Lecture Notes in Networks and Systems 992

The series "Lecture Notes in Networks and Systems" publishes the latest developments in Networks and Systems—quickly, informally and with high quality. Original research reported in proceedings and post-proceedings represents the core of LNNS.

Volumes published in LNNS embrace all aspects and subfields of, as well as new challenges in, Networks and Systems.

The series contains proceedings and edited volumes in systems and networks, spanning the areas of Cyber-Physical Systems, Autonomous Systems, Sensor Networks, Control Systems, Energy Systems, Automotive Systems, Biological Systems, Vehicular Networking and Connected Vehicles, Aerospace Systems, Automation, Manufacturing, Smart Grids, Nonlinear Systems, Power Systems, Robotics, Social Systems, Economic Systems and other. Of particular value to both the contributors and the readership are the short publication timeframe and the worldwide distribution and exposure which enable both a wide and rapid dissemination of research output.

The series covers the theory, applications, and perspectives on the state of the art and future developments relevant to systems and networks, decision making, control, complex processes and related areas, as embedded in the fields of interdisciplinary and applied sciences, engineering, computer science, physics, economics, social, and life sciences, as well as the paradigms and methodologies behind them.

Indexed by SCOPUS, INSPEC, WTI Frankfurt eG, zbMATH, SCImago.

All books published in the series are submitted for consideration in Web of Science.

For proposals from Asia please contact Aninda Bose (aninda.bose@springer.com).

Ivan Ostroumov · Maksym Zaliskyi
Editors

Proceedings of the 2nd International Workshop on Advances in Civil Aviation Systems Development

Springer

Editors
Ivan Ostroumov
Air Navigation Systems Department
National Aviation University
Kyiv, Ukraine

Maksym Zaliskyi
Department of Telecommunication and Radio
Electronic Systems
National Aviation University
Kyiv, Ukraine

ISSN 2367-3370 ISSN 2367-3389 (electronic)
Lecture Notes in Networks and Systems
ISBN 978-3-031-60195-8 ISBN 978-3-031-60196-5 (eBook)
https://doi.org/10.1007/978-3-031-60196-5

This Springer imprint is published by the registered company Springer Nature Switzerland AG
The registered company address is: Gewerbestrasse 11, 6330 Cham, Switzerland

Paper in this product is recyclable.

Preface

This volume contains the papers presented at the 2nd International Workshop on Advances in Civil Aviation Systems Development (ACASD 2024) held on March 26, 2024, organized by National Aviation University, Kyiv, Ukraine. ACASD 2024 received 35 research submissions from four different countries; among them, 32 papers were accepted for presentation. There are contributors from Ukraine, Ireland, China, Georgia, and Japan. ACASD was formed to bring together outstanding researchers and practitioners in the field of civil aviation systems including avionics systems, communication, navigation and surveillance systems, aerospace systems reliability development, and engineering from all over the world to share their experience and expertise. ACASD plays a unique role in the representation of new innovative technologies in the aerospace domain to researchers and industry practitioners to share complex problems and their solutions. Workshop covers major research areas focused on advances in civil avionics system design, modern trends in communication, surveillance and navigation systems development, human factor and safety system analysis, civil avionics system maintenance questions, and air traffic management systems. The presentation of the papers was carried out in two tracks: Development of Civil Aviation Systems; Reliability and Efficiency Analysis of Civil Aviation Systems.

Workshop was in hybrid format. Most reports are presented during two tracks of in-person mode in facilities of the National Aviation University. However, due to the war in Ukraine, participants from foreign countries have used virtual platform.

The organizers also wish to thank our publisher, Springer, for their acceptance to publish the proceedings, and for their kind assistance and cooperation during the editing process. The organizing committee wishes to thank CMT Research Conference Management System as it is a wonderful tool for easy organization and compilation of conference documents.

Fig. 1. Participants of ACASD workshop.

All our team invites you to visit the ACASD workshop website http://www.ans.nau.edu.ua/acasd where you can find all the relevant information about this exciting aviation event.

March 2024

Ivan Ostroumov
Maksym Zaliskyi

Organization

The 2nd International Workshop on Advances in Civil Aviation Systems Development (ACASD 2024) was organized by the National Aviation University of Ukraine.

General Chairs

Ivan Ostroumov	Department of Air Navigation Systems, National Aviation University, Ukraine
Maksym Zaliskyi	Department of Telecommunication and Radioelectronic Systems, National Aviation University, Ukraine

Program Committee Chairs

Vitalii Larin	National Aviation University, Kyiv, Ukraine
Olha Sushchenko	National Aviation University, Kyiv, Ukraine
Tetana Shmelova	National Aviation University, Kyiv, Ukraine
Roman Odarchenko	National Aviation University, Kyiv, Ukraine
Karen Marais	Purdue University, West Lafayette, USA
Neelakshi Majumdar	University of Arkansas, USA
Caner Ozcan	Karabuk University, Engineering Faculty, Karabuk, Turkey
Valeriu Vitan	Eurocontrol, Brussels, Belgium
Antonio Chialastri	Sapienza University of Rome, Rome, Italy
Minglei Fu	Zhejiang University of Technology, Hangzhou, China
Abdel-Badeeh M. Salem	Ain Shams University, Cairo, Egypt
Rauf Oglu Jafarzade	National Aviation Academy, Azerbaijan
Nina Rizun	Gdansk University of Technology, Poland
Viktoriia Ivannikova	Dublin City University Business School, Ireland
Vladimir Pavlikov	National Aerospace University "KAI", Kharkiv, Ukraine
Simeon Zhyla	National Aerospace University "KAI", Kharkiv, Ukraine
Kostiantyn Dergachov	National Aerospace University "KAI", Kharkiv, Ukraine

Sponsoring Organizations

ACASD 2024 could not been possible without support of the National Aviation University, Ukraine (https://www.nau.edu.ua) and Springer International Publishing AG, Germany (https://www.springer.com). Organizers and all participants of the International Workshop on Advances in Civil Aviation Systems Development express a lot of thanks.

Contents

Global Navigation Satellite Systems

Inertial Navigation System

Instrument Landing System

Artificial Intelligence in Civil Aviation Systems

Editors and Contributors

About the Editors

Dr. Ivan Ostroumov has been a faculty of Air Navigation Systems Department of the National Aviation University of Ukraine since September 2007. He obtained his Ph.D. degree of Engineering in Navigation and Traffic Control in 2009 and D. Sc. in 2020 from the National Aviation University of Ukraine. Since then, he has been a Senior research scientist and professor at National Aviation University. In 2017/2018, he has been a Fulbright scholar in the School of Aeronautics and Astronautics at Purdue. His research interests are Alternative Positioning, Navigation, and Timing (APNT); ADS-B; Safety of air transportation; air traffic analysis; efficient flight planning; navigation algorithms of FMS. He is head of bachelor's educational and professional program "Systems of Air Navigation Services" which is fully compatible with ICAO ATSEP Training program.

Dr. Maksym Zaliskyi is a professor of the Department of Telecommunication and Radioelectronic Systems at the National Aviation University. He received his B.Sc. and M.Sc. degrees in radio engineering from the National Aviation University, Ukraine, in 2005 and 2007, respectively. He obtained a PhD and D.Sc. degrees in aircraft equipment operation from the National Aviation University, in 2012 and 2021, respectively. He has 16 years of teaching experience and his areas of interest include data processing, equipment operation, and reliability analysis. The total number of research publications is 240.

Contributors

Veniamin Antonov National Aviation University, Kyiv, Ukraine

Maryna Arkhyrei National Aviation University, Kyiv, Ukraine

Oleksandr Avramenko National Defence University of Ukraine named after Ivan Cherniakhovskyi, Kyiv, Ukraine

Vadym Avrutov National Technical University of Ukraine "Igor Sikorsky Kyiv Polytechnic Institute", Kyiv, Ukraine

Yuliya Averyanova National Aviation University, Kyiv, Ukraine

Oleksandra Bankova National Aviation University, Kyiv, Ukraine

Yurii Bezkorovainyi National Aviation University, Kyiv, Ukraine

Liudmyla Blahaia National Aviation University, Kyiv, Ukraine

Dmytriy Bondarev National Aviation University, Kyiv, Ukraine

Ihor Bovdui Anatolii Pidhornyi Institute of Mechanical Engineering Problems of the National Academy of Sciences of Ukraine, Kharkiv, Ukraine

Mikle Burichenko National Aviation University, Kyiv, Ukraine

Kostiantyn Cherednichenko National Aviation University, Kyiv, Ukraine

Bohdan Chumachenko National Aviation University, Kyiv, Ukraine

Oleksii Chuzha National Aviation University, Kyiv, Ukraine

Anastasiia Dmytruk National Aviation University, Kyiv, Ukraine

Valentyn Dyptan National Defence University of Ukraine named after Ivan Cherniakhovskyi, Kyiv, Ukraine

Oleksii Hehelskyi National Technical University of Ukraine "Igor Sikorsky Kyiv Polytechnic Institute", Kyiv, Ukraine

Oleksii Holubnychyi National Aviation University, Kyiv, Ukraine

Yurii Hryshchenko National Aviation University, Kyiv, Ukraine

Oksana Ishchenko National Aviation University, Kyiv, Ukraine

Olga Ivanets Dublin City University, Dublin, Ireland

Viktoriia Ivannikova Dublin City University, Dublin, Ireland

Maxim Ivanytskyi National Aviation University, Kyiv, Ukraine

Oleg Ivashchuk National Aviation University, Kyiv, Ukraine

Volodymyr Kharchenko National Aviation University, Kyiv, Ukraine

Rimvidas Khrashchevskyi National Aviation University, Kyiv, Ukraine

Volodymyr Klymchuk National Aviation University, Kyiv, Ukraine

Volodymyr Kolotusha National Aviation University, Kyiv, Ukraine

Georgiy Konakhovych National Aviation University, Kyiv, Ukraine

Valery Konin National Aviation University, Kyiv, Ukraine

Lesia Konoplianyk National Aviation University, Kyiv, Ukraine

Olena Kozhokhina National Aviation University, Kyiv, Ukraine

Oleksandr Kutsenko National Aviation University, Kyiv, Ukraine

Borys Kuznetsov Anatolii Pidhornyi Institute of Mechanical Engineering Problems of the National Academy of Sciences of Ukraine, Kharkiv, Ukraine

Nataliia Ladohubets National Aviation University, Kyiv, Ukraine

Oleksandr Lavrynenko National Aviation University, Kyiv, Ukraine

Anzhela Lelechenko National Aviation University, Kyiv, Ukraine

Haoyang Li National Aviation University, Kyiv, Ukraine

Elvira Luzik National Aviation University, Kyiv, Ukraine

Tetiana Maliutenko National Aviation University, Kyiv, Ukraine

Maryna Maloied National Aviation University, Kyiv, Ukraine

Oleg Melnykov National Aviation University, Kyiv, Ukraine

Serhii Migel National Aviation University, Kyiv, Ukraine

Aria Nazarparvar Georgian Aviation University, Tbilisi, Georgia

Tatyana Nikitina Educational Scientific Professional Pedagogical Institute Ukrainian Engineering Pedagogical Academy, Universitetska Str, Kharkiv, Ukraine

Pavlo Openko National Defence University of Ukraine named after Ivan Cherniakhovskyi, Kyiv, Ukraine

Valerii Orlianskyi National Aviation University, Kyiv, Ukraine

Sergei Osadchy Flight Academy of the National Aviation University, Kropyvnytskyi, Ukraine

Alina Osipchuk National Aviation University, Kyiv, Ukraine

Ivan Ostroumov National Aviation University, Kyiv, Ukraine

Svitlana Pavlova Shanxi Agricultural University, Xiaodian, Taiyuan, China

Ivan G. Petrovski II IP Solutions Satellite Applications Inc., Tokyo, Japan

Zarina Poberezhna National Aviation University, Kyiv, Ukraine

Olexiy Pogurelskiy National Aviation University, Kyiv, Ukraine

Vasyl Polishchuk National Defence University of Ukraine named after Ivan Cherniakhovskyi, Kyiv, Ukraine

Ihor Prokopenko National Aviation University, Kyiv, Ukraine

Kostiantyn Prokopenko National Aviation University, Kyiv, Ukraine

Dmytro Prosvirin Antonov Company, Kyiv, Ukraine

Iryna Prykhodko National Aviation University, Kyiv, Ukraine

Oleksii Romanenko National Aviation University, Kyiv, Ukraine

Sergii Rupich National Technical University of Ukraine "Igor Sikorsky Kyiv Polytechnic Institute", Kyiv, Ukraine

Lev Ryzhkov National Technical University of Ukraine "Igor Sikorsky Kyiv Polytechnic Institute", Kyiv, Ukraine

Olexandr Salyuk National Aviation University, Kyiv, Ukraine

Olga Shcherbyna National Aviation University, Kyiv, Ukraine

Bohdan Shershen National Aviation University, Kyiv, Ukraine

Tetiana Shmelova Flight Academy of the National Aviation University, Kropyvnytskyi, Ukraine

Leonid Sibruk National Aviation University, Kyiv, Ukraine

Viktor Sibruk National Aviation University, Kyiv, Ukraine

Yuliya Sikirda Flight Academy of the National Aviation University, Kropyvnytskyi, Ukraine

Olena Sokolova National Aviation University, Kyiv, Ukraine

Oleksandr Solomentsev National Aviation University, Kyiv, Ukraine

Vadym Stratonov Flight Academy of the National Aviation University, Kropyvnytskyi, Ukraine

Olha Sushchenko National Aviation University, Kyiv, Ukraine

Oleksii Sushych National Aviation University, Kyiv, Ukraine

Serhii Tarasiuk National Aviation University, Kyiv, Ukraine

Nina Volianska National Technical University of Ukraine "Igor Sikorsky Kyiv Polytechnic Institute", Kyiv, Ukraine

Roman Voliansky National Technical University of Ukraine "Igor Sikorsky Kyiv Polytechnic Institute", Kyiv, Ukraine

Petro Yablonsky National Defence University of Ukraine named after Ivan Cherniakhovskyi, Kyiv, Ukraine

Alexandr Yakovchenko National Aviation University, Kyiv, Ukraine

Yaroslav Yakovlev Antonov Company, Kyiv, Ukraine

Maxim Yatsko National Aviation University, Kyiv, Ukraine

Serhii Yehorov National Aviation University, Kyiv, Ukraine

Ihor Zakutynskyi National Aviation University, Kyiv, Ukraine

Maksym Zaliskyi National Aviation University, Kyiv, Ukraine

Alexander Zamorsky National Technical University of Ukraine "Igor Sikorsky Kyiv Polytechnic Institute", Kyiv, Ukraine

Alexey Zhalilo National Aviation University, Kyiv, Ukraine

Oleksander Zhdanov National Aviation University, Kyiv, Ukraine

Yevheniia Znakovska National Aviation University, Kyiv, Ukraine

Valerii Zozulya State University of Trade and Economics, Kyiv, Ukraine

Computational Intelligence in Air Traffic Management

Spatial Indexing of Airplane Trajectory with Open Location Code and Hexagonal Hierarchical Index

Ivan Ostroumov(✉)

National Aviation University, Liubomyra Huzara Avenue, 1, Kyiv 03058, Ukraine
ostroumov@ukr.net

Abstract. Methods of vehicle navigation have been developing rapidly throughout the long history of aviation. Positioning is an important component of airplane navigation process. Performance-Based Navigation (PBN) principle is basic for civil aviation. PBN considers airplane location as an area of uncertainty based on a confidence band of 95%. Area of uncertainty has an ellipsoidal shape. Global spatial indexing systems use a global grid with a unique index of each cell, which could be zoomed for a particular level of precision. An Open Location Code (OLC) grounds on square cell. Hexagonal Hierarchical Spatial Index (H3) uses a combination of hexagons and pentagons to till the ellipsoidal surface of Earth. Nowadays spatial indexing proposes rich possibilities for addressing, data visualization, and navigation as well. In this paper we consider application of global spatial indexing for airplane flight path representation. An algorithm for coding airplane trajectory to OLC has been given. In numerical study airplane trajectory obtained under Automatic Dependent Surveillance-Broadcast (ADS-B) has been used for coding by OLC and H3 of different precision levels. Levels of precision that meets uncertainty area associated with airplane trajectory have been identified for both coding systems.

Keywords: Air Navigation · Data Processing · OLC · H3 · PBN · Spatial Indexing

1 Introduction

Air traffic control grounds on airplane trajectory analysis and management to meet requirements of aviation safety [1, 2]. Trajectory processing begins with validation of the requested flight plan. Then, at the stage of airspace usage, the trajectory of each airplane is recalculated based on flight scene development. Also, algorithms of mid-air and terrain awareness are continuously operated on both sides of airframe and ground surveillance data processing units. Finally, after airplane landing, trajectory is analyzed to meet the requirements of normative documents in order to identify the potential gaps in safety [3, 4].

A flight plan includes a detailed description of requested usage of airspace in the form of a sequence of waypoints that specify flight trajectory as linear movement between

I. Ostroumov and M. Zaliskyi (Eds.): ACASD 2024, LNNS 992, pp. 3–16, 2024.
https://doi.org/10.1007/978-3-031-60196-5_1

each pair of waypoints [5]. At the stage of airspace usage, services of air traffic control are using ground radars and multilateration systems to measure the exact location of airspace users. Primary radars in addition to airplane localization also measure current weather, based on quantity of water in the air [6]. An Automatic Dependent Surveillance-Broadcast (ADS-B) provides sharing of each airspace user position, identified by on-board equipment with other participants of air traffic and ground facilities of air traffic control [7]. Errors and technical state of aviation equipment have significant influence on accuracy of provided data [8, 9]. The surveillance data proceeding unit collects data from all sensors and provides effective error filtering and prediction of air traffic scene development [10, 11].

Trajectory data usually is presented as a set of geodetic coordinates that consist of matrixes of latitude, longitude, and pressure altitude accompanied by timestamp. Unfortunately, methods of trajectory data processing and analysis could not be realized in raw data of latitude and longitude, due to its spherical nature. During trajectory analysis, data are transformed from geodetic coordinates to some cartesian reference frame [12]. Global Earth-Centered, Earth-Fixed (ECEF) or local NED/NEU are used for most calculations of data.

Nowadays global spatial indexing provides various possibilities for data addressing and visualization [13–15]. One of the perspective technology development ways is navigation based on global spatial indexes. Operation with indexes of particular equal areas makes navigation computation effectively simplified, which is important for modern computational equipment. Global indexing systems represent a reference ellipsoid as a set of elementary cells, which could be zoomed iteratively. There are four basic types of cells: triangular, rectangular, pentagonal, and hexagonal [16]. Each type has a particular cell configuration and orientation in the space. Advantages and disadvantages of cell configurations are studied in a variety of papers [17]. Open Location Code (OLC) and Hexagonal Hierarchical Spatial Index (H3) are the most frequently used global indexing systems. OLC is associated with "Plus codes" and grounds on the rectangular form of elementary cells [18]. OLC supports addressing of cells with sides up to 1 dm (for 10th precision level). H3 uses a hexagonal form of elementary cell with an area of up to 0.895 m^2 (for 15th resolution) [19]. Spatial index provides a computational cheap solution for a variety of tasks for low-performance hardware [20]. Also, spatial indexing is widely used for geodata visualization [14, 21].

In the paper, we study the application of global spatial indexing for airplane flight path representation. OLC and H3 are used for coding airplane trajectory data. Also, a specific algorithm for trajectory data coding by OLC is proposed. Global spatial indexes propose multiple advantages for data representation which include minimization of space for data storage, due to the usage of a one-word index instead of two numbers of latitude and longitude and minimization of computation time for navigation and awareness algorithms. Spatial indexing makes it possible to use graph analysis for positioning, navigation, and awareness for various applications including aviation.

The rest of this paper is organized as follows: In Sect. 2 we consider airplane trajectory coding with Open Location Code. Section 3 describes coding with H3 library. Numerical study includes results of coding of airplane trajectory with OLC and H3 of different precision levels.

2 Open Location Spatial Index

OLC was issued in 2014 for automatically addressing any area in the world with precision up to a rectangular area [18]. Any point of location with particular coordinates could be represented as a cell of the spatial index with a precision level of OLC which meets the accuracy of navigation sensor. OLC supports the World Geodetic System (WGS) 84 ellipsoidal model of Earth for latitude and longitude data coding. The full range of latitude and longitude is coded by base of 20. The first level of encoding by OLC is presented in Fig. 1. Then, each square cell could be coded in the same way by 20 × 20 grid. There are five basic levels of precision based on 20 × 20 grid which forms short codes with precision up to 13.9 m side length. Also, a precise code uses additional five digits after a short code for indexing area up to a rectangle with sides 4 × 14 mm.

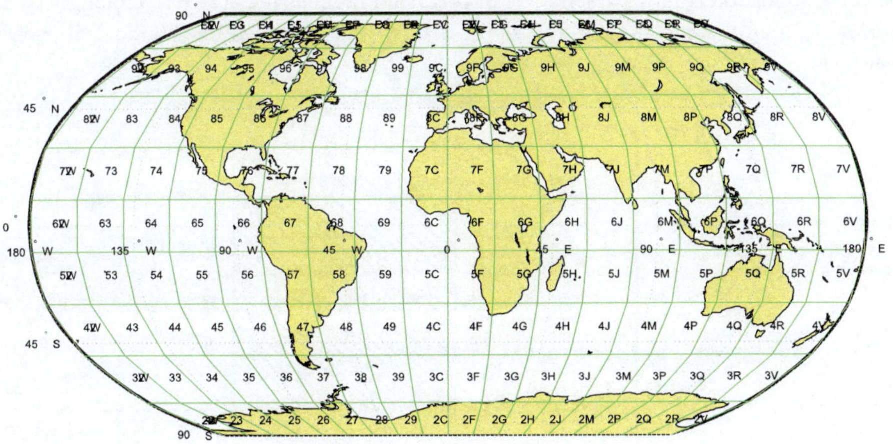

Fig. 1. First-level coding of OLC.

Shortcodes of OLC have the following format "AABBCCDD + EE", where AA is referred to as the first level of precision; BB is the second level of precision; CC is the third one; DD is the fourth; and EE is the fifth one [18, 22]. A sight " +" is used to differentiate code format from other addressing numbers. Results of precision calculation for point of crossing the Prime meridian and equator are presented in Table 1.

Each level of precision is coded by two characters. Aggregation of coded characters for each precision level forms OLC or "Plus code". The first character code Latitude index. Longitude index is the second one. OLC uses a specific sequence to form coding pattern to avoid word forming in codes. Character pattern for 20 numbers from 0 to 19 is the following [18]:

$$P = [2, 3, 4, 5, 6, 7, 8, 9, C, F, G, H, J, M, P, Q, R, V, W, X]. \quad (1)$$

Table 1. Parameters of rectangular area for different precision levels of shortcodes in OLC.

Level of Precision	Place in OLC format	Weight of precision level	Width of square Side	Average area of cell
1	AA	20°	2213.4 km	4.89×10^6 km^2
2	BB	1°	110.94 km	1.23×10^4 km^2
3	CC	0.05°	5.54 km	30.77 km^2
4	DD	0.0025°	277.36 m	0.076 km^2
5	EE	0.000125°	13.86 m	192.32 m^2

A rectangular cell is formed by two parallels and two meridians is used at a particular level of precision (Fig. 2). Each level of precision reduces the area of a rectangle by 20 times. In reality, a rectangular area is formed by four arches at the WGS84 ellipsoidal surface.

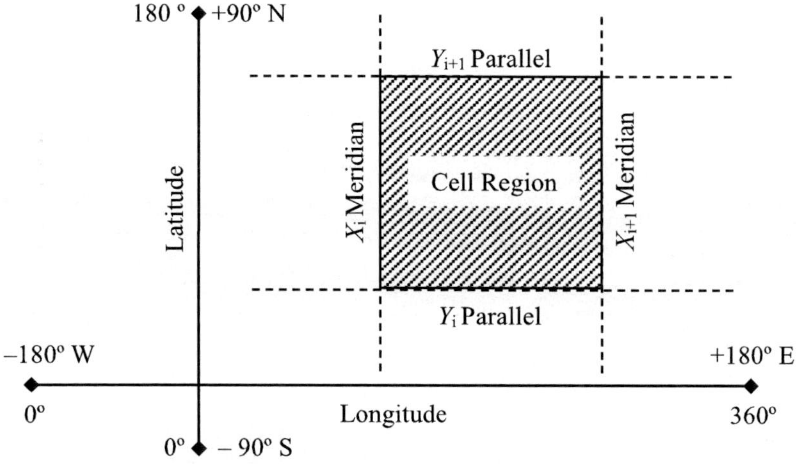

Fig. 2. A size of OLC cell.

Algorithm for coding data for short OLC codes is the following:

1. Align latitude and longitude sizes to the positive ones:

$$Lat = lat + 90°;$$
$$Lon = lon + 180°,$$

where $lat \in [-90°, +90°]$; $lon \in [-180°, +180°]$;

2. Setting codding range:

$$Lat = floor\left(Lat \times 20^3\right); \quad (2)$$

$$Lon = floor\left(Lon \times 20^3\right), \tag{3}$$

where $floor(x)$ is rounds argument x to the nearest integer less than x.

3. Coordinate transformation from decimal to base 20 system and code obtained decimal numbers by OLC pattern.

$$Lat_{10} \to Lat_{20}, Lat_{20} = A_{Lat}B_{Lat}C_{Lat}D_{Lat}E_{Lat}, \tag{4}$$

$$Lon_{10} \to Lon_{20}, Lon_{20} = A_{Lon}B_{Lon}C_{Lon}D_{Lon}E_{Lon}. \tag{5}$$

4. Reshape (4) and (5) to meet format rule:

$$A_{Lat}A_{Lon}B_{Lat}B_{Lon}C_{Lat}C_{Lon}D_{Lat}D_{Lon} + E_{Lat}E_{Lon}.$$

Based on righthand, the coded value could be represented with a reduced precision level by a group of two characters.

Decoding algorithm also includes only four basic steps:

1. Reshape OLC into two sets of latitude and longitude codes:

$$G_{Lat} = \{A_{Lat}, B_{Lat}, C_{Lat}, D_{Lat}, E_{Lat}\}, \tag{6}$$

$$G_{Lon} = \{A_{Lon}, B_{Lon}, C_{Lon}, D_{Lon}, E_{Lon}\}. \tag{7}$$

2. Decode sets of C_{Lat} and C_{Lon} with pattern (1)

$$G_{Lat} \to N_{Lat}; G_{Lon} \to N_{Lon}. \tag{8}$$

3. Multiply sets N_{Lat} and N_{Lon} with weight matrix of precision level w and getting a sum:

$$Lat_1 = \sum\nolimits_{i=1}^{p} N_{Lat}(i)w(i) - 90, \tag{9}$$

$$Lon_1 = \sum\nolimits_{i=1}^{p} N_{Lon}(i)w(i) - 180, \tag{10}$$

where w is a matrix of precision level weight, $w = [20, 1, 0.05, 0.0025, 0.000125]$; p is a maximum precision level, which corresponds to the number of chars in G_{Lat} set.

4. Calculation coordinates of cell center. Estimated by (9–10) Lat_1 and Lon_1 are coordinate of the bottom, left corner of a rectangular area. To get central point coordinates Lat and Lon the following equation could be used:

$$Lat = \frac{2Lat_1 + w(p)}{2}, \tag{11}$$

$$Lon = \frac{2Lon_1 + w(p)}{2}. \tag{12}$$

During the coding, we identify the cell index, which includes a particular point. However, after decoding we have a central point of a rectangular area. Thus, decoded coordinates are going to be different from input geodetic data. Also, the difference between obtained coordinates is proportional to the precision level p which is used in the decoding process.

An uncertainty area of a rectangular cell could be approximated by a rectangle with sides estimated by the arch length of a circle [23], which could be calculated as follows:

$$L = dasin\left(\sqrt{cos(Lat)cos(Lat \pm \Delta Lat)sin^2\left(\frac{\Delta Lon}{2}\right) + sin^2\left(\frac{\Delta Lat}{2}\right)}\right), \quad (13)$$

where d is the diameter of a sphere; Lat is latitude; ΔLat is latitude length of a rectangular cell; ΔLon is longitude width of a rectangular cell.

Representation of airplane trajectory as a sequence of OLC codes adaptive to the requested precision level requires to use algorithm of coding data for available sequences of coordinates. There are two basic approaches for evaluation of cell indexes in the linear parts between available points. The first one is grounded on the sequential brute force method for all neighbor cells and comparison with holding the linear part of trajectory. Logic of this method is presented in Fig. 3. Another approach is grounded in trajectory data interpolation to the level of required precision p with further coding of obtained data by OLC algorithm. After coding by data interpolation cells with identical codes are minimized for reducing the size of output data.

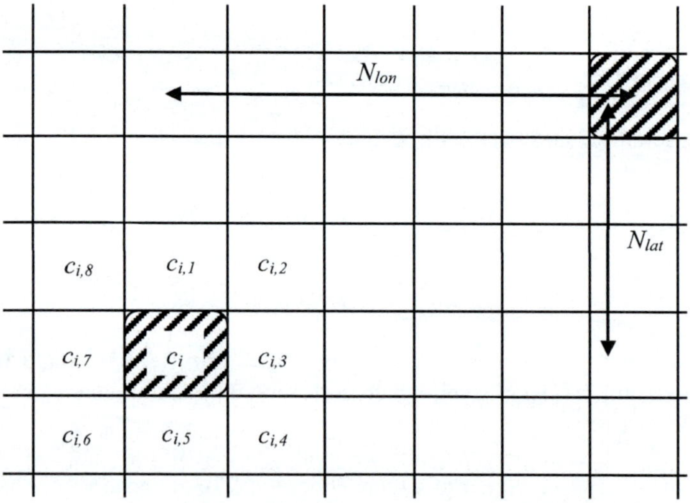

Fig. 3. Getting indexes at the line between any two cells.

We propose to use a brute-force method based on the following algorithm:

1. Identification indexes of all neighbor cells at the level of precision of the initial cell. Square or rectangular cell in OLC has only 8 neighbor cell indexes which could

be identified by logical operations in the set of cells of particular precision level. Identified indexes form a set of neighbor cells (Fig. 3):

$$C_i = \{c_{i,1}, c_{i,2}, c_{i,3}, c_{i,4}, c_{i,5}, c_{i,6}, c_{i,7}, c_{i,8}\}, \quad (14)$$

where i is an index of cell; $c_{i,1}$ is the first neighbor to i-th cell.

2. Estimation of a heading angle of a line as

$$H = \mathrm{tg}\frac{N_{lon}}{N_{lat}}, \quad (15)$$

where N_{lon} is a number of cells in longitude direction between the initial and final points of the linear part; N_{lat} is a number of cells on the latitude side.

3. Identification index of a neighbor cell from the set C_i for which H is corresponded.
4. Repeat steps 1–3 sequentially until the desired cell will not be reached.

Proposed method helps to identify each cell index which is placed in the linear path between two given cell indexes. Missing data and multiple gaps are common problems in surveillance data processing. Therefore, using spatial indexing helps to speedy interpolate the missing data from the given set.

3 Hexagonal Hierarchical Spatial Index

Another useful global spatial indexing system is the Hexagonal Hierarchical Spatial Index (H3). H3 uses a regular hexagonal shape to represent a cell. H3 grounds on gnomonic projection applied for each side of a regular icosahedron which consists of 20 equilateral triangles [19]. A Hexagonal grid is applied to the 2D projection. Also, H3 uses 12 pentagons in the vertexes of icosahedron to minimize distortions.

An example of a hexagonal grid at 10th level of precision in the vicinity of 50.438852 N, 30.428532 E is presented in Fig. 4. Edge length of hexagon at 10th level of precision is 75.8 m.

Fig. 4. Hexagonal grid at 10th level of precision.

A regular hexagon has multiple advantages for usage as a grid for a global indexing system:

- Hexagonal shape gives smaller distortion in projection in comparison to a square or triangle;
- Hexagon has only 6 neighbors;
- Centers of all neighbor hexagons are equally distanced from the center of the central hexagon;
- Size of hexagon could be easily zoomed based on precision level.

H3 supports 16 levels of precision. Indexing system of H3 supports division of each hexagonal cell into 7 child cells at the next level of precision (each pentagon is formed by 6 cells, which includes 1 pentagon and 5 hexagons) [19]. H3 supports indexing with hexagons of approximately 0.9 m^2 area at the highest level of precision. Indexing system of H3 uses 15 characters of hexadecimal numbers placed in 64 bits of space.

Open access H3 library proposes a variety of functions for operation with Hexagonal Hierarchical Spatial Index [19].

4 Numerical Demonstration

In the numerical calculation, we demonstrate the possibility of trajectory representation as a set of cell numbers of a particular global spatial indexing system. As an example, we consider airplane flight path coding of DLH 923 flight with a connection between London Heathrow (EGLL) and Frankfurt (EDDF) international airports. Flight data was extracted from the air traffic surveillance database for December 4, 2023. Data obtained under ADS-B network of ground stations includes 129 raw measurements of latitude, longitude, and altitude. A flight path of DLH 923 is represented in Fig. 5.

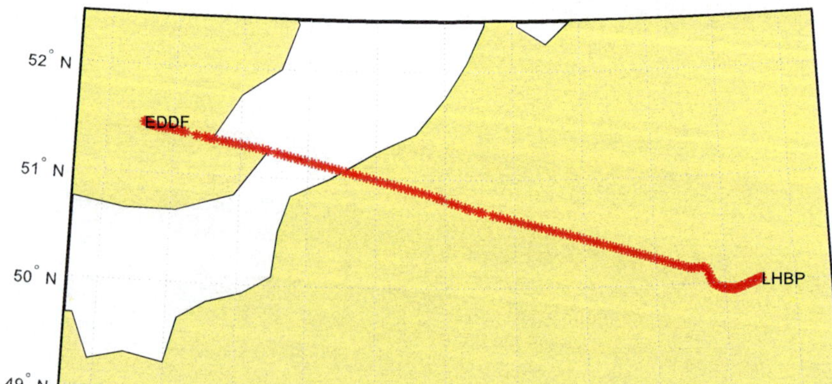

Fig. 5. Flight path of DLH 923 flight on December 4, 2023.

A trajectory data of DLH 923 flight is coded by OLC algorithm based on input precision level with the help of specially developed software in Matlab environment. Configuration of coded airplane trajectory for precision levels from second to fourth are presented in Fig. 6, 7 and 8.

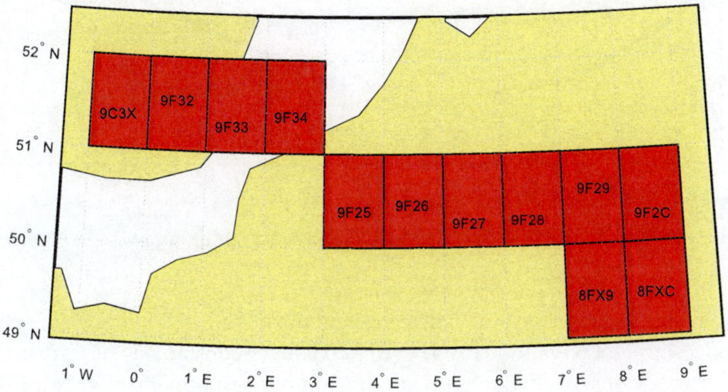

Fig. 6. Flight path coded by OLC to the second level of precision.

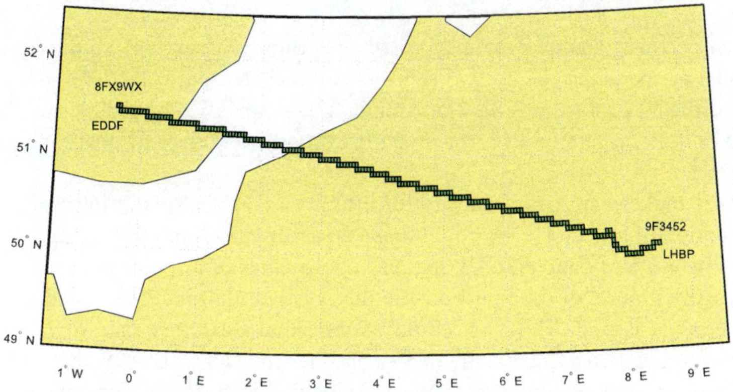

Fig. 7. Flight path coded by OLC to the third level of precision.

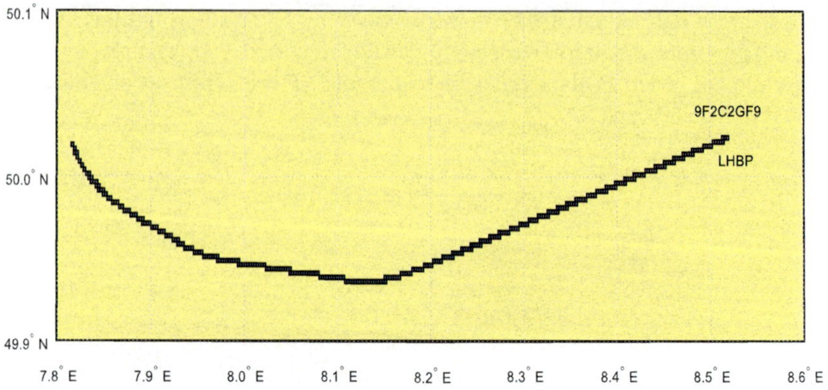

Fig. 8. Flight path coded by OLC to fourth level of precision.

We associate airplane location with a particular squared cell. It makes sense in case of performance-based navigation (PBN). PBN considers coordinates of airplane together with the accuracy of sensor which provides position measurements [24, 25]. PBN works with areas of uncertainty or areas of possible airplane location, based on 95% confidence band. Coordinate in PBN is a central point of the ellipsoidal area of possible airplane position. Coordinates and accuracy of measurements represent possible location [26, 27].

Actually, there are two main types of uncertainty areas: ellipsoidal and circular. Ellipsoidal is characterized by two semi-axes associated with standard deviation error on the North and East sides [28]. The circular area is specified by radius, which could be equal to double dispersion of error or standard deviation error of positioning powered by two $2\sigma_p^2$.

Thus it makes sense to associate uncertainty area of airplane location with OLC representation of trajectory. We could make an assumption for positioning performance provided by ADS-B data. ADS-B data use coordinates of airplane measured by GNSS receiver. Any reports of on-board positioning system malfunction related to DLH 923 flight were not issued. Thus, we could assume that trajectory data of DLH 923 has been measured with a precision of GNSS constellation at a particular time on December 4, 2023 and at a particular place. We use special software to simulate precision of coordinates measurements along considered trajectory for GPS navigation constellation only. An average navigation system error is equal to 12 m. Total uncertainty area for this trajectory was 15.6 km^2 which is estimated based on 95% confidence band.

Results of area analysis for trajectory coding by OLC are presented in Table 2. It includes a number of unique cells that are used to code airplane trajectory, the average area of the cell, and the total area of uncertainty associated with trajectory representation.

Table 2. Performance of trajectory coding by OLC.

Level of precision	Number of cells	Average area of cell	Average area of coded trajectory
2	12	1.23×10^4 km^2	1.47×10^5 km^2
3	214	30.77 km^2	6.58×10^3 km^2
4	3884	0.076 km^2	314.18 km^2
5	13444	1.92×10^{-4} km^2	2.58 km^2

Obtained results indicate, that the total area of uncertainty region in OLC of 5th precision level meets the approximate uncertainty area of data provided by on-board positioning sensor.

Then, trajectory data has been coded by H3 for different precision levels with specially developed software in JavaScript. Results of coding for 4th and 5th precision levels are presented in Fig. 9 and Fig. 10 in projection on OpenStreetMap. Performance of coded trajectory data with H3 is presented in Table 3.

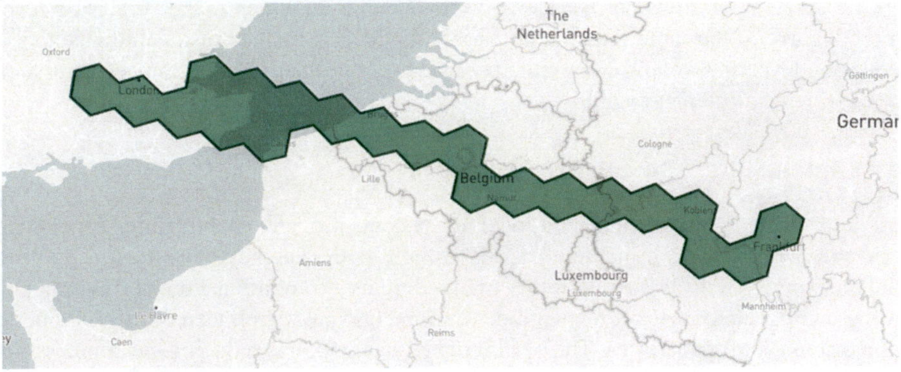

Fig. 9. Flight path coded by H3 to the fourth level of precision.

Fig. 10. Flight path coded by H3 to the fifth level of precision.

Table 3. Performance of trajectory coding by H3.

Level of precision	Number of cells	Average area of cell [km^2]	Average area of coded trajectory [km^2]
3	10	12393	123930
4	21	1770	37170
5	52	252	13104
6	146	36	5256
7	355	5	1775
8	958	0.73	699
9	2328	0.1	232
10	3281	0.015	49.21
11	9702	0.0021	20.4
12	20166	0.0003	6.04

H3 grounds on division of each cell into 7 children, which makes precision level more flexible to the input data in comparison to OLC. Obtained results indicate that a precision level of 14 could satisfy the precision requirements of input uncertainty area of considered airplane trajectory.

5 Conclusions and Future Works

Modern standards of global spatial indexing systems provide reach features for vehicle trajectory representation including airplane flight path. The key benefit of usage spatial indexing systems is the possibility of avoiding usage coordinate transformation and using complicated navigation equations. Cell navigation is grounded on logical operations, graph, and set theories. The benefit of cell navigation is reducing the computation performance required for solution navigation equations.

Airplane trajectory is an important navigation element. Proposed algorithm of trajectory coding helps to obtain a set of cells to represent in a global spatial indexing system. Trajectory representation with a particular level of precision makes it possible to consider uncertainty area connected with a given trajectory.

In future work, codded trajectories could be used for mid-air collision detection and avoidance with the help of logical functions only. That stimulates more efficient computation facility usage and open perspective for deep integration of different safety methods in aviation.

References

1. Zhang, S., Zhang, Y., Tay, T., Shankar, J.: Learning-based aircraft trajectory analysis tool for holding and vectoring identification with ADS-B data. In: 25th International Conference on Intelligent Transportation Systems, Macau, China, pp. 1100–1105. IEEE (2022). https://doi.org/10.1109/ITSC55140.2022.9921823

2. Ostroumov, I., Kuzmenko, N.: Statistical analysis and flight route extraction from automatic dependent surveillance-broadcast data. In: Integrated Communication, Navigation and Surveillance Conference (ICNS), Dulles, VA, USA, pp. 1–9. IEEE (2022). https://doi.org/10.1109/ICNS54818.2022.9771515
3. Ostroumov, I., Kuzmenko, N.: Risk assessment of mid-air collision based on positioning performance by navigational aids. In: 6th International Conference on Methods and Systems of Navigation and Motion Control (MSNMC), Kyiv, Ukraine, pp. 34–37. IEEE (2020). https://doi.org/10.1109/MSNMC50359.2020.9255506
4. Ostroumov, I.V., Kuzmenko, N.S.: Accuracy assessment of aircraft positioning by multiple Radio Navigational aids. Telecommun. Radio Eng. **77**(8), 705–715 (2018). https://doi.org/10.1615/TelecomRadEng.v77.i8.40
5. Simorgh, A., et al.: Comprehensive survey on climate optimal aircraft trajectory planning. Aerospace **9**(3), 146 (2022)
6. Popov, A., Tserne, E., Zhyla, S., Volosyuk, V., Pavlikov, V., Ruzhentsev, N.: Invariant polarization signatures for recognition of hydrometeors by airborne weather radars. In: Gervasi, O., et al. (eds.) Computational Science and Its Applications 2023, vol. 13956, pp. 201–217. Springer, Cham (2023). https://doi.org/10.1007/978-3-031-36805-9_14
7. Technical Provisions for Mode S Services and Extended Squitter, Doc 9871, 1st edn. ICAO, Monreal (2008)
8. Solomentsev, O., Zaliskyi, M., Nemyrovets, Y., Asanov, M.: Signal processing in case of radio equipment technical state deterioration. In: Signal Processing Symposium (SPS 2015), Debe, Poland, pp. 1–5. IEEE (2015). https://doi.org/10.1109/SPS.2015.7168312
9. Zaliskyi, M., Solomentsev, O., Kozhokhina, O., Herasymenko, T.: Reliability parameters estimation for radioelectronic equipment in case of change-point. In: Signal Processing Symposium (SPSympo 2017), Jachranka Village, Poland, pp. 1–4. IEEE (2017). https://doi.org/10.1109/SPS.2017.8053676
10. Zhang, X., Yu, W.: Research on the application of Kalman filter algorithm in aircraft trajectory analysis. In: 7th International Conference on Intelligent Computing and Signal Processing (ICSP), pp. 196–199. IEEE (2022)
11. Zeng, W., Chu, X., Xu, Z., Liu, Y., Quan, Z.: Aircraft 4D trajectory prediction in civil aviation: a review. Aerospace **9**(2), 91 (2022)
12. Osen, K.: Accurate conversion of earth-fixed earth-centered coordinates to geodetic coordinates. Doctoral dissertation, Norwegian University of Science and Technology (2017)
13. Azri, S., Ujang, U., Anton, F., Mioc, D., Rahman, A.A.: Review of spatial indexing techniques for large urban data management. In: International Symposium & Exhibition on Geoinformation (ISG), pp. 24–25 (2013)
14. Lv, Z., Hu, Y., Zhong, H., Yu, B., Wu, J., Li, B.: Spatial indexing of global geographical data with HTM. In: International Conference on Geoinformatics, pp. 1–6. IEEE (2010)
15. Aini, A.N., Dewantari, O.A., Mandala, D.P., Bisri, M.B.: An enhanced earthquake risk analysis using H3 spatial indexing. In: IOP Conference Series: Earth and Environmental Science, vol. 1245, no. 1, p. 012014. IOP Publishing (2023)
16. Uher, V., Gajdos, P., Snasel, V., Lai, Y.C., Radecky, M.: Hierarchical hexagonal clustering and indexing. Symmetry **11**(6), 731 (2019)
17. Lei, Y., et al.: Global multi-scale grid integer coding and spatial indexing: a novel approach for big earth observation data. ISPRS J. Photogramm. Remote Sens. **163**, 202–213 (2020)
18. Open location code specification. https://github.com/google/open-location-code/blob/main/docs/specification.md. Accessed 08 Dec 2023
19. Hexagonal hierarchical geospatial indexing system specification. https://h3geo.org. Accessed 08 Dec 2023

20. Voliansky, R., Sadovoi, O., Shramko, Y., Volianska, N., Sinkevych, O.: Arduino-based implementation of the dual-channel chaotic generator. In: 3rd International Conference on Advanced Information and Communications Technologies (AICT), Lviv, Ukraine, pp. 278–281. IEEE (2019). https://doi.org/10.1109/AIACT.2019.8847904
21. Kuznetsov, B., et al.: Method for design of magnetic field active silencing system based on robust meta model. In: Shukla, S., Sayama, H., Varghese, J., Kumar, D. (eds.) Data Science and Security, IDSCS 2023. LNNS, vol. 922, pp. 1–12. Springer, Singapore (2023, in press)
22. Open location (plus code). https://www.dcode.fr/open-location-code. Accessed 08 Dec 2023
23. Maria, E., Budiman, E., Taruk, M.: Measure distance locating nearest public facilities using Haversine and Euclidean Methods. J. Phys. Conf. Ser. **1450**(1), 012080 (2020)
24. Performance-Based Navigation (PBN) Manual. Doc 9613, 4th edn. ICAO (2013)
25. Kuzmenko, N.S., Ostroumov, I.V., Marais, K.: An accuracy and availability estimation of aircraft positioning by navigational aids. In: 5th International Conference Methods and Systems of Navigation and Motion Control, Kyiv, pp. 36–40. IEEE (2018). https://doi.org/10.1109/MSNMC.2018.8576276
26. Dergachov, K., Havrylenko, O., Pavlikov, V., Zhyla, S., Tserne, E., Volosyuk, V.: GPS usage analysis for angular orientation practical tasks solving. In: 9th International Conference on Problems of Infocommunications, Science and Technology (PIC S&T), Kharkiv, pp. 187–192. IEEE (2022)
27. Ostroumov, I.V., Kuzmenko, N.S.: Compatibility analysis of multi signal processing in APNT with current navigation infrastructure. Telecommun. Radio Eng. **77**(3), 211–223 (2018). https://doi.org/10.1615/TelecomRadEng.v77.i3.30
28. Ostroumov, I.V., Kharchenko, V.P., Kuzmenko, N.S.: An airspace analysis according to area navigation requirements. Aviation **23**(2), 36–42 (2019). https://doi.org/10.3846/aviation.2019.10302

Air Traffic Management with Hierarchical Hexagonal Geospatial Index

Ivan Ostroumov(✉) and Oleg Ivashchuk

National Aviation University, Liubomyra Huzara Avenue, 1, Kyiv 03058, Ukraine
ostroumov@ukr.net

Abstract. Nowadays global spatial indexes provide reach features for geographic data analysis. A variety of standards of spatial indexes with global coverage provide computational cheap processing and visualization of data associated with latitude and longitude. In the paper we consider potential benefits of Hierarchical Hexagonal Spatial Index (H3) implementation to support air traffic management (ATM). H3 could be useful in automatic data processing algorithms for representation of airspace, particular zones, and sectors of airspace with different shape complexity. Hierarchical hexagonal cell provides different levels of precision which could be associated with an area of uncertainty in positioning system used on board airplanes. Airspace users could be associated with a particular cell of geospatial index. H3 could be a complete environment to support ATM on the basis of aviation safety, which is required for the wide implementation of free route airspace globally. An airplane trajectory obtained by Automatic dependent Surveillance-broadcast technique has been classified to use different parts of airspace during departure from UKBB International airport based on a hexagonal representation of airspace user and airspace elements.

Keywords: Air Traffic Management · Spatial Index · Air Navigation · Aviation · Airspace · Geodata · Free Route Airspace

1 Introduction

Nowadays global air transportation network is in an ongoing process of development. Free Route Airspace (FRA) is planned for wide integration into the air transportation network [1, 2]. As an example, Ukraine is going to use rules of FRA for whole areas during airspace recovery after military restrictions [3]. FRA is based on free route planning with no more usage of flight route network [4, 5]. At the initial stage, FRA in Ukrainian airspace has been successfully implemented in upper airspaces. FRA provides multiple advantages for airspace users which include usage of effective cost-minimized trajectories which leads to minimized time of flight, increased airspace capacity, minimized risk of mid-air collision, and minimized carbon pollution as well [6].

Spatial index provides a reach feature to support air traffic management (ATM). Global spatial index is an addressing system that partitions ellipsoidal shape of WGS-84 into a set of cells with different sizes based on the level of precision. Each cell has

I. Ostroumov and M. Zaliskyi (Eds.): ACASD 2024, LNNS 992, pp. 17–30, 2024.
https://doi.org/10.1007/978-3-031-60196-5_2

a particular index which is a hierarchical address [7, 8]. There are different types of spatial indexes based on cell geometry. The most useful of them grounds on rectangular [9, 10] and hexagonal cells [11]. Rectangular spatial indexing partitions a range of latitude and longitude into a particular number of rectangles or squares based on a specific rule. At each step of iteration, each rectangular area is also partitioned with the same rule until required precision has not been reached. Hexagonal shape gives multiple advantages for navigation and simplifies relations for geometry recalculation during manipulation with data. Hexagonal Hierarchical Spatial Index (H3) is widely used in different implementations. A hexagonal spatial index could be useful for representation of data from a remote sensing system or mapping of the Earth's surface [12–14].

Area navigation specifications (RNP/RNAV) of Performance-Based Navigation (PBN) grounds on specific requirements for accuracy of on-board positioning system [15]. Total system error (TSE) in 1 NM (RNAV 1) is a basic specification for continental flights. Performance of primary positioning systems or methods of positioning by navigational aids (which are used as stand-by equipment) specifies a Navigation system error which is a main component of TSE [16, 17]. TSE provides a circular or ellipsoidal area of airspace user uncertainty according to confidence band of 95%. PBN operates with uncertainty areas of airplane position which is set up with a reference point (which coincides with measured airplane coordinates) and radius (equal to TSE) or semi-axes of ellipse (given by measured errors in the North and East directions). Also, action of degradation factors may affect the nominal positioning system operation [18, 19]. Action of these factors mostly depends on radio equipment technical state deterioration [20, 21].

H3 is well structured for representation of uncertainty areas in PBN and has great potential for implementation in ATM.

In the paper, we consider application of H3 for ATM to support automatic control algorithms with particular airspace representation. Also, we use representation of a particular airspace user as a hexagonal cell of the particular level of resolution which corresponds to performance of on-board positioning system. Another important problem in ATM is classification of airspace user presence in a particular airspace part. We propose a logical algorithm to identify airspace user position within the limited area to use in ATM.

The structure of the paper includes a description of the main airspace structural elements presented in Sect. 2. Section 3 describes the basic features of spatial index system for representation airspace areas of complicated shapes. Section 3 includes numerical demonstration of representation of structural elements of Ukrainian airspace with a hexagonal spatial index.

2 Airspace Structure

Nowadays civil aviation uses a wide network of flight routes to organize air traffic in terms of flight safety. Flight route network includes waypoints and routes which connect all waypoints in the network. Airspace users have to follow cleared trajectories within the network of flight routes and have to meet multiple requirements to airspace users [22]. Air transportation system includes many elements for organizing management and optimization. Airspace includes everything from the Earth's surface to 100 km above (the Karman line) and belongs to the administrative state that owns the territory below it, this also applies to territorial waters and economic zones.

Any usage of airspace forms air traffic which is managed by specialized authority. As an example, UkSATSE is an authorized company to support and manage air traffic in Ukraine. In general, airspace is a complicated structure that is partitioned into various components and elements for convenience of management, ensuring the required level of flight safety, and efficient air traffic management.

In general, airspace is classified into controlled and uncontrolled [22]. Controlled is the airspace where the air traffic control authority provides air traffic control services. Uncontrolled is the airspace where the air traffic control authority does not provide such service. Also, airspace volume is divided into seven classes called by alphabetic characters from A to G [22].

Each class except A has a specification for limitations in its space for Instrument (IFR) and visual (VFR) Flight Rules. These restrictions include requirements for the use of radio communications, speed limits, the need for a dispatcher's permission, and the types of services and services provided therein. Classes from A to E (in class E, only IFR) refer to controlled airspace, class E according to VFR, F, and G refer to uncontrolled airspace. Each country specifies particular geometry and type of airspace classes used. For example, Ukraine has C, D, and G classes only (Fig. 1) [23]. Class G from the Earth's surface to 1500 m. Class D from 1500 m to 2900 m. Class C is placed from 2900 m to 660 FL (flight level). Above 660 FL up to 100 km class G is used or unclassified airspace. Airspace around the airport is connected to class D to provide air traffic.

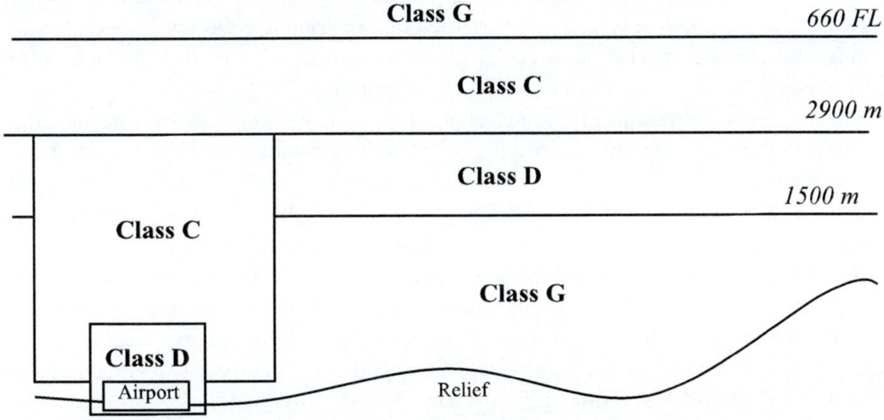

Fig. 1. Ukrainian airspace structure.

Airspace could include particular areas, zones, regions, and other elements. In case of wide areas to hold many users a flight information region (FIR) could be used. In Ukraine, there are five FIRs (Kyiv, Lviv, Dnipro, Odesa, and Simferopol) and one upper flight information region located above FIRs. FIR provides flight information service for any user from the ground to the upper limit. Also, flight information zones (FIZ) can be installed both within the controlled and uncontrolled airspace to provide particular flight information services. FIZ is much smaller in size than the FIR and has defined lateral and vertical perils. Airfield Flight Information Zone (AFIZ) provides flight information services in the airport vicinity. AFIZ has lateral and vertical limits that are determined by the air navigation provider.

Airspace elements that have air traffic control service include control area (CTA), terminal maneuvering area (TMA), and control region (CTR). CTA could be wide up to entire controlled airspace. However, the lower limit of the CTA is at a height of at least 200 m above ground level. TMA setting up around airports where a significant level of air traffic is expected. CTR is usually placed to connect the controlled airport with the CTA. Flights in the waiting area, landing, and climbing procedures must be taken in CTR. The lateral boundaries of CTR must be at the range of at least 9.3 km from the center of the airfield for which it is established. CTR is placed between the Earth's surface and lower limit of CTA.

Another important component of airspace is dangerous zones. There are three basic types of dangerous zones: prohibited, restricted, and dangerous areas.

Prohibited area is specified by geometrical perils and validity period which are set up by the air navigation services provider. Prohibited areas are placed over strategic objects where the risk of catastrophic situations is present. As an example, prohibited areas are placed under nuclear power plants.

Danger area holds a territory during flight within which there is a danger for the user due to the events that take place there. Parameters of this area are specified by authorized air navigation service provider. For example, an area with fireworks could be marked as dangerous.

Restricted area is a zone within which flights are restricted according to certain defined conditions. Air navigation chart includes restricted areas with their validity periods. Validity of restricted areas could be specified from a certain time in the morning to the evening or on certain days of the week or month.

There is also the possibility of reserving a part of the airspace for a certain time for carrying out some activity, such space is called a Temporary Reserved Area (TRA). Temporary Segregated Area (TSA) is a similar element to TRA, but it is more appropriate to consider it as a softer version of a Dangerous Area. Militaries may be required to segregate some part of airspace for their activities at any time and for any part.

Another important element of the airspace is a network of flight routes. Most flights are currently performed within a flight routes network. Flight routes could be controlled, uncontrolled, advisory, and conditional. During flight planning, airspace users prepare a detailed flight plan which includes a set of waypoints to specify a trajectory of future flight. Each flight plan is validated before approval for any potentially dangerous scenarios. During the flight, airspace users should strictly follow cleared flight plan and maintain a centreline of a predefined flight route. To guarantee airspace users' location within perils of flight routes a performance-based navigation (PBN) uses a various specification of area navigation (RNP/RNAV) [15]. Requirements of RNP/RNAV specify an acceptable level of airplane deviations from predefined trajectory in the form of total system error.

ATM could use dynamic sectorization of airspace management, which is used in cases where it is necessary to satisfy the demand of airspace users without involving additional human resources. Sectorization assumes the segregation of airspace into basic elements in order to combine them into air traffic control sectors to meet requirements of bandwidth and control. Sectorization is useful in some emergency cases, for example, in case of strong turbulence in controlled airspace, as a result, the problematic sector (CTR, TMA) could be partitioned into two children. Each created sector will have its operating frequency and its responsible controller. Different dangerous weather phenomena (tornado or volcanic ash) or increased airspace workload can lead to necessity sectorization.

The main reasons for dynamic sectorization are the following:

- Bad weather conditions;
- Exceeding the expected level of air traffic;
- A large number of aircraft that are in the sector at the same time;
- Difficulty of air traffic which includes many take-off/landing operations;
- Testing of air traffic control systems.

There are certain risks when implementing the procedure of separating or opening sectors:

- Low precision of traffic prediction leads to early or late controller enrolment;
- Incorrect transfer of the airplanes between sectors;
- Too many radio frequencies change the frequency of the sector (since when dividing, instead of the radio frequency of one large sector, we get two different frequencies).

When the situation has normalized, a decision to close or union of sectors could be reached, based on the following factors:

- Improvement of the weather;
- Air traffic levels are back to normal (or below normal);
- Reduction of the number of aircraft in the sector to the norm;
- Normalization of the number of operations.

There are certain risks associated with closing a section, for example, premature closing will lead to overloading of the combined sector, or underestimation of traffic. In some cases, it is possible to change not the number of sectors, but the size of the sector.

Safe air transportation could be guaranteed only in case of correct operation of each part of the airspace. Parts of airspace are specified with the help of geodetic coordinates of latitude and longitude defined for particular altitude perils. On the basis, airspace parts are defined in air navigation publication of ANSP as a set of geodetic points that could be marked in a digital air navigation chart. These basic airspace parts are specified and constantly present in a variety of airspace structure databases. At the same time, airspace volumes of specific geometry could be restricted based on some unpredicted factor action (weather or military) with no prior information. These emergency cases required to development of a way of immediately informing all air space users about segregated airspace geometry in some automatic way to minimize influence of human factors on both ATC and pilots' sides.

3 Global Spatial Index

Nowadays, spatial indexes propose a complex solution for flexible processing, representation, and storing of any data related to geographic coordinates. Global spatial index system considers the shape of ellipsoidal Earth model as a set of cells of different sizes based on resolution level. Hierarchical indexing proposes a unique addressing of each cell in the system. There are three different types of spatial index systems based on cell geometry: triangular, square (or rectangular), and hexagonal. Each type has particular benefits for defined applications. Rectangular geometry of cell is well-suited for mapping and addressing places. Hexagonal has multiple advantages for data processing and navigation.

Spatial index with a regular hexagonal shape of a cell makes any navigation computation simple due to equal ranges to the nearest hexagons. Centers of neighbor cells are placed in vortexes of another bigger hexagon (Fig. 2). Regular hexagonal shape makes computational cheap calculation coordinates of each vortex based on given coordinates of reference point.

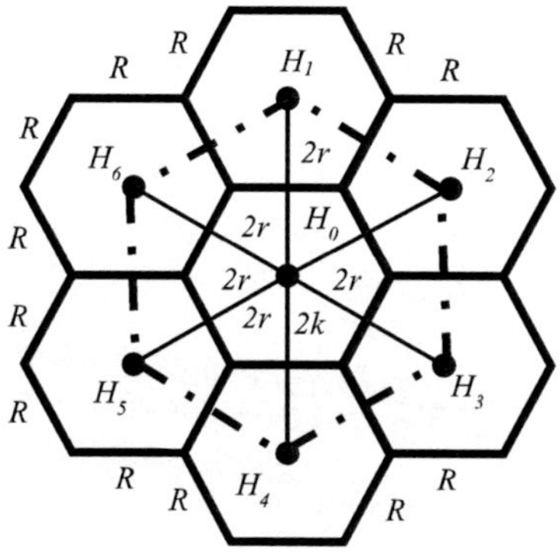

Fig. 2. Hexagonal spatial index.

Hierarchical hexagonal spatial index (H3) provides a spatial index system in a global scale that could be useful not only for statistical processing of geographic data, but also for vehicle navigation and traffic management (including air). The main advantage of H3 is to increase the speed of computation during solving particular tasks which is result of:

- using indexes (addresses) instead of geodetic coordinates;
- using logical operation and set theory instead of complicated trigonometric formulas.

Hierarchical structure of H3 proposes 16 levels of resolution. Basic level includes 122 cells with a 1281 km side length [24]. The highest resolution includes above 5.69 × 1014 cells of 0.58 m side length. Addressing in H3 includes hex code with fifteen chars. Hierarchical addressing makes addresses information of bigger cell present in the cell address of any child's cell. It makes identification of cell children only by comparison addresses. A cell of any precision, except last, has seven children.

We consider usage of a hexagonal spatial index to represent a complicated air space structure. Geometry of airspace parts like areas, zones, and sectors could be represented as a set of hexagonal cells. Hexagons of different resolution levels could be used to describe any geometry based on required precision. As an example, CTR of Lviv International Airport (UKLL) in hexagons is presented in Fig. 3. Hexagonal spatial index uses an address of cell and does not require computation which includes geodetic coordinates. It is the main advantage of H3 which helps to speed up data processing significantly.

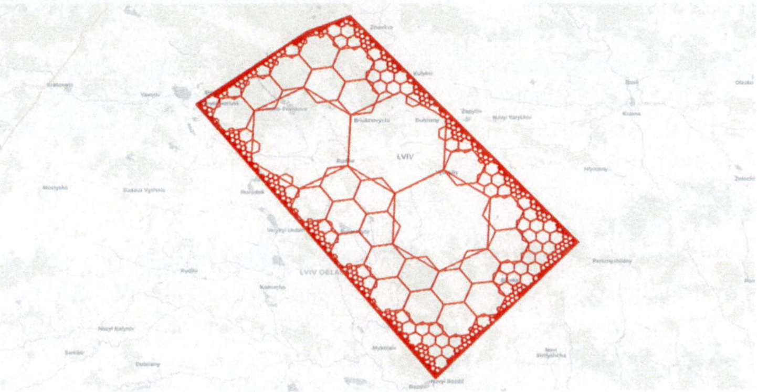

Fig. 3. CTR of Lviv International Airport (UKLL).

Airspace users also could be represented as hexagons. The size of hexagon could be selected dynamically based on precision level of on-board positioning system. It makes hexagons (or set of hexagons) be allocated with an area on uncertainty of each airspace user. Realization of that could be based on navigation category of accuracy which is transmitted together with airplane coordinates in digital messages of Automatic dependent surveillance-broadcast (ADS-B) in mode 1090ES [25].

Thus, both users and airspace could be represented in the space of hexagons. Procedures and operations required for air traffic management also could be done as action with cell data. ATM grounds on a particular set of tasks which includes: Identification of user presence in particular airspace volume; checking airspace users following a predefined trajectory; checking separation minimums between users; and terrain avoidance.

Identification of user entering or usage of particular airspace volume in hexagons is a simple comparison airspace user index with a set of hexagons used for representation of particular airspace structure:

$$P = \begin{cases} 1, & if\ acft \in H = \{h_1, h_2, \ldots, h_n\} \\ 0, & if\ acft \notin H = \{h_1, h_2, \ldots, h_n\} \end{cases}, \tag{1}$$

where P is a binary variable of presents airplane *acft* in the considered airspace; *acft* is an index of airspace user; H is a set of hexagon indexes of different resolutions which are used to describe particular airspace part; n is a number of hexagons used in data representation.

At the practical level, computation of (1) is much simpler due to comparing the indexes only. Search algorithm compares airplane index by each digit with an ordered set of hexagon indexes. Thus, child of parent hexagon is identified automatically as computationally cheap.

Check airspace users following a predefined trajectory could be realized based on two approaches. The first one could use a hexagonal representation of cleared flight path and its comparison with current user position. The second one could be grouted on identification of the two nearest waypoints (hexagons) of flight plan (previous and next one) and identification of deviation in metrics of hexagonal space from a line connecting these waypoints. Obtained value of deviation could be compared to the width of flight route. Airplane index placement out of cleared values generates an alarm based on risk values.

Checking the validity of separation minimums between users could be also easily done by comparing a grid distance between two indexes of airplanes. Grid distance could be calculated at the same level of precision as indexes of airplanes are given or could be calculated at the particular level defined in automatic air traffic control system. In this case, lateral and longitudinal separation are checked. Vertical separation minimums are validated based on barometrical altitude.

Terrain avoidance also could be easily done in H3. Digital elevation model could be represented in the space of hexagons. Thus, each hexagon could be associated with some mean value of relief height. This height usually is counted from global ellipsoidal model (WGS84). Thus, checking of dangerous flight into terrain could be realized as a simple comparison of current airplane altitude in WGS84 with relief height associated with the hexagon index of airplane.

4 Numerical Demonstration

In the numerical demonstration, we test representation of different airspace components in the hexagonal spatial index. A special software in Python has been created to develop a database of spatial indexes that identify airspace structured units in Ukrainian airspace. We use H3 library to operate with hexagonal space and leaflet mapping for visualization of data in OpenStreetMap. As an example, the CTR of Kyiv (Boryspil) International Airport (UKBB) coded by hexagonal spatial index is presented in Fig. 4. In geodetic coordinates this CTR could be represented as a set points: {(50.464 N, 30.706833 E), (50.5775 N, 30.872333 E), (50.573833 N, 30.997667 E), (50.103 N, 30.964667 E), (50.106167 N, 30.805667 E), (50.323167 N, 30.721333 E)}. This CTR is airspace of class D with vertical limits from ground to 900 m above mean sea level.

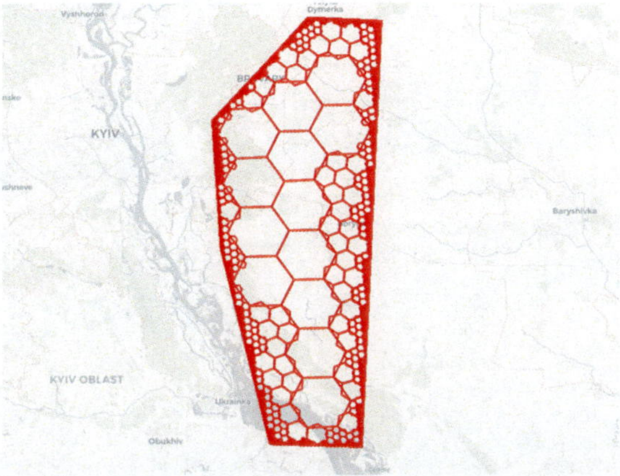

Fig. 4. CTR of Boryspil International Airport (UKBB).

Another example could be coding with hexagons a joined airspace of four training areas of Antonov facility in the Kyiv vicinity (UKT-500, UKT-501, UKT-502, UKT-503). Each of these areas has different altitude perils. Geometry of coded hexagons is presented in Fig. 5. Level of resolution has a significant influence on computation process in ATM. Performance of coding software with different precision levels is presented in Fig. 6. High-resolution levels require more computation time for processing different ATM tasks, thus minimization of cell numbers decreases the size of the hexagonal set. In H3 “compact_cells” helps to identify the best-fitting solution for particular regions. Thus minimizing the lower limit for resolution the following results represented in Table 1 could be obtained. Beginning from the 8th level of resolution minimization of cell numbers gives more than 90% of the saved set size.

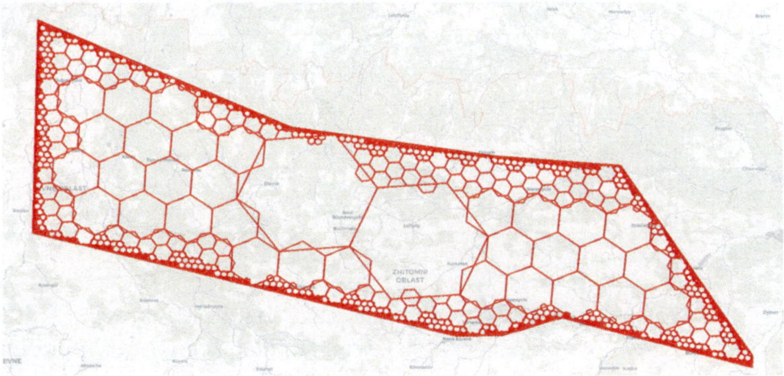

Fig. 5. Airspace of four training areas.

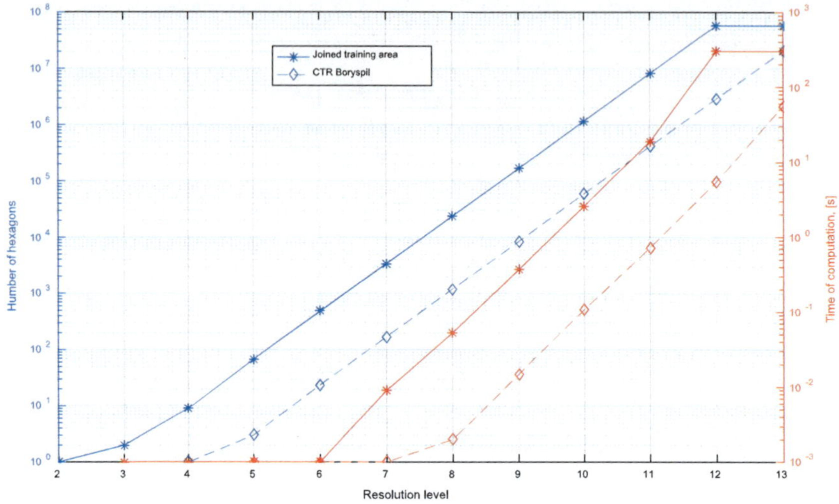

Fig. 6. Performance analysis of airspace representation.

Table 1. Results of effective cell minimization.

Level of resolution	Minimized number of hexagons for UKBB CTR	Saved number of cells for UKBB CTR, %	Minimized number of hexagons for training zone	Saved number of cells for training zone, %
5	3	0	50	26.4
6	18	25	138	71.6
7	73	56.8	434	87.3
8	200	83.4	1212	94.9
9	621	92.6	3302	98.0
10	1740	97.0	9108	99.2
11	4776	98.8	23793	99.7
12	12683	99.5	63277	99.8
13	34047	99.8	174258	99.9

In general case, the whole Ukrainian airspace could be represented as a set of hexagons. Minimized set of hexagons for Ukrainian airspace is shown in Fig. 7. This airspace representation required 1 hexagon for 2nd level of resolution, 33 hexagons for 3rd level, 294 for 5th, 103 for 4th, 817 for 6th, and 2158 for 7th level.

Also, we use airplane trajectory obtained by ADS-B. An open database of airplane trajectory data has been used to recover AUI 55 flight for connection between Boryspil

(UKBB) and Odesa (UKOO) International airports. We use historical flight from December 21, 2021, due to closed Ukrainian airspace now. Airplane trajectory is coded by hexagons. Presents of airplane within Boryspil CTR are simply detected by comparison indexes in hexagons by (1). Results of the detection of airplane flight within departure scheme of UKBB CTR are presented in Fig. 8 with green color. Also, in Fig. 8 a ninth resolution level of hexagons has been used to visualize a trajectory of AUI 55.

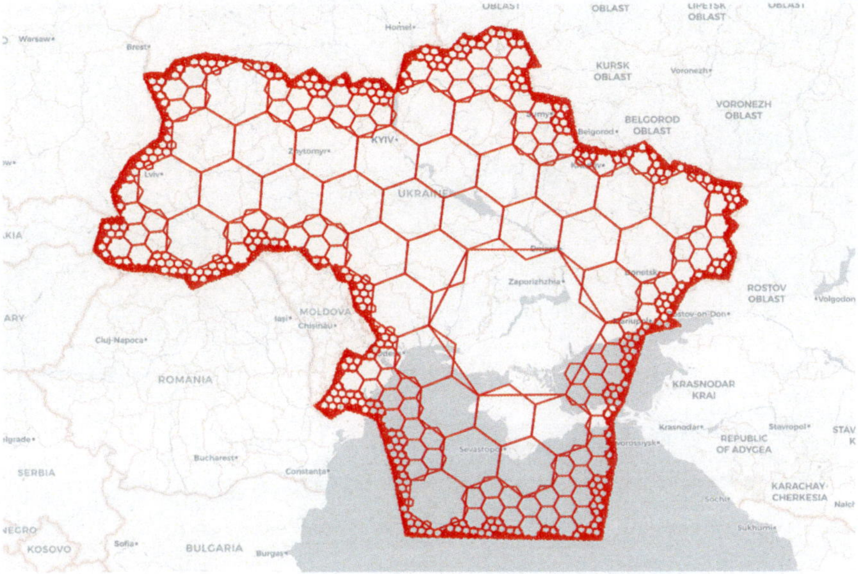

Fig. 7. Representation of Ukrainian airspace with H3.

The perils of CTR are not only lateral there are also vertical dimensions. Height of Boryspil CTR is 900 m about mean sea level [23]. Thus, an airplane could leave CTR early at 48 s of flight due to reach the altitude limit by climbing. Also, it should be noted that airplane trajectory used for simulation is available only for more than 100 m of relief due to poor SDR network capability.

Most ATM tasks could be easily solved with basic logical operations under elements of set in case of spatial index usage. Also, it opens multiple possibilities to integrate more operations in common ATM tasks. As an example, algorithms of automatic mid-air collision avoidance and terrain awareness.

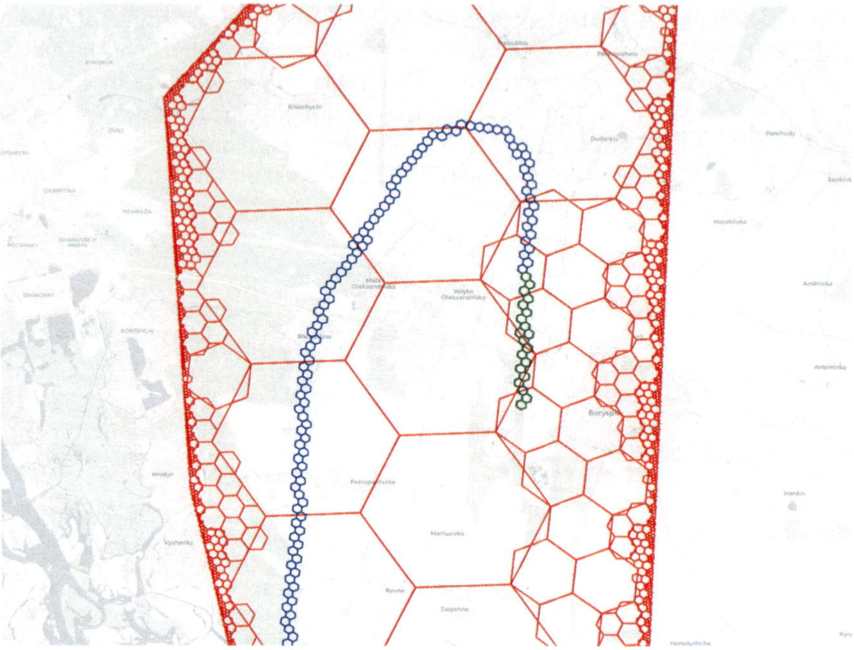

Fig. 8. Trajectory of AUI 55 within CTR Boryspil.

5 Conclusions and Future Works

Spatial indexes propose a reach feature for geodata processing and visualization. Spatial index with regular hexagonal cells proposes a new approach for airspace areas setting up and data processing. This approach is well structured for building an area of complicated shapes. H3 library includes multiple tools for area analysis and navigation. The main advantage of spatial indexing is switching to computational cheap logical operations with addresses or indexes of cell. This helps to simplify computation and uses a basis of set theory for solving particular air traffic control tasks.

Numerical demonstration shows a possibility of implementation of spatial index H3 for setting up a CTR and complicated union on training areas in Ukrainian airspace. On the other side, airspace users also could be associated with particular hexagons based on precision level of positioning sensor. Features of H3 library help to use simple algorithms of address comparison to identify presents of an airplane at a particular part of airspace.

The potential of special index systems proposes a huge step in air traffic management. Operation in hexagonal space makes it possible to integrate complicated navigation algorithms in cheap hardware side or could be run in equipment with limited computation performance. Efficient use of hardware is important for air traffic management within urban air mobility.

References

1. Majka, A., Pasich, A.: Cross-border Free Route Airspace concept and its impact on flight efficiency improvement. In: IOP Conference Series: Materials Science and Engineering, vol. 1226, no. 1, p. 012022. IOP Publishing (2022)
2. Szabo, P., Ferencova, M., Zeleznik, V.: Cloud computing in free route airspace research. Algorithms **15**(4), 123 (2022)
3. Ostroumov, I., Ivashchuk, O., Kuzmenko, N.: Preliminary estimation of war impact in Ukraine on the global air transportation. In: 12th International Conference on Advanced Computer Information Technologies (ACIT), Ruzomberok, Slovakia, pp. 281–284. IEEE (2022). https://doi.org/10.1109/ACIT54803.2022.9913092
4. Hirabayashi, H., Brown, M., Takeichi, N.: Feasibility study of free route airspace over the north pacific. J. Air Transp. **30**(2), 59–70 (2022)
5. Dudek, E.: ATM functionality development concept with particular analysis of free route airspace operations in Poland. In: Siergiejczyk, M., Krzykowska, K. (eds.) ISCT21 2019. AISC, vol. 1032, pp. 94–103. Springer, Cham (2020). https://doi.org/10.1007/978-3-030-27687-4_10
6. Ostroumov, I.V., Kuzmenko, N.S.: Risk assessment of mid-air collision based on positioning performance by navigational aids. In: 6th International Conference on Methods and Systems of Navigation and Motion Control (MSNMC), Kyiv, pp. 34–37. IEEE (2020). https://doi.org/10.1109/MSNMC50359.2020.9255506
7. Liu, Z., Chen, L., Yang, A., Ma, M., Cao, J.: Hiindex: an efficient spatial index for rapid visualization of large-scale geographic vector data. ISPRS Int. J. Geo Inf. **10**, 647 (2021)
8. Uher, V., Gajdos, P., Snasel, V., Lai, Y.C., Radecky, M.: Hierarchical hexagonal clustering and indexing. Symmetry **11**(6), 731 (2019)
9. Li, S., Pu, G., Cheng, C., Chen, B.: Method for managing and querying geo-spatial data using a grid-code-array spatial index. Earth Sci. Inf. **12**, 173–181 (2019)
10. Li, Z., Chen, L., Wang, Y.: G-Tree: an efficient spatial index on road networks. In: 35th International Conference on Data Engineering (ICDE), pp. 268–279. IEEE (2019)
11. Wang, R., Ben, J., Zhou, J., Zheng, M.: Indexing mixed aperture icosahedral hexagonal discrete global grid systems. ISPRS Int. J. Geo Inf. **9**(3), 171 (2020)
12. Ma, Y., et al.: A precision evaluation index system for remote sensing data sampling based on hexagonal discrete grids. ISPRS Int. J. Geo Inf. **10**(3), 194 (2021)
13. Birch, C.P., Oom, S.P., Beecham, J.A.: Rectangular and hexagonal grids used for observation, experiment and simulation in ecology. Ecol. Model. **206**(3–4), 347–359 (2007)
14. Nikitina, T., et al.: Algorithm of robust control for multi-stand rolling mill strip based on stochastic multi-swarm multi-agent optimization. In: Shukla, S., Sayama, H., Varghese, J., Kumar, D. (eds.) Data Science and Security, IDSCS 2023. LNNS, vol. 922, pp. 1–12. Springer, Singapore (2023, in press)
15. Performance-Based Navigation (PBN) Manual. Doc 9613, 4th edn. ICAO (2013)
16. Ostroumov, I.V., Kharchenko, V.P., Kuzmenko, N.S.: An airspace analysis according to area navigation requirements. Aviation **23**(2), 36–42 (2019). https://doi.org/10.3846/aviation.2019.10302
17. Kuzmenko, N.S., Ostroumov, I.V.: Performance analysis of positioning system by navigational aids in three dimensional space. In: 1st International Conference of System Analysis & Intelligent Computing (SAIC 2018), Kyiv, pp. 101–104. IEEE (2018). https://doi.org/10.1109/SAIC.2018.8516790
18. Solomentsev, O., Zaliskyi, M., Nemyrovets, Y., Asanov, M.: Signal processing in case of radio equipment technical state deterioration. In: Signal Processing Symposium 2015 (SPS 2015), Debe, Poland, pp. 1–5. IEEE (2015). https://doi.org/10.1109/SPS.2015.7168312

19. Solomentsev, O., Zaliskyi, M.: Correlated failures analysis in navigation system. In: 5th International Conference on Methods and Systems of Navigation and Motion Control, Kyiv, Ukraine, pp. 41–44. IEEE (2018). https://doi.org/10.1109/MSNMC.2018.8576306
20. Okoro, O.C., Zaliskyi, M., Dmytriiev, S., Solomentsev, O., Sribna, O.: Optimization of maintenance task interval of aircraft systems. Int. J. Comput. Netw. Inf. Secur. **14**(2), 77–89 (2022). https://doi.org/10.5815/ijcnis.2022.02.07
21. Solomentsev, O., Zaliskyi, M., Kozhokhina, O., Herasymenko, T.: Efficiency of data processing for UAV operation system. In: 4th International Conference on Actual Problems of Unmanned Aerial Vehicles Developments, Kyiv, Ukraine, pp. 27–31. IEEE (2017). https://doi.org/10.1109/APUAVD.2017.8308769
22. Air traffic management, Procedures for Air Navigation Services, Doc. 4444. ICAO (2016)
23. Aeronautical Information Publication of Ukraine. Ukrainian State Air Traffic Services Enterprise (2020)
24. Hexagonal hierarchical geospatial indexing system specification. https://h3geo.org. Accessed 08 Dec 2023
25. Ostroumov, I., Kuzmenko, N.: Statistical analysis and flight route extraction from automatic dependent surveillance-broadcast data. In: Integrated Communication, Navigation and Surveillance Conference (ICNS), Dulles, VA, USA, pp. 1–9. IEEE (2022). https://doi.org/10.1109/ICNS54818.2022.9771515

Aircraft Flight Path Correction During the Flight Execution Phase Using the Operative Information on Weather Hazards

Maxim Ivanytskyi, Yuliya Averyanova, Yevheniia Znakovska(✉), and Bohdan Shershen

National Aviation University, Liubomyra Huzara Avenue, 1, Kyiv 03058, Ukraine
zea@nau.edu.ua

Abstract. Modern aviation system operation is based on advanced technologies that allow to obtaining of crucial flight information in real-time. Systems of air navigation support share different data with all air space users to ensure the safety of air transportation. It plays an important role in operative flight trajectory correction when hazardous areas are detected along the pre-planned flight trajectory. However, this data may be unavailable at the stage of flight planning. In this paper, we consider weather-related hazards that influence the flight during the en-route flight phase. Also, we consider potential of modern information systems to obtain and share operative meteorological information as well as to use this information for in-flight trajectory correction. The Bellman algorithm was chosen to calculate the shortest part taking into account the operatively detected hazards. The simulation of trajectory correction taking into account different weather situation in the same area is done and results of simulation are discussed.

Keywords: Air Navigation · Flight Path Correction · Trajectory Correction · Weather-related Hazards

1 Introduction

The prospective aviation system aims to provide the safe and combined operation of air vehicles of different types including high-altitude, civil and transport, unmanned, space, as well as different missions that can be realized by them [1]. At the same time, the level of safety and security is expected to be improved. The possibility for operation of such a diverse aviation system is based on modern and prospective technologies that allow to obtain operative and crucial flight information in real-time, sharing this information with participants of air traffic, implementing such services as Air Traffic Flow Management, or concepts as Trajectory Based Operations [2] and Information-centered systems [3]. It is expected that mentioned improvements will serve to enhance the common situation awareness and view of all participants of air traffic and air traffic management. Enhancing abilities, in addition to the advantages of common information space and situation awareness are shown in Fig. 1.

I. Ostroumov and M. Zaliskyi (Eds.): ACASD 2024, LNNS 992, pp. 31–41, 2024.
https://doi.org/10.1007/978-3-031-60196-5_3

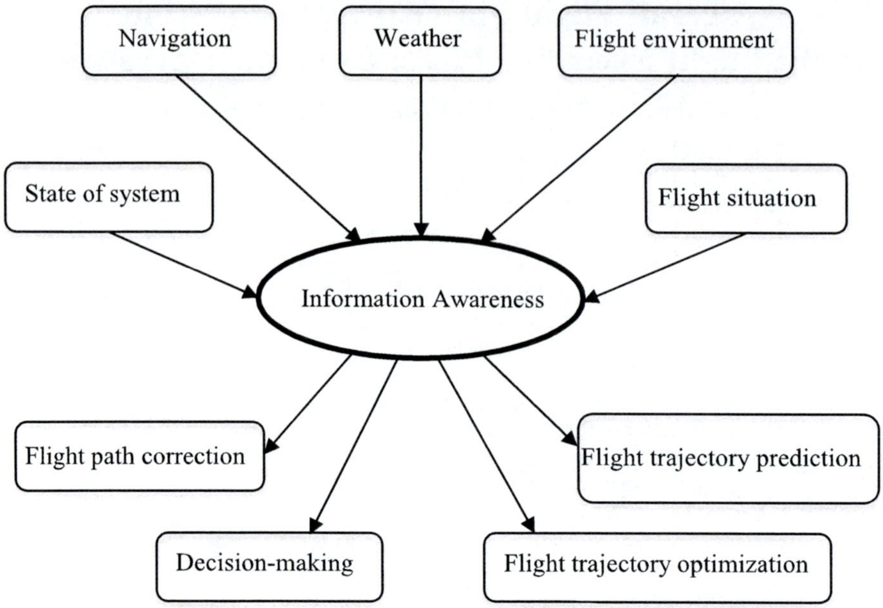

Fig. 1. Common information space and situation awareness advantages.

Common information awareness is based on fused and diverse information (Fig. 1) and, then, service-oriented products can be created to help in decision-making, close to optimal in-flight flight path correction and optimization as well as trajectory prediction.

In [4] it is indicated, that nowadays, the interacted flight trajectories are computed independently, and limited data is shared between involved participants. This can increase the disruption in air operations or influence flight safety. One of the important factors that influence flight safety and punctuality is weather [5]. The current and forecasted meteorological information about weather along the flight route is an important for flight planning. But, forecasted information has a probabilistic nature. Moreover, the atmosphere is a highly dynamic medium and weather phenomena are characterized by high variability and dynamics. Therefore, real-time data about current weather along the flight route is important for avoiding potentially hazardous areas during the flight. This becomes especially important under the conditions of increased numbers of convective weather events and turbulence during the last time [5, 6].

2 Motivation

Therefore, it was our motivation to study the weather conditions that influence the flight during en-route flight phase, consider abilities of modern systems to obtain and share operative meteorological information as well, as use this information for in-flight trajectory correction. The paper is organized in the next way: the short overview of the current research is presented in the 3rd section, the flight correction algorithm, simulation results, and discussions are presented in Sect. 4 and conclusions are given in Sect. 5.

3 Short Overview of Current Research and Achievements

Nowadays, there are a lot of research that investigates the influence of weather on flight operations. In paper [7] the phenomena that should be taken into account at pre-departure and sufficiently affect the take-off and landing are studied to quantified the limitations connected with specific airport equipment. Statistics given in the paper show that weather-related accidents cause fatal consequences during the stages of maneuvering and during flight along the flight route. At the same time, most accidents are found during the landing and take-off phases of flight. The estimation of flight delays caused by bad weather conditions is done in [5, 8]. The local weather impact on the airport operation is discussed as well as classification of airport performances on the base of weather data is made in [9]. In the paper [10] the dangerous for flight weather phenomena that are common in European airspace are analyzed. In this paper [11] the weather phenomena that are dangerous for low-level flights and Unmanned aerial systems operations are analyzed. The impact of weather on aviation transportation system proved by statistics [7, 12] shows the necessity to develop systems for detection of dangerous weather phenomena during the in-flight stage and mechanisms for improvement of meteorological service of flight.

The method of remote obtaining information about atmospheric processes and dangerous weather for aviation are considered in [13]. The prospective methods of obtaining information about dangerous wind-related phenomena during aircraft flight are presented in [14–18]. In these papers, the possibilities to detect dangerous weather phenomena using Doppler and polarimetric radars are presented and discussed. In paper [19] a method to detect turbulence by aircraft based on Mode surveillance downlink data. Sensors for detecting in-flight turbulence are presented in [20]. Instrument for detection of in-flight icing areas with satellite data is discussed in [21]. Ice detection methodology based on degradation of flight performances is presented in [22]. In the paper [23] the airborne sensor for measuring parameters of magnetic fields is presented. Method of hail zones detection using satellite-based radiometers is presented in [24]. Hail studies using the radar system are presented in [25].

The overview of new technologies can be continued. However, it is evidently, that new technology of operative obtaining information about weather hazards during the flight can sufficiently increase flight safety and regularity. On more key factor is the operative exchange of meteorological information between participants of air traffic. The modern aviation concepts that consider operative data exchange and on-demand access include system wide information management [26] and collaborative trajectory options program (CTOP) [27]. The concepts that allow the operative data obtaining and exchange are studied in [28, 29].

The combination of the achievements in operative data obtaining and information sharing opens the ability to improve algorithms for in-flight trajectory correction and optimization. Having a common awareness about current state of aviation system can assist in improvements of trajectory prediction as well. Papers [8, 30] also consider the utilization of operative data for in-flight path correction to avoid potentially dangerous regions. In this paper, we use the operative meteorological data to build the algorithm for in-flight path correction to avoid weather hazards. We are focusing on Bellman algorithm to find the shortest way and avoid hazardous areas detected during the flight.

4 Trajectory Correction Algorithm and Simulation

Bellman's algorithm computes the shortest paths from one source vertex to all other vertices in a weighted directed graph. Of course, it is slower than Dijkstra's algorithm for the same problem, but it is more versatile, as it can handle graphs in which the weight of some edges is negative [30].

As mentioned above, Bellman's algorithm is suitable for working with graphs that have negative weight edges. However, if a graph contains a "negative cycle", a cycle whose sum of edge weights is negative, then there is no shortest path tree for this graph (any path of this type can be improved by another pass through the edges that form the negative cycle). In this case, the Bellman algorithm can detect cycles of negative length and report their existence, but it cannot give the correct answer, and find the shortest path if a negative cycle is reachable from the source vertex.

Consider a directed graph with weighted edges that does not contain cycles of negative length. Also, let's suppose that we need to find the shortest paths from the selected vertex to all other vertices in graph.

All vertices of graph are considered untraversed and the edges are considered unviewed. During the execution of the algorithm, each vertex is assigned a number d_x equal to the length of the shortest path from vertex A to vertex X, which includes only traversed vertices. In the first step, we assume $d_a = 0$ and $d_x = 0$ for all, other than A. Also, at this step, the vertex is assigned the label "traversed" and $y = a$ (y is the last of the traversed vertices) is assigned.

For each of the vertices of the graph G, recalculate the following value d_x:

$$d_x = \min\{d_x, d_y + m_{yx}\}, \tag{1}$$

where m_{yx} is the weight of the edge (y, x).

If $d_x = \infty$ for all untraversed vertices X, the Bellman algorithm must be completed (there are no paths from vertex A to untraversed vertices in the original graph). Otherwise, the label "traversed" should be assigned to the vertex d_x with the smallest value. After that, by setting $y = x$, the iteration process continues. Note, if the value of d_x, decreases for some passed vertex, then the corresponding labels are removed from this vertex and its incident revised edge.

The procedure of the Bellman algorithm is completed only when all the vertices of the graph G have been traversed and when, after the next execution of step two, none of the numbers d_x has changed its value.

According to the algorithm, at the first step, we assign the label "passed" to vertex number one and assume that $d_1 = \infty$, $d_2 = \infty$, $d_3 = \infty$, $d_4 = \infty$, $d_5 = \infty$. After that, using $y = 1$, we recalculate the values of d_x.

$$d_2 = \min\{d_2, d_y + m_{y2}\} = \min\{\infty, 0 + 1\} = 1; \tag{2}$$

$$d_3 = \min\{d_3, d_y + m_{y3}\} = \min\{\infty, 0 + \infty\} = \infty; \tag{3}$$

$$d_4 = \min\{d_4, d_y + m_{y4}\} = \min\{\infty, 0 + \infty\} = \infty; \tag{4}$$

$$d_5 = \min\{d_5, d_y + m_{y5}\} = \min\{\infty, 0 + 3\} = 3. \tag{5}$$

Since the length value is less than, the label "traversed" is assigned to node "2" shown in Fig. 2. Along with it, the edge, which makes up the current shortest path tree, is also considered to have been traversed.

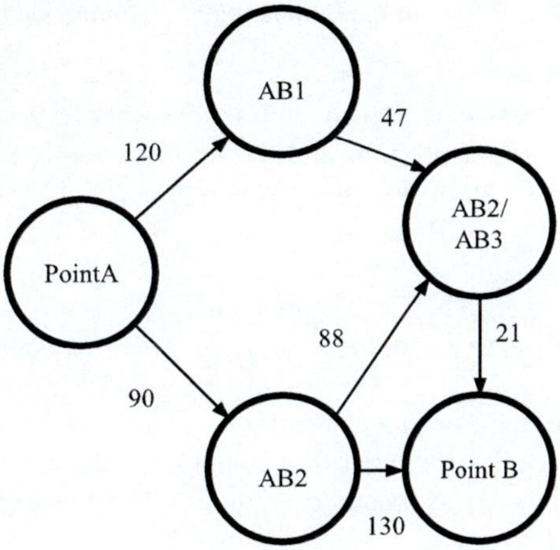

Fig. 2. Block diagram of the Bellman-Ford algorithm.

Therefore, for our situation, it is possible to demonstrate the efficiency of our aircraft on a certain section of the route in order to use fuel efficiently and save it during the flight in Fig. 3. This is achieved by pilots using headwinds that run parallel and in the direction of the aircraft, which makes it easier to pull and allows the pilot to use less fuel. In our situation, we took into account the weather conditions on the route and the clouds that are dangerous for the flight. Taking these factors into account, the flight correction is made using the Bellman-Ford algorithm. The two cases are simulated additionally and, then, considered. One is the correction using the pre-flight prognostic information. The other correction is realized using the operative data about meteorological hazards. The increase in efficiency when using the operative data is demonstrated.

The flight image shown in the figure illustrates a simulation of a flight from New York to Gander, taking into account weather conditions and potential risks. Figure 4 shows three different routes. During simulation, we use data about the forecasted weather situation over America from the National Oceanic and Atmospheric Administration.

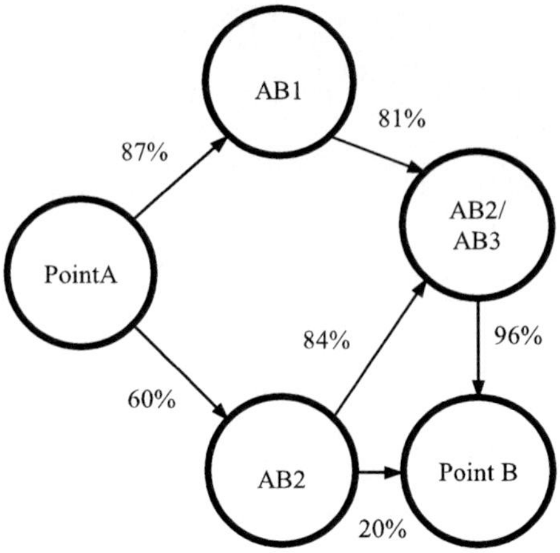

Fig. 3. Block diagram of the Bellman-Ford algorithm with effectivity.

The correction process takes into account the weather forecast at each stage of the route to avoid deteriorating conditions and ensure flight safety and operational data of the onboard systems. The algorithm also takes into account different route options to ensure optimal fuel efficiency. Random deviations from the standard route allow for the simulation of unforeseen situations such as wind direction changes, turbulence, or other factors that may affect the route. Each route is marked on the map and is accompanied by a legend: the length of the route, the altitude, and the fuel consumed to complete the route.

The duration of the flight from New York to Gander is an important planning aspect and software uses the Bellman's algorithm to optimize the route. It helps to take into account various factors such as weather conditions and fuel consumption to find the optimal path between two points.

The route starts from John F. Kennedy Airport (New York), where the runway is suitable for the start of the flight, the wind reaches 3 m/s and the temperature outside is -1 °C. The pilot sees that route he planned to take has obstacles in air between point of departure and landing point. At the Gander airfield, the pilot can see that the weather is clear, calm, and there is no turbulence nearby, which makes it easy to land the aircraft.

Figure 4 shows that the aircraft is passing through a zone of isolated cumulonimbus (CB) clouds. According to the requirements, intentional flight inside a CB cloud is prohibited. However, in the case of isolated clouds, they occupy less than 50% of the encircled area. Obviously, under such circumstances, inside the marked area, it is possible to find areas that may be free of CB clouds. The distribution of clouds is unknown. Therefore, up-to-date information on their current position, direction, and speed of movement can be crucial for the correct trajectory correction. It is important also for the trajectory correction algorithm operation.

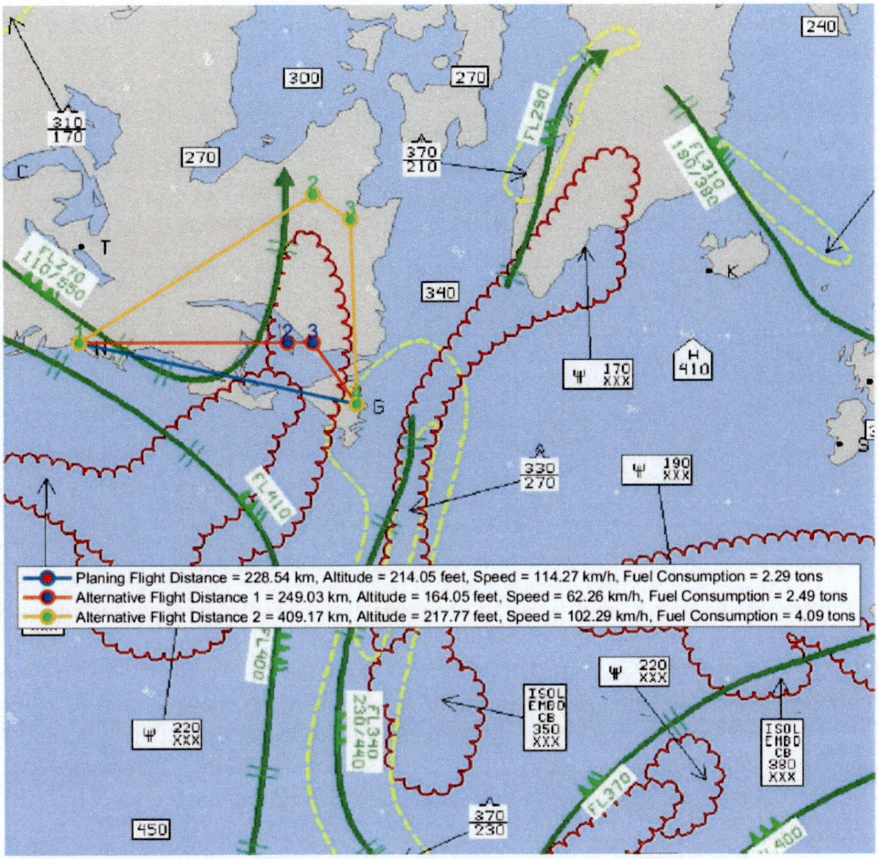

Fig. 4. An image of the flight paths of an aircraft (planned – line 1 and alternatives – lines 2 and 3, taking into account weather conditions.

The avoidance of potentially hazardous areas is possible when operative data are used in the flight trajectory correction algorithm. The example of correction can be seen in more detail in Fig. 4. The flight simulation shows the three flight paths. The first one that is shown with a blue line (line 1 in Fig. 4) is intentional flight. According to the rules, the area of potential hazards should be avoided. We can see the increased length of a flight path (shown with line 2 in Fig. 4). Although it is possible to fly straight in some areas that operatively detected as not dangerous. The correction of flight path taking into account the operative information about areas that are free from CB clouds (distance to the CB is less than 15 km measured with onboard weather radar) is shown with line 3 in Fig. 4. In this case, deviations from the intentional flight path are significantly less that help to reduce fuel consumption and keep the safety as priority.

In Table 1 the comparison of the flight parameters is shown. It is possible to see from the table, the significant economy in fuel consumption when flight correction using the operative data compared to the Alternative Flight (line 2) that is based on the pre-flight forecasted information.

Table 1. The comparison of the flight parameters.

	Distance, km	Altitude, feet	Speed, km/h	Fuel Consumption, tons	Efficiency, %
Planned Flight Distance 1	228.32	209.77	114.16	2.28	84
Alternative Flight Distance 3	253.29	180.46	84.43	2.53	88
Alternative Flight Distance 2	425.83	224.17	106.46	4.26	87

Results of simulation shown in Fig. 5 demonstrates another situation in the same area. In this case, if current and forecasted conditions in Gander airport allow pilot to make decision to flight, we analyze the forecasts along the flight route. The map shows that the clouds around this airport in the area of 50 km are characterized by a high probability of turbulence associated with isolated CB clouds. The simulated trajectory 2 is done with safety priority and using the information obtained during the preparation of the flight and operative data and finding the safe area close to the intentional trajectory 1.

In Table 2 the comparison of the flight parameters for another situation in the same area is shown. It is possible to see from the table the insignificant drop in total efficiency when weather hazards avoidance. But this is due to the keeping the safety.

Table 2. The comparison of the flight parameters.

	Distance, km	Altitude, feet	Speed, km/h	Fuel Consumption, tons	Efficiency, %
Planned Flight Distance	206.16	214.05	103.08	2.06	93
Alternative Flight Distance 1	245.62	164.05	61.4	2.46	94

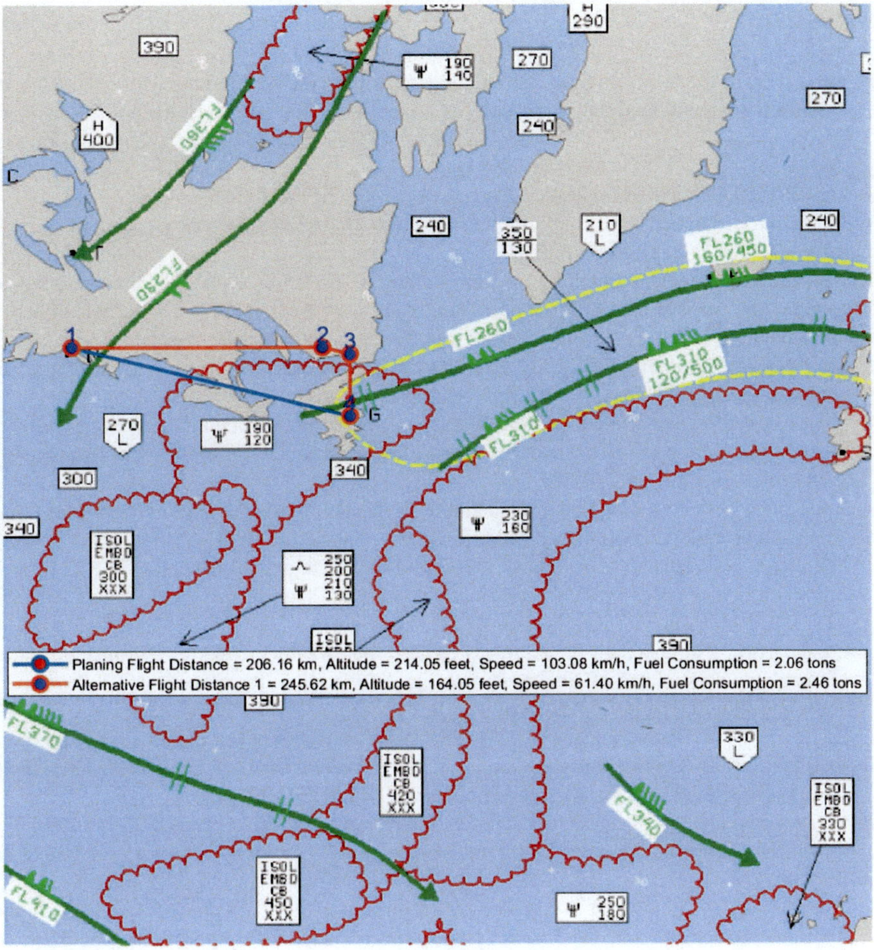

Fig. 5. An image of aircraft flight paths taking into account another weather conditions.

5 Conclusions

In this paper, we consider and simulate the aircraft trajectory correction using the Bellman algorithm and operative information about weather hazards along the flight routes. The benefits of operative meteorological information when in-flight path correction was demonstrated. The use of on-time information in the trajectory correction algorithm allows us to find the areas that are free from weather hazards inside zones that are marked as hazardous for flight in prognostic charts. The approach allows to increase the flight safety and reduces fuel consumption and flight time. It is our plan also to study the impact of operational information exchange between different aircraft flying in the same region into the awareness about precise positions of zones of potential hazards and their intensity.

References

1. Barney, L., Mallard, K., Buck, B.: Simulation of a representative future trajectory-based operations environment. National Aerospace and Space Administration (2020). https://ntrs.nasa.gov/api/citations/20205006885/downloads/NASA-TM-20205006885FINAL.pdf. Accessed 20 Jan 2024
2. Global TBO Concept. https://www.icao.int. Accessed 20 Jan 2024
3. Info-Centric National Airspace System. https://www.faa.gov/about/office_org/headquarters_offices/ang/icn. Accessed 20 Jan 2024
4. Network 4D Trajectory CONOP, 1st edn., Reference nr. 23/08/28/46, Eurocontrol (2023)
5. Skultety, F., Jarosova, M., Rostas, J.: Dangerous weather phenomena and their effect on en-route flight delays in Europe. Transp. Res. Procedia **59**, 174–182 (2021)
6. Storer, L.N., Williams, P.D., Gill, P.G.: Aviation turbulence: dynamics, forecasting, and response to climate change. Pure Appl. Geophys. **176**, 2081–2095 (2019)
7. Kwasiborska, A., Grabowski, M., Sedlackova, A.N., Novak, A.: The influence of visibility on the opportunity to perform flight operations with various categories of the instrument landing system. Sensors **23**, 7953 (2023)
8. Rodriguez-Sanz, A., Cano, J., Fernandez, B.R.: Impact of weather conditions on airport arrival delay and throughput. Aircr. Eng. Aerosp. Technol. **94**, 60–78 (2021)
9. Schultz, M., Reitmann, S., Alam, S.: Predictive classification and understanding of weather impact on airport performance through machine learning. Transp. Res. Part C Emerg. Technol. **131**, 103119 (2021)
10. Taszarek, M., Kendzierski, S., Pilguj, N.: Hazardous weather affecting European airports: climatological estimates of situations with limited visibility, thunderstorm, low-level wind shear and snowfall from ERA5. Weather Clim. Extremes **28**, 100243 (2020)
11. Averyanova, Y., Znakovskaja, Y.: Weather hazards analysis for small UASs durability enhancement. In: 6th International Conference on Actual Problems of Unmanned Air Vehicles Developments (APUAVD), Kyiv, Ukraine, pp. 41–44. IEEE (2021)
12. Serious incidents. https://skybrary.aero/articles/serious-incident. Accessed 20 Jan 2024
13. Doviak, R., Zrnic, D.: Doppler Radar and Weather Observations. Academic Press, San Diego (1993)
14. Averyanova, Y., Yanovsky, F., Averyanov, A.: Connection of reflected radar signal with liquid-hydrometeor deformation rate. In: 3rd Symposium on Microwaves, Radar and Remote Sensing Proceedings, Kiev, Ukraine, pp. 217–219. IEEE (2011)
15. Averyanova, Y., Larin, V., Kuzmenko, N., Ostroumov, I., Zaliskyi, M., Solomentsev, O.: Turbulence detection and classification algorithm using data from AWR. In: IEEE 2nd Ukrainian Microwave Week (UkrMW), Ukraine, pp. 518–522. IEEE (2022)
16. Averyanova, Y., Shcherbina, O., Yanovsky, F., Ostroumov, I., Kuzmenko, N., Solomentsev, O.: Model for wind-related phenomena estimation using polarization characteristics of microwave radar signal. In: 16th International Conference on Advanced Trends in Radioelectronics, Telecommunications and Computer Engineering Proceedings, Ukraine, pp. 173–177. IEEE (2022)
17. Averyanova, Y., Shcherbina, O., Yanovsky, F., Ostroumov, I., Kuzmenko, N., Solomentsev, O.: Polarimetric-radar drop size evaluation for wind speed estimate based on Weber criterion. In: Signal Processing Symposium Proceedings, Ukraine, pp. 17–22. IEEE (2021)
18. Popov, A., Tserne, E., Volosyuk, V., Zhyla, S., Pavlikov, V., Ruzhentsev, N.: Invariant polarization signatures for recognition of hydrometeors by airborne weather radars. In: Gervasi, O., et al. (eds.) Computational Science and Its Applications, ICCSA 2023, vol. 13956, pp. 201–217. Springer, Cham (2023). https://doi.org/10.1007/978-3-031-36805-9_14

19. Olive, X., Sun, J.: Detecting and measuring turbulence from mode-S surveilance downlink data. In: ICRAT Proceedings, pp. 1–8 (2020)
20. Microphone to detect in-flight turbulence passes initial flight tests. https://www.aerospacetestinginternational.com/news/flight-testing/microphone-to-detect-in-flight-turbulence-passes-initial-flight-tests.html. Accessed 20 Jan 2024
21. Zollo, A., Bucchignani, E.: A tool for remote detection and nowcasting of in-flight icing using satellite data. SAE technical paper 2023-01-1489 (2023)
22. Deiler, C., Fezans, N.: Performance-based ice detection methodology. J. Aircr. **57**(2), 209–223 (2020)
23. Sushchenko, O., Bezkorovalnyl, Y., Solomentsev, O., Kuzmenko, N., Golitsyn, V., Zaliskyi, M.: Airborne sensor for measuring components of terrestrial magnetic field. In: 41st International Conference on Electronics and Nanotechnology Proceedings, Ukraine, pp. 687–691. IEEE (2022)
24. Sante, L., Monte, G., Levizzani, V., Ferraro, R.R., Beauchamp, J.: A new method for hail detection from the GPM constellation: a prospect for a global hailstorm climatology. Remote Sens. **12**(21), 3553 (2020)
25. Fluck, E., Kunz, M., Geissbuehler, P., Ritz, S.P.: Radar-based assessment of hail frequency in Europe. Nat. Hazards Earth Syst. Sci. **21**, 683–701 (2021)
26. SWIM. https://www.eurocontrol.int/concept/system-wide-information-management. Accessed 20 Jan 2024
27. Collaborative trajectory options program. https://www.faa.gov/regulations_policies/advisory_circulars/index.cfm/go/document.information/documentID/1024987. Accessed 20 Jan 2024
28. Averyanova, Y.A., Rudiakova, A.N., Yanovsky, F.J.: Aircraft trajectories correction using operative meteorological radar information. In: International Radar Symposium Proceedings (IRS), Poland, pp. 256–259. IEEE (2020)
29. Ostroumov, I.: Relative navigation for vehicle formation movement. In: 3rd KhPI Week on Advanced Technology (KhPIWeek), Kharkiv, Ukraine, pp. 1–4. IEEE (2022)
30. Shortest Path: Dijkstra's and Bellman-Ford. https://courses.cs.duke.edu/spring18/compsci330/Notes/shortestpath.pdf. Accessed 20 Jan 2024

Data Processing in Civil Aviation

Enhancing Altitude Data Accuracy in Small Aircraft Systems Using Standard Kalman Filters

Olena Kozhokhina[1], Yaroslav Yakovlev[2], Liudmyla Blahaia[1](✉), Olga Shcherbyna[1], and Serhii Yehorov[1]

[1] National Aviation University, Liubomyra Huzara Avenue, 1, Kyiv 03058, Ukraine
b_ludmila@ukr.net
[2] Antonov Company, Akademika Tupoleva Street, 1, Kyiv 03062, Ukraine

Abstract. Accurate altitude estimation is a critical aspect of small aircraft system operational efficiency and design. This research studies pressure sensors and methods that could improve the accuracy of altitude data for small aircraft applications. Digital pressure sensors BMP180 and BMP280 have been used for data fusion. Altitude data derived from disparate sources have less uncertainty than if they were used individually. Despite the benefits of sensor fusion of the BMP180 and BMP280 sensors themselves, their sensitivity to noise presents a significant challenge requiring noise reduction strategies, such as filtering techniques. Standard Kalman filter (SKF) has been used for enhancing altitude data accuracy due to its robust real-time processing and ability to control linear systems noise. SKF reduces noise, increasing height measurement accuracy. The proposed methodology was substantiated through the integration of sensor fusion, on an Arduino Uno platform. Also, we consider application of these pressure sensors with SKF data processing in the altitude control systems of airplanes.

Keywords: Altitude Measurement · Pressure Sensors · BMP180 · BMP280 · Standard Kalman Filter · Noise Reduction · Sensor Redundancy · Compact Aircraft Systems · Autonomous Control · Data Processing

1 Introduction

1.1 Introduction to the Problem

Accurate altitude measurement is critical to aircraft operational efficiency. Despite being the standard method for measuring altitude, traditional barometric methods are prone to the following errors: instrument errors as a result of manufacture, installation, maintenance, instrument lag, position errors as a result of vent placement and ice blockage from static vents [1]. Moreover, because of their dimensions and weight, not all are suitable for use in aviation systems that require low weight and size.

The development of advanced pressure-measuring devices has resulted from computer technologies and integrated circuits.

I. Ostroumov and M. Zaliskyi (Eds.): ACASD 2024, LNNS 992, pp. 45–56, 2024.
https://doi.org/10.1007/978-3-031-60196-5_5

It should be noted, that these sensors are sensitive to noise. This may corrupt critical data, resulting in a degradation in navigation accuracy and the integrity of navigation systems.

As a result of piezo-resistant technology for enhanced electromagnetic compatibility robustness and high accuracy and linearity, the BMP180 micro-machined pressure sensors will deliver reliable readings over long periods. BMP280 is manufactured using Advanced Porous Silicon Membrane MEMS that offers lower noise levels.

However, both BMP180 and BMP280 sensors are not immune to errors that can occur due to temperature fluctuations, pressure variability, vibration and mechanical shock, humidity and moisture, drift, manufacturing variability, calibration issues, and other factors [2].

To overcome these errors that affect altitude sensors accuracy and to reduce noise, a Standard Kalman filter (SKF) could be useful. SKF is an analytical method that characterizes the linear data model of the sensors in order to provide real-time processing in resource-constrained systems with the highest computational efficiency. SKF was chosen as the primary filter in this study. The predictive capabilities of this system are critical for maintaining the stringent stability and control required for these systems.

As discussed in the following sections, a SKF will be integrated with the BMP180/280 sensors. This will provide more accurate and reliable altitude measurements for small aircraft systems.

1.2 Motivation

Altitude measurement accuracy is especially significant for small manned aircraft and is a key factor in flight safety and operational efficiency. Due to the increasing number of small aircraft applications, this is especially true. In this case, digital pressure sensors based on MEMS technology would be the best alternative to traditional barometric systems.

In such systems, altitude measurements are susceptible to noise inaccuracies, which create operational risks and reliability concerns. In order to improve pressure sensor data accuracy, advanced filtering methods, including Kalman filters, and especially SKFs, are being investigated and analyzed. Using sophisticated data processing techniques to improve altitude measurement in small aircraft systems motivates this research.

1.3 Contribution

A contribution of this study is to demonstrate how filtering methods of SKF can be used to obtain more accurate and less noisy altitude data for small aircraft.

Research increases the accuracy of altitude measurements as well as the operational reliability of system. Due to their computational efficiency and real-time processing capabilities, Kalman filters are effective in dealing with linear and Gaussian noises. In this study, the approach is validated using BMP180 and BMP280 sensors. Preliminary mathematical simulations were conducted using data obtained from the BMP180. Additionally, a microcontroller was used to perform real-time data processing with the BMP280.

As a result of the study and the filtering of real-time altitude data, SKF has demonstrated its suitability for environments in which sensor behaviour and environmental conditions can be more accurately described by linear models. In this study, SKF noise filtering is demonstrated to be effective in existing altitude control systems for compact aircraft. As a result, flight safety and efficiency will be significantly improved.

1.4 The Organization of the Paper

This paper has six sections, starting with an introduction. The second section summarizes existing research and technologies relevant to altitude data refinement in aircraft systems. The third section describes the experimental methodology and materials used to apply SKF to sensor data. The fourth section presents empirical evidence of SKF effectiveness. The conclusions and future scope of the study are presented in sections five and six.

2 Literature Review and Problem Statement

Previous studies provide a comprehensive overview of the challenges and technological advancements in navigation systems, emphasizing the critical importance of sensor accuracy in compact aircraft systems. Well-studied are correlated failure analyses within these systems, underscoring the need for reliable data to support system health monitoring and to enhance operational efficiency [1].

Research into statistical data processing for radio equipment addresses challenges arising from technical condition deterioration, highlighting the significance of precision in sensor data [2]. The optimal selection of navigational aids and the assessment of collision risks emphasize the necessity for precise altitude information in maintaining flight safety [2, 3].

The Kalman filter, a cornerstone of linear filtering and prediction in systems with uncertainty, provides the mathematical underpinning for refining sensor data [4]. Further exploration into the filter's applications makes it more accessible for various uses, including in compact aircraft systems [5].

Advances in sensor development, particularly for high-temperature and harsh-environment applications, are critical to the integration of sensors into compact aircraft systems [6, 7]. The progress of sensor technology, including flexible pressure sensors, informs the potential multifaceted uses of these sensors in aircraft systems [8].

Reviews of recent developments in MEMS capacitive pressure sensors and stretchable absolute pressure sensor networks underscore the importance of precision and reliability in these technologies, directly impacting the accuracy of aircraft systems [9, 10]. Estimation techniques for tracking and navigation highlight the role of merging sensor data for enhanced navigational accuracy [11–13]. Compensation for MEMS gyroscope drift relates closely to the challenges of altitude measurement in compact aircraft systems [14, 15].

Data processing techniques for monitoring reliability parameters and condition-based maintenance are crucial for maintaining the integrity of compact aircraft systems [16–18]. These studies highlight the necessity of employing advanced data processing techniques, such as the Kalman filter, to ensure precise operation and maintenance [19–27].

In summary, the literature review establishes the importance of accurate altitude data for compact aircraft systems and the potential of Kalman filtering techniques to enhance the reliability of sensor data. This study aims to contribute to this field by empirically demonstrating the effectiveness of a Kalman filter in improving sensor data accuracy, addressing a gap in the literature on the practical implementation of such filters in altitude measurement for compact aircraft systems.

3 Materials and Methods

The pressure sensors BMP180 and BMP280 were chosen based on this study's objectives. The BMP180 has been used due to its ultra-low power consumption, low voltage operation, and minimal altitude noise. Our selection was based on its ability to achieve high precision requirements in a compact aircraft system. In order to ensure accurate and linear pressure and temperature measurements, the device was calibrated using stored data from individual electrically erasable programmable read-only memory (EEPROM). Furthermore, the BMP280 sensor, which is also lightweight and compact, has improved performance features such as lower operating noise and advanced filtering modes. These features provide an additional set of measurements.

The practical assembly of the data acquisition system can be seen in Fig. 1, where the Arduino UNO microcontroller serves as the central component, interacting with both BMP180 and BMP280 sensors. This setup was required to collect raw sensor outputs from compact and lightweight aircraft systems under controlled conditions.

Fig. 1. A data acquisition setup using BMP180 and BMP280 sensors and an Arduino UNO.

SKF is used to process noisy sensor data. Initially, the parameters are initialized, followed by a prediction step based on the prior state and an update step that incorporates new measurements. According to the Kalman filter theory, the signal and observation models can be seen in Fig. 2.

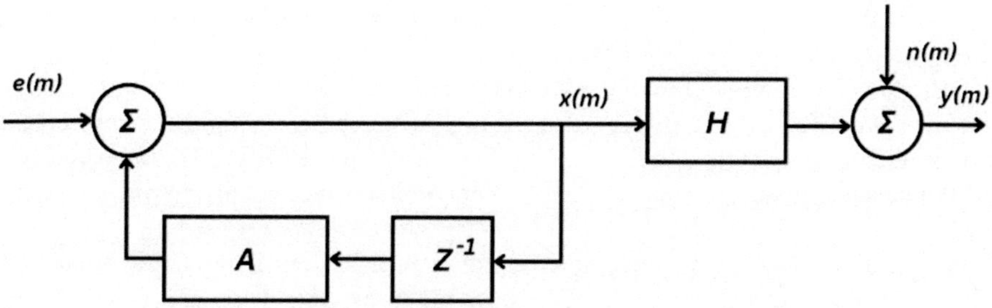

Fig. 2. An illustration of SKF signals and observation models.

In the case of a signal $x(m)$ and a noisy observation $y(m)$, the following equations will describe the state process model and the observation model:

$$x(m) = Ax(m-1) + e(m), \tag{1}$$

$$y(m) = Hx(m) + n(m), \tag{2}$$

where $x(m)$ is the state vector representing the P-dimensional signal or the system's state at the discrete time step m; A is the state transition matrix of size $P \times P$, which defines the relationship between the system's state at the previous time step $m-1$ and the current time step m; $e(m)$ is the process noise vector at time m, representing random fluctuations that affect the evolution of the state vector; $y(m)$ is the observation vector at time m, containing the M-dimensional noisy and distorted measurements obtained from the system sensors; H is the observation matrix of size $M \times P$, which maps the true state space into the observed space; $n(m)$ is the measurement noise vector at time m, accounting for random errors in the measurements, which is assumed to be normally (Gaussian) distributed.

Based on sensor data characteristics, SKF was initialized with parameters estimated from the data. Further prediction steps were based on the state transition matrix A, which represents the system dynamics.

As a recursive estimator, the SKF algorithm advances through a series of calculations to refine the estimation of the state of a dynamic system. These measurements are based on noisy data. The process begins by setting up the state estimate $\widehat{x_{0,0}}$ and the corresponding state error covariance $P_{0,0}$. These estimates encapsulate the initial guess of the system's state and the uncertainty associated with this guess. This algorithm re-initializes the state estimate and state error covariance as new measurements are acquired to ensure continuous estimation.

To predict the subsequent state of the system, the algorithm advances the current state estimate $\widehat{x_{n,n}}$ and error covariance $P_{n,n}$, to the next time step. This results in the expected estimate $\widehat{x_{n+1,n}}$ and the predicted error covariance $P_{n+1,n}$, respectively. This prediction phase incorporates process noise covariance Q, which represents the inherent uncertainty within system model.

In order to determine the Kalman Gain, the updated measurement must be weighed against the predicted state estimate. The Kalman Gain computation involves the predicted error covariance $P_{n,n-1}$ and the measurement noise covariance R_n. This characterizes the expected noise in the measurements.

Following the computation of the Kalman Gain, the algorithm updates the predicted state estimate $\widehat{x_{n,n-1}}$ by factoring in the actual measurement z_n and the discrepancy between this measurement and the predicted measurement $H\widehat{x_{n,n-1}}$. This step yields an updated state estimate $\widehat{x_{n,n}}$ which is more accurate due to the inclusion of the updated measurement information.

Finally, the algorithm updates the error covariance $P_{n,n}$, utilizing the Kalman Gain to adjust for the reduced uncertainty following the measurement update. It is critical to update the error covariance because it reflects the algorithm's increased confidence in the state estimate after measurement.

The algorithm described above for SKF can be expressed in formulas:

$$K_n = \frac{p_{n,n-1}}{p_{n,n-1} + \sigma^2}, \tag{3}$$

$$x_{n,n} = x_{n,n-1} + K_n\left(z_n + x_{n,n-1}\right), \tag{4}$$

$$p_{n,n} = (1 - K_n)p_{n,n-1}, \tag{5}$$

$$p_{n+1,n} = p_{n,n} + Q. \tag{6}$$

Experiment setups were designed to simulate small and compact aircraft systems. An environment controlled for altitude changes, temperature changes, and pressure differentials was used to test the sensors. By using this setup, data was collected that reflected the range of conditions the sensors would encounter during actual operation.

Mathcad software was used to process the data, applying the SKF algorithm to the sensor data. Using the software's robust computational tools, extensive datasets were processed, and the filtering process was visualized.

Comparative analyses of benchmark datasets were conducted to validate altitude data accuracy after Kalman filtering. In order to quantify the improvement in measurement accuracy, the SKF output was compared to raw sensor data.

A statistical analysis was conducted to evaluate SKF noise reduction capability. To assess sensor reading consistency following filter application, measurement variance was calculated using the formula (5).

$$\sigma^2 = \frac{1}{N}\sum\nolimits_{n=1}^{N} (x_n - \mu)^2, \tag{7}$$

where μ is mean value of measurements and x_n are individual altitude readings.

In addition, metrics such as the mean squared error were used to evaluate the system's performance. Assuming a noise level of 0.01 m, a test was conducted to evaluate the effectiveness of the filter in improving altitude measurements.

Using the formula (5), the estimated variance was calculated at 0.018. This was essential for initializing the SKF, which processed the data to produce the refined altitude estimate.

For the purpose of initializing the system, 172.8 m altitude was selected. According to initial estimation, the altimeter variance is 25 m, assuming a standard deviation of 5 m for the altimeter class used. This stage of the algorithm initializes the variance and altitude estimates. In calculating the Kalman gain for the first measurement, variance played a significant role.

4 Results and Discussions

In this study, we analyze the basic characteristics of different filtering technologies that could be applied for noise reduction in altitude measurement. Table 1 illustrates results of SKF comparison with other filters, including the Extended Kalman Filter (EKF), Moving Average Filter (MAF), Alpha-Beta (αβ) Filter, Least Squares Filter (LSF), and Robust Kalman Filter (RKF).

Table 1. Comparison of Filtering Methods for Small Aircraft Systems.

Criteria	SKF	EKF	MAF	αβ	LSF	RKF
Computational efficiency	High	Moderate	High	Moderate	Moderate	Moderate
Handling nonlinearity	Low	High	Low	Low	Low	High
Predictive capability	High	High	Low	Moderate	Low	High
Robustness to noise	High	Moderate	Moderate	Moderate	Low	Very High
Ease of implementation	High	Low	High	High	High	Moderate
Real-time processing	Suitable	Less Suitable	Suitable	Suitable	Less Suitable	Suitable
Resource utilization	Low	High	Low	Low	Low	Moderate
Accuracy	Moderate-High	High	Low-Moderate	Moderate	Moderate	High
Suitability for linear models	Very High	Moderate	High	High	High	Moderate

According to this comparison, SKF offers high computation efficiency. It should be mentioned that this will be essential when integrating with small aircraft systems in a real-time setting and for a system that requires fast and accurate data processing.

Also, the filtering algorithm ensures accuracy due to its robustness to Gaussian noise and predictability. Despite its inherent ability to thrive in linear model conditions, the SKF can be considered a suitable choice for this study and small aircraft applications. This is because of its ease of implementation and low resource utilization.

Table 1 shows SKF strikes the right balance between complexity and performance. This enhances compact aircraft systems navigational capabilities without excessive computational burdens.

A crucial parameter in SKF is the Kalman gain, which is used to weigh uncertainty in estimation and measurement. Based on the calculations provided, the filter updates the estimate as it iterates over each altitude measurement to reduce uncertainty.

Table 2 shows the Kalman gain, altitude estimates, and estimated variances by (1), (2), and (3) throughout the filtering process. Data measured with BMP180 was collected and obtained results demonstrate a consistent decrease in variance, indicating an improvement in altitude measurement precision.

Based on Table 1, Fig. 3 shows a graphical representation of measured raw and refined altitude values processed by SKF. Based on the graph, it can be seen that the filter effectively reduces noise in the altitude data. This is done with the refined estimates (x_n) trending closer to the mean altitude, demonstrating a lower variance than the raw measurements (z_n) updated every second.

According to the study results it can be highlighted that employing SKF can improve the accuracy and reliability of altitude data. This provides stable and precise navigation.

Table 2. Kalman Gain and Altitude Estimates.

Number of measures	Altitude measures (m) z_n	Kalman gain K_n	Altitude estimation (m) $x_{n,n}$	Estimated variances $p_{n,n}$
0	–	–	172.800	25
1	172.31	0.999	172.310	0.026
2	172.45	0.591	172.392	0.01163
3	172.49	0.393	172.430	0.00807
4	172.38	0.310	172.415	0.00657
5	172.60	0.267	172.464	0.00581
6	172.43	0.244	172.456	0.00539
7	172.78	0.230	172.531	0.00515
8	172.60	0.222	172.546	0.00500
9	172.63	0.217	172.564	0.00491
10	172.54	0.214	172.559	0.00485
11	172.31	0.212	172.506	0.00482
12	172.57	0.211	172.520	0.00480

Figure 4 shows a comparative analysis of altitude readings prior to and following SKF application during real-time processing. The blue line represents raw data from the BME 280 sensor, while the orange line represents refined data. Using the program downloaded onto the Arduino UNO microcontroller board, refined data was calculated.

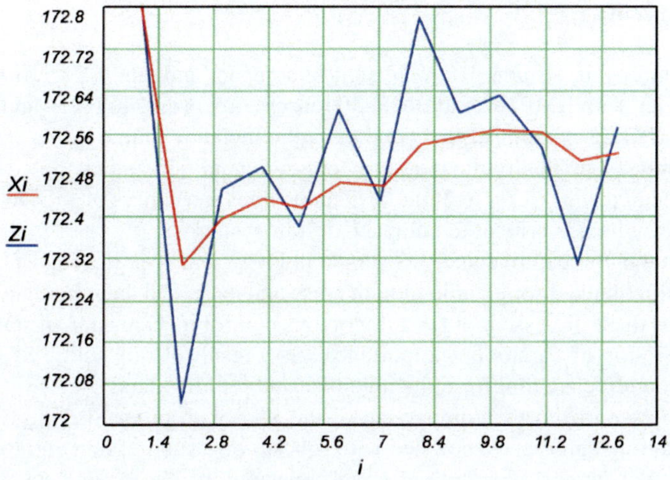

Fig. 3. A one-dimensional Kalman filter graphical representation of measurements and their refinement.

A new update of this data was made every 0.005 s. The visual comparison of processed data in Fig. 4 illustrates SKF effectiveness in noise reduction, which is essential to compact aircraft system precision.

These results indicate that due to its ability to balance efficiency and effectiveness without excessive computational demand, the SKF filter is suitable for small aircraft systems and applications requiring precision and operational constraints.

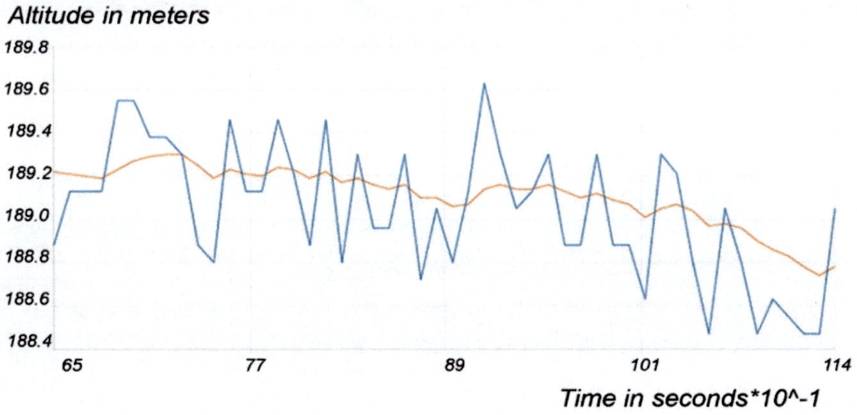

Fig. 4. Altitude data during real-time processing with raw and Kalman filter-refined data.

5 Conclusions

This study improved altitude data precision in compact and small aircraft systems by integrating SKF with BMP180 and BMP280 sensors. According to Fig. 3 and 4, the one-dimensional SKF has demonstrated good enough efficiency in mitigating sensor noise, enabling highly accurate altitude measurements. A quantified and tabulated analysis of the experimental results (refer to Table 2) demonstrates robustness of SKF in processing linear models, which is critical to compact aircraft systems.

Based on the comparative analysis shown in Fig. 3 and 4, performance of SKF can be visually articulated through reduction in noise and increased data accuracy. Enhancements such as these are essential for autonomous control mechanisms in small aircraft, where precision is of paramount importance. As a result of this study, SKF has been validated as an effective tool for reducing noise and refining data.

As can be seen from the comparison of the SKF with other filtering methods, its high computational efficiency, coupled with robust noise suppression and low resource utilization, underscores its suitability for real-time applications in compact aviation systems. As a result of the study, SKF was recommended as an optimal choice for small aircraft altitude control. It is achieved by striking a balance between performance needs and cost considerations.

6 Future Scope

Future research directions include:

- long-term field testing to confirm SKF accuracy and effectiveness;
- consolidating GPS, accelerometers, and gyroscope data with SKF could provide a more robust solution for altitude control;
- studies on the impact of improved altitude measurements on aircraft maintenance and energy consumption could lead to more cost-effective systems.

References

1. Halit, E, Roocke, P.: Altimeters. Encycl. Electr. Eng. **24**, 7–20 (2000)
2. Matyja, T., Kubik, A., Stanik, Z.: The MEMS-based barometric altimeter inaccuracy and drift phenomenon. Sci. J. Silesian Univ. Technol. Ser. Transp. **116**, 141–162 (2022). https://doi.org/10.20858/sjsutst.2022.116.9
3. Solomentsev, O., Zaliskyi, M.: Correlated failures analysis in navigation system. In: 5th International Conference on Methods and Systems of Navigation and Motion Control, pp. 41–44. IEEE, Kyiv (2018). https://doi.org/10.1109/MSNMC.2018.8576306
4. Shaukat, S., Katscher, M., Wu, C.L., Delgado, F., Larrain, H.: Aircraft line maintenance scheduling and optimisation. J. Air Transp. Manag. **89**(101914), 1–11 (2020). https://doi.org/10.1016/j.jairtraman.2020.101914
5. Zaliskyi, M., Shcherbyna, O., Zuiev, O., Solomentsev, O., Kozhokhina, O., Petrova, Y.: Statistical data processing for radio equipment in case of technical condition deterioration. In: 11th International Conference on Advanced Computer Information Technologies (ACIT), pp. 91–94. IEEE, Deggendorf (2021). https://doi.org/10.1109/ACIT52158.2021.9548454

6. Ostroumov, I., Kuzmenko, N., Marais, K.: Optimal pair of navigational aids selection. In: 5th International Conference on Methods and Systems of Navigation and Motion Control, pp. 32–35. IEEE, Kyiv (2018). https://doi.org/10.1109/MSNMC.2018.8576293
7. Ostroumov, I., Kuzmenko, N.: Risk assessment of mid-air collision based on positioning performance by navigational aids. In: 6th International Conference on Methods and Systems of Navigation and Motion Control, pp. 34–37. IEEE, Kyiv (2020). https://doi.org/10.1109/MSNMC50359.2020.9255506
8. Kalman, R.: A new approach to linear filtering and prediction problems. J. Basic Eng. **82**, 34–45 (1960)
9. Welch, G., Bishop, G.: An introduction to the Kalman filter, pp. 1–16. University of North Carolina at Chapel Hill (2006)
10. Mansoor, M., et al.: An SOI CMOS-based multi-sensor MEMS chip for fiuidic applications. Sensors **16**(11), 1608 (2016). https://doi.org/10.3390/s16111608
11. Werner, M., Fahrner, W.: Review on materials, microsensors, systems and devices for high-temperature and harsh-environment applications. IEEE Trans. Industr. Electron. **48**(2), 249–257 (2001)
12. Zang, Y., Zhang, F., Di, C.-A., Zhu, D.: Advances of flexible pressure sensors toward artificial intelligence and health care applications. Mater. Horiz. **2**(2), 140–156 (2015). https://doi.org/10.1039/C4MH00147H
13. Eswaran, P., Malarvizhi, S.: MEMS capacitive pressure sensors: a review on recent development and prospective. Int. J. Eng. Technol. **5**, 2734 (2013)
14. Guo, Y., Li, Y., Guo, Z., Kim, K., Chang, F., Wang, S.: Bio-inspired stretchable absolute pressure sensor network. Sensors **16**(1), 55 (2016). https://doi.org/10.3390/s16010055
15. Bar-Shalom, Y., Li, X., Kirubarajan, T.: Estimation with Applications to Tracking and Navigation: Theory Algorithms and Software, p. 547 (2001)
16. Deng, Z., Hu, Y., Yu, J., Na, Z.: Extended Kalman filter for real time indoor localization by fusing WiFi and smartphone inertial sensors. Micromachines **6**, 523–543 (2015). https://doi.org/10.3390/mi6040523
17. Shin, E.: A Quaternion-based unscented Kalman filter for the integration of GPS and MEMS INS. In: 17th International Technical Meeting of the Satellite Division of the Institute of Navigation (ION GNSS 2004), Long Beach, CA, USA, pp. 1060–1068 (2004)
18. Diao, Z., Quan, H., Lan, L., Han, Y.: Analysis and compensation of MEMS gyroscope drift. In: 2013 Seventh International Conference on Sensing Technology (ICST), pp. 592–596. IEEE, Wellington (2013). https://doi.org/10.1109/ICSensT.2013.6727722
19. Guo, Y., Han, F., Du, S., Ma, G., Zhu, L.: Performance analysis of MEMS gyro and improvement using Kalman filter. In: 34th Chinese Control Conference, pp. 4789–4794. IEEE, Hangzhou (2015). https://doi.org/10.1109/ChiCC.2015.7260380
20. Solomentsev, O., Zaliskyi, M., Herasymenko, T., Kozhokhina, O., Petrova, Y.: Data processing in case of radio equipment reliability parameters monitoring. In: Advances in Wireless and Optical Communications (RTUWO), pp. 219–222. IEEE, Riga (2018). https://doi.org/10.1109/RTUWO.2018.8587882
21. Backer, A.: Kalman Filter from the Ground Up, p. 436 (2023)
22. Zaliskyi, M., Solomentsev, O., Kozhokhina, O., Herasymenko, T.: Statistical data processing for condition-based maintenance. In: Signal Processing Symposium (SPSympo), pp. 232–235. IEEE, Krakow (2019). https://doi.org/10.1109/SPS.2019.8882103
23. Bansal, D., Bajpai, A., Kumar, P., Kaur, M., Rangra, K.: Low actuating voltage spring - free RF MEMS SPDT switch. J. Electr. Comput. Eng. 7984548 (2016). https://doi.org/10.1155/2016/7984548

24. Ahmed, A., Al-Gayem, Q.: Micro accelerometer built-in self-test and calibration using genetic algorithm and interpolation method. In: 2022 IEEE International Conference on Semiconductor Electronics (ICSE), pp. 49–52. IEEE, Kuala Lumpur (2022). https://doi.org/10.1109/ICSE56004.2022.9863170
25. Zhang, H., Wei, X., Ding, Y., Jiang, Z., Ren, J.: A low noise capacitive MEMS accelerometer with anti-spring structure. Sens. Actuat. A **296**, 79–86 (2019). https://doi.org/10.1016/j.sna.2019.06.051
26. Han, C., Li, C., Zhao, Y., Li, B.: High-stability quartz resonant accelerometer with micro-leverages. J. Microelectromech. Syst. **30**(2), 184–192 (2021). https://doi.org/10.1109/JMEMS.2020.3036121
27. Zhao, Y., Xia, G., Shi, Q., Qiu, A.: Expanding bias-instability of MEMS silicon oscillating accelerometer utilizing AC polarization and self-compensation. Sensors **20**(5), 1455 (2020). https://doi.org/10.3390/s20051455

Information Technology for Identification of the Dynamics Model of Multidimensional Moving Object

Valerii Zozulya[1](✉) and Sergei Osadchy[2]

[1] State University of Trade and Economics, Kyoto Street, 19, Kyiv 02156, Ukraine
bazi.v41@gmail.com

[2] Flight Academy of the National Aviation University, Stepana Chobanu Street, 1, Kropyvnytskyi 25005, Ukraine

Abstract. The purpose of this paper is the development of a suitable information technology for identifying the model of the dynamics of a multidimensional moving object and the disturbances affecting it during operation, based on the data of vectors of control signals and signals at the output of the object, aimed at increasing the reliability of obtaining calculation results. The Stateflow Matlab application was used to create the information technology of identification. A distinctive feature of information technology is the increased accuracy and reliability of performing calculations, which are achieved due to introduction of a new approach to the factorization of polynomial matrices, improvement of algorithms for multiplying polynomial, fractional-rational matrices, respectively, to reduce the loss of true significant figures due to ordering and ranking of elementary operands. To illustrate the operation of information technology, a multidimensional moving object was chosen a spatial mechanism of parallel kinematic structure based on the Stewart platform. Such mechanisms can be used as simulators for aircraft, such as the A320 and Mi-17-1V, as well as a platform for dynamic testing or calibration of inertial navigation systems. On the basis of the developed information technology of identification for the mechanism based on Stewart platform, its transfer function was determined, as well as the transfer function of the shaping filter, which, in the input signal in the form of white noise, has a random process with given statistical characteristics at the output.

Keywords: Identification · Stewart Platform · Information Technology

1 Introduction and Analysis of Research and Publications

The creation of guaranteed competitive motion stabilization systems for complex multidimensional moving objects that function under the influence of random uncontrolled disturbances, with minimal design costs, is one of the main requirements for achieving success in the market of this class of devices. One of the most effective and proven ways to meet this condition in the creation of aviation and space technology is the use of experimental and analytical approaches to the design of optimal stochastic stabilization

I. Ostroumov and M. Zaliskyi (Eds.): ACASD 2024, LNNS 992, pp. 57–70, 2024.
https://doi.org/10.1007/978-3-031-60196-5_6

systems [1]. One of the factors in the successful application of this approach is determined, on the one hand, by the correspondence of algorithms for identifying structure and parameters of the model in conditions close to the real mode of operation of the prototype of the object, and on the other hand, by the presence of computer mathematics system that allows the application of algorithms for the automated design of optimal automatic control systems.

The spatial mechanisms of the parallel kinematic structure based on Stewart platform, as an object of identification, which has six identical kinematic chains (actuators) is of particular interest [2]. Such mechanisms can be used as simulators for flight vehicle, such as the A320 [3] and Mi-17-1V [4], as well as a platform for dynamic testing or calibration of inertial navigation systems [5, 6]. Stewart platform has six degrees of freedom of the moving platform. By programmatically adjusting the length of the actuators of Stewart platform, it is possible to control the position of the motion base, move it in vertical and horizontal directions, and rotate it in three planes. Consideration of Stewart platform as a control object reveals a number of features, the main of which is that, for many technological tasks, the parameters of disturbing influences that are applied to the operating part, individual axes are not determined in advance and the difficulty of constructing an adequate, analytical, mathematical model [2, 7].

Today, a fairly wide range of methods for identifying models of the dynamics of control objects, which function under the conditions of uncontrolled centered stationary random effects [1, 8, 9], according to "input-output" data, is known. Input-output method [9] involves combining the signals acting at the input and output of the system into a single signal vector. This vector is considered to be the output of some imaginary dynamic system, at the input of which there is a virtual test signal with the known dynamics, for example, "white noise". However, almost all of them are designed for use in the conditions where there is no mutual influence between the external disturbance and the control signal. At the same time, the presence of feedback, which cannot be unlocked during identification, makes it impossible to accept the above hypothesis even with different sources of disturbances and control signals. For example, paper [10] presents algorithms for identifying the dynamics of elements of a multidimensional stabilization system, where it is claimed that in the conditions when the signals in the control loop belong to a set of centered stationary random processes, and the noise of the sensors has different sources of origin with disturbances to the system, then it is possible unambiguously determine the matrices of fractional-rational functions of the disturbance generator and find the system of ordinary differential equations of minimum order that characterizes the dynamics of the control object. At the same time, there are no information technologies of identification for evaluating the model of dynamics of control objects, which would help in solving the problem of the complexity of performing such calculations.

The paper is structured into main sections:

- introduction and analysis of research and publications. An analysis of the current state of problems and solutions to the task of identifying the dynamic model of a complex multidimensional moving object has been conducted;

- the purpose and objectives of research. The goal is formulated as the creation of an information technology for identifying the dynamic model of the Stewart platform. The research objectives are outlined as well;
- research methods. The algorithm for solving the problem of identifying the dynamic model of the Stewart platform is justified;
- dynamics of Stewart platform based on the results of structural Identification. The results of identifying the dynamic model of the Stewart platform, obtained using the developed information technology, are presented;
- conclusions.

2 The Purpose and Objectives of Research

The purpose of the research is the development of a suitable information technology for the identification of the model of the dynamics of a multidimensional moving object – Stewart platform and the disturbances affecting it during operation, based on the data of vectors of control signals and signals at the output of the object, aimed at increasing the reliability of obtaining calculation results. The task of identifying the dynamics of Stewart platform, which functions in a closed control system, arises when feedback cannot be opened.

As a result of the structural transformation of the working surface (WS) motion control scheme of Stewart platform and taking into account that we can measure the m-dimensional vector of control signals r_0 and the vector of real values of the motion of the working surface x_1, and in the case of the impossibility of opening the feedback, this scheme can be converted to the form presented in Fig. 1. Moreover, most often, the source of the appearance of disturbances Δ and the vector of control signals r_0 have different physical nature, therefore, the hypothesis of their independence is accepted.

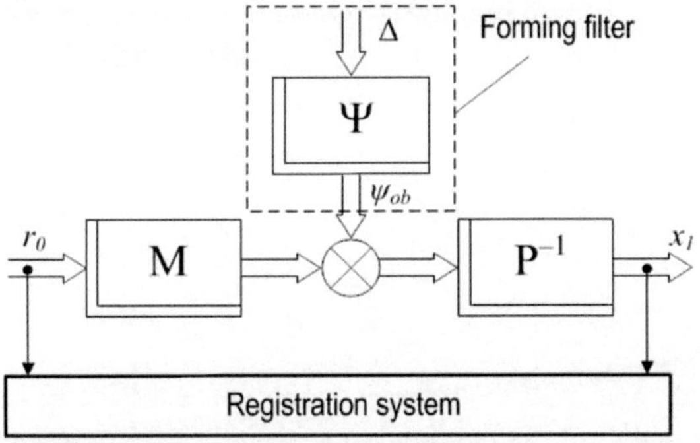

Fig. 1. Structural diagram of the identification object.

Figure 1 shows P, M are polynomial matrices from the differentiation operator s of the appropriate size, which characterize the dynamics of a closed system, Ψ is the unknown matrix of dimension $n \times n$, the elements of which are function operators from s, ψ_{ob} is a vector of centered stationary random disturbances, which characterizes the action of all disturbances and noises in the control system and n is a measurable stationary random process with zero mathematical expectation and the unknown matrix of spectral densities $S_{\psi_{ob}\psi_{ob}}$.

Based on the analysis of the types of tasks that are solved by mechanisms based on Stewart platform [2, 7], it can be argued that the working surface performs small movements around the center of rotation. Then, in the first approximation, an ordinary differential equation can be written that characterizes the dynamics of the control object, that is, the relationship between the system inputs r_0, ψ_{ob} and the output x_1 (see Fig. 1):

$$Px_1 = Mr_0 + \psi_{ob}. \tag{1}$$

The output x_1 and one of the inputs r_0 can be measured, but the input ψ_{ob} cannot be measured.

The vector ψ_{ob} is formed from the noise vector Δ by a linear stable filter with a matrix of transfer functions Ψ and can be represented as:

$$\psi_{ob} = \Psi\Delta. \tag{2}$$

The task of obtaining the dynamic characteristics of a multidimensional moving object and a disturbance vector is as follows: based on the data of the measured input r_0 and output vectors x_1, it is necessary to find the order and parameters of the matrices P, M and Ψ that minimize the next indicator of the identification quality:

$$J = \frac{1}{j}\int_{-j\infty}^{j\infty} tr\left(S'_{\varepsilon\varepsilon}R\right)ds, \tag{3}$$

where j is a complex unit; $tr(x)$ is trace of the matrix [11]; s is a complex variable ($s = j\omega$); R is a symmetric positive definite weight matrix; $S'_{\varepsilon\varepsilon}$ is the transposed matrix of spectral densities of the vector of identification errors ε.

The vector of identification errors ε is equal to the difference between the vector of measured actual values of the movement of the working surface x_1 and the estimate of the vector of movement of the working surface x_{id} obtained as a result of identification:

$$\varepsilon = x_1 - x_{id}. \tag{4}$$

Also, the vector of identification errors ε can be defined as:

$$\varepsilon = x_1 - \Phi\zeta, \tag{5}$$

where Φ is the block matrix of the transfer functions of the identification object and is defined as $\Phi = [W_1, W_2]$, W_1 is the optimal structure of the transfer function matrix of the identification object, W_2 is the optimal structure of the filter transfer function matrix Ψ; $\zeta = \begin{bmatrix} r_0 \\ \Delta \end{bmatrix}$ is an extended vector of input signals.

Then the task of identifying the Stewart platform dynamics model is reduced to finding two matrices of transfer functions W_1, W_2 in such a way that the error ε (5) or the indicator of the quality of identification (3) are minimal.

The solution to the task was found as a result of three stages of work:

– development of the algorithm for structural identification of a multidimensional dynamic object with random input signals;
– collection and processing of experimental data of vectors r_0 and x_1;
– analysis of the change in the variance of the vector of the output coordinates of the WS of Stewart platform x_1, with random changes in the vectors r_0 and ψ_{ob}.

3 Research Methods

To achieve the objective, according to the work [1], the task of building an algorithm for solving the problem of identification and its information technology was set and solved. The algorithm of preparation for solving the problem of identification of the model of the dynamics of a multidimensional moving object, that is Stewart platform, consists of the following stages:

1. Finding the spectral and mutual spectral densities $S_{x_1x_1}$, $S_{r_0r_0}$, $S_{x_1r_0}$, $S_{r_0x_1}$ of the vector of control signals r_0 and the vector of real values of the movement of the working surface x_1, which have already been obtained as a result of processing experiments.
2. Determination of the estimate of the spectral and mutual spectral densities $S_{x_1x_1}$, $S_{r_0r_0}$, $S_{x_1r_0}$,$S_{r_0x_1}$ vector of control signals r_0 and vector of real values of movement of the working surface x_1.
3. Finding the relationship of the transposed matrix of spectral densities of identification errors $S_{\varepsilon\varepsilon'}$ with the block matrix Φ, based on Wiener-Khinchin theorem in vector form [11], applied to vector (5):

$$S'_{\varepsilon\varepsilon} = S'_{x_1x_1} - S'_{\zeta x_1}\Phi_* - \Phi S'_{x_1\zeta} + \Phi S'_{\zeta\zeta}\Phi_*, \tag{6}$$

where $S'_{x_1x_1}$ is a transposed matrix of spectral densities of random process x_1; $S'_{x_1\zeta}$ is a transposed matrix of mutual spectral densities between vector random processes ζ and x_1;

$$S'_{\zeta\zeta} = \begin{bmatrix} S_{r_0r_0} & O_{6\times6} \\ O_{6\times6} & S_{\Delta\Delta} \end{bmatrix}, \tag{7}$$

where $O_{6\times6}$ is a zero matrix with dimension 6×6;

$$S'_{\zeta x_1} = \left(S'_{r_0x_1}, S'_{\Delta x_1}\right). \tag{8}$$

4. Determination of the matrix of mutual spectral densities $S'_{\Delta x_1}$ through the output data. For this, taking into account the matrices $S'_{\zeta x_1}$, $S'_{\zeta\zeta}$ and performing some transformations, we obtain the connection equation

$$S_{x_1\Delta}S^{-1}_{\Delta\Delta}S_{\Delta x_1} = S_{x_1x_1} - S_{x_1r_0}S^{-1}_{r_0r_0}S_{r_0x_1}. \tag{9}$$

To find the function $S_{\Delta x_1}$, it is necessary to factorize the matrix $S_{x_1\Delta}S^{-1}_{\Delta\Delta}S_{\Delta x_1}$, taking into account the specificity of the vector Δ as a vector of single white noises [12].

5. Determination of the first variation of the identification quality functional δJ, under the condition of minimization on the class of stable and physically realized matrix Φ:

$$\begin{array}{c}\delta J = \frac{1}{j}\int_{-j\infty}^{j\infty} tr\Big[R_{0*}\Big(-R_0 S'_{\zeta x_1} D_*^{-1} + R_0 \Phi D\Big) D_* \delta\Phi_* \\ + R_0\Big(-S'_{x_1\zeta} R_0 D^{-1} + S'_{\zeta\zeta}\Phi_* R_{0*} D_*\Big)\delta\Phi D\Big] ds\end{array}, \tag{10}$$

where δJ, $\delta\Phi$ are variations of the corresponding functions; R_0 is the result of matrix factorization $R = R_{0*}R_0$.

6. Definition of the complex variable matrix D as a result of Wiener factorization:

$$DD_* = S'_{\zeta\zeta}. \tag{11}$$

$$D = \begin{bmatrix} d_u & O_{6\times 6} \\ O_{6\times 6} & E_{6\times 6} \end{bmatrix}, \tag{12}$$

where $E_{6\times 6}$ is unit dimension matrix 6×6.

7. Calculation of $T_0 + T_+$ as a result of separation:

$$T = T_0 + T_+ + T_- = S'_{\zeta x_1} D_*^{-1}. \tag{13}$$

where T_0 are the elements of the polynomial matrix s; T_+ are matrix elements, regular fractions with poles only in the left half-plane; T_- are matrix elements, regular fractions with poles only in the right half-plane.

8. Calculation of the blocks of matrix Φ (optimal structures of the matrices of the transfer functions of the identification object W_1 and the filter Ψ, which forms the dynamic characteristics of the disturbance brought to the output of the system W_2):

$$\Phi = \begin{bmatrix} W_1 & W_2 \end{bmatrix} = R_0^{-1}\big[T_0 + T_+\big]D^{-1}. \tag{14}$$

9. Determination of the dynamic properties of matrix P using the operation of one-sided removal of the poles of matrices W_1 and W_2 [12, 13], after which matrix M is found:

$$M = \tilde{P} W_1. \tag{15}$$

10. Calculation of the matrix of spectral densities of disturbing influence:

$$S'_{\psi_{ob}\psi_{ob}} = \tilde{P} W_1 W_2 \tilde{P}_*. \tag{16}$$

The above-grounded algorithm enables proposing information technology for identifying the model of dynamics of a multidimensional moving object, which should include procedures for adding and multiplying polynomial and fractional-rational matrices, Wiener factorization, Wiener separation of fractional-rational matrices, and finding dispersion integrals.

To create information technology of identification, the Stateflow Matlab application was used, which showed high efficiency in solving the problems of designing multidimensional automatic control systems [14]. Stateflow is an interactive development tool for modeling complex, event-driven systems with complex behavior. The Stateflow software is an extension of the Matlab Simulink system and provides possibility of forming finite state machines by building hierarchical models in the form of subsystems.

In the Stateflow system, a corresponding flow diagram is created, which implements the interactive process of identifying Stewart platform dynamics model (see Fig. 2). Rachet block in Stateflow is included in the Simulink system model.

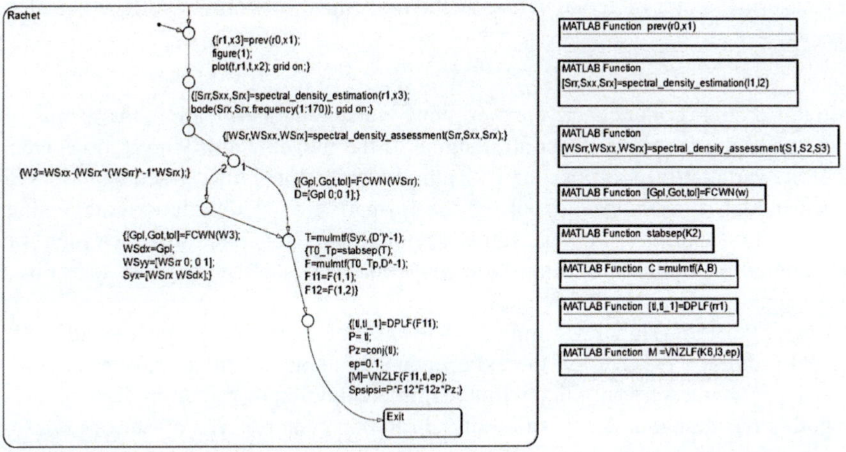

Fig. 2. Flow diagram of the process of identification of the dynamics model of a multidimensional moving object in Stateflow.

A distinctive feature of the information technology for performing stages of identification of the model of the dynamics of a multidimensional moving object is the increased accuracy and reliability of the calculations, which are achieved due to the introduction of a new approach to the factorization processes of polynomial matrices (the `FCWN` function (see Fig. 2)) and the computational algorithm built on it based on the improvement of the algorithms for multiplying polynomial matrices (the `Mulmtf` function (see Fig. 2)) from the Stateflow Matlab calculation system to reduce the loss of true significant figures due to the appropriate ordering and ranking of elementary operands (the `Stabsep` function (see Fig. 2)), multiplication of fractional-rational matrices (function `Mulmtf` (see Fig. 2)) to reduce the growth of the order of results based on the use of the operations of removing the corresponding poles to the right and left (functions `DPRT` and `DPLF` (see Fig. 2)) and entering corresponding zeros to the right and left (functions `VNZRT` and `VNZLF` (see Fig. 2)).

4 Dynamics of Stewart Platform Based on the Results of Structural Identification

In order to illustrate the work of information technology of identification of the dynamics model of a multidimensional moving object, let us consider only the move-ment of the rotation center of WS mechanism based on Stewart platform [7, 10] and perform identification of only the dynamics that characterizes the movement of this center under the influence of the control influence vector:

$$r_0 = \left[\xi_0 \; \eta_0 \; \zeta_0 \right]', \tag{17}$$

and measuring vector of actual values of the movement of working surface is equal to:

$$x_1 = \left[\xi_1 \; \eta_1 \; \zeta_1 \right]'. \tag{18}$$

In this case, to conduct active experiment, with such identification, the formation of the vector of the program, the control signal of the movement of the rotation center of WS of Stewart platform, is possible with a multidimensional filter based on the standard Dryden model, as a universal tool for the formation of stochastic external influence [15]. Based on this, the software signal vector r_0 was generated using Dryden model implemented in Matlab/Simulink. After pre-processing (`prev` function (see Fig. 2)), control signals r_0 are shown in Fig. 3.

The range of this signal was limited by the size of the working area Stewart platform physical prototype [7]. During the experiment, the inputs of the control system of the physical prototype were supplied with the implementation of vector program signals r_0, simulating the change in the position of the rotation center of WS of Stewart platform.

Obtaining the displacement vector of WS of Stewart platform on the basis of a full-scale experiment was performed using a physical prototype of Stewart platform, with an orientation and navigation sensor installed in the center of rotation of WS. The orientation and navigation sensor are built based on 3D MEMS sensors: accelerometer, gyroscope, and magnetometer. It was previously assessed by the method of structural identification of the matrix of transfer functions and spectral densities of interference measurements of the developed orientation and navigation sensor [16], which contributed to increasing the accuracy with optimal stochastic filtering of signals [17].

After preprocessing (`prev` function (see Fig. 2)), a selection of records of the signal vector of the displacement of the rotation center of WS of Stewart platform x_1 was obtained, the graph of the change of which is shown in Fig. 4.

As a result of the full-scale experiment, two collections of points of the input signal r_0 and the response of Stewart platform to this signal x_1 were obtained, which were entered into the MATLAB software. These signals are the input parameters of the Rachet block in Stateflow (see Fig. 2).

Vectors of useful signals and interference are multidimensional stationary random processes with zero mathematical expectations and fractional-rational matrices of spectral and mutual spectral densities$S_{x_1x_1}$, $S_{r_0r_0}$, $S_{x_1r_0}$, $S_{r_0x_1}$ which have already been obtained as a result of experimental processing. Measurement interferences are uncorrelated with each other and with the useful signal, and the disturbance ψ_{ob} is uncorrelated with the control signal r_0 .

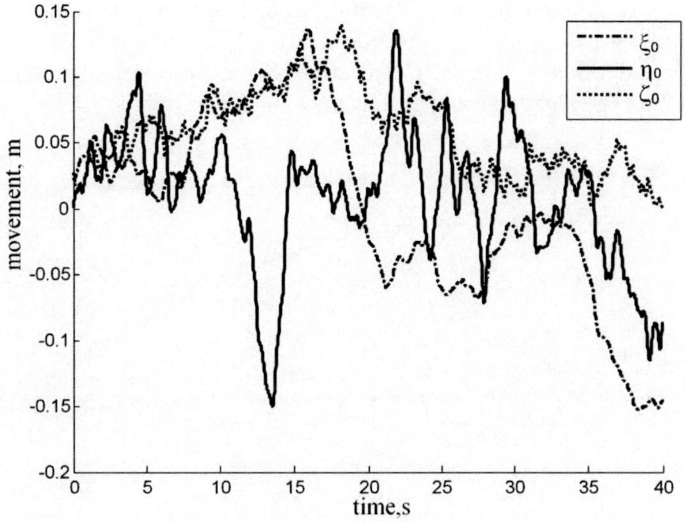

Fig. 3. The values of the input signal r_0.

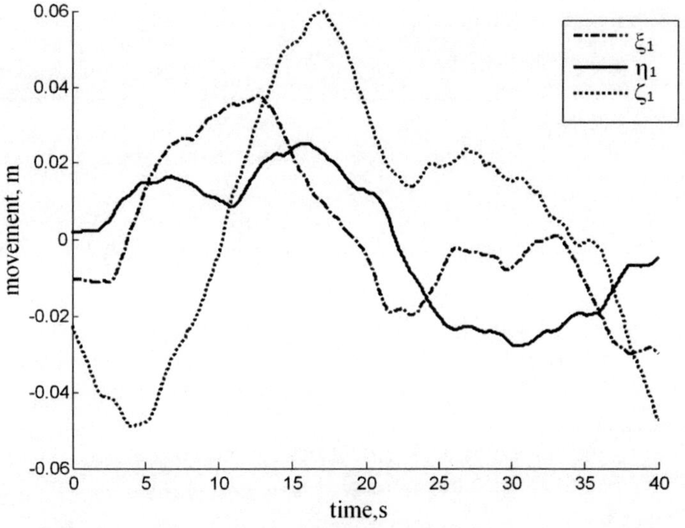

Fig. 4. Value of the output signal x_1.

According to the 1st point of the identification algorithm and with the help of the `spectral_density_estimation` function (see Fig. 2), it was determined: spectral density of the output signal $S^{/}_{x_1x_1}$, spectral density of the input signal $S^{/}_{r_0r_0}$, mutual spectral density of the signal $S^{/}_{r_0x_1}$.

According to the 2nd point of the identification algorithm and thanks to the `spectral_density_assessment` procedure (see Fig. 2), estimates of these densities were determined. As an example of execution, Fig. 5 shows the spectral density of element (2, 2) of the input signal of matrix $S_{r_0r_0}$ and its evaluation.

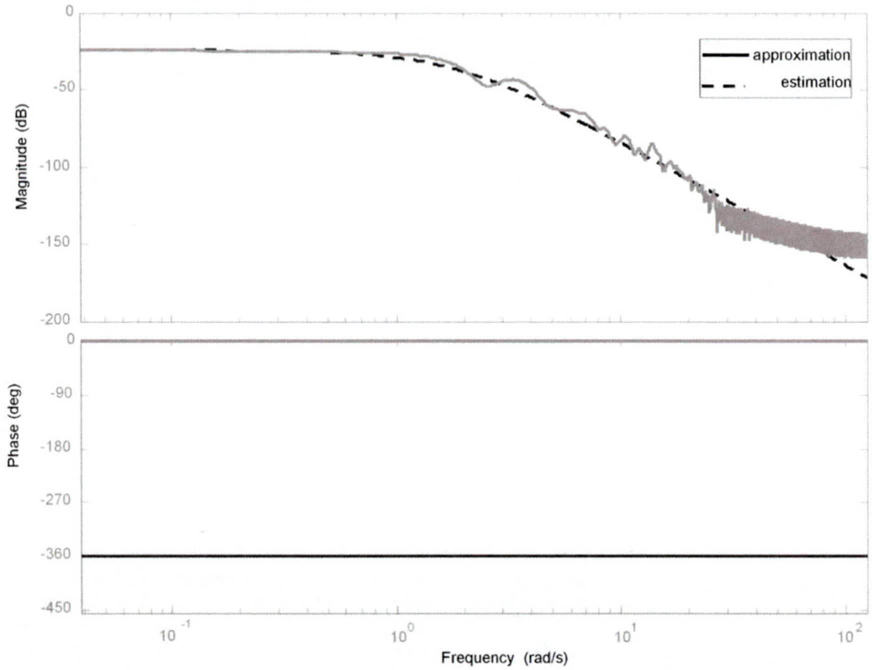

Fig. 5. Contact Spectral density of element (2, 2) of matrix $S_{r_0r_0}$ and its assessment.

According to the 3rd point of the identification algorithm (7) and the corresponding flows in the Rachet block in Stateflow, using Wiener factorization of polynomial matrices (`FCWN` function (see Fig. 2)) at $S_{\Delta\Delta} = 1$, by substituting the spectral density of the input signal $S_{r_0r_0}$, one can get $S'_{\zeta\zeta}$.

According to the 4th point of the identification algorithm, after factoring matrix $S_{x_1\Delta}S^{-1}_{\Delta\Delta}S_{\Delta x_1}$, using the `FCWN` function (see Fig. 2). After that, we substitute $S'_{\Delta x_1}$ and $S'_{r_0x_1}$ in (8) and get $S'_{\zeta x_1}$.

According to the 6th point of the identification algorithm, after factoring matrix $S'_{r_0r_0}$ using the `FCWN` function (see Fig. 2), we obtained matrix D (12).

Further, according to the 7th point of the identification algorithm (13), having fractional-rational matrices $S'_{\zeta x_1}$ and D, they are multiplied using the `Mulmtf` function (see Fig. 2), and then Wiener separation is performed (`Stabsep` function (see Fig. 2)), as a result of which the matrix is obtained $T_0 + T_+$.

Entering the notation $R_0 = 1$, in accordance with the 8th point of the identification algorithm (14), multiplying the fractional-rational matrices $(T_0 + T_+)$ and D^{-1}, using the `Mulmtf` function (see Fig. 2), the matrix $\Phi = [W_1\ W_2]$ is obtained, where W_1 is the optimal structure of the matrix of the transfer function of the identification object:

$$W_1 = \frac{10^{-2}}{z_1}\begin{bmatrix} 1.3(s+5.4)(s+0.8)(s+0.15)\left(s^2+0.3s+0.067\right) & -1.6(s-2)(s+0.14)(s-0.03)\left(s^2+1.9s+1\right) & -0.8(s-2.1)(s+0.76)(s+0.19)\left(s^2+0.25s+0.08\right) \\ (s+2.1)\left(s^2+0.22s+0.03\right)\left(s^2+1.35s+0.83\right) & -0.4(s+9)\left(s^2+1.95s+1\right)\left(s^2+0.07s+0.086\right) & 0.4(s+1.96)\left(s^2+0.22s+0.055\right)\left(s^2+1.06s+0.67\right) \\ 4.9(s+0.96)\left(s^2+0.5s+0.14\right)\left(s^2+0.018s+0.2\right) & 0.6(s+0.19)\left(s^2+1.43s+0.54\right)\left(s^2+1s+4\right) & 2.5(s+0.94)\left(s^2-0.2s+0.068\right)\left(s^2+0.59s+0.17\right) \end{bmatrix}, \quad (19)$$

where $z_1 = \left(s^2+0.11s+0.04\right)$.

W_2 is the optimal structure of the matrix of transfer function of the filter, which forms the dynamic characteristics of the disturbance brought to the output of the system:

$$W_2 = \begin{bmatrix} \frac{0.03}{(s+0.5)(s+0.55)} & 0 & 0 \\ \frac{0.018}{(s+0.5)(s+0.55)} & \frac{0.024}{(s+0.5)(s+0.55)} & 0 \\ \frac{0.01}{(s+0.5)(s+0.55)} & \frac{0.02}{(s+0.5)(s+0.55)} & \frac{0.054}{(s+0.5)(s+0.55)} \end{bmatrix}. \quad (20)$$

Further, according to the 9th point of the identification algorithm, applying the operation of one-sided removal of the poles of matrices W_1 and W_2, using the `DPLF` function (see Fig. 2), a polynomial matrix P. After that, there is matrix M according to (15).

According to the 10th point of the identification algorithm (16), applying the operation of entering zeros to the left using the `VNZLF` function (see Fig. 2), the matrix of spectral densities of the disturbing influence was obtained:

$$S'_{\psi_{ob}} = z_2\begin{bmatrix} 9 & 5.3 & 3 \\ 5.3 & 9 & 6.7 \\ 3.2 & 6.7 & 34.4 \end{bmatrix}, \quad (21)$$

where $z_2 = \left|\frac{\left(s^2+0.11s+0.04\right)\left(s^2+0.63s+0.15\right)\left(s^2+2s+1\right)}{(s+0.45)(s+0.55)}\right|^2$.

Analysis of the structure of matrices W_1 and W_2 and logarithmic frequency characteristics show that the considered mechanism is a multidimensional stable mechanical filter both for control signals and for disturbances in the operating zone of the mechanism based on Stewart platform. The natural frequency of such a filter is determined in element (1, 2) of matrix W_1, which has the highest cutoff frequency and is close to 0,52 s^{-1} (Fig. 6), and depends on the moments of inertia and the masses of the moving elements of the mechanism. It can also be argued that the disturbance is much less than the controlling influence, so the disturbance can be replaced by white noise due to the fact that it has a low intensity.

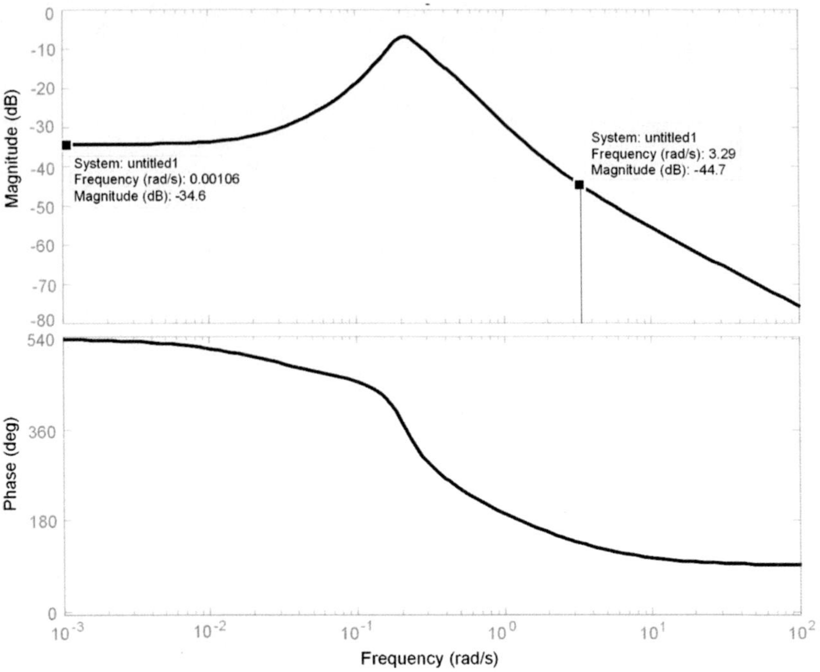

Fig. 6. Logarithmic frequency response of element (1, 2) of matrix W_1.

5 Conclusions

On the basis of well-known frequency algorithms of structural identification, a new information technology was obtained, which allows finding the order and parameters of the linearized system of ordinary differential equations of a multidimensional object and the matrix of spectral densities of disturbances that affect it under operating conditions close to the real mode of functioning of the experimental object.

A distinctive feature of the information technology of the stages of identification of the model of the dynamics of a multidimensional moving object is the increased accuracy and reliability of the calculations, which were achieved due to the introduction of a new approach to the factorization of polynomial matrices and the computational algorithm built on its basis, the improvement of algorithms for multiplying polynomial matrices from the system Stateflow Matlab calculations to reduce the loss of true significant figures due to the appropriate ordering and ranking of elementary operands, multiplication of fractional-rational matrices to reduce the growth of the order of results based on the use of the operations of removing the corresponding poles on the right and left and entering the corresponding zeros on the right and left.

The paper identified Stewart platform dynamics model and determined its transfer function, as well as the transfer function of the shaping filter. It is determined that the considered mechanism is a multidimensional stable mechanical filter both for control signals and disturbances in the mechanism's operating zone. It can also be argued that the disturbance is much less than the controlling influence, so it can be replaced by white noise because it has a low intensity.

Prospects for further research involve the development of an Information Technology for synthesizing the control system of the motion of the WS of the Stewart platform. These should allow for determining the structure and parameters of a multidimensional controller of the system control moving object.

References

1. Azarskov, V.N., Blokhin, L.N.: Methodology for designing optimal systems of stochastic stabilization. NAU, Kyiv (2006)
2. Taghirad, H.D.: Parallel Robots. Mechanics and Control. Taylor & Francis Group, CRC Press (2013)
3. Saudi Pilot Simulator Training Center Holds Airbus A320 Full Flight Simulator Acceptance. https://afm.aero/saudi-pilot-simulator-training-center-holds-airbus-a320-full-flight-simulator-acceptance. Accessed 17 Jan 2024
4. Comprehensive helicopter flight simulators Mi-8MTV, Mi-171. http://avia-ts.ua/production/8-kompleksnyy-trenazher-vertoletov-mi-8mtv-i-mi-171.html. Accessed 02 May 2021
5. Sushchenko, O.A., Bezkorovainyi Y.M., Golitsyn, V.O.: Fault-tolerant inertial measuring instrument with neural network. In: 40th International Conference on Electronics and Nanotechnology (ELNANO), pp. 797–801. IEEE, Kyiv (2020). https://doi.org/10.1109/ELNANO50318.2020.9088779
6. Sushchenko, O.A., Bezkorovainyi, Y.N., Novytska, N.D.: Nonorthogonal redundant configurations of inertial sensors. In: 4th International Conference on Actual Problems of Unmanned Aerial Vehicles Developments (APUAVD 2017), pp. 73–78. IEEE, Kyiv (2018). https://doi.org/10.1109/APUAVD.2017.8308780
7. Osadchy, S., Zozulya, V., Timoshenko, A.: The dynamic characteristics of the manipulator with parallel kinematic structure based on experimental data. In: Advances in Intelligent Robotics and Collaborative Automation – Robots, chap. 2, pp. 27–48. River Publishers (2015). https://doi.org/10.1201/9781003337119
8. Sushchenko, O., Bezkorovainyi, Y., Ostroumov, I., Zaliskyi, M., Averyanova, Y., Larin, V.: Identification of disturbed stabilization systems based on experimental data. In: 4th KhPI Week on Advanced Technology, pp. 1–6. IEEE, Kharkiv (2023). https://doi.org/10.1109/KhPIWeek61412.2023.10312938
9. Ebadat, A.: Experiment design for closed-loop system identification with applications in model predictive control and occupancy estimation. Doctoral thesis, Royal Institute of Technology (KTH), Stockholm, Sweden (2017)
10. Melnichenko, M.M., Osadchy, S.I., Zozulya, V.A.: Identification of the signals in position control circuits of a hexapod platform. In: 4rd International Conference Methods and Systems of Navigation and Motion Control (MSNMC 2016), pp. 51–57. IEEE, Kyiv (2016). https://doi.org/10.18372/1990-5548.50.11387
11. Korn, G., Korn, T.: Handbook of Mathematics (for scientific workers and engineers). Nauka, Kyiv (1977)
12. Aliev, F.A., Bordyug, V.A., Larin, V.B.: Time and frequency methods of synthesis of optimal regulators. Institute of Physics of the Academy of Sciences of the Azerbaijan SSR (1988)
13. Kucera, V.: The H2 control problem: a general transfer-function solution. Int. J. Control. **80**(5), 800–815 (2007). https://doi.org/10.1080/00207170701203590
14. Patankar, P., Kulkarni, S.: MATLAB and Simulink In-Depth: Model-Based Design with. BPB Publications (2022)

15. Tunik, A.A., Rydlo, K., Savchenko, O.V., Melnik, K.V.: Practical experience of intellectual UAV attitude stabilization system computer-aided design. Electron. Control Syst. **1**(43), 17–25 (2015). https://doi.org/10.18372/1990-5548.43.8842
16. Osadchy, S.I., Zozulya, V.A., Rudiuk, G.I.: The Dynamics of 3-dimensional micro-mechanic sensor of angle motions of a robot-hexapod. In: 8th International Conference on Intelligent Data Acquisition and Advanced Computing Systems: Technology and Applications (IDAACS 2015), pp. 908–912. IEEE, Warsaw (2015). https://doi.org/10.1109/IDAACS.2015.7341435
17. Osadchy, S.I., Zozulya, V.A.: Optimal filtering of hexapod acceleration data obtained under action of electromagnetic interference. In: 3rd International Conference Methods and Systems of Navigation and Motion Control (MSNMC 2014), pp. 21–23. IEEE, Kyiv (2014). https://doi.org/10.1109/MSNMC.2014.6979719

Automatic Control in Civil Aviation Systems

Mathematical Description of System for Stabilization of Aviation Equipment in Problems of Synthesis and Simulation

Olha Sushchenko(✉), Yurii Bezkorovainyi, Olexandr Salyuk, and Serhii Yehorov

National Aviation University, Liubomyra Huzara Avenue, 1, Kyiv 03058, Ukraine
sushoa@ukr.net

Abstract. This paper deals with developing the mathematical description of the system for stabilization of the aviation equipment. The set of equations describing separate units of the above-mentioned system is proposed. The united model of the stabilization object and motor is represented. The features of mounting the stabilization system with payload on the moving vehicle are described. The non-linearities inherent in real moving vehicles are taken into consideration. Features of the united model of the electric motor are listed. The elastic connection between the stabilization object and the electric motor and the drive backlashes are taken into consideration in the mathematical model. The influence of the hysteresis in the model of the gyroscopic tachometer is mathematically described. The block diagram of the mathematical model is represented. The control law is proposed. Simulation of transient processes of the control system synthesized using the developed mathematical model has been carried out. The simulation results are represented. The obtained results can be useful for designing stabilization systems of the wide class.

Keywords: Stabilization System · Moving Vehicle · Control Law · Mathematical Model · Simulation · Aviation Equipment

1 Introduction

Recently, the urgency of creating new promising systems for stabilizing aviation equipment has been growing. Modern trends in design of complex technical systems lie in significant shortening of design terms with simultaneous expansion of the set requirements. Therefore, the creation of new modern systems of stabilization requires a mathematical description, which allows us to study the stabilization system of aviation equipment already at the early stages of development.

This research considers the stabilization system of the equipment unit installed on some rotating base, which, in turn, is placed on an aircraft or is used as a part of the hardware for aircraft instrumentation tests. In general, the system of stabilization of equipment consists of vertical and horizontal channels. These channels are mutually independent, so they can be considered separately, moreover, from the point of view of

I. Ostroumov and M. Zaliskyi (Eds.): ACASD 2024, LNNS 992, pp. 73–85, 2024.
https://doi.org/10.1007/978-3-031-60196-5_7

the methodological approach to creating a mathematical model, they have no differences. In the given research, vertical channel is considered, since it is much more complicated.

During the development of the proposed model, moments of resistance, moments of imbalance, moments of inertia, as well as the presence of an elastic balancing system were taken into account. This approach ensures that created model matches real equipment. For the same purpose, the characteristics of reducers are introduced into the mathematical description, taking into account their stiffness and backlash. It is worth noting that all moments included in the model are reduced to the axis of rotation of motor.

Considerable attention is now being paid to the creation of new promising systems of stabilizing equipment for various purposes. General problems of navigation of moving objects and a specific navigation system are described in [1, 2], which emphasize the use of autonomous navigation. Ways to improve the accuracy of navigation systems are discussed in [3–5], which deal with the improvement of navigation information processing algorithms. Consideration of issues of mathematical description of stabilization systems for equipment of moving vehicles including aircraft requires a separate study. The general principles of creating models of drives of stabilization systems are considered in many publications, for example in [6, 7] but the solution of a specific problem requires some additions and new proposals.

The purpose of this research is to create a complete model of the system for stabilizing equipment of aircraft, taking into account the models of all the components, as well as the features inherent in real systems. Such a model is created to study the dynamics of the system and its errors. The finite goal is to carry out the synthesis of the stabilization system.

In the paper, we propose an approach to developing the mathematical description and synthesis of the system for the stabilization of equipment assigned for operation on aircraft. The feature of the approach is taking into consideration nonlinearities inherent in real stabilization systems of the considered type including hysteresis in the gyroscopic angular rate and backlashes of the drive. Such an approach ensures the quality of the simulation of the system, which is synthesized with the help of the proposed mathematical description.

The paper is organized as follows. In Sect. 2 we consider features of the stabilization system. A mathematical description of measurement instruments and electronics is given in Sect. 3. The united model of the stabilization plant and motor taking into consideration their elastic connection is represented in Sect. 4. The numerical demonstration of the results of the synthesized system is shown in Sect. 5.

2 Features of System Assigned for Stabilization of Equipment on Moving Vehicles

The mathematical model of the system for the stabilization of equipment on moving vehicles is presented in the form of differential and algebraic equations since such a presentation is the most visual for theoretical description of a complex dynamic system. Direct software implementation is expedient to be carried out by means of Simulink using the transfer functions of the basic devices of the system. Such an approach will allow us to bring the model closer to the real system and, accordingly, to obtain an effective means of research for the newly created system of stabilization of aviation equipment at the initial stages of the development [8, 9]. In addition, it facilitates the comparison of the results of the model and the results of tests of real system components, for example, their mock-up samples, or the results of modeling the electronic devices of the system by one of the well-known systems for their development (Workbench, MultiSIM).

The main function of the stabilization system is to reproduce the spatial movement and stabilize the module installed on the moving vehicle [10]. The module, in which the equipment for a certain purpose is located, is installed on a rotary base and can rotate in the horizontal plane. In turn, the equipment is installed on the unit that ensures the rotation in the vertical plane. The module and equipment are connected to the horizontal and vertical drives of tracking and stabilization systems with the help of reducers as shown in Fig. 1.

The implementation of rotational movements of stabilization objects, i.e. the module and the equipment, can be implemented using a shoulder strap with ball bearings and an axis installed in a rolling or sliding bearing [4], as shown in Fig. 2 and Fig. 3.

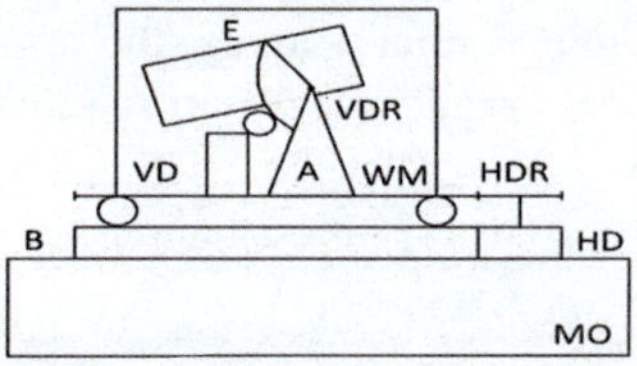

Fig. 1. Stabilization system installed on the vehicle: MO is the moving object; B is the base; HD is the horizontal drive; HDR is the horizontal drive reducer; WM is the working module; A is the axle; VD is the vertical drive; VDR is the vertical drive reducer; E is equipment.

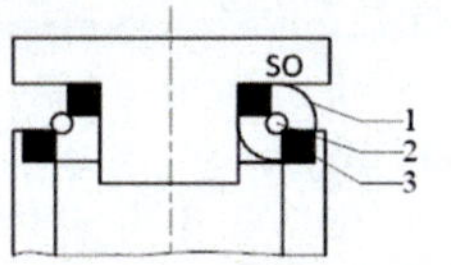

Fig. 2. Installation of a stabilization object on a shoulder strap with ball-bearing supports (1 is shoulder strap; 2 is shoulder strap balls; 3 is support ring).

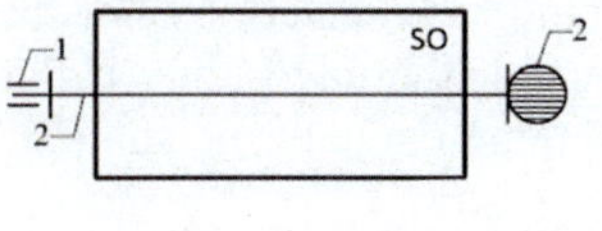

Fig. 3. Installation of a stabilization object in a bearing (1 is axis; 2 is supports with bearings).

3 Mathematical Description of Measuring Instruments and Electronics

Gyroscopic tachometers, which provide measurement of the absolute angular speed of the aviation equipment in the horizontal and vertical planes, and gyroscopic angular speed sensors of another type can be used as angular motion sensors of the system under study [11–13]. As a drive motor, a direct current motor is usually used, which has high control, dynamic, and energy characteristics, although its mass and dimensions are significant [7]. For systems of the studied type, it is important to achieve high accuracy in tracking and stabilization processes. Traditionally, the solution to this problem is based on the use of gyroscopic stabilization tracking systems. The most common types include a high-speed tracking system, which is characterized by feedback on the absolute angular speed of the control object, obtained with the help of a gyroscopic angular speed sensor [7]. The mathematical model of the system of stabilization of moving vehicle equipment is a set of mathematical models of the devices included in this system. The main devices of the stabilization system include sensors, electronic devices, and the motor. A block of equipment is an object of stabilization.

The measurement of absolute angular velocity (the sum of the relative velocity with the installed equipment and the translational velocity of moving object) can be performed using a gyroscopic tachometer.

The gyroscopic tachometer dynamics model about the output angle of rotation and output voltage are follow [7]:

$$\begin{gathered} J_g\,\ddot{\alpha} + 2\nu\sqrt{J_g c_t}\dot{\alpha} + M_\alpha - M_{gt}^1 + M_{gt}^2 - M_{fr} = M_{im} - \dot{\omega}_\eta J_g; \\ \ddot{U} = k_{adt}\alpha + U_0, \end{gathered} \tag{1}$$

where J_g is the moment of inertia of gyroscopic tachometer gimbals; α is the output signal of the gyroscopic tachometer; ν is the damping coefficient; c_t is torsion rigidity; M_α is the moment taking into account the torsion's hysteresis; $M_{gt}^1 = S\omega_\xi cos\alpha$ is the gyroscopic moment caused by the influence of measured angular rate ω_ξ, here S is the kinetic moment of the gyroscopic device; $M_{gt}^2 = S\omega_\varsigma sin\alpha$ is the gyroscopic moment caused by the influence of the angular rate ω_ς normal to the measured one; $M_{fr} = -M_f sign\dot{\alpha}$ is the moment of friction in gimbals bearings; $M_{im} = \delta W$ is the moment of imbalance, here δ is the displacement due to imbalance; W is acceleration, which leads to imbalance; $\dot{\omega}_\eta$ is the angular acceleration relative to the longitudinal axis of the device; k_{adt} is the transfer factor of the angle data transmitter; U_0 is the zero signal of the angle data transmitter.

The necessity to take into consideration the influence of the hysteresis is a feature of the model of the gyroscopic tachometer (1). For this goal, it is proposed to form the moment M_α according to such an algorithm. If the condition $\dot{\alpha} \le 0$ is satisfied, the moment M_α depending on the range of the angle α measurement is formed in the following way:

$$\begin{cases} M_\alpha = c_t\alpha_0, & \text{if } \alpha_0 - k_{gis}\alpha_0 \le \alpha \le \alpha_0; \\ M_\alpha = c_t\alpha, & \text{if } \alpha \le \alpha_0(1 - k_{gis}), \end{cases} \tag{2}$$

where k_{gis} is a coefficient, which characterizes a value of the hysteresis.

Expression (2) supplements model (1). The signal of the gyroscopic tachometer enters an electronic unit, which can include the limit device, demodulator, and filter.

The model of the limit device is sufficiently simple:

$$\begin{cases} A_\omega = U, & \text{if } |U| \le U_{lim}; \\ A_\omega = 12\text{sign}(U), & \text{if } |U| \ge U_{lim}, \end{cases} \tag{3}$$

where U_{lim} is the limit voltage; U is some nominal voltage; sign(x) is signum function.

The model of a semi-period demodulator looks like:

$$\begin{cases} U_{dm} = 0, & \text{if } \sin 2\pi ft \le 0;; \\ U_{dm} = A_\omega sin 2\pi f, & \text{if } \sin 2\pi ft \ge 0, \end{cases} \tag{4}$$

where f is the demodulator frequency. The filter model looks like:

$$\begin{cases} T_f = T_1, \text{ if } U_{dm}\cos 2\pi ft \ge 0; \\ T_f = T_2, \text{ if } U_{dm}\cos 2\pi ft \le 0; \\ U_f = (kU_{dm} - U_f)/T_f, \end{cases} \tag{5}$$

where T_f is the filter time constant; k is the transfer constant.

The control unit is assigned to forming a control, which must compensate for a signal of the absolute angular rate [14–16]. The input signal of the control unit is an angle of the turn, which can be given in some way. The mathematical model of the control unit represents the dependence of the control console signal on the input angle of the turn. The dependence looks like:

$$U = f(\gamma) = U_0 \frac{1 - \cos(\frac{2\alpha_0}{\varphi_0} i\gamma)}{1 - \cos\alpha_0} \text{sign}(\gamma), \tag{6}$$

where U is the output voltage; U_0 is the voltage relative to the middle point; γ is a current turning angle; φ_0, α_0 are working area parameters; and i is a transfer ratio; $sign(x)$ is signum function.

The stabilization system model also includes a tachogenerator that measures the angular speed of the equipment and a location sensor that determines an angle of the rotation:

$$\begin{aligned} U_{tg} &= k_{tg}\omega_{wm}; \\ U_{ls} &= k_{ls}\varphi_{wm}, \end{aligned} \tag{7}$$

where k_{tg}, k_{ls} are appropriate transfer coefficients of tachogenerator and location sensor; ω_{wm}, φ_{wm} are angular rate and angle of turn of working module.

The control law is formed on the basis of the output signals of the aforementioned sensors, taking into account the relevant coefficients depending on the mode of operation of the stabilization system [17–19]. In the general case, the control influence can be formed by the total signal of the control panel and gyroscopic tachometer after integration, the total signal of the gyroscopic tachometer and tachogenerator, and the feedback signal by the current. In some cases, the signal from the location sensor can be used [8, 20].

The output signal of the adder is level-limited, demodulated, amplified, and fed to the input of the pulse-width modulator. At the output of this device, a sequence of pulses is formed, the amplitude U_{PWM} and duration of which are determined based on the following relationships:

$$\begin{aligned} U_{PWM} &= (27 - k_{PWM}|I_a|)\text{sign}U_{in}; \\ t_{\text{PWM}} &= \frac{U_{in}}{U_{ref}}\tau, \end{aligned} \tag{8}$$

where k_{PWM} is the transfer constant of the pulse-width-modulator; I_a is the current in the armature's circuit; U_{in} is the input signal of the adder; U_{ref} is the reference voltage of the pulse-width-modulator; τ is the clock frequency of the pulse-width-modulator. Expressions (4)–(8) define the mathematical description of electronics of the system.

4 The United Mathematical Model of Stabilization Plant and Motor

The formation of the current in the armature circuit of the motor is described by the following expression:

$$I_a = \frac{U_{PWM} - c_e\dot{\varphi}_m}{R_w} \tag{9}$$

where c_e is the constant of electromotive force; R_w is the resistance of windings in the motor armature circuit; $\dot{\varphi}_m$ is the motor's angular rate, which represents the feedback signal.

The generated current flows to the motor and to the current sensor. The time constant of the motor is taken into account through an aperiodic unit with a unit gain coefficient, which can be presented in the following form:

$$T_m\dot{J}_{cor} + J_{cor} = J, \tag{10}$$

where T_m is the motor's time constant; J_{cor} is the input signal of the aperiodic unit (the current after correction, which enters the mechanical part of the motor).

The mathematical model of the motor can be developed based on [3]. It represents a differential equation of the second order:

$$J_m\ddot{\varphi}_m + M_r\text{sign}\,\dot{\varphi} = c_mJ_{cor}\frac{c_r}{n_r}(\varphi_b - \varphi_a), \tag{11}$$

where J_{cor} is the input signal of an aperiodic unit; c_m is the constant coefficient of the moment of loading on the motor shaft; M_r is the moment of resistance to motor rotation; c_r is reducer rigidity; n_r is the gear ratio; J_m is the moment of the motor's inertia; φ_m is an angle of the motor's turn; φ_b is an angle of the motor's turn taking into consideration the nonlinearity caused by the drive's backlashes; φ_a is an angle of apparatus turn.

The proposed model of the mechanical part of the engine is characterized by the following features:

1. The model takes into account the nonlinearity caused by the backlash of the drive, for which the output signal of the linear model, i.e. the angle of rotation of the motor φ_m, is passed through the nonlinear unit.
2. Interconnection of the mechanical part of the engine and the equipment block is carried out through a reducer taking into account its rigidity. Mathematically, such a relationship is ensured by entering into Eq. (11) the angle of rotation of the equipment block.
3. To establish the adequacy of the model and the real motor, zeroing of the acceleration of the engine rotation is introduced under the condition of its zero speed and the fulfilment of a ratio between the moments acting on the motor:

$$\left| c_m J_{cor} - \frac{c_r}{n_r}(\varphi_b - \varphi_a) \right| \leq |M_r|. \tag{12}$$

Models of stabilization objects and the motor play a significant role in creating a mathematical description of the stabilization system of information and measurement devices. From the point of view of the mathematical approach, these models are characterized by some features. In systems of the studied type, the drive, i.e., the motor with a gearbox, together with the object of stabilization can be considered as a single control signal reproduction device [7]. The output signal of such a device is the angular speed of rotation of the stabilization object, and the control signal is the output voltage of the pulse width modulator. With this approach, it can be assumed that the drive and the object of stabilization are connected by an elastic connection, that is, the gearbox is characterized by some rigidity. The kinematic scheme of the device, taking into account the elastic connection, under the condition of the monolithic structure of the stabilization object, is presented in Fig. 4.

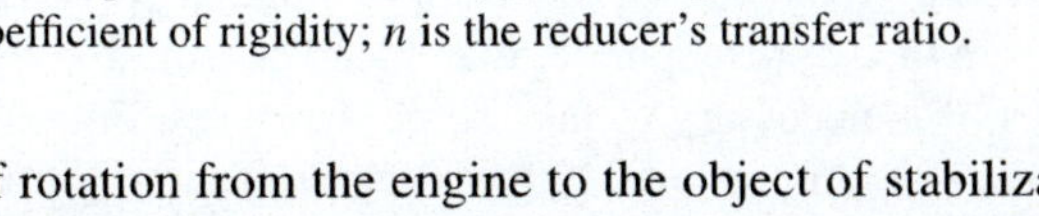

Fig. 4. Kinematical scheme of the unit, consisting of the motor, reducer, and stabilization object: M_m is the dynamic moment of the motor; I_m is the moment of inertia of the motor; I_{so} is the moment of inertia of the stabilization plant; φ_m is an angle of the turn of the motor shaft; $\varphi_r = \varphi_m/n$ is an angle of the turn of the output reducer's shaft; φ_{so} is an angle of the turn of the stabilization plant; c_r is the reducer's coefficient of rigidity; n is the reducer's transfer ratio.

The transfer of rotation from the engine to the object of stabilization is carried out through an elastic reducer. To create mathematical models of stabilization objects, it is necessary to take into account the effect of moments of friction, moments of imbalance, and moments of inertia. The presence of an elastic balancing system should be taken into account in the mathematical model of the equipment block. The elastic connection between the stabilization objects and the engine is determined by taking into account the stiffness and gear ratio of the gearbox [8–10].

The mathematical model of the stabilization plant must take into account not only the mechanical transmission, but also drive backlashes [21–24]. The mathematical model of the stabilization object of the vertical channel taking into account its elastic connection with the motor can be presented in the form of an equation of moment equilibrium [3, 4]. The mathematical model of the equipment can be given in the form of a second-order differential equation:

$$J_a\ddot{\varphi}_a + M_{fr}\text{sign}\dot{\varphi}_a - M_{im}\text{sin}\varphi_a + k_s\varphi_a + c_r\varphi_a = k_sA + \frac{c_r\varphi_b}{n_r} - M_{im}, \quad (13)$$

where M_{fr} is the friction moment; M_{im} is the moment of imbalance; k_s is the elastic rigidity; A is an angle of the setting angle deviation; J_a is the moment of apparatus of inertia; φ_a is an angle of apparatus turn.

For the model of the equipment, a requirement similar to the requirement for the motor model must also be met:

$$\left| -M_{im}\text{sin}\varphi_a + k_s\varphi_a + c_r\varphi_a - k_sA - \frac{c_r\varphi_b}{n_r} + M_{im} \right| \le \left| M_{fr} \right|. \quad (14)$$

The final differential equation of the vertical channel allows us to determine the absolute angular velocity of the equipment $\dot{\varphi}_{abs}$ based on the angular velocity of the equipment (relative velocity) $\dot{\varphi}_a$ and the angular velocity of the object (translational velocity) $\dot{\varphi}_{car}$:

$$\dot{\varphi}_{abs} = \dot{\varphi}_a + \dot{\varphi}_{car}. \quad (15)$$

The received absolute speed signal $\dot{\varphi}_{abs}$ is the output signal for the gyroscopic tachometer, which is part of the stabilization system of the equipment in the vertical plane.

5 Block Diagram of Stabilization System and Control Law

The block diagram of the mathematical model of the stabilization system, which shows the relationship of all its components (1)–(15), is shown in Fig. 5. For systems of studied type, it is important to achieve high accuracy of tracking and stabilization processes. Traditionally, the solution to this problem is based on the use of gyroscopic stabilization and tracking systems. The most common types of them include the speed tracking system, which is characterized by feedback on the absolute angular speed of the control plant, obtained with the help of a gyroscopic angular speed sensor. To improve the accuracy of control processes, it is necessary to take some measures. First, in such systems, in many cases, the control is proportional to the integral of the error, which ensures high accuracy characteristics in stable modes.

Secondly, achieving small dynamic errors for powerful drives is impossible without using invariant approaches. The formulation of conditions for the compensation of disturbances and the main provisions of the principle of invariance, as well as the extension of this principle to several applied problems, are presented in [15].

Features of the design of invariant drives are highlighted in papers [25, 26]. The main feature of invariant systems is the combined control [15]. This approach includes the creation of control systems, where, along with regulation by deviation or error, regulation

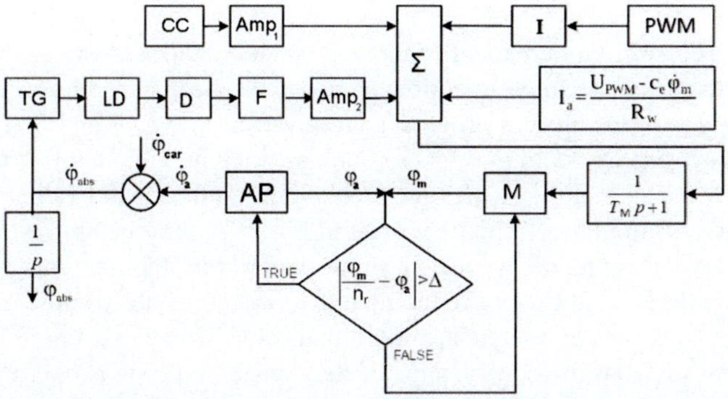

Fig. 5. The block diagram of the stabilization system: CC is the control console; A is the amplifier; LD is the limit device; D is the demodulator; A is the adder; I is the integrator; PWM is the pulse-width-modulator; M is the motor; A is apparatus; *Δ* is the dead zone.

by satisfying or disturbing influence is used. Taking into account the need to ensure high accuracy of guidance processes, in this case, it is advisable to use adjustment based on the desired effect. The local feedback must be used to improve control accuracy. The system has the local feedback on the voltage or current of the motor armature circuit. It is known that, depending on the selected parameter – voltage or current, a direct current motor has the properties of a speed or torque controller [27]. For direct current motors, the motor armature current can be taken as this value [28]. Gyro angular velocity sensors, tachogenerators, and motor armature circuit voltage and current signals can be used to organize feedback in tracking systems [29]. Depending on the way local feedback is organized, there are several approaches to organizing control in systems of the type under study. The basic principles of system functioning are as follows. The tracking angular rate is set on the control console. The gyroscopic tachometer provides the measurement of the absolute angular speed of the equipment. The compensating gyro tachometer measures the translational speed. The tachogenerator ensures the formation of a signal proportional to the angular speed of the motor. The system control loop uses feedback from the motor armature circuit current. The control law is formed by signals of the control console, ordinary and compensation gyro tachometers, and position sensor:

$$\begin{gathered} U_{con} = R_2 \int_0^{\Delta t} \left(R_1 K_{cor} U - K_{\dot{\varphi}_{abs}} \dot{\varphi}_{abs}\right) dt \\ + K_{kp} \varphi_{abs} + R_4 \left(R_3 K_{\dot{\varphi}_{so}} \dot{\varphi}_{so} + K_{\dot{\varphi}_m} \dot{\varphi}_m\right) + R_5 K_{I_a} I_a, \end{gathered} \tag{16}$$

where U_{con} is control signal; R_i, $i = \overline{1, 5}$ are adjustment coefficients; K_{cor} is the coefficient of the correction unit; U is a signal of the control console; K_{cor}, $K_{\dot{\varphi}_{abs}}$, K_{kp}, $K_{\dot{\varphi}_{so}}$, $K_{\dot{\varphi}_m}$, K_{I_a} are transfer coefficients by U, $\dot{\varphi}_{abs}$, $\dot{\varphi}_{so}$, $\dot{\varphi}_m$, I_a; Δt is the control cycle.

6 Modelling Results

Using the developed mathematical model of the stabilization system, the robust parametrical synthesis of the above-mentioned system has been carried out. The simulation results in the form of transient processes are given in Fig. 6, 7 and 8. The considered stabilization system works in two modes such as tracking and stabilization. In the first mode, the constancy and the small time of reducing to the steady value of the tracking rate are of great importance. This is proved in Fig. 6, where the above-mentioned time is 0.75 s. In the second mode, we need a small error of the stabilization of the platform's position, and also a small time of reducing to the constant stabilization error. As we can see from Fig. 7, these parameters are more than satisfactory – 2.5 arcmins and 0.75 s. The feature of the stabilization systems of the considered type is operation in difficult conditions caused by irregularities of the road and terrain. Figure 8 represents the transient process similar to Fig. 7 but in conditions of simulated disturbances, which have been obtained using experimental results. And we can see that the stabilization error and time of reducing to the steady state are acceptable.

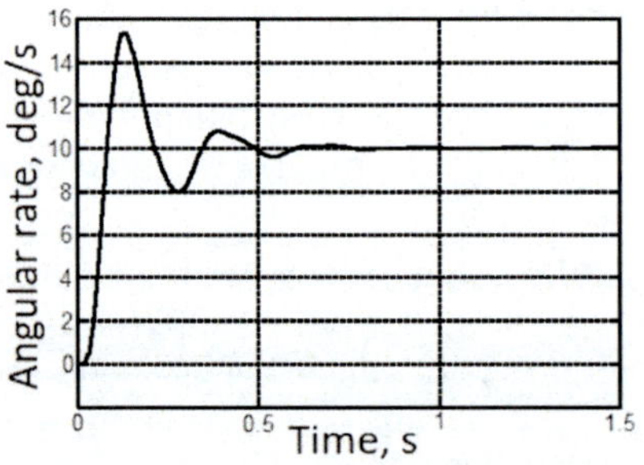

Fig. 6. Stabilization of the angular rate.

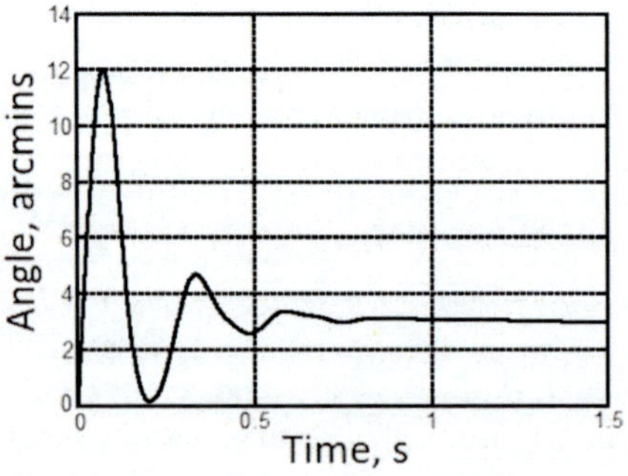

Fig. 7. The location of the stabilized platform.

The represented model has two main features. Firstly, the model's characteristics are maximally close to real systems. Secondly, the model represents the so-called two-mass system. This is of great importance when it is necessary to stabilize a payload with significant mass and dimensions. Of course, a sufficiently powerful motor must be used in this case. Therefore, the represented model can be useful for any moving object with the above-mentioned payload. In particular, the obtained results can be useful for tested equipment of aviation systems during manufacturing and checks.

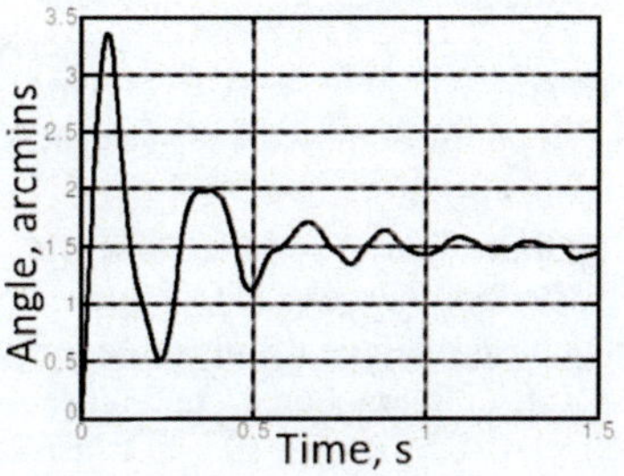

Fig. 8. The location of the stabilized platform with disturbances.

Another area of the usage of the developed model are stationary powerful telescopes and antennas mounted on the stabilized platform, which are used in research centers. The obtained model can be adapted to another areas by neglecting some parameters, namely, coefficient of elastic connection between the stabilized platform and electric motor.

7 Conclusions

A complete model of the system for the stabilization of equipment assigned for the air-moving object has been developed that ensures the possibility of modeling newly created stabilization systems, which are subjected to strict accuracy requirements. The model was created taking into account nonlinearities inherent in real systems. The proposed model allows for studying the dynamics of the system, as well as determining stabilization errors. The obtained simulation results prove the efficiency of the proposed approach.

In future work, it is proposed to improve procedures of designing systems for the stabilization of equipment for moving vehicles based on the developed mathematical description and learning machine methods to predict the states of the stabilization system.

References

1. Hilkert, J.M.: Inertially stabilized platform technology concepts and principles. IEEE Control. Syst. Mag. **28**(1), 26–46 (2008). https://doi.org/10.1109/MCS.2007.910256
2. Sushchenko, O., et al.: Integration of MEMS inertial and magnetic field sensors for tracking power lines. In: XVIII International Conference on the Perspective Technologies and Methods in MEMS Design, pp. 33–36. IEEE, Polyana (2022). https://doi.org/10.1109/MEMSTECH55132.2022.10002907

3. Sushchenko, O.A., Bezkorovainyi, Y.N., Novytska, N.D.: Theoretical and experimental assessments of accuracy of nonorthogonal MEMS sensor arrays. Eastern-Eur. J. Enterp. Technol. **39**(93), 40–49 (2018). https://doi.org/10.15587/17294061.2018.131.945
4. Sushchenko, O.A., Tunik, A.A.: Robust stabilization of UAV observation equipment. In: 2nd International Conference on Actual Problems of Unmanned Air Vehicles Development (APUAVD), pp. 176–180. IEEE, Kyiv (2013). https://doi.org/10.1109/APUAVD.2013.6705318
5. Sushchenko, O., et al.: Airborne sensor for measuring components of terrestrial magnetic field. In: 41st International Conference on Electronics and Nanotechnology (ELNANO), pp. 687–691. IEEE, Kyiv (2022). doi: https://doi.org/10.1109/ELNANO54667.2022.9926760
6. Scogestad, S., Postlathwaite, I.: Multivariable Feedback Control. Analysis and Design. Wiley, London (2005)
7. Zerrik, El.H., Castillo, O.: Stabilization of Infinite Dimensional Systems. Springer, Cham, (2021). https://doi.org/10.1007/978-3-030-68600-0_2
8. Osadchyi, S., Zozulia, V.: Synthesis of optimal multivariable robust systems of stochastic stabilization of moving objects. In: 5th International Conference Actual Problems of Unmanned Aerial Vehicles Developments (APUAVD), pp. 106–111. IEEE, Kyiv (2019). https://doi.org/10.1109/APUAVD47061.2019.8943861
9. Sushchenko, O.A., Bezkorovainyi Y.M., Golitsyn, V.O.: Fault-tolerant inertial measuring instrument with neural network. In: 40th International Conference on Electronics and Nanotechnology (ELNANO), pp. 797–801. IEEE, Kyiv (2020). https://doi.org/10.1109/ELNANO50318.2020.9088779
10. Buhaiov, D.V., Shelever, V.M., Avrutov, V.V.: Artificial neural networks application to MMG temperature calibration. In: 5th International Conference Actual Problems of Unmanned Aerial Vehicles Developments (APUAVD), pp. 44–47. IEEE, Kiev (2019). https://doi.org/10.1109/APUAVD47061.2019.8943908
11. Kuznetsov, B.I., Nikitina, T.B., Bovdui, I.V.: Multiobjective synthesis of two degree of freedom nonlinear robust control by discrete continuous plant. Tech. Electrodyn. **5**, 10–14 (2020). https://doi.org/10.15407/techned2020.05.010
12. Sanfelice, R.G.: Hybrid Feedback Control. Princeton University Press (2021)
13. Wang, L.: PID Control System Design and Automatic Tuning using MATLAB/Simulink. Wiley, Hoboken (2020)
14. Sushchenko, O., Averyanova, Y.: Algorithms for design of robust stabilization systems. In: Gervasi, O., Murgante, B., Hendrix, E.M.T., Taniar, D., Apduhan, B.O. (eds.) ICCSA 2022. LNCS, vol. 13375, pp. 198–213. Springer, Cham (2022). https://doi.org/10.1007/978-3-031-10522-7_15
15. Sushchenko, O.A., Shyrokyi, O.V.: H2/H∞ optimization of system for stabilization and control by line-of-sight orientation of devices operated at UAV. In: International Conference Actual Problems of Unmanned Aerial Vehicles Developments (APUAVD), pp. 235–238. IEEE, Kyiv (2015). https://doi.org/10.1109/APUAVD.2015.7346608
16. Niu, S.S.: Xiao, D: Process Control: Engineering Analysis and Best Practices. Springer, New York (2022). https://doi.org/10.1007/978-3-030-97067-3
17. Asadi, F.: State-Space Control Systems: The MATLAB/Simulink Approach. Morgan & Claypool, Sun Rafael (2021). https://doi.org/10.2200/S01050ED1V01Y2020.9CRM006
18. Hernandez-Gurman, V.M., Silva-Ortigoza, R.: Automatic Control with Experiments. Springer, Berlin (2019). https://doi.org/10.1007/978-3-319-75804-6
19. Lia, Y., Huang, N.Z., Jiang, T.: Selective Maintenance Modelling and Optimization. Springer, Cham (2023). https://doi.org/10.1007/978-3-031-17323-3
20. Voliansky, R.S., Sadovoi, A.V.: The transformation of linear dynamical object's equation to Brunovsky canonical form. In: 4th International Conference Actual Problems of Unmanned

Aerial Vehicles Developments (APUAVD), pp. 196–199. IEEE, Kyiv (2017). https://doi.org/10.1109/APUAVD.2017.8308808
21. Sushchenko, O., Chikovani, V., Tsiruk, H.: Redundant information processing techniques comparison for differential vibratory gyroscope. Eastern-Eur. J. Enterp. Technolog. **4**(7–82), 45–52 (2016). https://doi.org/10.15587/1729-4061.2016.75206
22. Hryshchenko, Y., Skripets, A., Tronko, V.: Mathematical description of phenomenon of amplification of integral-differential motive dynamic stereotype. In: 3rd International Conference on Methods and Systems of Navigation and Motion Control (MSNMC), pp. 71–74. IEEE, Kiev (2014). https://doi.org/10.1109/MSNMC.2014.6979733
23. Zhiteckii, L.S., Azarskov, V.N., Solovchuk, K.Y., Sushchenko, O.A.: Discrete-time robust stead-state control of nonlinear multivariable system. IFAC Proc. Vol. 8140–8145 (2014). https://doi.org/10.3182/20140824-6-ZA-1003.01985
24. Avrutov, V., et al.: Wavelet de-noising and Kalman filtering of mems sensors for autonomous latitude determination. Int. J. Sens. Wirel. Commun. Control **12**(5), 344–351 (2022). https://doi.org/10.2139/ssrn.4042780
25. Wright, S.J., Recht, B.: Optimization for Data Analysis. Cambridge University Press, Cambridge (2022). https://doi.org/10.1017/9781009004282
26. Zoppoli, R., Sanguineti, M., Gnecco, G., Parisini, T.: Neural Approximations for Optimal Control and Decision. Springer, Cham (2020). https://doi.org/10.1007/978-3-030-29693-3
27. Osadchy, S.I., Zozulya, V.A., Bereziuk, I.A., Melnichenko, M.M.: Stabilization of the angular position of hexapod platform on board of a ship in the conditions of motions. Autom. Control. Comput. Sci. **56**(3), 221–229 (2022). https://doi.org/10.3103/S0146411622030051
28. Bittanti, S.: Model Identification and Data Analysis. Wiley, New Jersey (2019)
29. Avrutov, V.V., Ryzhkov, L.M.: About one method of autonomous determination of the navigation parameters. In: 5th International Conference Actual Problems of Unmanned Aerial Vehicles Developments (APUAVD), pp. 214–217. IEEE, Kiev (2019). https://doi.org/10.1109/APUAVD47061.2019.8943835

Hierarchical-Correlation Method for Designing of an Adaptive Neural Flight Control System in Compliance with European and US Standards

Dmytro Prosvirin[1(✉)] and Volodymyr Kharchenko[2]

[1] Antonov Company, Akademika Tupoleva Street, 1, Kyiv 03062, Ukraine
dmytro.prosvirin@antonov-airlines.aero

[2] National Aviation University, Liubomyra Huzara Avenue, 1, Kyiv 030058, Ukraine

Abstract. Insufficient knowledge about the object being modeled and the conditions of its functioning poses a challenge in mathematical and computer modeling of dynamic systems. To address this issue, a proposed solution involves combining the strengths of theoretical and neural network modeling. By employing this approach, the performance is showcased through the simulation of a civil aircraft's motion. The theory of building a hierarchical-correlation neural network is developed. It includes the solution of problems of structural and parametric synthesis, which are based on precise mathematical models, algorithms, and programs for calculating aerodynamic, energy, temperature characteristics, take-off mass, stability and controllability, and efficiency indicators. This method belongs to the category of synergistic networks. The network starts with only input and output neurons. In the learning process, neurons are selected from the pool of candidates and added to the hidden layer. The hierarchical-correlation method of building an adaptive neural network has a number of advantages over multilayer perceptron networks.

Keywords: Adaptive Neural Flight Control · Hierarchical Correlation Method · Mathematical Modeling

1 Adaptive Neural Network Modeling of Dynamic Systems

Ordinary differential equations are extensively employed as a modeling tool for controlled dynamical systems. These models, along with the corresponding numerical methods, find widespread application in addressing the challenges associated with the synthesis and analysis of controlled motion for aircraft (AC) across different classes.

To date, detailed methods have been developed for creating and utilizing traditional models, which have proven effective in solving a wide range of problems. However, when it comes to modern and promising complex technical systems, several challenges arise that cannot be addressed by conventional approaches. These challenges stem from the uncertainties present in the system's properties and operating conditions, which can only be overcome if the system possesses adaptability. This adaptability refers to the ability to adjust the control algorithm and model of the system in response to changing

I. Ostroumov and M. Zaliskyi (Eds.): ACASD 2024, LNNS 992, pp. 86–97, 2024.
https://doi.org/10.1007/978-3-031-60196-5_8

circumstances. Furthermore, the accuracy requirements imposed on models, based on the specific problem at hand, often surpass the capabilities of traditional methods.

The current situation necessitates the use of methods and tools based on artificial neural networks as the most suitable modeling apparatus. This approach offers an alternative to traditional methods of modeling dynamic systems and allows for the development of adaptive models. However, traditional neural network models for dynamic systems, commonly employed in modeling controlled dynamic systems, are essentially empirical or "black box" models. These models rely solely on experimental data regarding the behavior of the object. Unfortunately, when dealing with the complexity inherent in aviation technology, these empirical models often fail to achieve the required level of accuracy necessary for solving problems related to aircraft motion control. Furthermore, due to the structural organization of such models, they are unable to address the issue of identifying the characteristics of a dynamic system, such as the aerodynamic properties of an aircraft. This limitation represents a significant drawback of this particular class of models.

The inefficiency of traditional neural network models in dealing with complex technical systems can be attributed to the formation of a purely empirical model, commonly referred to as a "black box". This model aims to encompass all the intricacies of the dynamic system's behavior. To achieve this, neural network models with a high dimensionality, containing a large number of adjustable parameters, are required. However, it is well-known from modeling experience that as the dimensionality of the neural network model increases, a larger amount of training data is necessary for its setup. Consequently, due to the limited availability of experimental data for complex technical systems, it becomes impossible to train such models and achieve the desired level of accuracy.

In order to address such challenges commonly encountered in conventional models, including those formulated as differential equations and neural network models, a solution known as the Hierarchical-correlation method is suggested. This approach relies on neural network modeling and employs specialized mathematical modules with adaptive control through the utilization of dedicated software. Notably, it has been effectively validated at the beer stand of the Antonov State Enterprise.

2 Adaptive Neural Control

Let us consider the system of differential equations written as a ratio in [1]. Here, the trajectory angle θ and engine thrust P are the control parameters. θ is also called the elevator deflection angle. We assume that both control parameters are normalized:

$$\theta = \tilde{\theta} \cdot \theta_0, -1 \leq \tilde{\theta} \leq 1, \tag{1}$$

$$P = \tilde{P} \cdot P_0, \qquad 0 \leq \tilde{P} \leq 1. \tag{2}$$

The control signals will be $\tilde{\theta}$, $\tilde{P}$ within the specified ranges. This explains the choice of two activation functions in the elevator control channel and the trim control channel:

$$h_\theta(x) = \frac{2}{1+e^{-x}} - 1, -1 < h_\theta(x) < 1, \tag{3}$$

$$h_{\mathrm{P}}(x) = \frac{2}{1 + e^{-x}}, 0 < h_{\mathrm{P}}(x) < 1. \tag{4}$$

The first step is to generate a population of chromosomes using functions:

$$\begin{aligned} f_{\theta}(x) &= h_{\theta}\left(\sum\nolimits_{j=1}^{2} w_j^{\theta} x^{(j)} + w_0^{\theta}\right), \\ f_P(x) &= h_{\mathrm{P}}\left(\sum\nolimits_{j=1}^{2} w_j^{P} x^{(j)} + w_0^{\mathrm{P}}\right), \end{aligned} \tag{5}$$

where $x = (x(1), x(2)) \in \mathrm{R}$, $x(1) = v$ is the airspeed; $x(2) = H$ is the flight height.

Synaptic weights w_j^{θ}, w_j^{P} and threshold displacements $w_0^{\theta}, w_0^{\mathrm{P}}$ are selected using a random number generator. They are assumed to vary from -1 to 1. Each of the neural network parameters is encoded with 32 bits.

Next, set the angle of deflection of the elevator:

$$\theta = f_{\theta}(v, H)\theta_0 = h_{\theta}\left(w_1^{\theta} v + w_2^{\theta} H + w_0^{\theta}\right)\theta_0, \tag{6}$$

and engine thrust

$$P = f_P(v, H)P_0 = h_{\mathrm{P}}\left(w_1^{P} v + w_2^{P} H + w_0^{\mathrm{P}}\right)P_0. \tag{7}$$

After that, we solve the above system of differential equations using the 4th order Runte-Kutta method. The system is integrated in steps of 0.01 s. At each step, the value of θ, P is updated according to formulas (1) and (2) [2].

For each chromosome the control quality criterion:

$$J = \int_0^T [(\theta_0 - \theta_t)^2 + (v_0 - v_t)^2]dt, \tag{8}$$

where θ_0 is the set elevator deflection angle; v_0 is the set speed.

A population consists of 15 chromosomes. They are ranked by value J:

$$J_{i_1} \leq J_{i_2} \leq \cdots \leq J_{i_{15}}, \tag{9}$$

where $i_1, i_2, \ldots, i_{15}$ are the corresponding chromosome numbers.

Modify the functionality J, so that it has a more transparent physical meaning than the functionality (3).

Let the accuracy thresholds for θ: δ_{θ} and velocity: δ_v. If simultaneously following condition are true then such control accuracy is considered acceptable:

$$\frac{1}{T}\int_0^T (\theta_0 - \theta(t))^2 dt \leq \delta_{\theta}^2, \tag{10}$$

$$\frac{1}{T}\int_0^T (v_0 - v(t))^2 dt \leq \delta_v^2. \tag{11}$$

Therefore, the dimensionless functional can be used as a criterion for control quality:

$$J = \max\left(\frac{1}{T\delta_{\theta}^{2}}\int_{0}^{T}(\theta_0 - \theta(t))^2 dt;\ \frac{1}{T\delta_{v}^{2}}\int_{0}^{T}(v_0 - v(t))^2 dt\right). \tag{12}$$

If $J \le 1$, then this quality is acceptable. Therefore, the chromosomes are ranked by the values of: $J_{i_1} \le J \le \cdots \le J_{i_{15}}$. If $J_{i_1} \le 1$, then the evolution ends, and the best chromosome with the number i_1 sets the control according to formulas (6), (7).

If the criteria for stopping evolution are not met, i.e. $J_{i_1} > 1$, , then a new generation of 15 chromosomes is created by chromosome selection and genetic operations. Then the values of the functional (12) are ranked for it: J_1 is chromosome sets the desired control. If. $J_{i_1} > 1$, then evolution continues.

The criteria for ending the evolution are the following factors:

1. Completion of a given number of epochs (for example, 100);
2. The algorithm does not improve the minimum already achieved.

In situation 2, we have consistent values of the minimum functional for different epochs:

$$\mu_1 \ge \mu_2 \ge \mu_3 \ge \ldots, \tag{13}$$

where μ_k are minimum values of the functional for the k-th generation.

So, if $\mu_i = \mu_{i+1} = \cdots = \mu_{i+9}$, then we can stop the evolution at the step $i + 9$. Or if the relation $\frac{\mu_i}{\mu_{i+9}} \le 1 + \delta$, 0 is the threshold is set. The accuracy of the algorithm can be improved by introducing several layers of neurons for each chromosome.

Reproduction Algorithm. At the first stage, we have a generation with $N_1 = 15$ chromosomes. Each chromosome consists of two neurons, each of which sets the function of two variables (5) or (6), respectively. A chromosome is characterized by a vector of parameters:

$$Z = \left(W_0^{\Theta}, W_1^{\Theta}, W_{2,}^{\Theta} W_0^{P}, W_1^{P}, W_2^{P}\right). \tag{14}$$

The parameters change from -1 to 1. The population is ranked according to the criterion (12). The smaller J, the higher the rank of the chromosome in the population.

We are building a reproduction algorithm in which each subsequent k-th generation will also have $N_k = 15$ chromosomes. Next, we write $N = 15$ is the number of chromosomes in a generation.

To move to the next generation, you specify numbers: Elite count is the number of elite children in the new generation, by default Elite count = 2;

$$F = \frac{n}{N - \mathrm{j}}, \tag{15}$$

where F is Crossover function; n is the number of persons created by Crossover; j is Elite count.

Crossover children are new chromosomes that are created from two parents by exchanging components of the respective Z vectors.

Let's set the Crossover function to 0.8. Then the number of crossover children is equal to $(15 - 2) \times 0.8 = 10.4$. So, we assume that there will be 10 crossover children in each generation. Accordingly, three mutated children are born from the three individuals with the lowest rank.

Further, the two people with the highest rank are passed on to the next generation without any changes (these are elite children). Three people with the lowest rank produce three mutation-children. Let's describe the process of mutation.

Let $z = (z_1, \ldots, z_6)$ be the vector of parameters for one of the three people with the lowest rank. The new vector z for the mutation child will be:

$$z_{mut} = (z_1 + \sigma_k \gamma_1, \ldots, z_6 + \sigma_k \gamma_6), \tag{16}$$

where $\gamma_1, \ldots, \gamma_6$ are independent Gaussian random variables with distribution $N(0, 1)$; k is the generation number to which the chromosome with vector z belongs.

Next, we assume

$$\sigma_k = \frac{\sigma_1 (N - k)}{N - 1}, k = 2, 3, \ldots, N - 1, \tag{17}$$

where N is the maximum number of generations allowed.

Value N is given by the value Generations. It is suggested that $N = 100$.

We put $3\sigma_1 = 1$, $\sigma_1 = 1/3$. This is done to ensure that the magnitude of the disturbance $\sigma_1 \gamma_j$ almost never goes beyond $3\sigma_1 = 1$, where the number 1 is associated with the range of change in the components of the z vector. If it happens that some coordinate $z_{mut} : z_j + \sigma_k \gamma_j$ falls outside the range $[-1; 1]$, then we continue to generate the Gaussian value $y_j \sim N(0, 1)$ as long $z_j + \sigma_k \gamma_j$ falls withinthe range $[-1; 1]$. Thus, we achieve that all coordinates of the perturbed vector z_{mut} lie on the interval $[-1; 1]$.

Let's describe the process of reproduction of ten chromosomes with ranks from 13 to 4 inclusive; they give rise to crossover children. These chromosomes are assigned new ranks:

$$4 \to 1, 5 \to 2, 6 \to 3, 7 \to 4, 8 \to 5, 9 \to 6, 10 \to 7, \tag{18}$$

$$11 \to 8, 12 \to 9, 13 \to 10.$$

So, we have ranks: 10, 9, 8, 7, 6, 5,…, 1 - in descending order of the functional value. Next, we assign these chromosomes mating probabilities proportional to the new ranks. We have:

$$1 + 2 + \cdots + 10 = \frac{11 \cdot 10}{2} = 55, P10 = \frac{10}{55} = \frac{2}{11}, P9 = \frac{9}{55}, P8 = \frac{8}{55}, \ldots, P1 = \frac{1}{55}. \tag{19}$$

These probabilities are successively plotted on the interval [0, 1]. Next, we generate a random number $U \in [0; 1]$ using a uniform distribution on $[0; 1]$.

Consider the numbers in the interval $[0; 1]$:

$$\gamma_k = \{U + k_h\}, k = 0, 1, 2, \ldots \tag{20}$$

where { } denotes the fractional part of a number.

It is possible to put $h = 0.25$. Each γ_k corresponds to the segment of lengthP_j, that we have placed on the interval [0;1], then we correspond to the number γ_k to the chromosome with (new) rank j. So, we have a sequence of ranks:j_0, j_1, j_2,... From this sequence, we select the parents $(jj_0, 1), (jj_2, 3), (jj_4, 5)$, and so on. In this case, two different chromosomes must be mated. Therefore, for example, if $j_2 = j_3$, we select the following pairs: $(jj_0, 1)$, $(jj_3, 4)$, $(jj_5, 6)$,... And then, if necessary, we adjust the selected pairs so that both parents are different. Now let's describe the process of creating Crossover children. Let $Z = (Z_1, \ldots, Z_6)$, $U = (U_1, \ldots, U_6)$ are two different chromosomes. Then the algorithm creates a child $V = (V_1, \ldots, V_6)$, where $V_k = \begin{cases} Z_k \\ U_k \end{cases}$ with a choice at each coordinate. If it happens that in the end $V = Z$, or $V = U$, then we rebuild the vector V until the child differs from each parent by at least one coordinate.

Stopping Algorithm. We have 5 conditions for stopping the process of creating new generations:

- Generations is the algorithm stops when the number of generations reaches the value Generations. You can set Generations = 100.
- Time limit. Stop when the time in seconds reaches the Time limit value.
- Fitness limit. We stop when the best chromosome of the generation has $J \leq 1$.
- Stall generations. Stops if

$$\frac{\mu_i}{\mu_{i+9}} \leq 1 + \delta, \tag{21}$$

  where $\delta > 0$ is a specified threshold (for example, $\delta = 0.001$); μ_j is the minimum value of J in the $i + 9$-th generation; stalling is performed for the $i + 9$-th generation. We chose the parameter Stall generations = 10, but other choices are possible.
- Stall time limit. We stop if

$$\frac{\mu_i}{\mu_{i+k}} \leq 1 + \delta \tag{22}$$

  on the time interval from the occurrence of the i-th generation to the occurrence of the $i + k$-th generation, while k is equal to Stall time limit.

The output of the propagation algorithm is the best chromosome of the form (12) from the last generation where the stop occurred. This chromosome sets the optimal control according to (5) and (6).

3 Hierarchical-Correlation Method for Building an Adaptive Neural Network

This approach falls under the classification of synergistic networks. The network commences with solely input and output neurons. Throughout the training process, neurons are chosen from the pool of candidates and incorporated into the hidden layer. Hierarchical-correlation network an adaptive neural network (HCNANN) has a number of advantages over multilayer perceptron networks:

1. There is no need to concern oneself with determining the number of layers and neurons to use in the network, as they are self-organizing and develop a hidden layer during training.
2. The training time for HCNANN is remarkably fast, often being two orders of magnitude quicker than a perceptron network. This attribute makes it well-suited for handling large training sets.
3. Typically, feed-forward ANNs are relatively small, with the hidden layer consisting of fewer than twelve neurons. On the other hand, probabilistic neural networks necessitate a hidden layer of neurons for each training case.
4. HCNANN training is highly dependable and satisfactory outcomes can usually be achieved with minimal or no adjustment of parameters.

It is less prone to getting stuck in local minima compared to multilayer reception networks. The disadvantages of this approach include the following:

1. Overfitting training data can lead to exceptional accuracy during training, but it often results in poor accuracy when dealing with new and unseen data due to their extraordinary potential for overfitting.
2. It is worth exploring the possibility that these networks may exhibit lower accuracy compared to probabilistic and general regression neural networks when dealing with small and medium-sized tasks (less than a couple of thousand training series). Further research is required in this field to gain a deeper understanding. However, it is important to note that hierarchical correlation excels in solving larger problems more effectively than probabilistic or general regression networks.

The HCNANN architecture is characterized by a unique structure where hidden neurons are incrementally added to the network and remain unchanged once added. This hierarchical approach involves feeding the output of all existing neurons in the network to the newly added neurons. When new neurons are introduced to the hidden layer, the learning algorithm aims to maximize the correlation between the output of the new neuron and the residual error of the network, with the goal of minimizing this error.

HCNANN has three layers: input, hidden and output. The input layer represents a vector of predictor variables ($x_1 \ldots x_p$). Its purpose is to distribute these values to the neurons in the hidden and output layers without performing any additional actions. In addition to the predictor variables, a constant input of 1.0, known as the offset, is also provided to each hidden and output neuron. This offset is multiplied by a weight and added to the sum that is sent to the neuron.

In the hidden layer, each value from the input neurons is multiplied by weight when it reaches a neuron. These weighted values are then added together to create a combined value. This weighted sum is then passed through a transfer function, which produces an output value. The outputs from the hidden layer are then distributed to the output layer.

The output layer consists of a single neuron for regression tasks, while for classification tasks, there is a neuron for each category of the target variable. Each output neuron receives values from all input neurons (including biases) and hidden layer neurons. Similar to the hidden layer, each value presented to the output neuron is multiplied by a weight and the resulting weighted values are added together to create a combined value. This combined value is then passed through a transfer function, which produces the final output value. These output values represent the outputs of the network. In regression tasks, a linear transfer function is used for the output neurons, while a sigmoid transfer function is used for classification tasks.

Learning Algorithm HCNANN. At the beginning, a cascade correlation neural network is composed solely of input and output layer neurons, without any hidden layer neurons. Each input is connected to each output neuron through a connection with an adjustable weight.

The weight value between the input and output neuron is represented by "x". The values on the vertical line are multiplied by their weights and then added together. Therefore, each output neuron receives a weighted sum from all input neurons, including the offset. The output neuron applies its transfer function to this weighted input sum in order to generate the final result. Even a simple HCNANN without hidden neurons possesses significant predictive capability. For many problems, a cascade correlation network with well-balanced input and output layers offers excellent predictions. The input weights for the hidden neuron are depicted as square boxes, indicating that they remain fixed once the neuron is added. On the other hand, the weights of the output neurons, represented by "x", continue to be adjusted. Here is a diagram of a network with two hidden neurons. It is worth noting how the second neuron receives inputs from external inputs as well as existing hidden neurons.

Theoretical knowledge regarding the object of modeling, which is typically represented by ordinary differential equations (such as traditional models of aircraft movement), is incorporated into a neural network model of a combined type (known as a semi-empirical neural network model) in a specialized manner. In this process, a portion of the neural network model is constructed using existing theoretical knowledge and does not require further adjustment or training. Only the elements that involve uncertainties, such as the aerodynamic characteristics of an aircraft, undergo adjustment and/or structural modification during the training process of forming the neural network model. In order to comply with the requirements of the EASA FAA, the development of a simulation complex for the implementation of an adaptive neural flight control system can be carried out within a Safety Critical Application Development Environment (SCADE) [3, 4]. The design technology for the Adaptive Control System (ACS) has significantly advanced at present.

There have been three major developments in the design technology:

1. Development of software for modeling the dynamics of controlled aircraft movement.
2. Software development for onboard equipment.
3. Development of a bench complex for semi-natural modeling, which serves as the final stage of development before conducting flight tests.

The software enables the selection and justification of the structure of the Adaptive Flight Control System (AFSC), as well as the analysis of system stability and the verification of compliance with the technical specifications for the control system with a specified probability. The software development process concludes with the synthesis of control algorithms based on a comprehensive mathematical model of a closed control loop, which includes statistical modeling (Fig. 1).

Software development involves creating programs that implement synthesized algorithms. Once the stand complex is developed, a closed control loop needs to be established. This control loop includes real computers, drives, and a control panel. The stand complex itself is a combination of hardware and software, offering versatility and the ability to easily reconfigure it for different aircraft with varying aerodynamics, traction, inertial characteristics, control wiring, steering systems, and other unique features. During the process of mathematical modeling, both deterministic and statistical modeling blocks are utilized.

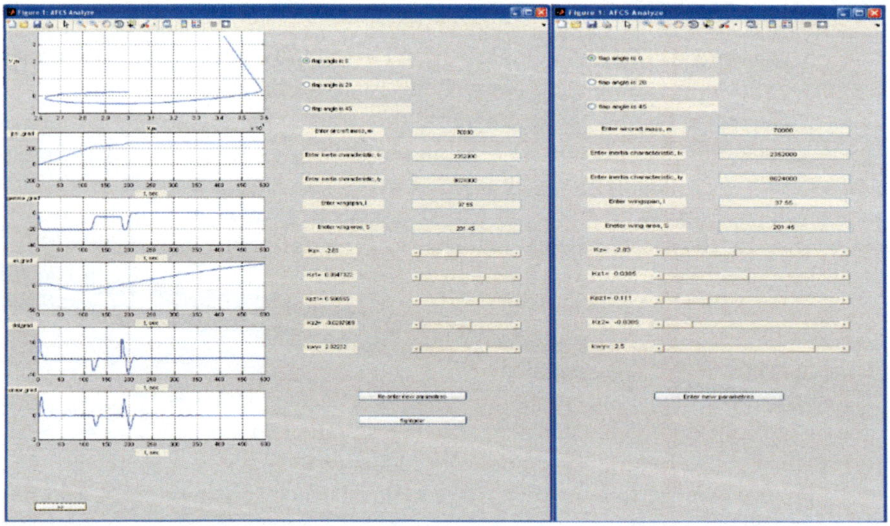

Fig. 1. Software for automatic flight control system development and analyzing.

The aerodynamic characteristics block contains numerical data regarding the components of these characteristics. These data are obtained from the aircraft designer and are interpolated using a program that operates in two or three variables, namely mass-director and combined trajectory control.

The trajectory algorithms produce signals that are then transmitted to the control elements of the aircraft, such as the elevator, ailerons, rudder, spoilers, and stabilizer. Alternatively, these signals can be directly sent to the drives [5].

The mathematical model consists of two blocks: one for graphical representation of the results and another for statistical processing of simulation results (Fig. 2). This allows for the evaluation of the system's compliance with the technical specification requirements.

The simulation complex for adaptive neural flight control system is based on neural network simulations. It utilizes software and mathematical modules that employ adaptive control. These modules have been successfully tested at the semi-natural stand of Antonov as part of the development of AN-158 and AN-178 aircraft (Fig. 3). The proposed neural adaptive control system incorporates a hierarchical-correlation neural grid (HCNG), which is a direct-distribution artificial neural mesh with a variable growing architecture. The HCNG is formed as a multilayer grid with a special hierarchical structure. Notably, it can be incrementally taught, allowing for the gradual addition of new neurons when convergence to the neural grid is achieved. Additional connections are then established for the newly added neurons during further training.

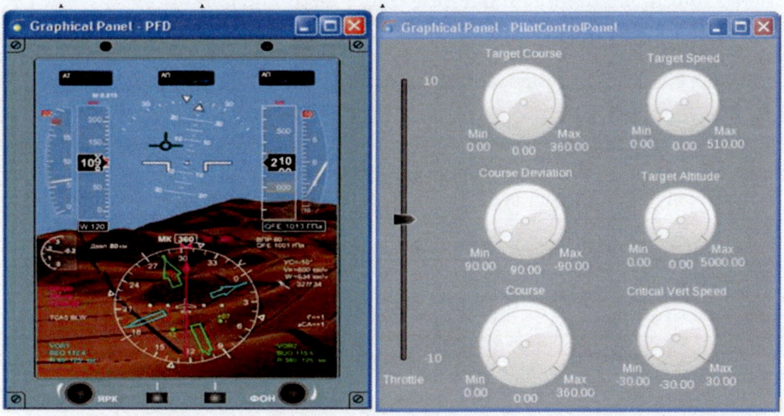

Fig. 2. Display of flight data processing results and control panel for the virtual flight test bench display.

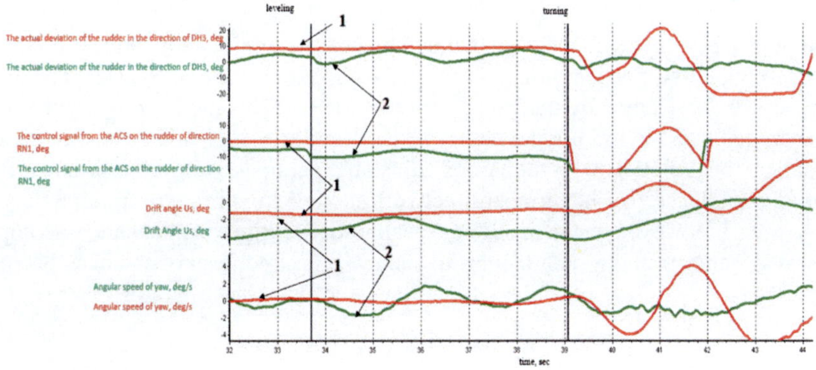

Fig. 3. Comparison of the simulation results with existing models (Flight No1, red color) and with pro-posed algorithm (Flight No2, green color).

Excessive transients at leveling altitude before touch down (Flight No. 1). The path channel algorithm has been corrected (Flight No. 2). The rudder response to deviations is less. Damping in the travel channel has been reduced, the values of the coefficients have been reduced: by angular speed of yaw, Drift angle (Us).

In accordance with the above methodology, the results of the modelling and flight tests at the points of contact were compared. The comparison results are presented in Table 1. The table shows that the mathematical model is quite close to the real processes at the points of contact when reproducing the conditions of real flights. The discrepancy between the estimated parameters does not exceed the tolerances specified in the methodology. In addition to comparing the main flight parameters at the landing points, the dynamic characteristics of the landing transients were compared based on the results of flight tests and mathematical modelling under adequate conditions.

Table 1. Comparison of modelling and flight test results. Flight 393 (landing 1).

Landing characteristics	Flight tests	Mathematical modelling	Permissible difference in methodology	Actual discrepancy
L_{pos}, m	520	494	75	26
Z_{pos}, m	1.71	1.39	1	0.32
V_{ypos}, m/sec	–0.18	–0.34	0.2	0.16
γ, deg	–0.046	–0.060	1	0.014
ϑ, deg	4.44	5.21	1	0.77

4 Conclusions

This approach effectively manages dynamic objects and significantly reduces computational learning costs. The learning parameters, such as architecture, NG size, and learning speed, are determined by the learning process itself. This methodology overcomes the challenges faced by traditional neural network models, particularly in constructing high-dimensional models with numerous adjustable parameters. It is well-known from modeling experience that larger neural network models require more training data for accurate setup. However, due to the limited availability of experimental data for complex technical systems, training such models to achieve a desired level of accuracy becomes impractical.

References

1. Haupt, R., Haupt, S.: Practical Genetic Algorithms, 2nd edn. Wiley, New Jersey (2004)
2. Ceylan, H., Bell, M.: Traffic signal timing optimization based on genetic algorithm approach, including drivers' routing. Transp. Res. Part B Methodol. **38**(4), 329–342 (2004)

3. Efficient Development of Safe Avionics Display Software with DO-178B Objectives Using SCADE Suite. Methodology Handbook. France, Esterel Technologies (2012)
4. RTCA DO-178C Software Considerations in Airborne Systems and Equipment Certification, cRTCA, Inc. Washington, DC USA (2011)
5. Prosvirin, D., Kharchenko, V.: Improvement of "aircraft-automatic flight control system" control loop quality electronics and control systems. Electron. Control Syst. **4**(47), 22–28 (2016)

Global Navigation Satellite Systems

Revealing the Criteria for Detecting the Spoofing and Premediated Interference of GNSS Signals Using the Experimental Simulation Model

Valery Konin[1], Oleksiy Pogurelskiy[1](✉), Iryna Prykhodko[1], Tetiana Maliutenko[1], Alexey Zhalilo[2], and Alexandr Yakovchenko[2]

[1] National Aviation University, Liubomyra Huzara Avenue 1, Kyiv 03058, Ukraine
pogurelskiy@gmail.com
[2] Kharkiv National University of Radio Electronics, Nauky Avenue 14, Kharkiv 61166, Ukraine

Abstract. The results of the research on the detection of meaconing spoofing or intentional interference of Global Navigation Satellite Systems (GNSS) signals at the initial stage when the user enters the area of the indicated interference are presented. The research was carried out using an experimental simulation model (prototype) of the effects of GNSS/GPS signal spoofing. The prototype included a dual frequency GNSS receiver, two antennas, a combiner and attenuator, a computer for accumulation of measurement information. The impact of spoofing was analyzed using the output measurement information of the receiver. During the experiment, the signals from the first antenna were fed to the input of the receiver through a combiner. After 75 min of observation, the input of the receiver was fed with satellite signals from the second antenna. The amplitudes of the signals from both antennas did not differ much, which can be characterized as the effect of intentional interference or spoofing attack at the initial stage. The receiver output data were recorded, processed and analyzed. The results of the whole experiment have been analyzed. It was found that the main criteria for the presence of interference/spoofing are the signal-to-noise ratio, code and phase pseudorange differences, which change significantly and rapidly. The additional criteria can be the pseudorange standard deviations and a multiple increase of the positioning errors. The data obtained make it possible to indicate the influence of spoofing and to recommend switching to alternative navigational aids.

Keywords: Global Navigation Satellite Systems (GNSS) · Spoofing · Premeditated Interference · Signal-to-Noise Ratio · Pseudorange Differences

1 Introduction

Global Navigation Satellite Systems (GNSS) technologies are being intensively implemented in all sectors where coordinate and time determination are required [1]. In [2], sixteen strategically important state infrastructures were identified for which the failure of GNSS equipment due to interference of various types can have significant or even catastrophic effects. The aviation industry is also among such infrastructures.

I. Ostroumov and M. Zaliskyi (Eds.): ACASD 2024, LNNS 992, pp. 101–113, 2024.
https://doi.org/10.1007/978-3-031-60196-5_9

GPS L1 uses a 1559–1610 MHz frequency band. Usage of this frequency band should be strictly controlled by the relevant international and national services. As the specified signal strength of GNSS satellites is very low (ranging from -153 to -161 dBW), it is susceptible to distortion by unintentional or intentional interference. For aviation users acceptable signal level of interference in the form of a sinusoidal signal in signal search mode can be 6 dB more than the useful signal, and in the tracking mode by 10–12 dB [3]. Signal levels for pulse interference are somewhat higher and depend on the spectrum width of the interfering signal.

GNSS interference is classified as follows:

- Jamming. When interference is produced by extraneous signals that exceed the levels of navigation satellite signals, resulting in an increase in the interference/satellite signal ratio and complete or partial inoperability of the navigation receiver;
- Spoofing of GNSS signals (simulated interference), when the navigation receiver receives the imposed signals as standard and determines the position, velocity vector, position and time (PVT) according to the data included in the interfering signals;
- retransmission/transmission (meaconing spoofing) of a full or partial copy of GNSS satellite signals observed at the point where the repeater-spoofer antenna is installed.

Spoofer signals can be quite diverse. They can be generated in a simulator where satellite-like signals are generated. It is also possible for a spoofer to receive signals from several satellites, increase their power and broadcast them to the object whose navigation receiver is to be interfered with. It is possible to use co-application, for example, of a powerful signal that jams the normal functioning of the receiver, then feed the signals of a spoofer in which an intruder-imposed trajectory is generated.

The development of electronic components, given that the characteristics of GNSS signals are publicly available, has created conditions for the production of both "cheap" interferers of navigation equipment and imitators of fake satellite signals, which creates a threat potentially affecting the safety of aircraft flights. International Civil Aviation Organization (ICAO) pays great attention to flight safety issues, but the measures provided for to improve navigation performance, in particular the procedures for autonomous control of satellite signal-in-space integrity (RAIM, ARAIM), do not fully protect satellite navigation equipment from spoofing.

Taking into account that GNSS is a primary navigation system in aviation, equipping aircrafts with satellite navigation hardware, introduction of automatic dependent surveillance technologies, which use the data received from satellite navigation systems to transmit the position of aircraft, the issues of creation and development of methods and technologies mitigating the impact of spoofing on navigation characteristics are certainly relevant, and a priority.

2 Literature Review and Problem Statement

The process of receiving, signal processing and obtaining a navigation solution in the receiver is as follows: in the upper hemisphere, the signal of each satellite is searched and detected by comparing a copy of the signal code available in the receiver with the satellite signal. Once the signal is detected and captured by the receiver, the signal is tracked. The received microwave signal is down-converted, digitized, correlated with the code copy for possible code delays and the Doppler frequency. As a result, a correlation peak is formed, which is a time reference for pseudorange calculation. When spoofing is occurring, the correlation peak is distorted, or a second spoofing signal peak appears. If the amplitude of the spoofing signal peak is sufficient to be captured, the receiver begins to interact with the spoofed signal and produces a false navigation solution [4].

Counteraction to spoofing consists in its timely detection and, if possible, weakening or elimination, and if it is impossible to prevent the spoofing attack, the aircraft should switch to alternative means of navigation.

Interference mitigation techniques and in particular spoofing have received much attention in the scientific and technical literature [5–7].

Interference mitigation of GNSS signals can be broadly divided into:

- Methods of signal processing directly using software in the navigation receiver at the signal level;
- Methods and algorithms to defeat spoofing by an autonomous stand-alone module that rejects the interfering signal and delivers an uninterfered signal to the receiver.

Detection and protection against the effects of spoofing in a navigation receiver can be performed in several ways.

To detect spoofing and jamming type interference, algorithms for tracking changes in received signals [8], and changes in the signal-to-noise ratio (spike decrease, increase, comparison with signals from different satellites [9] are used.

A combined anti-interference and spoofing mitigation algorithm is proposed in [10]. The possibility of effective mitigation of jamming type interference in the first step and spoofing in the second step by modeling methods on a 9-element antenna array is shown. Spoofer signals, obtained by receiving and retransmitting the signals of four satellites, are amplified and radiated by a single antenna. The signals from the same satellites are received by the receiver antenna system.

It is possible to detect and determine the direction of arrival of the fake signal. If the receiving navigation equipment has an antenna array with a controlled radiation pattern, then at the formation of zero in the direction of arrival of spoofer signals on the receiver monitor will disappear all satellites [11–13] and the navigation receiver will receive satellite signals that are not affected by signals from an external source.

The method of detection and localization of a spoofer using a mobile five-element antenna array is proposed in [14, 15]. To separate the spoofer signal from the satellite signals, the analysis of the double phase difference of the carrier frequency is used, and when the antenna array is moved, the coordinates of the spoofer are determined.

Obtaining correlation peaks of multibeam propagation, spoofing and selection of useful satellite signals allows the method [16], which for GPS signals is tested according to the results of signal processing in Software-Defined Receiver (SDR). The efficiency of the method was verified using test files of TEXBAT measurements [17].

In the article [18] spoofing detection is carried out by analyzing the distortion of the correlation characteristic of the received signals.

The use of four navigation receivers with a special arrangement of four antennas made it possible to determine the direction of arrival of spoofer signals in a 5x5 degree sector, to attenuate multipath and to mitigate the impact of spoofing due to special processing of navigation receiver signals [19].

In [20], the possibility of spoofing detection using statistics of signal-to-noise ratio differences in a navigation receiver with an antenna array was researched.

A spoofing protection method by monitoring pseudoranges to satellites is proposed in [21].

In [22], the occurrence of spoofing is determined by testing the statistical hypothesis of the minimized probability of missed detections based on the directions of the signals arriving at the navigation receiver.

In [23] method of interference detection with application of neural networks to satellite signals including those from low-orbit satellite systems is considered.

In [24] results of modelling the impact of interference on the equipment provides a solution to the navigation task for multi-system GNSS observations, including three-frequency observations GPS, Galileo, and two-frequency observations GLONASS. Experiments were conducted in an anechoic chamber with a NovAtel PwrPak 7 navigation receiver supporting GPS, GLONASS, Galileo, and BeiDou with all frequencies and high-performance GNSS signal and trajectory simulators. The jamming type interference was generated at frequency $L_1 = 1575.42$ MHz. The interference was applied at 15-min intervals for two minutes. The experiments showed that the use of multi-system and multi-frequency observations gives reliable position determination at low and even high level of jamming power.

Satellite signal authentication technology is also being developed [25, 26], which involves embedding special keys into navigation signals that are not known to generate spoofing signals (cryptographic digital signatures from satellites).

It follows from the above brief literature review that detection of the effects of spoofing can be organized using software in the navigation receiver, for example, by correlation characteristics of the received signals and outside the receiver by the data produced by the receiver to solve the navigation problem. The second method is favored in some cases because it does not affect the receiver's signal processing structure and software.

The research does not analyze the issue of spoofing detection at the initial stage. If this point is missed, the receiver will capture false signals and the navigation solution will not be correct.

The aim of this research is to identify stable criteria and intervals of reliable detection of the impact of meaconing spoofing on non-specialized GPS navigation receivers based on the results of the analysis of the output measurement information (raw GPS observations and positioning results) on the basis of the created experimental simulation model (mock-up).

The established criteria of GNSS signal spoofing impact and the criteria that will be identified and verified in future studies are the basis for the development of technologies (experimentally verified methods, algorithms, and software) for reliable detection of spoofing impact in real time with subsequent embedding of the relevant software and mathematical software into onboard processors of GNSS receivers.

3 Research Method

The proposed method for identifying the criteria for reliable detection of spoofing impact is based on the data generated by the navigation receiver during positioning. The hardware structure for obtaining experimental data is shown in Fig. 1.

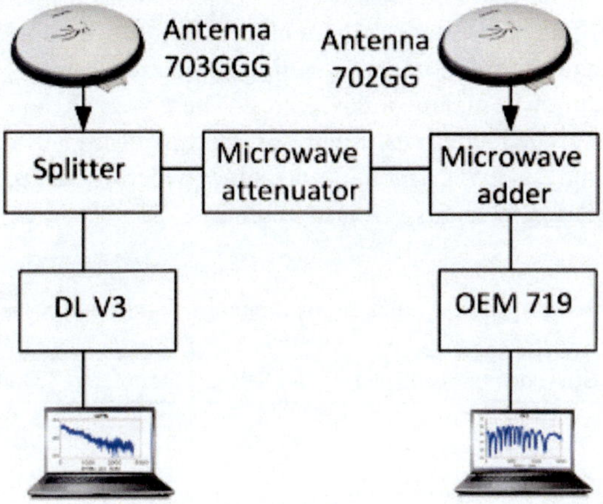

Fig. 1. Interconnection of equipment during the experimental study.

The DL V3 and OEM 719 navigation receivers were tuned to receive GPS signals at L1 and L2 frequencies at a rate of 1 Hz. The DL V3 receiver performed the control function. The antennas were geodetically referenced and spaced ~1.5 m apart. The receivers recorded: code and phase pseudoranges, standard deviations of pseudorange fluctuation errors, signal-to-noise ratios, location (latitude, longitude, altitude), standard deviations of latitude, longitude, altitude, GPS ephemeris, and receiver time variations.

Record Procedures. The microwave attenuator was set in the full attenuation mode. In this case, the inputs of the receivers received signals only from the antennas to which they were connected.

During 13416 s the above parameters were recorded. Then, without disassembling and without switching off equipment, approximately at 4500 s, the attenuator was switched for 20–30 s to the zero-attenuation mode. During 8900 s the recording of output parameters continued. At the same time, the DL V3 receiver remained in the mode of transmitting the results of observations of signals from the 703GGG antenna, while the OEM 719 receiver received signals from the two antennas at almost the same level.

Thus, OEM 719 receiver data from signal observations from 702GG antenna alone were recorded at 0–4500 s, and at 4500–13416 s, these data from simultaneous signal observations from 702GG and 703GGG antennas were recorded. Any of the antennas can be considered as a spoofer, and by processing the received data to determine 1) the presence of spoofing in the signal mixture and 2) the moment of its occurrence.

Section 4 presents the results of the data processing.

4 Research Results

4.1 Processing and Analyzing the Message Range

The purpose of the analysis is to identify changes in the signals arriving at the OEM719 receiver from two antennas compared to the signal arriving from a single antenna. The signals coming from the two antennas can be considered as a form of spoofing at the most critical moment when the navigation receiver smoothly enters its area of coverage. If spoofing is detected at this moment, spoofing mitigation can be initiated or the user should switch to alternative means of navigation. There were 20 satellites in view at the observation interval and 12 satellites in view at the time of transition.

A generalized analysis from which it is possible to identify the parameters of change of two signals in relation to one is given in Table 1.

Table 1. Identifying the changing in signal by data from message range.

PRN	Range	Std (cod)	Std (phase)	S/N	Doppler	Delta range	Total
2	0	0	2	5	0	5	12
5	0	5	2	5	0	5	17
7	0	5	0	3	0	5	13
13	0	0	0	5	0	5	10
14	0	0	1	5	0	5	11
15	0	5	0	5	0	5	15
17	0	0	3	3	0	5	11
19	0	0	0	5	0	4	9
22	0	5	1	5	0	5	16
23	0	0	1	5	0	5	11
24	0	0	0	5	0	5	10
30	0	2	1	5	0	5	13

In Table 1 the numbers denote the priorities of parameters from 0 to 5 ("5" is maximum priority, "0" is parameter can be ignored in the analysis). Parameters Range, Doppler can be ignored when evaluating the presence of an interfering signal. Standard deviations (Std) of code and phase pseudorange errors have the lowest priority and can be taken into account only additionally. The highest priority is given to signal-to-noise ratios (S/N) and differences of code and phase pseudorange (Delta range), changes of which take place for all satellites. The data in Table 1 are valid for observations at GPS frequencies L1 and L2.

Typical features of signals, obviously affected by spoofing, are shown in Fig. 2 for GPS satellite №15 at frequencies L1, L2. Horizontal axis shows time in seconds in Fig. 2, vertical axis shows signal-to-noise ratio in dB and linear combinations of Delta range in meters. As can be seen from Fig. 2, at 4500 s the signal-to-noise ratio and Delta range change. This behavior of the pseudorange components is typical for all satellites.

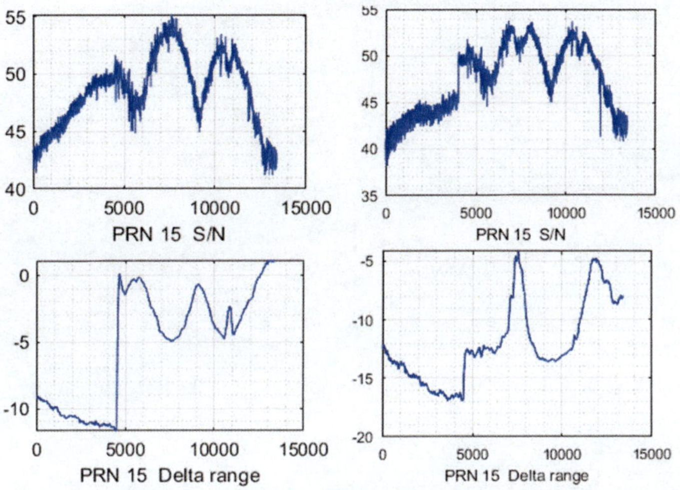

Fig. 2. Parameters analyzed: graphs on left obtained for frequency L1, diagrams on right for L2.

The detailed behavior of the parameter-characteristics of the pseudorange components in the transition area is shown in Fig. 3, Fig. 4, and Fig. 5. Delta range parameters varied within 6–10 m (Fig. 3).

Of the 12 satellites, the signal-to-noise ratio varied within 1–2 dB for three satellites, within 3- 4 dB for the fourth satellite, and within 5–8 dB for the fifth satellite (Fig. 4). All satellites are characterized by changes in the values of standard deviations of phase and code pseudorange errors at the moment of transition.

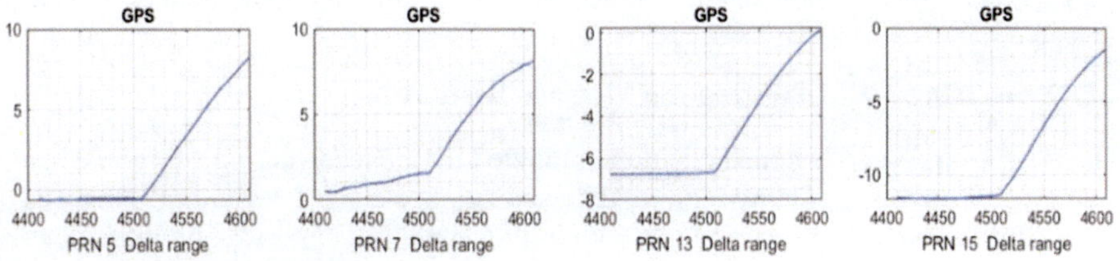

Fig. 3. Delta range changes of some satellites in the transition area.

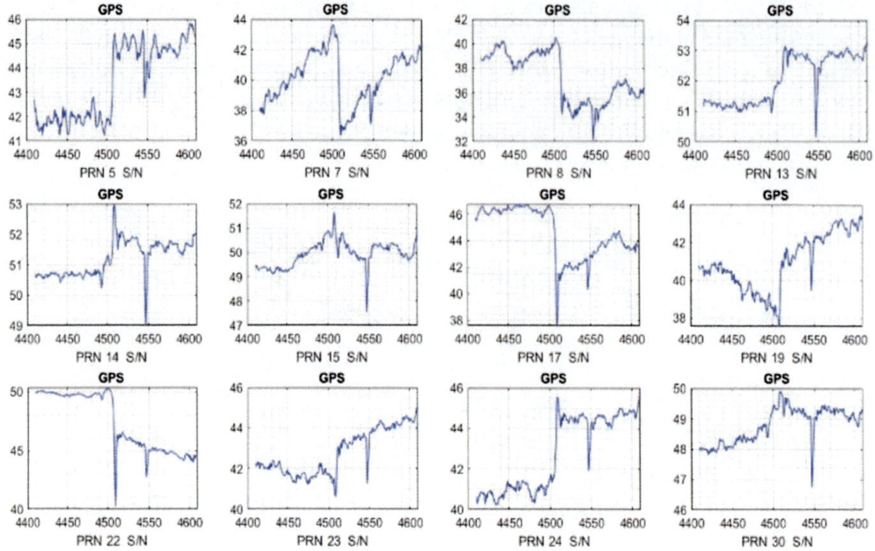

Fig. 4. Changes in the signal-to-noise ratio in the transition area.

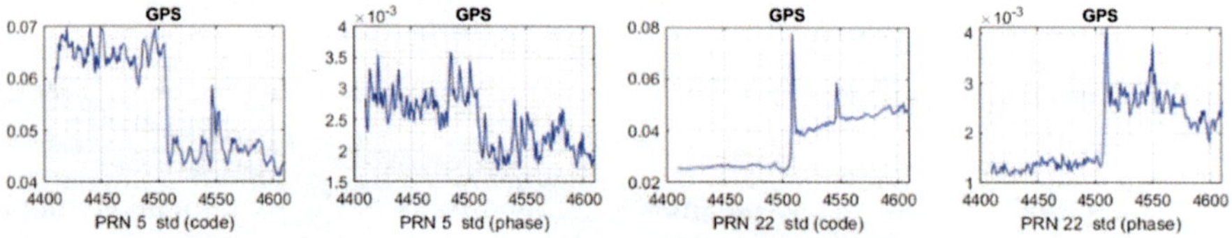

Fig. 5. Typical characteristics of the variation of the standard deviation of fluctuations of code and phase pseudorange errors.

The results obtained allow us to conclude that a jump-like change in the signal-to-noise ratio can be a criterion for detecting the impact of spoofing. An additional criterion can be a jump-like change in the difference of phase and code pseudorange.

Let's also consider how the position parameters change at the moment of transition.

4.2 Processing and Analyzing the Message Bestpos

The Bestpos message contains position computed by the receiver and additional statistical evaluations such as latitude, longitude and height standard deviations. Analysis of the Bestpos message was carried out both in the whole time interval and in the area of transition from the data of signal arrival in the receiver from one antenna to the data of signal arrival in the receiver from two antennas.

In the first case for the entire time interval was obtained that the error in determining the location of signals from two antennas in the horizontal plane in 99% of cases reaches 5 m, in the vertical plane was 10 m. The following are the results of location estimation in the transition area. The transition area was chosen to be 400 s with the center at the 4510 s. This point is determined by analyzing the range message and the 400 s is related to the smoothing interval of the positional data. The results are shown in Fig. 6. From the data in Fig. 6, it can be concluded that for at least 200 s receiver has received extraneous signals and the navigation system should be avoided until it is cleared.

The data in Fig. 6 are presented in a topocentric coordinate system. The data confirm that at the moment of transition there are changes in coordinates, but it is not so significant to be the main criterion. These parameters can be considered as an addition to the criteria obtained in the study of pseudoranges. A number of results characterizing the extended scope of the research are presented in Appendix.

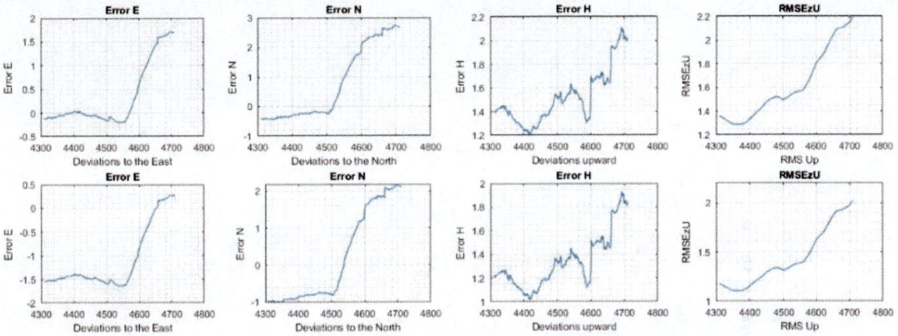

Fig. 6. Position deviations: on 4 upper graphs data referenced to antenna 702 GG; on 4 lower graphs data referenced to antenna 703 GGG.

Obviously, it is necessary to determine the most effective criteria for detecting spoofing, and in the development of algorithms and software for detecting the action of spoofed signals it is necessary to use simultaneously several (a combination of) criteria.

In the future, it is also expected to develop technologies to determine the location (direction finding) of spoofing sources in critical areas in the presence of groups (networks) of receivers with information integration of their measurement and other information.

5 Conclusions

The performed studies have shown that at the early stage of spoofing interference its impact can be detected by analyzing the data produced by the navigation receiver. Priority data are signal-to-noise ratios, code, and phase pseudorange differences. These characteristics are almost immediately apparent when interfering signals appear, with amplitudes comparable to the standard satellite signals, and can be criteria for detecting spoofing. Additional criteria include changes in the user's location along the east-north-up coordinate axes. It is important to highlight that when spoofing is detected by the proposed method, there is no need to interact with the software of GNSS receiver. Only data processing is required, which is produced by practically most GNSS navigation equipment.

In the future, in order to identify additional criteria for detecting interference of false and "real" (authentic) signals, as well as the impact of spoofing of GNSS signals is expected to be performed experimental research:

- with analysis of an extended set of linear combinations of observations, characteristics, and quality control of navigation solutions, analyses of velocity vector components, etc.;
- with the use of multi-system measurements of signal parameters of different GNSS (GPS, GLONASS, Galileo, BeiDou) to compare combinations of navigation solutions (position, velocity vector, and time (PVT)) and assessment of data consistency from different GNSS.

Appendix

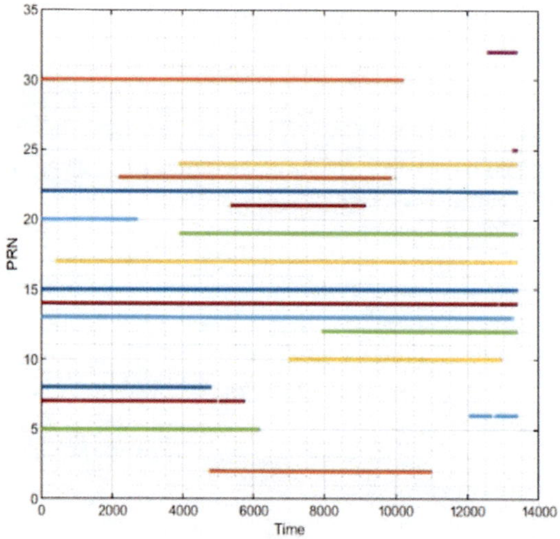

Fig. 7. Diagram of satellite visibility.

The visibility diagram (Fig. 7) shows the number of visible satellites at an arbitrary observation interval.

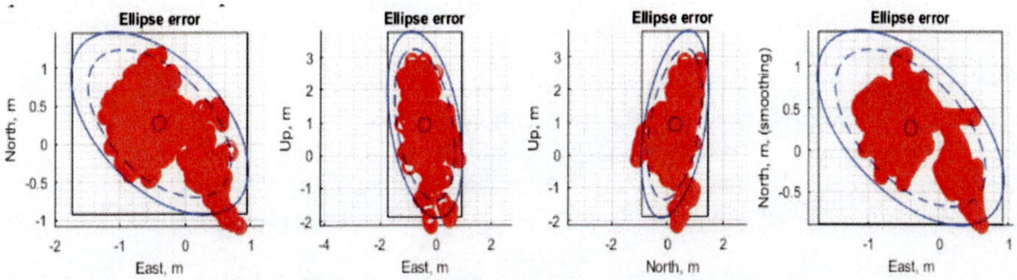

Fig. 8. Ellipses of errors for DLV3 signal from a single antenna.

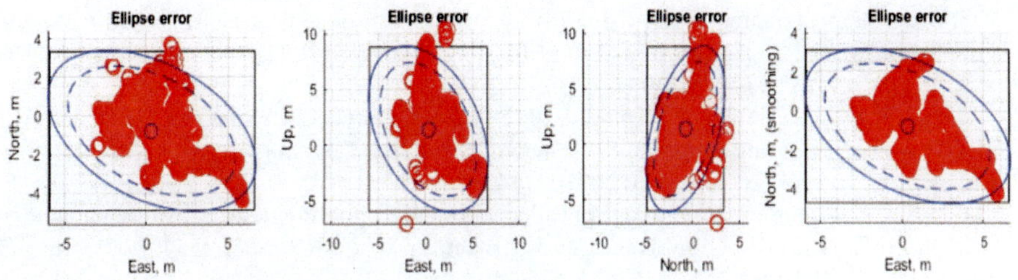

Fig. 9. Ellipses of errors OEM 719 during spoofing.

In Fig. 8, over the entire observation interval, the signal arrives from antenna 703GGG. In Fig. 9, over the entire observation interval the signal comes from antenna 703GGG and 702GGG. In Fig. 8, Fig. 9, the solid line corresponds to 95%, the dashed line to 99.5%.

References

1. EUSPA EO and GNSS Market Report, Issue 1. Luxembourg: Publications Office of the European Union (2022). https://doi.org/10.2878/94903
2. O'Connor, A.C., et al.: Economic benefits of the global positioning system (GPS). RTI Report Number 0215471 (2019)
3. Aeronautical Telecommunications. Annex 10 to the Convention on International Civil Aviation. vol. I (Radio Navigation Aids), 8th edn. ICAO, Montreal (2023)
4. GNSS Signal Spoofing How to Evaluate the Risks to Safety-Critical and Liability-Critical Systems, White paper, Spirent Communications, Inc (2020)
5. Morales-Ferre, R., Richter, P., Falletti, E., Fuente, A., Lohan, E.: A survey on coping with intentional interference in satellite navigation for manned and unmanned aircraft. IEEE Commun. Surv. Tutorials **22**(1), 249–291 (2020). https://doi.org/10.1109/COMST.2019.2949178
6. Chamola, V., Kotesh, P., Agarwal, A., Naren, N., Gupta, N., Guizani, M.: A comprehensive review of unmanned aerial vehicle attacks and neutralization techniques. Ad Hoc Netw. **111**, 102324 (2020). https://doi.org/10.1016/j.adhoc.2020.102324
7. Schmidt, E., Ruble, Z., Akopian, D., Pack, D.: Software-defined radio GNSS instrumentation for spoofing mitigation: a review and a case study. Trans. Instrum. Measur. **68**(8), 2768–2784 (2019). https://doi.org/10.1109/TIM.2018.2869261

8. Akos, D.M.: Who's afraid of the spoofer? GPS/GNSS spoofing detection via automatic gain control (AGC). Ann. Navig. **59**, 281–290 (2012)
9. Liu, J., Chen, F., Xie, Y., Ge, B., Lu, Z., Sun, G.: Robust spoofing detection for GNSS array instrumentation based on C/N0 difference measurements. Trans. Instrum. Measur. **72**, 1–11 (2023). https://doi.org/10.1109/TIM.2023.3328684
10. Qiong, Y., Zhang, Y., Tang, C., Lian, J.: A combined antijamming and antispoofing algorithm for GPS arrays. Int. J. Antennas Propag., 8012569 (2019). https://doi.org/10.1155/2019/8012569
11. Averyanova, Y., Kutsenko, O., Konin, V.: Interference suppression at cooperative use of GPS, GLONASS, GALILEO, BEIDOU. In: Ukrainian Microwave Week (UkrMW), pp. 1–5. IEEE, Kharkiv, Ukraine (2020). https://doi.org/10.1109/UkrMW49653.2020.9252766
12. Kutsenko, O., Averyanova, Y., Konin, V.: Simulation of four-directional spoofing suppression with five-elements antenna array. In: 3rd Ukraine Conference on Electrical and Computer Engineering (UKRCON), pp. 213–216. IEEE, Lviv, Ukraine (2021). https://doi.org/10.1109/UKRCON53503.2021.9575929
13. Averyanova, Y., Kutsenko, O., Konin, V.: Method of GPS, GLONASS, GALILEO, and BEIDOU systems spoofing suppression. Telecommun. Radio Eng. **80**(7), 51–64 (2021). https://doi.org/10.1615/TelecomRadEng.2021037162
14. Merwe, J.R., Rugamer, A., Lipka, M.: Enhanced spatial spoofing detection with and without direction of arrival estimation. Trans. Aerosp. Electron. Syst. **59**(5), 5530–5540 (2023). https://doi.org/10.1109/TAES.2023.3262455
15. Liu, R., Yang, Z., Chen, Q., Liao, G., Zhu, Q.: Localization of GNSS spoofing interference source based on a moving array antenna. Remote Sens. **15**(23), 5497 (2023). https://doi.org/10.3390/rs15235497
16. Ahmed, S., Khanafseh, S., Pervan, B.: GNSS spoofing detection based on decomposition of the complex cross ambiguity function. In: 34th International Technical Meeting of the Satellite Division of The Institute of Navigation (ION GNSS+ 2021), pp. 3569–3580. St. Lou-is, Missouri (2021). https://doi.org/10.33012/2021.17987
17. Humphreys, T., Bhatti, J., Shepard, D., Wesson, K.: The Texas spoofing test battery: toward a standard for evaluating GPS signal authentication techniques. In: 25th International Technical Meeting of the Satellite Division of The Institute of Navigation (ION GNSS 2012), pp. 3569–3583. Nashville (2012)
18. Khan, A.M., Ahmad, A.: Global navigation satellite systems spoofing detection through measured autocorrelation function shape distortion. Int. J. Satell. Commun. Network. **40**(2), 148–156 (2021). https://doi.org/10.1002/sat.1425
19. Mao, P., et al.: GNSS spoofing detection and direction-finding method based on low-cost commercial board components. Remote Sens. **15**(11), 2781 (2023). https://doi.org/10.3390/rs15112781
20. Liu, J., Xie, Y., Chen, F., Ni, S., Sun G.: Research on spoofing detection based on C/N0 measurements for GNSS array receivers. In: Yang, C., Xie, J. (eds.) China Satellite Navigation Conference (CSNC 2024), LNEE, vol. 1093, pp. 556–566. Springer, Singapore (2024). https://doi.org/10.1007/978-981-99-6932-6_46
21. Zhang, X., Liang, T., Tian, J., Wu, J., Wang, C., Chen, M.: Anti-spoofing method for improving GNSS security by jointly monitoring pseudo-range difference and pseudo-range sum sequence linearity. Sensors **23**(20), 8418 (2023). https://doi.org/10.3390/s23208418
22. Rothmaier, F., Chen, Y.H., Lo, Sh., Walter, T.: GNSS spoofing detection through spatial processing. NAVIGATION: J. Inst. Navig. **68**(2), 243–258 (2023). https://doi.org/10.1002/navi.420
23. Mehr, I.E., Dovis, F.: A deep neural network approach for detection and classification of GNSS Interference and Jammer. TechRxiv. Preprint (2023).https://doi.org/10.36227/techrxiv.22212121.v

24. Elmezayen, A., Karaim, M., Elghamrawy, H., Noureldin, A.: Enhanced GNSS reliability on high-dynamic platforms: a comparative study of multi-frequency, multi-constellation signals in jamming environments. Sensors **23**, 9552 (2023). https://doi.org/10.3390/s23239552
25. Chen, X., Luo, R., Liu, T., Yuan, H., Wu, H.: Satellite navigation signal authentication in GNSS: a survey on technology evolution, status, and perspective for BDS. Remote Sens. **15**, 1462 (2023). https://doi.org/10.3390/rs15051462
26. Fernandez-Hernandez, I., et al.: SBAS message authentication: a review of protocols, figures of merit and standardization plans. In: International Technical Meeting of the Institute of Navigation, pp. 111–124 (2021). https://doi.org/10.33012/2021.17829

Performance Analysis of Intentional Interference on Multi-GNSS Receivers

Oleksii Sushych, Olexiy Pogurelskiy(✉), Valery Konin, Oleksandr Kutsenko, Iryna Prykhodko, and Tetiana Maliutenko

National Aviation University, Liubomyra Huzara Avenue, 1, Kyiv 03058, Ukraine
pogurelskiy@gmail.com

Abstract. Nowadays Global Navigation Satellite System (GNSS) has been used in a variety of applications, however number of interference actions on the frequency band of GNSS has increased rapidly during the last decades. Research of GNSS resistance methods to interference arose in the late 1990s. This became evident when the first prototypes of jammers targeting GNSS were available in the market. This situation persisted until the suppression of electronic warfare systems was achieved. Currently, available jamming equipment can almost entirely nullify the effectiveness of weaponry and military equipment utilizing GNSS signals. The problem is also arising with multiple privacy protection devices, basic jammers, and GNSS transmitters available at low cost, as the use of such devices leads to unintentional disruption of GNSS signals. Interference and GNSS spoofing are periodically observed in different Flight Information Regions, especially in conflict zones and areas of increased tension. Their impact on aircraft and ground systems is a potential risk to the safety of civil aviation. The paper presents the results of experimental research on impact of intentional interference on multi-GNSS receivers of satellite navigation systems GPS, Glonass, Galileo, Beidou, QZSS.

Keywords: Multi GNSS · GPS · Glonass · Galileo · Beido · QZSS · Intentional · Interference · Receiver

1 Introduction

Multiple navigation satellite systems are currently in various stages of deployment and operational readiness. Technologies based on Global Navigation Satellite Systems (GNSS) are widely used in areas related to position determination and high-precision time synchronization. This is, first of all, transport, in particular, civil aviation [1, 2], control systems for complex technological processes (for example, nuclear power stations), unmanned vehicles (aerial vehicles, ground vehicles, boats) [3].

The annual report of the European Union Space Program Agency (EUSPA) reflects the prospects for the development of GNSS in 17 different market segments [3]. According to EUSPA's forecast, annual deliveries of GNSS receivers in the world will grow steadily over the next ten years (from 1.8 billion units in 2021 to 2.5 billion units in 2031). The vast majority of them are associated with the consumer solutions, tourism,

I. Ostroumov and M. Zaliskyi (Eds.): ACASD 2024, LNNS 992, pp. 114–124, 2024.
https://doi.org/10.1007/978-3-031-60196-5_10

and healthcare segments, which account for approximately 92% of all global annual deliveries due to the huge number of smartphones and portable devices, that are sold annually. It should be noted that aviation and unmanned aerial vehicles represent a significant market, which is expected to grow from 42 million units in 2021 to 49 million in 2031 [3].

2 Literature Review and Problem Statement

The introduction of GNSS technologies is accompanied by the creation of consumer equipment operating in the spectrum of GNSS. Nowadays GNSS is available in GPS, Glonass, Galileo, Beidou, QZSS. GNSS signals received in the users segment have a low signal power. According to the GPS interface control document, the minimum level of GPS L1 C/A signal reaching the Earth's surface should not fall below -160.0 dBW [4].

Very low power of signals from GNSS satellites makes GNSS vulnerable to intentional and unintentional interference, which can impact the flights of numerous aircraft within a wide area [5].

User technology report on GNSS indicates many cases of GNSS interference, with the overwhelming majority caused by jammers (privacy protection devices, which are illegal in most countries) [6].

Easy access to low-power GNSS signal jammers which are out of the law in many countries can lead to the disruption of GNSS equipment over a wide area.

Common jammers generate low-power (10–30 dBm) and narrowband interference (frequency band 20 MHz) in the civilian L1/E1 frequency range. Interferences often represented as continuous signals or linearly frequency-modulated signals of varying difficulty (typical sweep time is 8–9 μs). Jammers degrade positioning accuracy by deteriorating the signal/noise ratio of GNSS signals. This is similar to receiving weak and distorted signals in urban terrain.

A simple GNSS signal repeater, freely available to provide uninterrupted indoor coverage, can become a source of spoofed GNSS signals in its vicinity when misused [6].

On December 25, 2023, in many areas of Poland, Lithuania, southern Sweden, and some Baltic countries, GPS signals were unavailable. Many users in these areas were left without reliable GNSS service.

Also, serious disruptions of GPS signals were observed in the territory of Finland. The most notable impact of interference on the eve of the festive events was reported in aviation reports and low navigational integrity of ADS-B systems [7].

GNSS spoofing and meaconing are increasing significantly. Research from the University of Texas at Austin has shown that: "Spoofing is more effective for denial of service than jamming: a 1-W spoofer is more effective than a 1 kilowatt narrowband/broadband jammer over the same distance".

Introduction of new satellite constellations and the use of multiple frequencies for GNSS can greatly reduce the probability of losing service due to unintended interference. This is achieved through the diversity of used frequencies and the increased number of satellites in range. However, certain types of intentional interferences still remain [5].

We investigate the problem of experimental evaluation of noise immunity of modern navigation multi-GNSS equipment under the influence of deliberate broadband interference.

3 Experimental Study of GNSS Jamming

In the first part of the experiment, the performance of the multi-GNSS navigation receiver was evaluated under the disruption of the integrity of the radio navigation field by radio interference of various intensities in the L1 and L2 frequency ranges of the GPS.

We use multi-GNSS receiver, a three-channel antenna [8, 9], and a low-power interference generator of pseudo-noise signals at L1 and L2 frequencies. Measurements included signal/noise ratios in the receiver channels and obtained coordinates. The intensity of the intentional radio interference was varied by changing the distance between the interference generator and the navigation receiver antenna.

The hardware component of the experiment consisted of an OEM719 Multi-Frequency GNSS Receiver [8], a GPS-703-GGG antenna [9], and a deliberate radio interference generator. The software component of the experiment consisted of the NovAtel Connect navigation receiver [10].

Without intentional interference, the receiver operates in normal mode: it detects and utilizes 12 GPS satellites, 8 Glonass satellites, 5 Galileo satellites, and 3 Beidou satellites for navigation measurements (see Fig. 1).

Fig. 1. Satellites visibility during operation without interference.

The signal-to-noise ratio in the receiver channels across the entire frequency range ranged from 31 dB to 54 dB. The receiver used all available frequencies, including: GPS – L1C/A, L2P(Y), L2C, L5Q; Glonass – L1C/A, L2P, L2C/A; Galileo – E1C, E5A, E5B, E5AltBOC; Beidou – B1 (see Fig. 2).

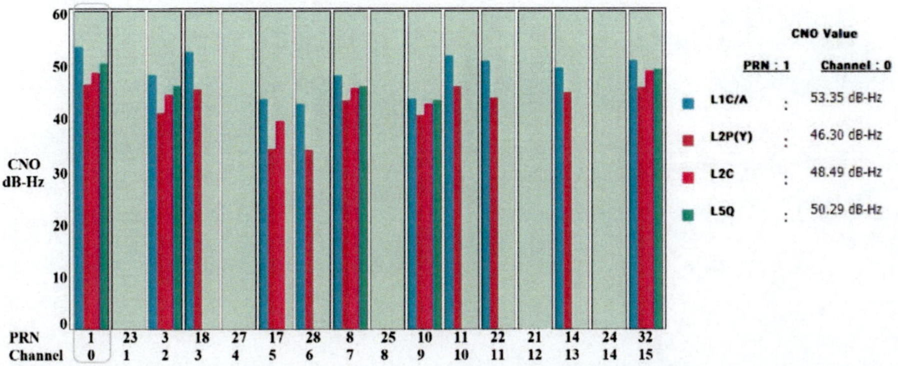

Fig. 2. Signal-to-noise ratio of GPS signals in the receiver channels (without interference).

The interference generator (output power 10 dBm) creates wideband interference in the L1 frequency range: 1562 MHz-1598 MHz (see Fig. 3) and the L2 frequency range: 1205 MHz-1243 MHz (see Fig. 4).

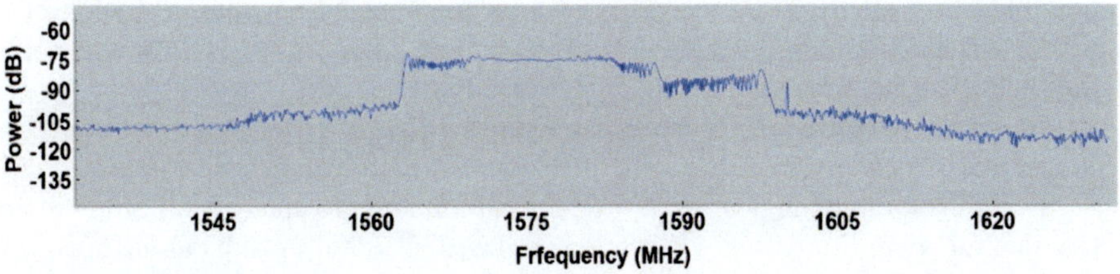

Fig. 3. Frequency spectrum of the interference generator signal in the L1 GPS band.

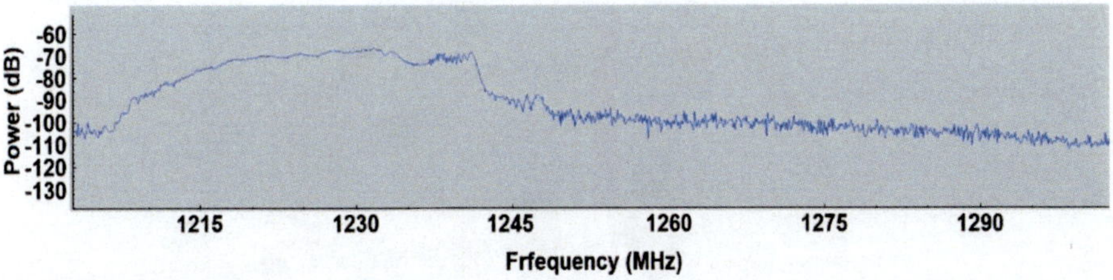

Fig. 4. Frequency spectrum of the interference generator signal in the L2 GPS band.

Below is the frequency band of the interference generator according to the GNSS frequency plan [11]. The interference region is highlighted by a red dashed line (see Fig. 5).

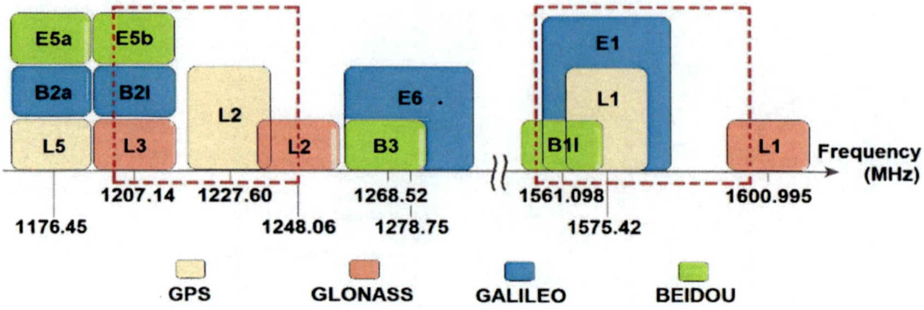

Fig. 5. The frequency plan of multi-GNSS.

The operation of the interference generator at a distance of 20 m significantly degraded the performance of the receiver but did not stop it. In navigation measurements, 2 out of 5 visible GPS satellites, 6 out of 8 visible Glonass satellites, 0 out of 3 visible Galileo satellites, and 0 out of 1 visible Beidou satellite were used (see Fig. 6).

It is worth noting that when operating in interference conditions, the receiver uses satellites with a high signal/noise ratio. These satellites typically have high elevation angles, and the receiver uses frequencies both within and outside the interference generator's frequency band. For satellites with a low signal/noise ratio, usually characterized by low elevation angles, the receiver only uses frequencies outside the interference generator's frequency band: GPS – L5Q frequencies, Glonass – L1C/A frequencies, Galileo – E5A, E5B, E5AltBOC frequencies (see Fig. 7).

It should be noted that the receiver captures satellites only with an acceptable signal/noise ratio. Typically, satellites with the highest elevation angles are captured, while those with low elevation angles are not. This situation leads to an artificial limitation of GNSS satellite visibility with low elevation angles. This in turn results in the operation of the multi-GNSS receiver under conditions of limited availability [11].

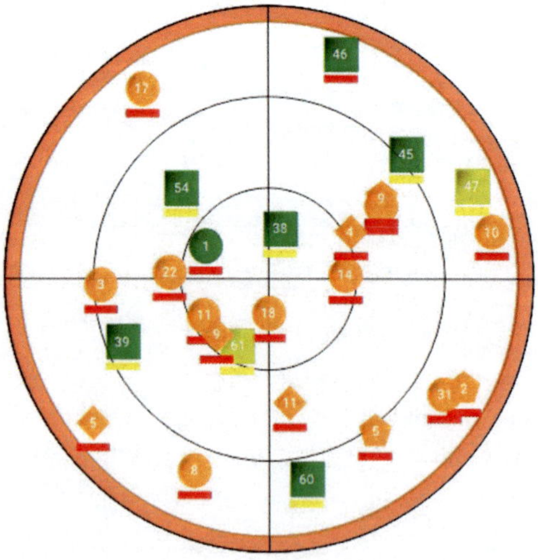

Fig. 6. Satellite visibility during the operation of the interference generator at a distance of 20 m.

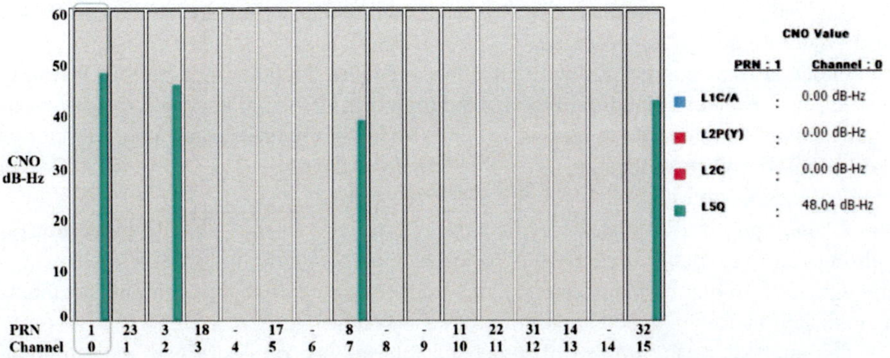

Fig. 7. Signal/noise ratio of GPS signals in the receiver channels (with interference).

Depending on the angle between the directions to the satellites relative to the user position, the uncertainty area of the user's position can vary, being either larger or smaller. In other words, the larger the angle between the directions to the satellites, the better the accuracy of the coordinate measurement. As a criterion for qualitative and quantitative assessment of the impact of satellite visibility on the expected accuracy of coordinate measurements, Dilution of Precision (DOP) factors are used.

In the absence of interference to the multi-GNSS receiver, the angles between the directions to the satellites relative to the user position are large (multi-GNSS receiver uses up to 30 satellites in measurements), the dilution of precision factors have low values, and the accuracy of multi-frequency coordinate measurements is high (see Table 1).

In the presence of interference on the multi-GNSS receiver, the angles between the directions to the satellites relative to the user's position are small (the multi-GNSS receiver uses up to 10 satellites in measurements), DOP factors have high values, and the accuracy of coordinate measurements is low (see Table 1).

Table 1. Evaluation of the user's position without interference/with interference.

Radio interference	No	Yes
Number of SVs used	30	10
Latitude standard deviation, [m]	1.13	6.44
Longitude standard deviation, [m]	1.04	5.31
Height standard deviation, [m]	2.39	14.23

In the second part of the experiment, the performance of the multi-GNSS navigation receiver was assessed under the disruption of the integrity of the radio navigation field by radio interference in the L-band range of the GNSS.

Several multi-frequency satellite navigation receivers from NovAtel were used in the experiments, including a GPS/Glonass receiver, a GPS/Glonass/Galileo receiver, a GPS/Glonass/Galileo/Beidou receiver, and multi-GNSS antennas from NovAtel.

During the measurements, all satellite navigation receivers were subjected to radio interference in the L-band range of GNSS, with frequencies ranging from 1 GHz to

2 GHz. The software component of the experiment consisted of the standard software of the satellite navigation receivers.

Under normal conditions, without intentional interference and satellite visibility restrictions, receivers function in their standard mode, detecting and utilizing the maximum available number of navigation satellites for precision measurements. DOP factors are minimal, and the precision of multi-frequency coordinate measurements is high (Table 2, first column for each system).

When exposed to low-power, wideband interference in the L-band range, the signal/noise ratio gradually deteriorates for all available satellite navigation systems and frequencies (see Fig. 8). Consequently, this leads to degradation in positioning accuracy, as indicated in Table 2 (second column for each system).

The analysis of the impact of wideband interference on coordinate measurement accuracy is presented in Fig. 9. The figure displays the results of assessing the accuracy of measuring the longitude of the location using three GNSS receivers.

With a further increase in interference power, the receivers stop determining location coordinates until the interference is discontinued (see Fig. 10).

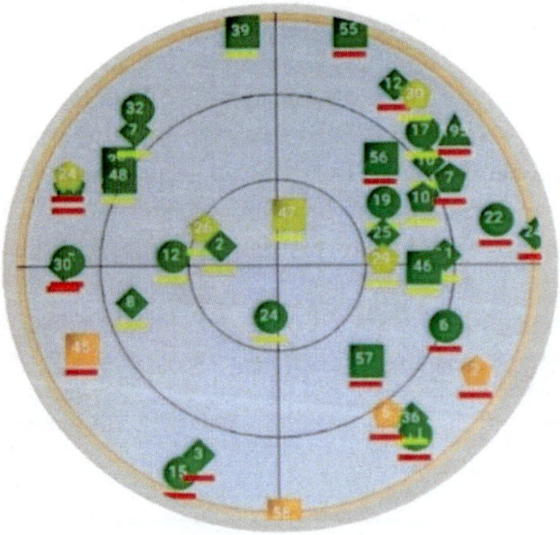

Fig. 8. Effect of low-power jamming interference in the L-band range.

The figure illustrates that during the occurrence of wideband interference, all receivers lose satellite signals (the GDOP factor reaches values of 9999) and leading to stop in determining coordinates.

It is worth noting that during the experiment, the four-system GPS/Glonass/Galileo/Beidou multi-frequency multi-GNSS demonstrated significantly more stable operation, remaining functional during secondary interference but with a notable degradation in positioning accuracy (see Fig. 11). Dual-system GPS/Glonass and triple-system GPS/Glonass/Galileo multi-frequency multi-GNSS receivers stopped determining coordinates during secondary interference (GDOP reaches values of 9999).

Table 2. Evaluation of the user's position without interference/with interference.

GNSS	GPS/Glonass	GPS/Glonass	GPS/ Glonass/Galileo	GPS/ Glonass/Galileo	GPS/ Glonass/ Galileo/ Beidou	GPS/ Glonass/ Galileo/ Beidou
Radio interference	No	Yes	No	Yes	No	Yes
Number of satellites, min/max	15/ 19	7/ 9	25/ 30	11/ 13	28/ 33	14/ 18
GDOP, min/max	1.2/ 1.5	2.9/ 4.0	1.2/ 1.4	3.2/ 3.8	1.5/ 1.6	3.5/ 4.6
PDOP, min/max	1.1/ 1.3	2.4/ 3.3	0.9/ 1.0	2.1/ 2.4	0.9/ 1.0	2.1/ 2.7
Latitude standard deviation, [m]	1.6	3.0	1.1	2.3	1.1	2.3
Longitude standard deviation, [m]	1.2	2.0	0.8	1.4	0.8	1.4
Height standard deviation, [m]	2.9	7.7	2.2	5.4	2.2	5.5

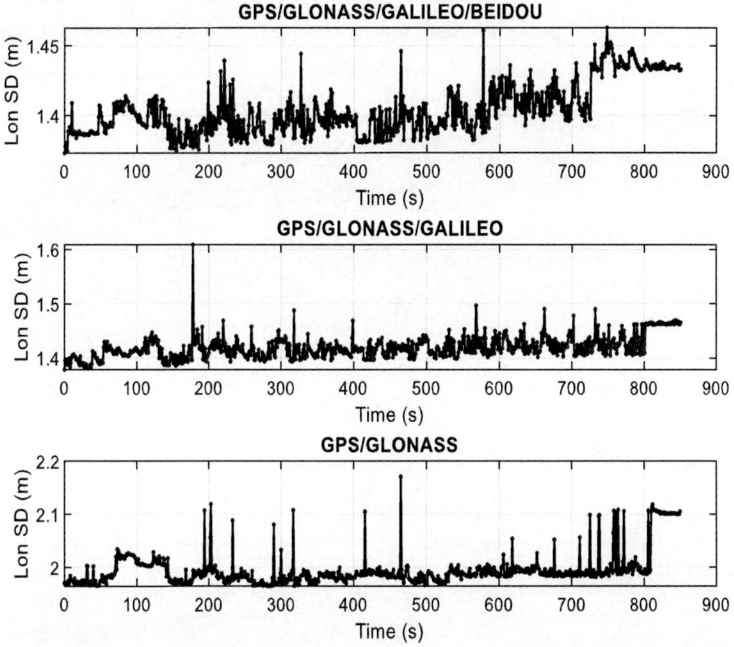

Fig. 9. The influence of low-power interference on the precision of coordinate measurements.

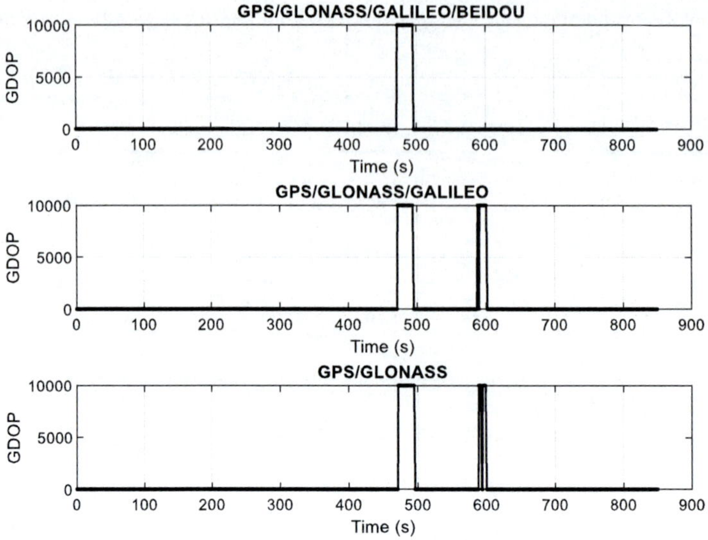

Fig. 10. The influence of powerful interference on the precision of coordinate measurements.

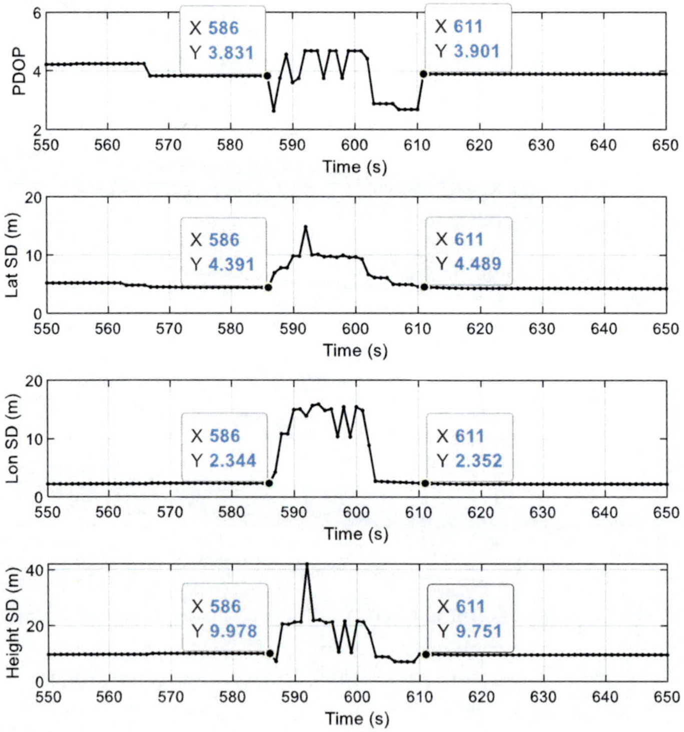

Fig. 11. Measurements accuracy of GPS/Glonass/Galileo/Beidou with interference.

4 Conclusions

The rapid growth in the number of satellite navigation systems jammers raises reasonable concerns. Commonly used low-power jammers and transmitters in the L1 GPS/E1 Galileo frequency range can cause significant deterioration in positioning and lead to a complete loss of satellite signal acquisition in the single-frequency equipment segment. Interference has different effects on receivers, depending on the characteristics of the navigation receiver and its antenna, as well as the characteristics of the interference itself, its power, and the relative positioning of the jammer and the receiver. Intentional interference degrades the signal/noise ratio of signals received by the receiver. This, in turn, leads to the fact that the receiver cannot capture weak satellite signals, and coordinate measurements get worse or become impossible.

The use of multi-frequency multi-GNSS receivers in conditions of jamming interference within the operational range of one or more satellite navigation systems provides a significant advantage compared to single-frequency receivers, enabling positioning, but with reduced accuracy. The higher the interference power the worse the positioning accuracy.

The advantage of multi-GNSS receivers is provided by the ability to select navigation satellites from different navigation systems based on the quality of received signals (signal/noise ratio). In the case of wideband interference in the L-band GNSS, the advantages of multi-GNSS receivers are less noticeable because the interference affects all frequencies of all satellite navigation systems. The use of multiple GNSS constellations provides an advantage in terms of availability and reliability of positioning, especially in conditions of marginal signal-to-noise ratio levels.

The low power of GNSS signals under jamming conditions makes civilian/unencrypted multi-GNSS receivers vulnerable to jamming. Minimizing the impact of intentional interference is achieved by using signal encryption mechanisms, statistical signal analysis methods, and adaptive antenna systems.

References

1. Global Air Navigation Plan 2016–2030. ICAO Doc. 9750 AN/963 (2016)
2. European ATM Master Plan (2020)
3. EUSPA EO and GNSS Market Report iss. 1 (2022). https://www.euspa.europa.eu/sites/default/files/uploads/euspa_market_report_2022.pdf. Accessed 01 Feb 2024
4. NAVSTAR GPS Space Segment/Navigation User Segment Interfaces. IRN-IS-200N. https://www.gps.gov/technical/icwg/IS-GPS-200N.pdf. Accessed 01 Feb 2024
5. Global navigation satellite system (GNSS) implementation issues. AN-Conf/12-WP/21. https://www.icao.int/Meetings/anconf12/WorkingPapers/ANConfWP21.6.1.EN.pdf. Accessed 01 Feb 2024
6. GNSS User Technology Report. https://www.euspa.europa.eu/sites/default/files/uploads/technology_report_2020.pdf. Accessed 01 Feb 2024
7. From Russia with love for Christmas: Jamming Baltic GPS. https://www.gpsworld.com/from-russia-with-love-for-christmas-jamming-baltic-gps. Accessed 01 Feb 2024
8. OEM719 Multi-Frequency GNSS Receiver. https://novatel.com/en-gb/products/receivers/gnss-gps-receiver-boards/oem719. Accessed 01 Feb 2024

9. GPS-703-GGG. https://novatel.com/support/previous-generation-products-drop-down/previous-generation-products/gps-703-ggg-antenna. Accessed 01 Feb 2024
10. NovAtel Connect. https://novatel.com/support/support-materials/software-downloads. Accessed 01 Feb 2024
11. Konin, V., Pogurelskiy, O., Prykhodko, I., Maliutenko, T., Sushych O., Ishchenko, O.: Multi-GNSS in limited navigation satellite availability. In: Ostroumov, I., Zaliskyi, M. (eds.) Proceedings of the International Workshop on Advances in Civil Aviation Systems Development. LNNS, vol. 736, pp. 126–140. Springer, Cham (2023). https://doi.org/10.1007/978-3-031-38082-2_10

Simulation of the Functional and Performance Behavior of Multi-GNSS Constellation

Olexiy Pogurelskiy(✉), Valery Konin, Iryna Prykhodko, Tetiana Maliutenko, Oleksii Sushych, and Oksana Ishchenko

National Aviation University, Liubomyra Huzara Ave., 1, Kyiv 03058, Ukraine
pogurelskiy@gmail.com, ans@nau.edu.ua

Abstract. The paper presents the results of forming a current almanac of Global Navigation Satellite System (GNSS) in Yuma format. This type of format use in various scenarios focused on simulating availability, geometrical dilution of precision (GDOP) and positional navigation system precision when using Galileo, GPS, Beidou systems standalone and all together inside a multi-system GNSS constellation. The Yuma format was originally used for analyzing and simulating the orbit of the GPS system, and there is a large amount of software available for analyzing the availability of satellite navigation systems based on data in this format. Due to the lack of almanacs in the Yuma format for most satellite navigation systems, specialized software was used in the research to convert data into a unified format. The evaluated availability and accuracy are important performances of GNSS that provides global coverage of signals for positioning, navigation and timing (PNT). These performances depend on structure of space segment which is number of satellite systems provide transmitting of navigation signals. The better level of PNT can be achieved by receiving data from all available systems and augmentations such as Wide Area Augmentation System, European Geostationary Over-lay Service, Indian Regional Satellite Navigation System and Qazi-Zenith Satellite System.

Keywords: Global Navigation Satellite Systems · GNSS · GPS · Galileo · Beidou · Almanac · Availability · Geometry · Accuracy

1 Introduction

Positioning, navigation and timing (PNT) based on Global Navigation Satellite System (GNSS) signals is used in aviation along with data from traditional navigation aids, but in the future will become the main source of navigational information during flight on route, approach and landing. As stated in Global Air Navigation Plan [1] with the continuous evolution of aircraft navigation capability through performance-based navigation, and the widespread use of GNSS positioning, regions of high traffic density no longer need as high a density of navigation aids. Until 2023, the GPS and GLONASS satellites were considered as the core-constellations of the GNSS [2]. However, starting with the 4th edition of [2], the BeiDou system was also added to them. In the future, they will be joined by the European satellite navigation system Galileo.

I. Ostroumov and M. Zaliskyi (Eds.): ACASD 2024, LNNS 992, pp. 125–136, 2024.
https://doi.org/10.1007/978-3-031-60196-5_11

The GNSS also includes regional satellite systems, such as Indian Regional Satellite Navigation System (IRNSS), Qazi-Zenith Satellite System (QZSS), wide-area augmentations (WAAS), European Geostationary Overlay Service (EGNOS) [3, 4]. The coverage of these space-based augmentation systems (SBAS) is shown in Fig. 1. The satellites of these systems transmit additional information to the data from the core-constellations. The use of data from these satellites is optional and depends on the ability of the user's equipment to receive the relevant signals.

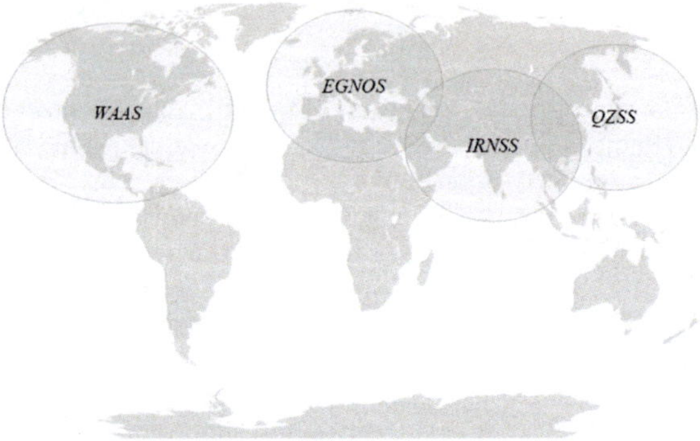

Fig. 1. Existing coverage of the space-based augmentation systems.

Obviously, the characteristics of GNSS depend on the combination of satellite constellations, which may include all systems available in a given region, or part of them in various combinations. A particular scientific interest is the estimation of the characteristics of satellite navigation systems when low-orbit satellite constellations are used or when they are added to the existing GNSS configuration. Such research can be oriented either to evaluate performances at a specific geographical location or covering the entire Earth's surface.

To analyze the combination of satellite constellations, it is necessary to establish a unified representation of orbital constellation data for each satellite navigation system. One of the widely used formats for representing almanac data is YUMA format, which is an ASCII message containing almanac element for each GPS satellite. GPS data began being stored in YUMA format after 1990 and is used in a large number of software to analyze GPS system performance. To analyze the combination of satellite constellations, it is necessary to convert the almanacs of GPS, Glonass, Galileo, and BeiDou systems into a unified GNSS almanac in the YUMA format.

2 Literature Review and Problem Statemen

The main performance characteristics of GNSS are accuracy, integrity, reliability and availability. The corresponding definitions as well as the values of these parameters for specific GNSS systems are given in [2, 5–7].

Possibility of implementation satellite navigation on the Low Earth Orbit (LEO) constellation of existing or planned for deployment communication satellite systems such as OneWeb, SpaceX are discussed in [8, 9]. LEOs have the advantage of being closer to the Earth compared to GNSS systems in Medium Earth Orbit (MEO), thus experiencing less path loss and potentially delivering stronger signals. This makes them more resilient to jamming. LEO spacecraft also have much faster motion in the sky, passing overhead in minutes instead of hours in MEO. This gives rise to more multipath rejection, as reflections are no longer static over short averaging times. The addition of these constellations adds a wealth of geometric diversity [8].

Possibility of using the Starlink system for navigation was investigated [10]. A presented research method includes the transformation satellites almanac into YUMA format. The proposed method of forming Yuma almanac can be applied for set of Keplerian orbital elements set specified in so-named two-line element (TLE) format.

The algorithm and software for transforming almanacs of GPS, GLONASS, Galileo and BeiDou in one common GNSS almanac were described in [11]. The common almanac allows to consider the multi-component GNSS space segment as a single constellation and further use it with the help of existing software tools.

The one of possible example is software tool Galileo Satellite Simulation Facility (GSSF) [12]. The GSSF described as "simulation environment that reproduces the functional and performance behavior of the Galileo system to support the simulation needs during the entire Galileo program lifecycle." It is capable of simulating, besides the nominal system, also its various degraded modes, in a deterministic or probabilistic manner. Space segment and user segment are not limited to Galileo and provide GPS and EGNOS related simulation models, and support also user-defined satellites and user-defined satellite constellations [12]. The option to import user-defined constellation into GSSF environment gives possibility to apply this tool for simulation of the functional and performance behavior of real constellations that included to GNSS now or can be added in future and potentially interesting for such type research.

3 Research Method

Information on the current status of the operating GPS, Galileo, Beidou, QZSS systems is available in almanacs. It is broadcasted inside navigation message from space. Alternatively, it can be downloaded from Internet services such as available on [13]. The almanac data is used as initial data for calculation of satellite positions with coarse level of accuracy enough for accelerating signal searching, acquisition and making shorter time to first fix. Also, almanac data used for prediction satellites geometry relative specified Earth location and calculation intervals of their observing.

The common GNSS almanac in December 5, 2023 contains information on a total of 152 satellites with a range of numbers from 2 to 390. The satellites of each navigation systems have their own range of numbers according to Table 1. The data on the GLONASS system is not considered in this research.

The format of this file should be converted to Yuma view, which is more common in GNSS data processing software. For conversion we use the Matlab software `2023_Convert_Trimble` developed by the Navigation Systems Laboratory of the

Table 1. Ranges of satellite numbers in almanac.

Satellite number in Trimble almanac	Satellite system
1–37	GPS
38–64	GLONASS
111–118	QZSS
201–263	Galileo
264–283	BeiDou

National Aviation University. The scheme to explain the usage of particular and common GNSS almanac in simulations is shown in Fig. 2.

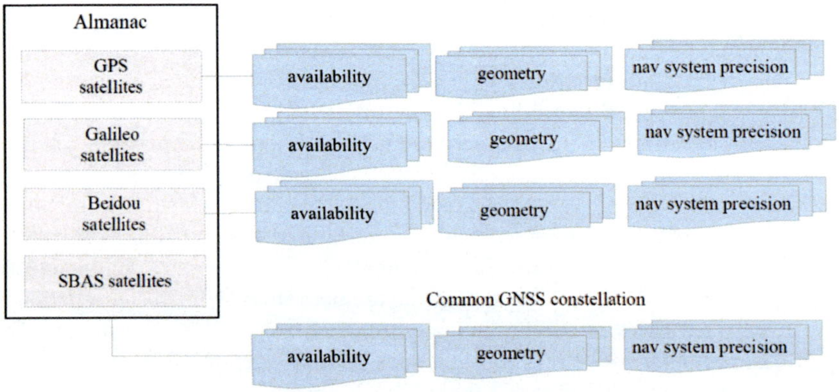

Fig. 2. Scheme of the common GNSS almanac data.

The names of the output files are specified in the program script. These files are almanacs in Yuma format for each of the systems separately as a result of the `ReadAlmTrGGGGBX` script and a common almanac of all systems as a result of running the `ReadAlmTrFed`. Recorded and stored in Yuma-format almanacs of GPS, Galileo, Beidou and SBAS systems can be opened in a text editor and analyzed for its correctness by comparing the week number and total number of satellites for each system with corresponding data received in alternative way. The GPS week number inside the almanac used for research is 2291. We use, online services to validate obtained results [14].

The common GNSS almanac contains data for satellites from all available constellations. This data set can be used for creating scenarios of estimation satellite system performances either with any single system or full-scale multi-system constellation.

The analyzed characteristics are availability of satellites estimated in numbers of visible satellites, their geometry relative ground located users and predicted navigation system precision. Simulation is performed for global covering by estimated characteristic and visualizing on graphs minimal, maximal and average value (Fig. 3).

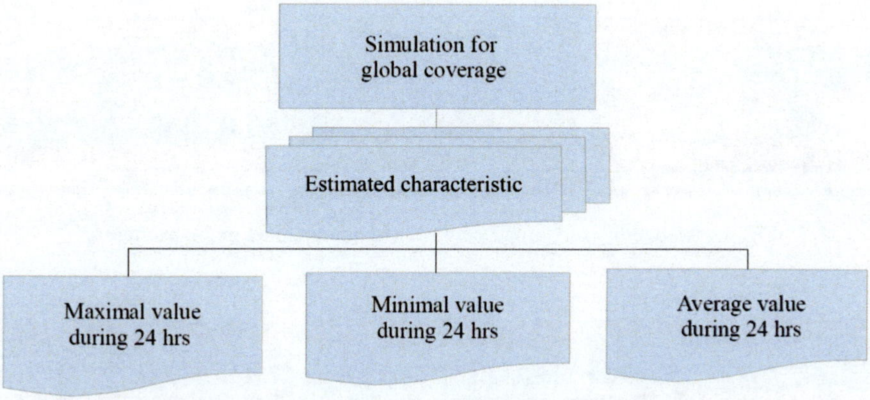

Fig. 3. Scheme of data processing during GNSS analysis.

Section 4 presents the results of simulation performed with the help of common multi-system almanac of GNSS.

4 Research Results

4.1 Availability Analysis

Availability is determined by the probability of the consumer receiving reliable information at a given time with the required accuracy. Alternatively, availability refers to the number of satellites available to the user.

To evaluate the availability of satellite navigation services were created scenarios with current almanacs of systems Galileo, GPS, BeiDou and type of analysis – Visibility. The visibility between satellites and users or elements of the ground segment. It returns the number of satellites at each time step, and the number of satellites in view for each user or ground segment element, during the simulation time period. Results of constellation availability are presented in Fig. 4, 5, 6 and 7 for the maximum, minimum, and average number of satellites.

From Fig. 4, 5, 6 and 7, it can be concluded that the average availability for Galileo system is about 6.8 to 9.1 satellites. The average availability of the GPS system is 8.5 to 33.5 satellites, which is higher than the availability of the Galileo system. As seen, the maximum number of GPS satellites is accessible at the equator and polar region. Average availability of BeiDou system is 7.4 to 24.3, and higher for Asian regions. It influents on results obtained for GNSS constellation which includes all Galileo, GPS and BeiDou satellites. In addition, the combination of different GNSS systems gives much more availability than using only one satellite navigation system.

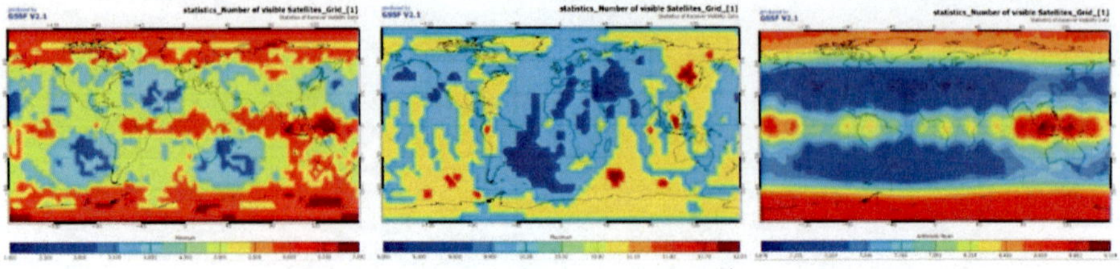

Fig. 4. Availability of Galileo satellites (minimum, maximum, average).

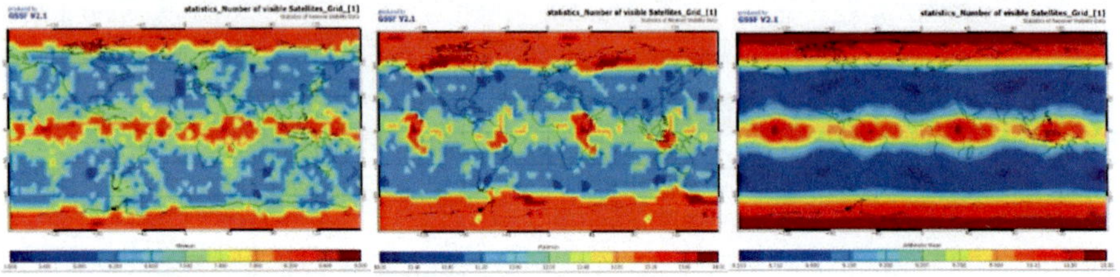

Fig. 5. Availability of GPS satellites (minimum, maximum, average).

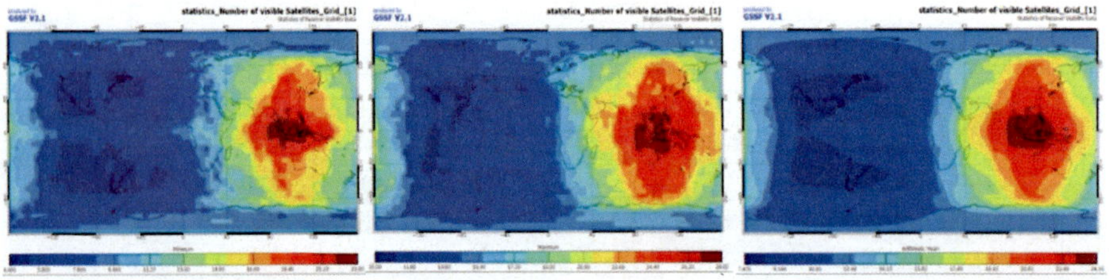

Fig. 6. Availability of BeiDou satellites (minimum, maximum, average).

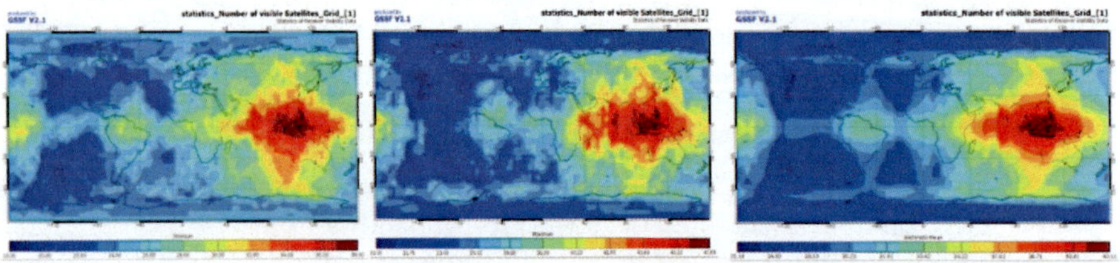

Fig. 7. Availability of GNSS satellites (minimum, maximum, average).

4.2 GDOP Analysis

The accuracy of measured coordinate depends both on factors related to the type of used equipment and the method of coordinate determination (dual-frequency or single-frequency measurements, phase or code ranges), and on factors common to all users, which are included in the total parameter User Equivalent Range Error (UERE). Results of GDOP analysis are presented in Fig. 8, 9, 10 and 11.

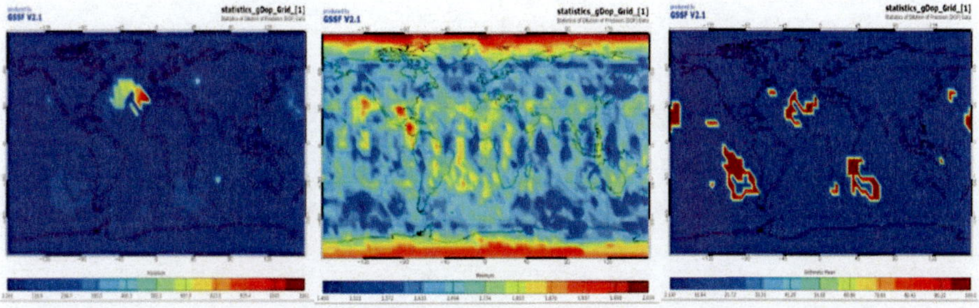

Fig. 8. GDOP of Galileo (minimum, maximum, average).

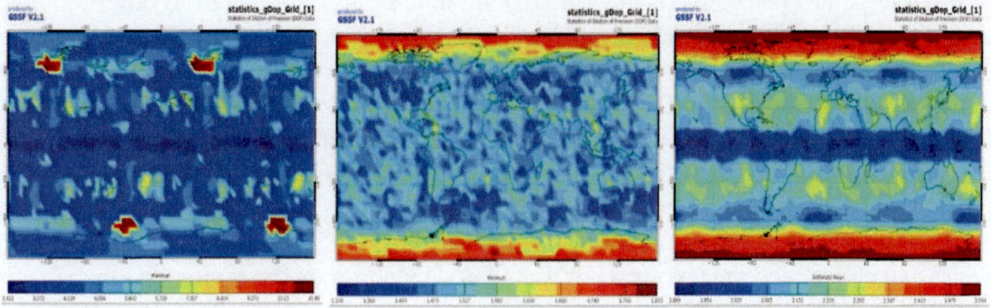

Fig. 9. GDOP of GPS (minimum, maximum, average).

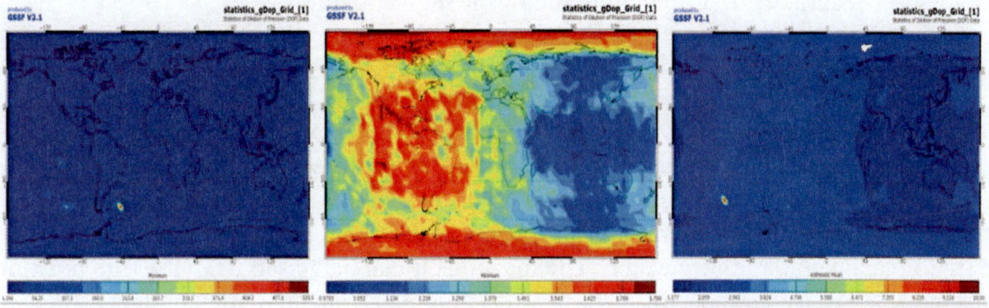

Fig. 10. GDOP of BeiDou (minimum, maximum, average).

The final relation between a particular UERE (which for simulation purposes can be set at a certain level) and the accuracy of the navigation system is established through the geometric dilution of precision factor (GDOP).

Figure 8 shows the level of GDOP values for Galileo from 3.16 and more, which indicates a decrease in the accuracy of coordinate determination and makes it impossible to use Galileo for accurate coordinate determination. GDOP in the range from 1.45 to 2.059 indicates sufficient accuracy of coordinate determination when using one Galileo satellite system.

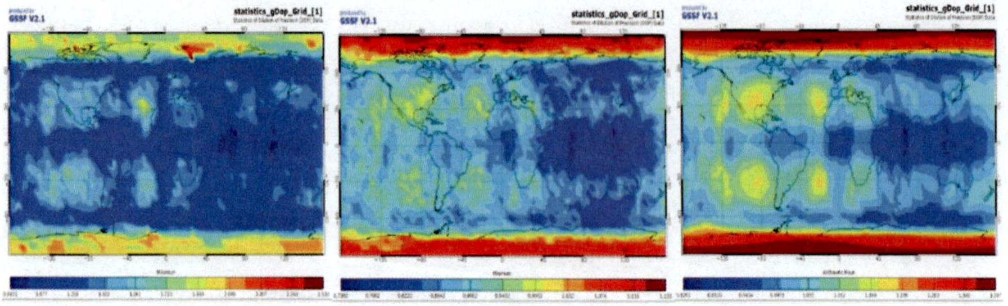

Fig. 11. GDOP of GNSS (minimum, maximum, average).

GDOP values near the equator (Fig. 9) are much lower compared to other regions, where the GDOP increases. This is due to the greater availability of navigation satellites than in other regions. In turn, the minimum GDOP value in the range from 1.31 to 1.853 indicates the maximum availability of GPS navigation satellites. The average values GDOP in the range from 1.889 to 2.544, which is a sufficient level for determining coordinates and indicates different availability of GPS navigation satellites, depending on the region.

Figure 10 shows minimum GDOP from 0.9705 to 1.788, maximum GDOP from 1.386 and above, and an average GDOP 2.059, depending on the region and the area of availability to BeiDou navigation satellites.

The average level of GDOP for GNSS is between 0.826 and 1.397. It corresponds to minimal influence of satellite geometry on accuracy of PNT. The results obtained for evaluation accuracy are presented in the following subsection.

4.3 Accuracy Analysis

The Navigation System Precision (NSP) describes the dispersion of user receiver estimated position around the mean position. NSP is estimated for each user or ground segment element over the simulation time period. It returns the mean, maximum, and minimum NSP for each user or ground segment element as well as the instantaneous data for each simulation time step. The mean itself can be biased from the true position but the NSP does not reflect this. The NSP is computed from the UERE budget estimated by environment model.

$$C = \left(A^T W A\right)^{-1}, \tag{1}$$

$$W = \begin{pmatrix} w_1 & \cdots & 0 \\ \vdots & \ddots & \vdots \\ 0 & \cdots & w_n \end{pmatrix}, \tag{2}$$

where W is weight matrix; A is the observation matrix.

If the coordinate coefficients are expressed in the receiver local reference frame, then the positional navigation system precision (PNSP) can be calculated as:

$$PNSP = \sqrt{C_{1,1} + C_{2,2} + C_{3.3}}. \tag{3}$$

where $C_{1,1}$, $C_{2,2}$, $C_{3,3}$ are elements of matrix C.

The diagonal elements of the weight matrix W are calculated as follows:

$$w_i = \sigma_i^{-2} = \left(\sigma_{ODTS}^2 + \sigma_{iono}^2 + \sigma_{tropo}^2 + \sigma_{multipath}^2 + \sigma_{rnoise}^2\right)_i^{-1}, \tag{4}$$

where σ_{ODTS} is component of pseudorange error caused by overall difference of time scales between user and satellite; σ_{iono} and σ_{tropo} are components of pseudorange error caused by ionosphere and troposphere; $\sigma_{multipath}$ is pseudorange error component caused by multipath propagation; σ_{rnoise} is pseudorange error component caused by other noises inside user receiving equipment.

In the research the individual contribution in Eq. (4) is generated by UERE model. The values of σ_{iono}^2, σ_{tropo}^2, σ_{rnoise}^2 need to take into account the corrections made inside a user receiver (ionosphere, troposphere and other corrections in a real user receiver).

The results of accuracy analyses presented in Fig. 12, 13, 14, 15 and 16 for positional navigation system precision (PNSP).

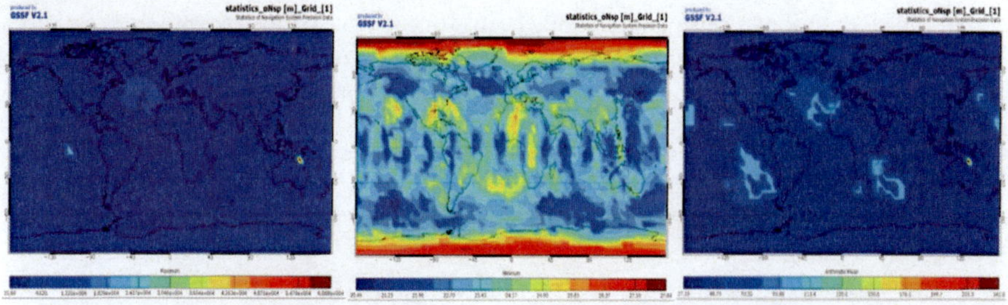

Fig. 12. PNSP of Galileo (minimum, maximum, average).

The results for estimation accuracy of Galileo system (Fig. 12) show relation between geometrical dilution of precision and positional navigation system precision. It concluded on base of the presence of the same shape areas on graphs for both evaluated characteristics. The arithmetic mean of PNSP value for simulated period changes near the 30 m.

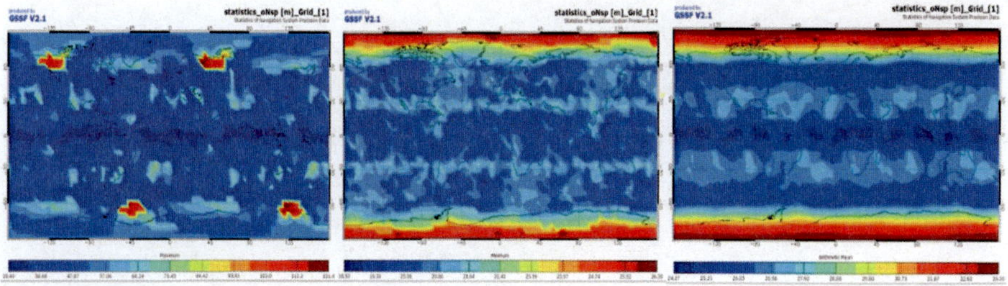

Fig. 13. PNSP of GPS (minimum, maximum, average).

The results for estimation accuracy of GPS system (Fig. 13) shows dependence of it from geographical latitude and seen on plots as horizontal lines that symmetrically marks regions near the Earths poles as regions with lower accuracy. The arithmetic mean of PNSP value for simulated period changes near the 25 m.

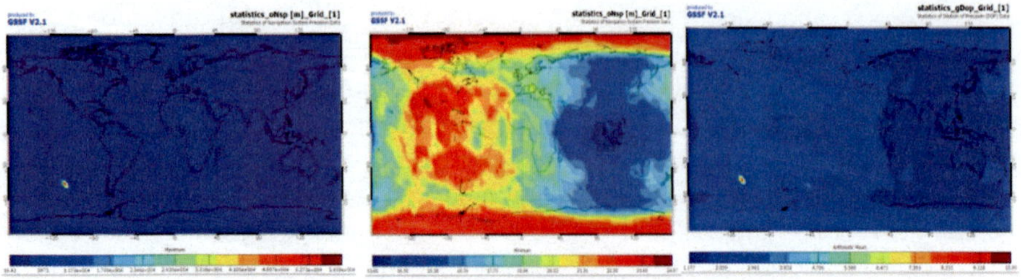

Fig. 14. PNSP of BeiDou (minimum, maximum, average).

In the case of BeiDou accuracy simulation the received plot (Fig. 14) for maximal level of PNSP that doesn't informative due to just one-color filling except one small area. Such can be interpreted as presence during 24 h moments of time than error achieve high value. The appearance of errors of high level influents on arithmetic mean changes in the range between 15 m and 35 m.

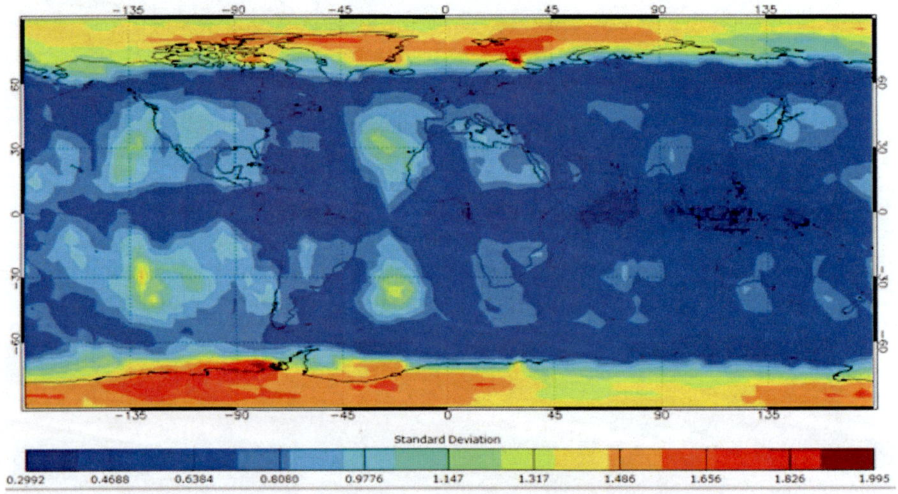

Fig. 15. Standard deviation of PNSP for GNSS.

Finally, the standard deviation of navigation system precision was simulated (Fig. 15). The obtained result does not exceed level of 2 m in higher latitudes and for most part of Earth surface ranges between 0.5 and 1.2 m.

The results for estimation accuracy of multi-system GNSS system (Fig. 16) shows highest level of positioning navigation system precision. The corresponding plots show

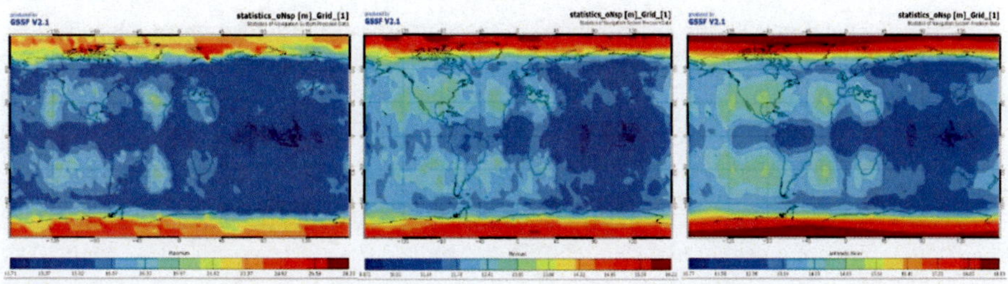

Fig. 16. PNSP of GNSS (minimum, maximum, average).

the minimal level in the range 10–16 m, maximal in the range 13–25 m, and average in the range 12–18 m dependently on the geographical region. Better results were achieved for equatorial region and for Asia where the influence of BeiDou is obvious.

5 Conclusions

The research has shown that it is possible to generate an actual almanac of GPS, Galileo and BeiDou systems in Yuma format using specially developed software. Yuma almanacs of single systems or the entire GNSS can be used for simulation of the functional and performance behavior of each constellation alone and in combination.

The best level of availability and accuracy performances was achieved for multi-system structure GNSS. The obtained results demonstrated non-uniform distribution of performance over the Earth surface and can be explained by the peculiarities of space segment construction, which is most clearly evident for Beidou satellites. Obviously, a similar distribution of simulated performances will be observed when GNSS include geostationary satellites of space-based augmentations continuously visible only from geographically restricted regions.

The further improvement of this research can be achieved by obtaining Yuma almanac of LEO constellations of communication satellites and providing with the help of them the simulations by the same scenarios of availability, GDOP and accuracy analysis.

References

1. Global Air Navigation Plan 2016–2030. Doc 9750-AN/963, 5th edn. ICAO, Montreal (2016)
2. Global Navigation Satellite System (GNSS) Manual. Doc 9849, 4th edn. ICAO, Montreal (2023)
3. Gulam, R.: ICAO Global Provisions and Regional Developments related to GNSS. Joint ACAC/ICAO MID Workshop on GNSS. ICAO MID Regional Office, Cairo (2016)
4. Aeronautical Telecommunications. Vol. I (Radio Navigation Aids). Annex 10 to the Convention on International Civil Aviation, 8th edn. ICAO, Montreal (2023)
5. Global Positioning System Standard Positioning Service Performance Standard, 5th edn. Washington (2020)
6. European GNSS Agency: Galileo Open Service: signal-in-space interface control document (OS SIS ICD), issue 2.1, Publications Office of the European Union (2023). https://doi.org/10.2878/39727. Accessed 10 Feb 2024

7. BeiDou navigation satellite system signal in space interface control document: Open service signal B2b (Version 3.0). China Satellite Navigation Office (2022)
8. Reid, T., Neish, A., Walter, T., Enge, P.: Leveraging commercial broadband LEO constellations for navigating. In: the 29th International Technical Meeting of the Satellite Division of The Institute of Navigation (ION GNSS+ 2016), pp. 2300–2314. Portland, Oregon, (2016). https://doi.org/10.33012/2016.14729
9. Reid, T., Walter, T., Enge, P., Sakai, T.: Orbital representations for the next generation of satellite-based augmentation systems. GPS Solut **20**, 737–750 (2016). https://doi.org/10.1007/s10291-015-0485-x
10. Konin, V., Pogurelskiy, O., Turovska, A.: Simulation and estimation parameters of low orbit satellite navigation system. In: 2021 IEEE International Conference on Information and Telecommunication Technologies and Radio Electronics (UkrMiCo), pp. 160–163. Odesa, Ukraine (2021). https://doi.org/10.1109/UkrMiCo52950.2021.9716604
11. Konin, V., Pogurelskiy, O., Pryhodko, I., Maliutenko, T.: Methods for research and study GNSS in remote mode. In: 2021 IEEE International Conference on Information and Telecommunication Technologies and Radio Electronics (UkrMiCo), pp. 195–198. Odesa, Ukraine (2021). https://doi.org/10.1109/UkrMiCo52950.2021.9716655
12. Kalden, O., Schmidt-Brücken, F., Zimmermann, F., Pfister, J.: Galileo system simulation facility recent applications & developments. In: 6th ESA Workshop on Satellite Navigation Technologies (Navitec 2012) & European Workshop on GNSS Signals and Signal Processing, pp. 1–6. ESA, Noordwijk, Netherlands (2012). https://doi.org/10.1109/NAVITEC.2012.6423085
13. Information on the current status of GNSS. http://ftp.trimble.com/pub/eph. Accessed 19 Jan 2024
14. LabSat GPS Time Calculator Homepage. https://www.labsat.co.uk/index.php/en/gps-time-calculator. Accessed 19 Jan 2024

Operationalizing the Split-Beam Ionospheric Scintillation Model in GNSS RF Simulators: Geometrical Optics to Nakagami-M Distribution

Ivan G. Petrovski II(✉)

IP Solutions Satellite Applications Inc., Tokyo, Japan
ivan@ip-solutions.jp

Abstract. This paper introduces the Split-Beam model, developed to simulate ionospheric scintillation effects on Global Navigation Satellite Systems (GNSS). The model dissects a radio signal into two components upon encountering ionospheric irregularities, cooperatively contributing to amplitude and phase scintillation. Through Monte Carlo simulations and scintillated GNSS RF signal simulation the model demonstrates its capability to replicate the empirical distribution of signal amplitudes, closely aligning with the Nakagami-m distribution, and the behavior of the receiver tracking loops tracking scintillated live satellite signal. Leveraging a combination of off-the-shelf and in-house instruments, our approach integrates readily available technologies with highly customizable instruments. The Split-Beam model addresses the limitations of traditional weak scattering models, particularly in deriving and explaining the Nakagami-m distribution and cross-correlation properties of real data. This paper also presents a variation of the Split-Beam model, employing geometrical optics to provide insights into the physical processes underlying scintillation and allow for realistic simulation of the scintillated RF signal in GNSS signal simulator. The application of the model extends to areas including ionospheric tomography, radio wave propagation, wireless communications, and remote sensing. By accurately modeling the scattering and attenuation effects in ionospheric scintillation, the Split-Beam model offers a methodological approach for signal analysis in GNSS and related fields. This work provides a foundation for improved understanding and simulation of ionospheric scintillation, contributing to the advancement of signal processing in atmospheric and astrophysical research.

Keywords: Ionospheric Scintillation · Nakagami-m Distribution · Ionospheric Bubble · GNSS Receiver · GNSS RF Simulator

I. Ostroumov and M. Zaliskyi (Eds.): ACASD 2024, LNNS 992, pp. 137–150, 2024.
https://doi.org/10.1007/978-3-031-60196-5_12

1 Ionospheric Scintillation: Mechanisms and Impact on GNSS a Subsection Sample

1.1 Implications of Ionospheric Scintillation for Civil Aviation and Atmospheric Studies

Ionospheric scintillation refers to the rapid variations in the amplitude and phase of a radio signal as it passes through the ionosphere. These variations, caused by irregularities in the ionosphere's electron density, can lead to serious issues for Global Navigation Satellite Systems (GNSS) receivers, including positioning errors and signal loss.

The significance of scintillation is paramount for GNSS applications [1], particularly in civil aviation. This importance is further amplified as solar activity reaches its peak within the 11-year solar activity cycle. Beyond navigation, scintillation offers valuable insights into Earth's atmosphere and climate, presenting a dual challenge: while we strive to mitigate the effects of ionospheric scintillation on communication and navigation systems, we also seek to leverage these variations to study atmospheric dynamics.

Predicting and counteracting these effects are crucial for maintaining the reliability of GNSS receivers. Conversely, for atmospheric studies, enhancing the clarity of scintillation effects is desirable.

Our research in this domain commenced with the development of an ionospheric scintillation monitor for a Japan Aerospace Exploration Agency (JAXA) project initiated in 2007. This endeavor led to the creation of the JAXA Cosmode Software Defined Radio (SDR) Ionospheric Scintillation Monitor (ISM) and, subsequently, the Agile Receiver for Atmospheric Monitoring and Ionospheric Scintillation (ARAMIS) SDR GNSS receiver [2, 3].

Integrating the ionospheric scintillation monitor into a SDR has facilitated various coupling methods for integrating the receiver with Inertial Navigation Systems (INS). An ultra-tight coupling with JAXA INS has enabled on-board aircraft measurements during high scintillation periods when conventional ISMs struggle to maintain a lock.

This research also underscored the need for an RF scintillation simulator. Given the unpredictable nature of heavy scintillation and the challenges in acquiring relevant data sets for receiver testing and tuning, simulating such effects is of paramount importance. In our development work for the JAXA scintillation product, we initially employed phenomenological models and later introduced a physical split-beam model to more accurately represent the underlying physics.

To enhance practicality and implementation ease, this paper presents new variations of the split-beam model based on geometrical optics. This model not only provides a more profound physical explanation of scintillation effects but is also applicable to fields beyond GNSS.

As described in [4], phenomenological models have been highly successful in fitting experimental data. The model we used in simulating GNSS signals under scintillation conditions belongs to this category. The ITU-R Radio communication Study Group 3 currently recommends the Global Ionospheric Scintillation Model (GISM) [5], which predicts parameters such as the S4 index, depth of amplitude fading, and RMS phase and angular deviations due to scintillation. While GISM provides comprehensive scintillation maps and data, our focus was on examining the physical processes at the specific receiver baseband processor level, creating physical RF disturbances to induce identical effects on the receiver processor.

1.2 Phenomenological Scintillation Model

In this section, we focus on a phenomenological model for ionospheric scintillation, tailored for GNSS RF simulation and adjusted to align with experimental data. The off-the-shelf GNSS RF signal simulator Replicator, integrating this and subsequent models, is depicted in Fig. 1.

Fig. 1. IP-Solutions Off-the-Shelf GNSS RF Simulator "Replicator": This advanced simulator is utilized for accurately replicating GNSS signals under ionospheric scintillation conditions.

The simulator provides a robust and highly customizable platform for testing and analyzing the performance of GNSS receivers by simulating realistic scintillation scenarios, thereby contributing significantly to our understanding and mitigation strategies for scintillation effects in GNSS applications.

Both phase and amplitude scintillations can be described as random processes [6]. They can be described by two negatively correlated random processes with Gaussian and Gamma distribution functions describing phase and amplitude scintillation, respectively.

A GPS signal can be described by the following equation:

$$A = A_0 e^{j\varphi}, \tag{1}$$

where A is amplitude; φ denotes the phase.

Under the influence of scintillation, the signal is modified as:

$$A = A_0 \cdot \delta A \cdot e^{j(\varphi_0 + \delta\varphi)}, \tag{2}$$

where δA is amplitude scintillation, and $\delta\varphi$ is phase scintillation.

Phase scintillation is modeled as a zero-mean Gaussian random process with a standard deviation $\sigma_{\delta\varphi}$. This Gaussian process adheres to a normal distribution. Post-generation, these scintillation processes are processed through shaping filters to introduce necessary autocorrelation. Without such filtering, the processes would possess unlimited power, and their consecutive values would be entirely independent.

To obtain negative correlation properties, we consider phase and amplitude scintillations as a two-dimensional random process with a bivariate gamma distribution. The components of this vector are correlated, and we utilize the trivariate reduction method to construct such stochastic processes as two-dimensional random vector is defined as:

$$\left\{\delta A, \delta\phi'\right\} = \{G_1 + G_3, G_2 - G_3\}, \tag{3}$$

where G_1, G_2, G_3 are independent gamma random variables with parameters $\alpha_1, \alpha_2, \alpha_3$.

The marginal gamma distributions of δA and $\delta\varphi\prime$ have parameters $\alpha_A = \alpha_1 + \alpha_3$ and $\alpha_\phi = \alpha_2 - \alpha_3$ respectively. To account for negative correlation, the parameters for the marginal gamma distributions should be defined as follows:

$$\alpha_1 = \alpha_A - \rho_{A\varphi}\sqrt{\alpha_A\alpha_\varphi}, \tag{4}$$

$$\alpha_2 = \alpha_\phi + \rho_{A\phi}\sqrt{\alpha_A\alpha_\phi}, \tag{5}$$

$$\alpha_3 = \rho_{A\varphi}\sqrt{\alpha_A\alpha_\varphi}, \tag{6}$$

where $\rho_{A\phi}$ is correlation between components.

Because the phase scintillation is in fact better presented by a random process with zero-mean Gaussian distribution, the parameters of the second component of bivariate gamma distribution shall be adjusted to a special case when gamma distribution will be close to Gaussian.

Post-generation of the phase and amplitude scintillation processes as random values, we apply shaping filters to instill appropriate autocorrelation. An Infinite Impulse Response (IIR) filter serves as an effective shaping filter:

$$x_i = \phi_i - a_1 x_{i-1} - a_2 x_{i-2} - a_3 x_{i-3} \ldots - a_n x_{i-n}, \tag{7}$$

where n is the filter order;a_i are coefficients.

Filter order and coefficients are calculated to fit experimental spectra.

In line with the Nyquist theorem, the sampling rate should be at least double the maximum frequency of interest. Experimental findings peg the maximum scintillation frequency at 10 Hz. The standard ISM ARAMIS 50 Hz output rate is well within these requirements, when we use it to measure scintillation data. The ReGen control center, which runs Replicator simulator, has 1,000 Hz data generation rate, so it generates the scintillation process at one-millisecond signal loop, interpolating scintillation for each sample.

2 Split-Beam Model Development for Scintillation Simulation in GNSS RF Simulators

Phenomenological models show some inconsistencies when compared with experimental results. Primarily, they do not naturally account for Nakagami-m [7] distribution, nor do they elucidate the observed negative cross-correlation between scintillation phase and amplitude. In our previous work [2], we introduced the "Split-Beam" model, which naturally encapsulates both phenomena.

In GNSS applications, it is predominantly the strong scintillation that is of concern. In such scenarios, scintillation affects primarily the signal-to-noise ratio, rather than altering the code and phase observables significantly. This effect can lead to the degradation of the observables or, in more severe cases, to the loss of lock.

GNSS receiver tracking loops are robust, but may falter under intense scintillation conditions. Based on our analysis of GNSS data, the Nakagami-m distribution provides an optimal fit for the strong scintillation distribution [2], though it does not derive directly from theoretical constructs such as the Rytov approximation or solutions to the wave propagation problem [8, 9]. Instead, it is a statistical model that characterizes the amplitude fluctuations of the signal and can be associated with the physical parameters of the system via the Rytov variance or analogous metrics. As for phase perturbations, they are considered statistically independent and uniformly distributed [4, 8]. The sample of amplitude and phase distribution generated in accordance to (1)-(8) to fit real data is presented on Fig. 2. The scintillation data were recorded using IP-Solutions ARAMIS ISM SDR receiver with PORTOS front end, showcased in Fig. 3, which allows to operate on four user defined frequencies, such as for example L1, G1, B1 and L5. The data then were used for simulation of the scintillated signal with ReGen and processed again with GNSS receivers.

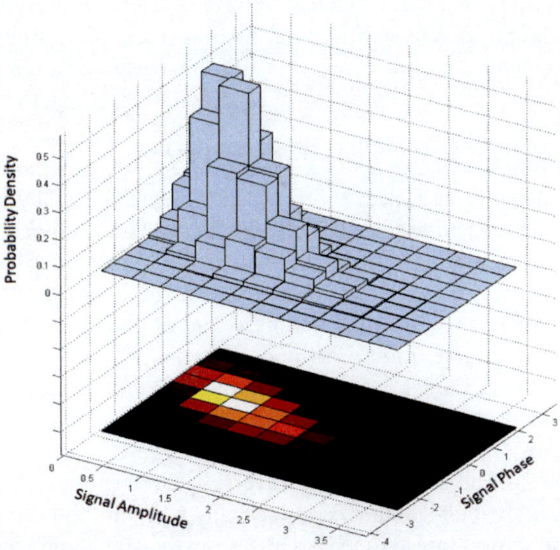

Fig. 2. Nakagami-m distribution of the experimental data, simulated using ReGen.

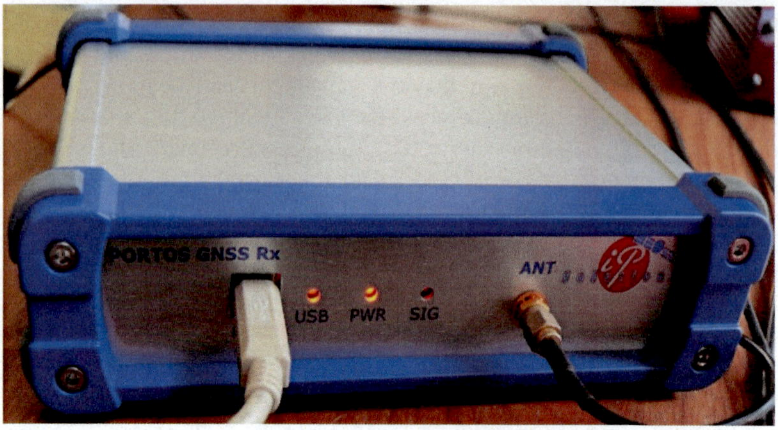

Fig. 3. IP-Solutions off-the-shelf PORTOS front end for SDR GNSS ionospheric scintillation monitor ARAMIS.

This paper presents further advancement and utilization of the Split-Beam scintillation model for high-fidelity simulation of GNSS signals. Beyond its application in simulation, the Split-Beam scintillation model potentially contributes to a deeper understanding of the physics of scintillation. Its implementation promises to introduce repeatability and transparency into receiver testing and research-development processes.

2.1 Split-Beam Scintillation Model. Weak Scattering.

While conventional models are adept at fitting experimental data, they often lack insights into the mechanisms underlying scintillation generation. This section introduces an alternative model grounded in the physical mechanisms of signal scintillation. The "Split-Beam" model [2], offers a novel perspective to elucidate the scintillation of GNSS signals as they traverse ionospheric depletion cavities or plasma bubbles populated with a multitude of eddies. Unlike traditional models that might employ Rician or Rayleigh statistics, the Split-Beam model boasts several unique features and potentials.

In the Split-Beam model, when the signal encounters a cavity, it bifurcates into two or more beams. One beam is scattered, with its related field describable by the Rytov approximation. The second beam, traversing through the bubble, is attenuated. Phase scintillation is attributed to the scattered beam, while amplitude scintillation in this model arises not only from interference but also from the attenuation of the direct signal. This model necessitates considering a complex refractive index due to the involved attenuation. The scattered signal is formulated as:

$$E_S = E_{S0} e^{i\Phi_S}, \tag{8}$$

where are E_{S0} and Φ_S represent the amplitude and phase of the scattered beam, respectively. Correspondingly, the direct signal can be described as:

$$E_D = E_{D0} e^{i\Phi_D}. \tag{9}$$

The total signal therefore is a sum of two components. The fluctuations due to source movement should satisfy the conservation of energy:

$$P_S + P_D = P, \tag{10}$$

where P_S is scattered signal; P_D is direct signal.

When the beams are combined the phase scintillation is defined by the first beam and the amplitude scintillation is defined by both beams.

The intensity, encapsulated as the square of the signal's absolute magnitude, can be depicted as:

$$I_t = |P_t|^2 = \left| E_{S0} e^{i\Phi_S} + E_{D0} e^{i\Phi_D} \right|^2. \tag{11}$$

Expanding this and taking the expected value gives:

$$E[I_t] = E\left[|E_{S0}|^2\right] + E\left[|E_{D0}|^2\right] + 2E[|E_{S0}||E_{D0}|cos(\Phi_S - \Phi_D)], \tag{12}$$

where $E[x]$ denotes the expected value.

This elucidates that the scintillation amplitude spectrum is sculpted by two signals, with amplitude scintillation emerging from two distinct mechanisms, thereby diverging from the normal distribution framework. Part of the amplitude scintillation originates from eddy-induced attenuation altering the path geometry, while the remainder is due to interferences of scattered signals. The simulation results, adhering to the model Eqs. (8)–(12), align well with the Nakagami distribution, reinforcing the model's efficacy.

The total energy of the signal is partitioned between the two beams. A larger portion scattered results in intensified phase scintillation, also reshaping the Nakagami-m distribution for amplitude scintillation. The amplitude scintillation contribution from the direct beam wanes correspondingly. The phase scintillation is caused only by the scattered beam. Therefore, the phase scintillation spectrum has normal distribution.

The total energy of the signal is divided in two beams. If larger part of the signal is scattered, then the phase scintillation become higher. The scattered signal is generated to have Rayleigh distribution. The part of amplitude scintillation coming from the direct beam is diminishing with its attenuation.

The Split-Beam model shares foundational assumptions with the Rice distribution. However, the Rice distribution does not accommodate an SI index exceeding unity [4]. Rino also noted that the log-normal distribution offers a superior fit for weak-to-moderate scintillation. Based on our work with GNSS scintillation data, for GNSS applications, the focus should primarily be on strong scintillation, as modern GNSS receivers are well-equipped to handle weak and moderate scintillation scenarios. For the strong scintillation of the GNSS signals in L- and S-bands the Nakagami-m distribution provides the best fit.

2.2 Split-Beam Scintillation Model. Geometrical Optics

In the Split-Beam Weak Scattering model, we conceptualized the signal as bifurcating into two beams upon encountering an eddy: a direct beam and a scattered beam. The scattered beam is chiefly accountable for the Nakagami-m distribution, while their amalgamation accounts for the amplitude and phase cross-correlation properties. Statistically, the model aligns well with experimental data, but our interest extends to the behavior of the RF signal at the antenna phase center, beyond mere intensity calculations over an area.

Weak scattering models, phenomenological models, including GSIM, the split-beam weak scattering model, and even experimental data used for signal generation, all inherit a limitation: they do not perfectly align with the behavior of scintillated signals as observed in the tracking loops of a receiver's baseband processor.

To bridge this gap, we have devised a new variant of the split-beam model, purely formulated using geometrical optics. This model not only captures the statistical distribution and cross-correlation properties of real data but also mirrors the behavior of live, satellite-scintillated signals as perceived within a receiver.

Scintillation in our test data is predominantly recorded near equatorial regions in Asia, often in conjunction with ionospheric depletion zones or ionospheric bubbles. When a satellite's initial radio signal encounters such an ionospheric depletion area, diffraction causes the signal to diverge into multiple beams. Each beam undergoes its own amplitude attenuation, code/phase alteration, and Doppler shift, akin to multipath effects commonly managed by receiver practitioners.

As depicted in Fig. 4, a collimated beam encountering a low-refractivity ionospheric depletion area splits into several beams. Analogous to handling multipath scenarios, we consider for our model a single additional ray, factoring in attenuation changes, code and phase delays, and excess Doppler, without significantly increasing the number of rays. An increase in ray count would induce normality in the process, moving away from the Nakagami-m distribution. Strong scintillation is likely to manifest when one or two rays with substantial Doppler shifts predominate.

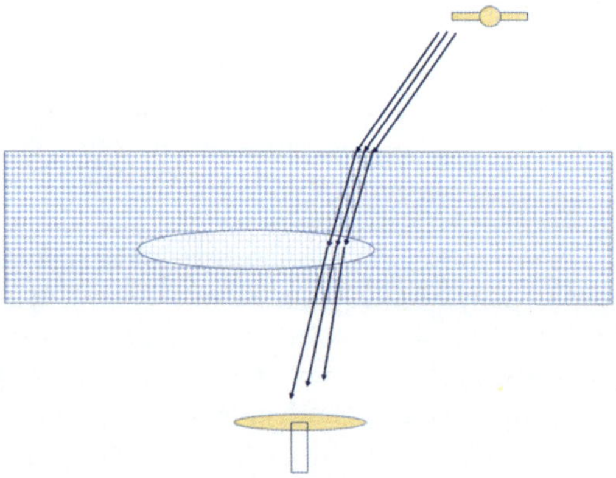

Fig. 4. Collimated beam encounters lower refractivity ionospheric depletion area.

It's noteworthy that rays traversing longer paths through the ionospheric depletion area experience lesser delay and phase advance. The excess Doppler for such a bubble is conjectured to be within a few Hz, leading to signal oscillations close to the original frequencies, but with significantly slower amplitude variations, at a frequency equal to the frequency difference of the original signals.

This phenomenon, akin to multipath effects, causes the beam to split into several beams with distinct Doppler shifts. It can also be conceptualized as the ionospheric depletion bubble disrupting the spatial coherence of beam waves, causing the beam to diverge instead of remaining collimated.

Demonstrated in the simplest multipath-like scenario encountered in GPS occultation that the phase rate difference between beams, contingent on the size of the refractivity perturbation, can reach 10 to 15 Hz for LEO occultation [10]. This is comparable to the scenario with a ground-based receiver and an ionospheric depletion bubble. The Doppler shift of the collimated beam due to satellite movement typically falls within 6 kHz.

The Split-Beam model [2], elucidates the intricate interplay of signals within the ionosphere. According to this model, a signal bifurcates upon encountering an eddy: one component traverses directly through the eddy, undergoing attenuation (contributing to amplitude scintillation), while the other component is scattered by the eddy (contributing to phase scintillation). These components subsequently interfere, culminating in the observed signal. The Split-Beam model, enhanced by geometrical optics, considers multiple beams, each characterized by variations in phase, amplitude, and Doppler with excess range up to 15 Hz. Utilizing a Monte Carlo approach, we generate numerous signal instances, 100,000 in this instance, based on the model. We then construct a histogram of the resulting amplitudes, serving as an empirical probability distribution for comparison with the Nakagami-m distribution.

Figure 5 exhibits the empirical distribution of signal amplitudes derived from the Monte Carlo simulation of the Split-Beam weak scattering and geometrical optics models, juxtaposed with the Nakagami-m distribution. The Split-Beam weak scattering uses normal distribution for direct beam and Rayleigh distribution for scattered beam, whereas geometrical optics model uses normal distribution for both beams, direct and multipath.

The red line shows target Nakagami-m distribution with parameters $m = 0.8$ and $\Omega = 2.0028$. The target shape parameter $m = 0.8$ corresponds to $S_4 = 1.25$, indicative of substantial scintillation intensity. This level of scintillation is high enough to induce loss of lock in GNSS receiver, a scenario especially critical for civil aviation or algorithms and methods, which use snapshot data [11]. In these receivers, the compounded effect of additional Doppler can significantly challenge tracking loop dynamics, exacerbating the likelihood of signal loss.

The green step histogram shows results of the Monte-Carlo simulation (empirical data) for the Normal-Rayleigh beam combination. The green line represents fitted Nakagami-m distribution for the Normal-Rayleigh beam combination.

The blue step histogram shows results of the Monte-Carlo simulation (empirical data) for the Normal-Normal beam combination (geometrical optics model). The blue line gives fitted Nakagami-m distribution for this model.

The shape parameter m of the Nakagami-m distribution correlates with the extent of multipath scattering. A diminished m signifies intensified scintillation, potentially due to an increased count of eddies in the bubble for the weak scattering model or elevated excess Doppler in the geometrical optics model. A m value below 1 denotes severe fading, indicative of substantial ionospheric-induced scintillation.

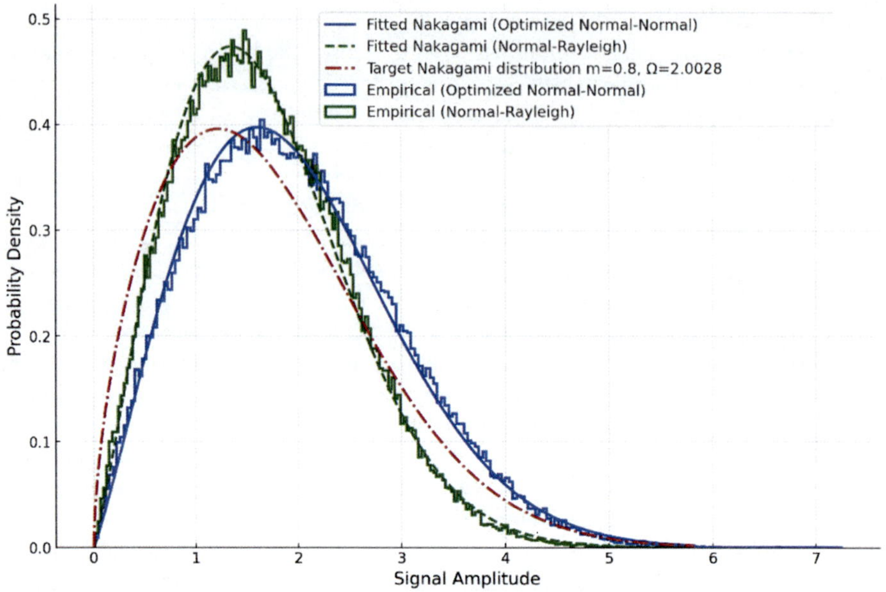

Fig. 5. The Split-Beam model distribution for weak scattering and geometrical optics vs. Nakagami-m distribution.

The scale parameter Ω pertains to the mean signal power. An augmented Ω implies enhanced signal power, possibly attributable to a more robust direct signal or diminished ionospheric attenuation.

The Split-Beam model, under the weak scattering paradigm, presumes the signal, upon encountering an ionospheric eddy, splits into two fractions: one scattered by the eddy (phase scintillation) and the other directly passing through the eddy but attenuated (amplitude scintillation). These components then amalgamate to form the observed signal.

An exponential distribution models the direct signal navigating the ionosphere. Its scale parameter embodies the average signal amplitude of the direct signal. Elevated values signify a stronger direct signal. A Rayleigh distribution typifies the scattered signal, deflected by the eddies in the bubble. Its scale parameter represents the root-mean-square signal amplitude of the scattered signal. Higher values indicate more pronounced scattering, potentially due to an increased eddy count within the bubble.

Conversely, the Split-Beam geometrical optics model posits that the signal, upon interacting with an ionospheric cavity of depleted TEC, generates a multipath ray. Both rays have Gaussian-distributed PDFs, but the resultant integrated PDF closely fits to Nakagami-m distribution.

Introducing a third scattered signal into the model could encapsulate more intricate scattering phenomena within the ionosphere, such as multifaceted scattering paths or a non-uniform eddy distribution.

The Normal-Normal Beam Combination has 88.45% Fitting Percentage to Target Nakagami-m distribution, Kolmogorov-Smirnov statistics 0.1155 with p-value of 0, which is an indicator of significant difference. The Normal-Rayleigh Beam Combination vs Target Nakagami gives Fitting Percentage: 94.30% with KS Statistic 0.0570 and KS p-value $\sim 4 \times 10^{-283}$, which also is indicating a significant difference.

If we look at the fitting to the fitted Nakagami-m distribution, then in both cases with 98% fitting and Kolmogorov-Smirnov statistics around ~ 0.01 with p-value $\sim 4 \times 10^{-10}$.

It's evident from the analysis that the process of fine-tuning parameters to fit experimental data is more challenging for the weak scattering model than for the geometrical optics model. The distribution parameters, including shape and scale, are meticulously adjusted to align with the model's specifics, highlighting the nuanced approach required for the weak scattering model.

The weak scattering model, particularly when employing the Normal-Rayleigh beam combination, demonstrates a commendable fit to the target Nakagami-m distribution, suggesting its effectiveness in representing the statistical characteristics of ionospheric scintillation. Nonetheless, the geometrical optics model has distinct advantages in providing insights into the physical processes underlying scintillation. Its ease of simulation and the clarity with which it represents phenomena in the baseband processor make it a valuable model for practical applications.

As we proceed to examine the baseband processor's response to these models, it's crucial to consider these factors. The baseband processor's interpretation of signals, influenced by the chosen scintillation model, can significantly impact the performance and reliability of GNSS receivers in scintillation-prone environments. The geometrical optics model's direct correlation with physical parameters offers an intuitive framework for understanding and simulating these effects, potentially leading to more robust and accurate receiver designs.

The ISM ARAMIS facilitates real-time RF GNSS signal processing or records the signal as a Digitized Intermediate Frequency for subsequent applications [12]. This versatility permits multiple analyses of the same signal using diverse algorithms and parameters. The ISM ARAMIS offers an insightful glimpse into the tracking loops at the baseband processor's core. Figure 6 portrays the ARAMIS tracking loop panel with an acquisition peak, tracking loop constellation IQ plot, and Q-time plot, depicting NAVDATA with noticeable shape degradation. Figure 6 presents results from processing actual live satellite scintillation, while Fig. 7 demonstrates the simulated baseband signal utilizing the split-beam geometrical model.

The bottom Q-time plot on the Fig. 6 is provided for comparison and represents a signal without scintillation. The demonstrated signal is GPS L1C/A. The time axis of the simulated signal is in seconds, contrasting with the 20 ms scale divisions on the receiver tracking panel's time scale. Both simulated and live satellite data demonstrate Doppler excess on the level of 10 Hz.

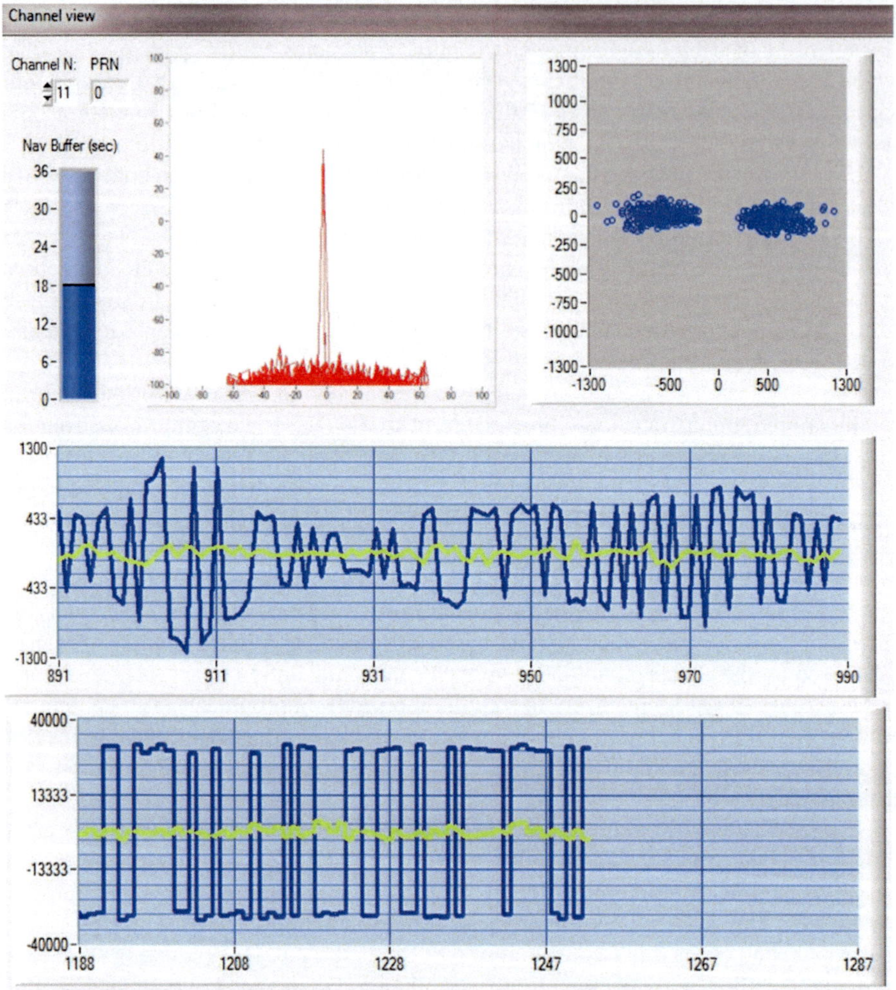

Fig. 6. ARAMIS ISM tracking panel with acquisition and tracking of live satellite scintillated signal. Q-plot for scintillated (top) and normal (signal).

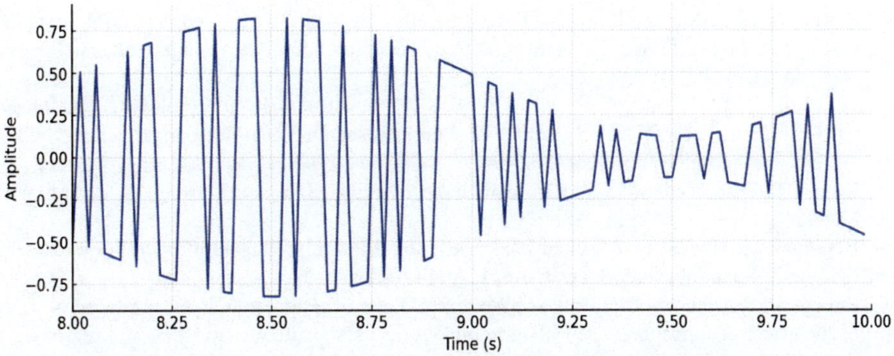

Fig. 7. The simulated baseband signal for high level of scintillation with Split-Beam Geometric Optics model.

3 Conclusions

The proposed split-beam model for simulating ionospheric scintillation extends its utility beyond GNSS applications, offering valuable insights and enhancements in several related fields. For ionospheric tomography the provided model could refine tomography techniques, enhancing the reconstruction accuracy of electron density distributions within the ionosphere. This could lead to improved predictions and understanding of ionospheric behavior.

Beyond GNSS signals, the model's principles could be applied to comprehend the scintillation effects on various types of radio waves as they traverse through random media, contributing to a broader understanding of wave propagation under diverse atmospheric conditions. In urban environments or scenarios where signals encounter scattering and attenuation, the model could offer predictive insights, potentially enhancing the reliability and quality of wireless communication links. The proposed split-beam model could significantly benefit remote sensing applications, particularly in interpreting signals from synthetic aperture radar (SAR) or other radar systems that penetrate the atmosphere or other random media. Enhancements in signal interpretation could lead to more accurate and detailed remote sensing data.

The proposed model's potential extends into the field of astrophysics, where it could be instrumental in elucidating the scintillation effects on light from celestial bodies as they pass through the interstellar medium. This could contribute to a deeper understanding of cosmic phenomena and the characteristics of the interstellar medium. By transcending its initial domain, the split-beam scintillation model paves the way for a wide array of advancements across multiple disciplines, enhancing our understanding of scintillation phenomena and improving the accuracy of systems reliant on signal propagation through random media.

References

1. Foken, T.: Springer Handbook of Atmospheric Measurements. Springer Handbooks, Springer Nature Switzerland AG (2021)

2. Petrovski, I.I., Ivan, G., Tsujii, T.: Digital Satellite Navigation and Geophysics, A Practical Guide with GNSS Signal Simulator and Receiver Laboratory. Cambridge University Press, New York (2012)
3. Petrovski II, I.G.: Ionosphere with GNSS SDR: constructing and employing specialized software-defined radio for in-depth ionospheric research with global navigation satellite system. Forthcoming. Springer (2024)
4. Rino, C.L.: The Theory of Scintillation with Applications in Remote Sensing. Wiley, Hoboken (2011)
5. Beniguel, Y., Hamel, P.: A global ionosphere scintillation propagation model for equatorial regions. J. Space Weather Space Clim. **1**, A04 (2011)
6. Pullen, S., Opshaug, G., Hansen, A., Walter, T., Enge, P., Parkinson, B.: A preliminary study of the effect of ionospheric scintillation on WAAS user availabilty in equatorial regions. In: Proceedings of ION GPS-98, Inst, of Navig., Alexandria, Va. (1998)
7. Nakagami, M.: The m-Distribution, a general formula of intensity of rapid fading. In: William, C.H. (ed.) Statistical Methods in Radio Wave Propagation: Proceedings of a Symposium held 18–20 June 1958, pp. 3–36. Permagon Press (1960)
8. Wheelon, A.D.: Electromagnetic Scintillation, vol. I. Cambridge University Press, Geometrical optics (2001)
9. Wheelon, A.D.: Electromagnetic Scintillation, vol. II. Cambridge University Press, Weak scattering (2003)
10. Melbourne, W.G.: Radio Occultations Using Earth Satellites: A Wave Theory Treatment. Wiley, Hoboken (2005)
11. Petrovski II, I.G.: RTK-B enhancing real-time kinematic (RTK) positioning with genetic algorithms and long-code GNSS signals for rapid ambiguity resolution and improved accuracy. In: Proceedings of the International Workshop on Advances in Civil Aviation Systems Development. Lecture Notes in Networks and Systems, vol. 736, pp. 116–125. Springer, Singapore (2023). https://doi.org/10.1007/978-3-031-38082-2_9
12. Desai, M.V., Darshna, J., Shweta, S.N.: Implementation and analysis of grid-based ionospheric correction technique and positioning errors of NavIC + GPS ARAMIS SDR receiver. IETE Tech. Rev. **40**, 663–670 (2022). https://doi.org/10.1080/02564602.2022.2152880

Regional Navigation Satellite Systems Exploration and Integration: Concept Evaluation with the Agile SDR Platform

Ivan G. Petrovski II(✉)

IP Solutions Satellite Applications Inc., Tokyo, Japan
ivan@ip-solutions.jp

Abstract. In the evolving landscape of global navigation, the strategic integration of Regional Navigation Satellite Systems (RNSS) like Japan's QZSS and India's NavIC is imperative for enhancing civil aviation. This paper presents a comprehensive examination of the potential of RNSS to serve as reliable navigational aids, addressing the unique challenges and requirements of regional and international flight operations. It elucidates the significance of orbital configurations, such as Geostationary Earth Orbit (GEO), Highly Elliptical Orbits (HEO) and Inclined Geosynchronous Orbit (IGSO), and their implications for satellite coverage, signal availability, and positional accuracy. Furthermore, the study delves into the specifics of Receiver Autonomous Integrity Monitoring (RAIM) for RNSS, ensuring the reliability and integrity of navigational data, a prerequisite for safe aviation practices. Through a simulated test flight from Tokyo to Delhi, employing Agile SDR Platform, which encapsulates GNSS receiver ARAMIS and state-of-the-art GNSS RF simulator Replicator, the research validates the operational efficacy of RNSS. It emphasizes the necessity of multi-system integration, robust satellite signal processing, and adherence to stringent accuracy requirements in diverse flight phases. The paper advocates for continued innovation, international collaboration, and policy harmonization to fully harness the potential of RNSS. By fostering the development and integration of RNSS, the paper contributes to envisioning a future where civil aviation is characterized by heightened safety, operational efficiency, and global interconnectedness.

Keywords: QZSS · NavIC · L1S Signal · SDR GNSS Receiver ARAMIS · GNSS RF Simulator Replicator

1 Emergence and Strategic Importance of Regional Navigation Satellite Systems

In recent years, the landscape of satellite navigation has witnessed a significant shift, marked by the emergence and accelerated development of Regional Navigation Satellite Systems (RNSS). This trend underscores a strategic move towards achieving enhanced navigational capabilities and autonomy in satellite-based navigation, particularly within specific countries or regions. The impetus behind this shift is multifaceted, driven by the need to establish a reliable backup for primary Global Navigation Satellite Systems (GNSS) and to cater to unique regional navigational requirements.

I. Ostroumov and M. Zaliskyi (Eds.): ACASD 2024, LNNS 992, pp. 151–165, 2024.
https://doi.org/10.1007/978-3-031-60196-5_13

RNSS are designed to provide high-precision navigational signals tailored to specific regions, offering a level of accuracy that is sometimes superior to that of global systems within those regions.

The enhanced accuracy of RNSS stems from the use of high-density regional reference station network, which can provide high accuracy ionospheric grid models, as for example for NavIC, or high accuracy real-time corrections, like in QZSS Multi-GNSS Advanced Orbit and Clock Augmentation (MADOCA) service, which allows for precise point positioning. It is very advantageous to use SDR GNSS receivers to develop, test and validate such advanced services (see application of ARAMIS SDR GNSS Receiver for research of NavIC ionospheric grid model in [1]). The additional options for inter-satellite links, time keeping design and system synchronization can also allow for some additional advantages for epheeris estimation and thus overall accuracy [2].

Countries are increasingly seeking to establish their own satellite navigation systems to maintain sovereignty over critical infrastructure and to enhance national security. The aspiration to leverage RNSS as backup navigation systems extends beyond national borders, encompassing the realm of international civil aviation. This strategic orientation is particularly pronounced in regions without proprietary access to global navigation systems, with Asia being a prime example.

RNSS also allow for the provision of customized services, including regional augmentation systems, to meet specific needs such as disaster response, agriculture, and civil aviation. For example, QZSS L1S signal provides such information, including earthquakes, tsunami, volcanic activity, ash fall, tornado and other information related to natural disasters early warning, status and parameters.

It is apparent that use of such information in addition to standard augmentation data provided by Space Based Augmentation System (SBAS) can be extremely valuable for civil aviation.

Integrating RNSS into the broader navigational framework of civil aviation promises to significantly enhance redundancy and reliability. The importance of having alternative navigation methods to GNSS as a backup for civil aviation aircraft systems, as well as the methods for their evaluating these alternatives, has been drawing more and more attention in recent years [3]. In scenarios where the primary GNSS might be compromised due to technical issues, environmental factors, or deliberate interference, RNSS can serve as a critical fallback option, ensuring uninterrupted navigation.

The effective use of RNSS in international flights necessitates a high degree of interoperability and standardization across different systems. This involves harmonizing technical specifications, signal formats, and operational protocols to ensure seamless transition and handover between RNSS and GNSS.

The integration of RNSS into international civil aviation must be underpinned by robust regulatory frameworks and compliance mechanisms. This involves aligning RNSS operations with international aviation standards, ensuring safety, security, and efficiency in air navigation.

2 The Dynamics and Implications of Regional Navigation Satellite System Orbits

The architecture and efficacy of RNSS are significantly influenced by the types of orbits they employ. Primarily, these systems utilize Geostationary Earth Orbits (GEO), Highly Elliptical Orbits (HEO), and/or Inclined Geosynchronous Orbits (IGSO), each offering distinct advantages and challenges for satellite-based positioning and navigation.

We use GNSS Development framework, which includes SDR GNSS receiver and GNSS simulator in this work. The simulator and receiver can work with each other in both RF and digital domain. The Fig. 1 shows simulation of GPS, NavIC, and QZSS orbit ground tracks. Please note that to enhance visibility the time scale for the constellations may differ.

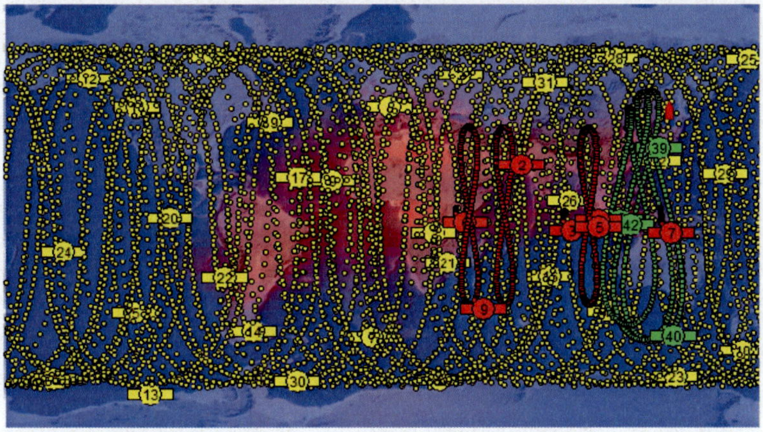

Fig. 1. IP-Solutions GNSS simulator ReGen screen with simulated GPS (yellow), NavIC (red) and QZSS (green) orbits.

GEO satellites orbit the Earth at an altitude of approximately 35.786 km, matching the Earth's rotation period. As a result, they appear stationary from any point on Earth's surface. Among the advantages of such orbits is constant visibility from a fixed location, ideal for communication, broadcasting, and regional augmentation services. However, GEO has disadvantages as well, such as limited coverage at higher latitudes, potential signal obstruction in mountainous or urban areas. They require station keeping and maintenance, which includes regular adjustments to maintain precise orbital position, known as station-keeping. The GEO satellites are used as part of SBAS for correction signals and system integrity information rather than primary positioning due to coverage limitations, less accurate and less stable ephemeris.

HEO satellites follow elongated elliptical paths, with long dwell times at high latitudes. They have first were implemented in Molniya orbits [4], which were developed and used in the former USSR to provide prolonged visibility over high-latitude regions for communication applications.

IGSO satellites follow a geosynchronous orbit but are inclined relative to the equator. This inclination allows the satellite to appear to oscillate in latitude from the perspective of an observer on Earth. Among the IGSO advantages enhanced coverage and signal availability, especially in mid-latitude regions where GEO satellites have limited visibility. In comparison to GEO, IGSOs offer improved visibility and signal strength at higher latitudes compared to GEO satellites, but inferior to that respect to HEO.

The QZSS have altitude similar to geostationary orbits, and similar to HEO in that they have noticeable inclination and non-zero eccentricity. The NavIC satellites have a circular but inclined orbit, maintaining a fixed distance from Earth and providing consistent coverage to specific mid-latitude areas, they do not feature significant non-zero eccentricity and correspondently have similar coverage in both hemispheres. Based on eccentricity and inclination parameters, we classify QZSS orbits as HEO [5] and NavIC orbits as IGSO. As HEO orbits QZSS orbits allows one to provide robust coverage and availability in the wider areas of Asia-Pacific region with more coverage in the Northern hemisphere, which is especially important for regional navigation satellite systems. The orbital design of QZSS with satellites following an analemma-like pattern over Japan ensures high availability and accuracy. This unique trajectory ensures that a satellite is almost directly overhead (near zenith) over Japan for most of the day [6]. The term "Quasi-Zenith" reflects this operational characteristic.

To demonstrate these types of the orbits, we use ReGen simulator and simulate HEO satellite using QZSS almanac parameters. We use eccentricity: $7.499032974 \times 10^{-2}$ and orbital inclination angle: 0.7119348647 rad. These parameters belong to QZSS satellite with ground track depicted on Fig. 1. This ground track is repeated on Fig. 2 for clarity. Figure 3 shows satellite orbit with the eccentricity set to 0, which is IGSO and Fig. 4 shows the satellite orbit (a dot on a map) with inclination set to 0, which gives is GEO.

When it comes to HEO and IGSO, we have to mention that unlike systems such as GPS, Galileo, GLONASS, and BeiDou incorporates GEO and IGSO satellites as integral parts of its GNSS constellation, not just as separate augmentation systems. This integration enhances the system's coverage and performance, particularly over the Asia-Pacific region.

While GEO satellites are invaluable for corrections, alarms, and augmentation, their use for precise positioning is limited. However, they can serve as effective backups in a comprehensive RNSS, providing critical data and services. The maintenance, station-keeping, and overall system integrity of GEO, HEO and IGSO satellites require meticulous management to ensure consistent performance and reliability.

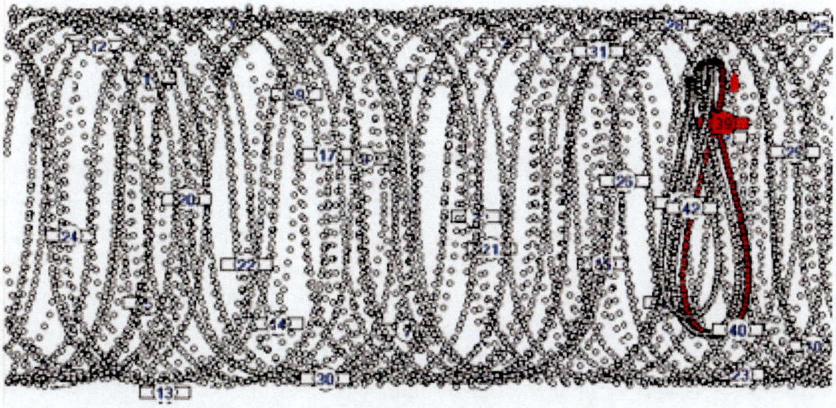

Fig. 2. Simulated QZSS HEO satellite ground track (red).

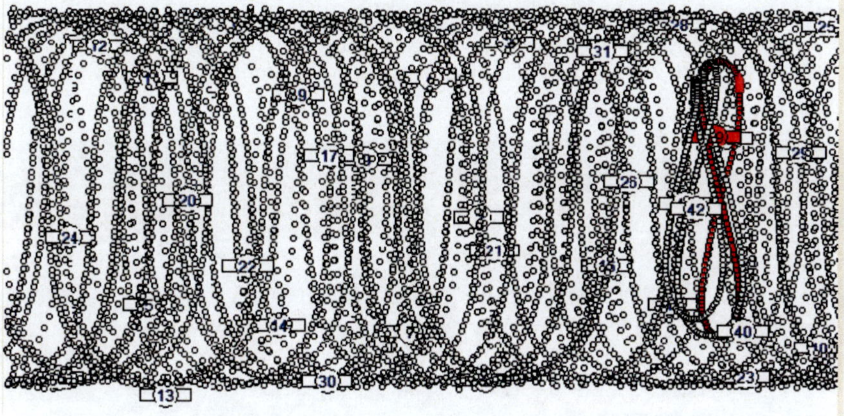

Fig. 3. Simulated QZSS HEO satellite ground track with eccentricity set to 0.

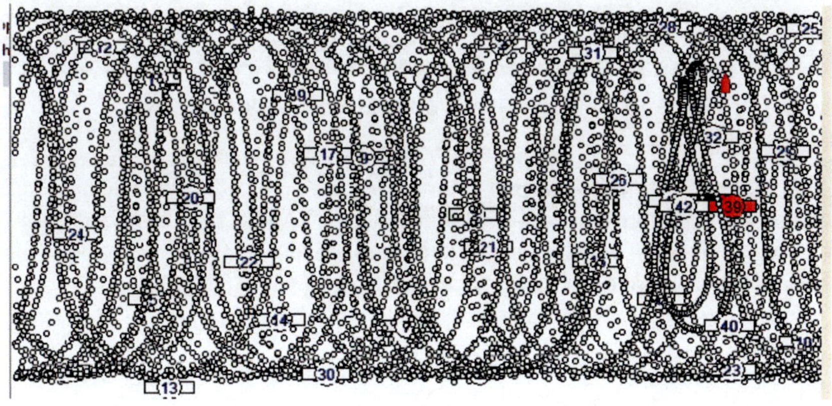

Fig. 4. Simulated QZSS HEO satellite ground track with inclination set to 0.

The ability to evaluate and utilize these diverse orbital configurations, especially with tools like SDR receivers, presents an opportunity to optimize RNSS design and operation for regional and global navigation needs. The main advantage of RNSS in terms of accuracy is more direct interactions with high density regional networks, which allows for better error models and more accurate corrections with low latency. The research of formation and usage of such corrections and models is extremely important and can be made easy with usage of SDR instruments. One of example of such application is a use of ARAMIS SDR receiver to get access to ionospheric grid data provided by NavIC to enhance positioning accuracy [1], which had implemented and compared the single-frequency ionospheric corrections models for Klobuchar and NavIC regional GIVD using L5-band NavIC ARAMIS SDR receiver from IP-Solutions. It was established that regional GIVD model is more efficient compared to Klobuchar model, and its performance is very similar to dual-frequency correction. To implements the GIVD model in [1] Navigation Application Program Interfaces (API) has been used, which provides access to navigation processor source code and raw data, to implement algorithms. Figure 5 depicts the experimental setup of NavIC + GPS ARAMIS SDR receiver at the Communication Research Lab of the Electronics and Communication Engineering Department, Sardar Vallabhbhai National Institute of Technology (SVNIT) in Surat.

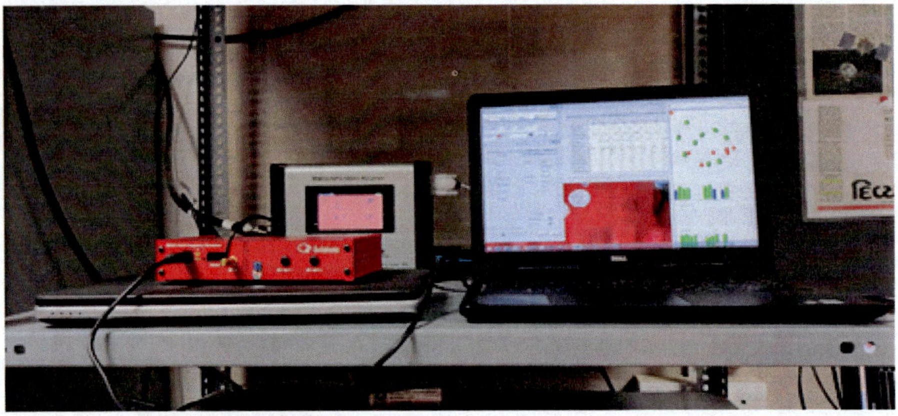

Fig. 5. IP-Solutions GNSS ARAMIS receiver setup at SVNIT.

The ARAMIS RF front-end (red color device in Fig. 5) uses the antenna to receive the NavIC L5 and GPS L1 composite signal. The ARAMIS receiver has the software API (Navigation API), which is run on the Visual Studio platform. Personal computer is running ARAMIS Window-based GUI with real-time user interface.

3 Validation and Testing of RNSS for Civil Aviation: A Case Study with QZSS and NavIC

RAIM is a technology used in GNSS receivers to assess the integrity of GNSS signals and the accuracy of the position solution. RAIM detects faults in the GNSS constellation by performing consistency checks on the navigation signals. The positioning with GNSS requires four satellites to determine three-dimensional user position. The RAIM requires a minimum of five satellites to detect a fault and six satellites for fault detection and exclusion.

The additional satellite is required to substitute the faulty satellite in the position calculation. However, the RAIM algorithm requires an additional satellite making minimum number of required satellites in two constellations to be seven. This additional satellite is to be used to estimate time difference between two satellite constellations, in this case NavIC and QZSS. Though this information can possibly be provided, and for some constellations is provided, as a part of navigation message, the accurate and reliable estimation of this value must be done as a part of navigation solution. Indeed, the standard navigation task should be reformulated as follows:

$$\overline{Z} = [A(\overline{X})], \tag{1}$$

$$\overline{Z} = \begin{bmatrix} \rho_{s_1} \\ \rho_{s2} \\ \dots \\ \rho_{s_n} \end{bmatrix}; \overline{X} \begin{bmatrix} x_r \\ y_r \\ z_r \\ \delta t_r \\ \Delta \tau \end{bmatrix}; \tag{2}$$

where ρ_{s_i} are pseudoranges or code phase measurements; x, y, z are coordinates of the antenna phase center; δt_r is receiver clock error; $\Delta\tau$ is time difference between system time frames.

In this test, we leverage our state-of-the-art GNSS SDR receiver ARAMIS with PORTOS front end (see Fig. 6), capable of processing NavIC L5 + S and QZSS signals, a capability matched only by select advanced receivers like the Novatel OEM-7 series. The platform we use includes our off-the-shelf simulator Replicator (Fig. 7). These instruments constitute Agile GNSS Development Platform offering the flexibility to work with both RF and Digitized Intermediate Frequency (DIF) files. The Replicator simulator allows us to provide both ARAMIS and Novatel OEM-7 receivers with simulated NavIC and QZSS RF signals at the waypoints to save test time. The ReGen control center allowed also simulation of DIF files, which ARAMIS has processed in the baseband deferred mode [7].

The ARAMIS receiver was also used to process recorded with ATOS front end live satellite signals in India and Japan. The recorded signals allow us to include only chosen satellite systems for positioning. The receiver by default also use GPS, GLONASS, BeiDou and Galileo signals, which were not considered for this study.

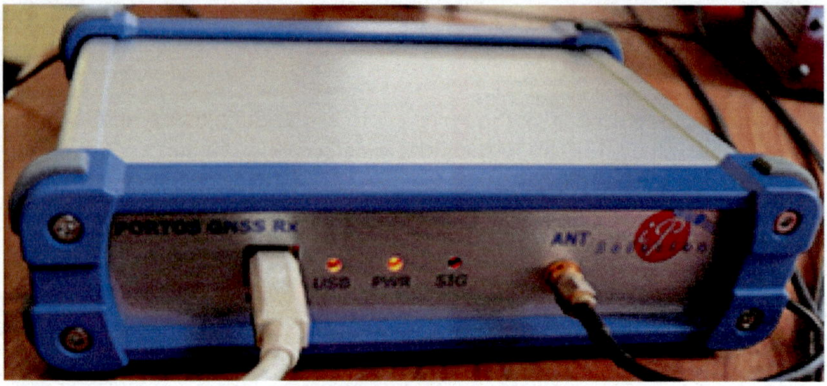

Fig. 6. IP-Solutions GNSS SDR receiver front end PORTOS (L- and S- band capable).

Fig. 7. IP-Solutions GNSS RF simulator Replicator for simulation GPS, Galileo, QZSS, GLONASS and NavIC on L1, L2, L5 and S frequency bands.

The ReGen had generated flight path shown on Fig. 8 using the following waypoints:

1. Sapporo, Japan (43.0621° N, 141.3544° E);
2. Tokyo, Japan (35.6895° N, 139.6917° E);
3. Xi'an, China (34.3416° N, 108.9398° E);
4. Kathmandu, Nepal (27.7172° N, 85.3240° E);
5. Delhi, India (28.6139° N, 77.2090° E).

The number of satellites visible at specific waypoints and the corresponding DOP values are calculated. While DOP is a crucial factor in determining positional accuracy, our primary focus is on satellite visibility, considering the relatively moderate accuracy requirements for en-route navigation. For an in-depth analysis of the effects of multi-constellation design on the geometrical factors, which affect navigation accuracy, such as dilution of precision [8]. In this work, we were specifically interested in number of satellites in relation to the 7 satellites threshold. Plots of 24-h satellite availability (slightly extended) and sky charts are generated to visualize coverage and signal consistency. The ReGen allows us to generate all constellation, but only GPS, NavIC, QZSS and SBAS were generated for clarity. The ReGen screenshot of all constellations over Tokyo is shown as a reference on Fig. 9. Figures 10, 11, 12 and 13 show the number of satellites over correspondent waypoint. The GPS satellites are depicted by blue line. They are shown for comparison. It is clear that there are enough GPS satellites (even an extra one) at each waypoint, taking into account that the threshold in case of GPS is 6 satellites.

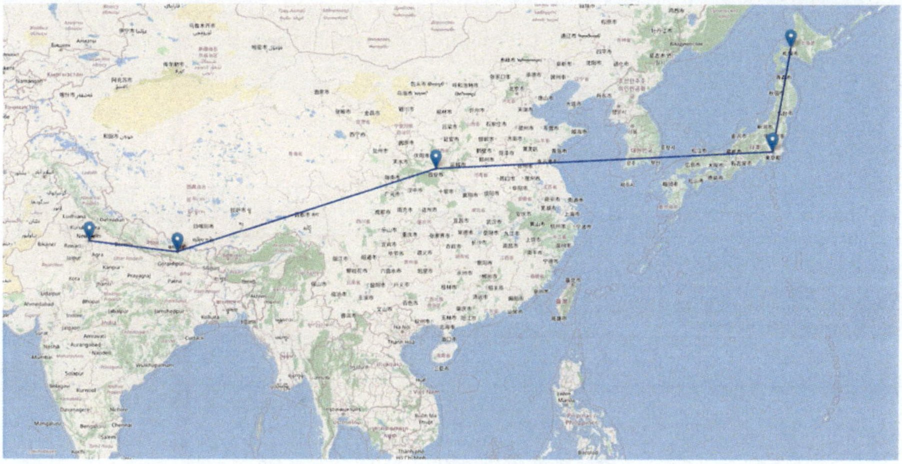

Fig. 8. Simulated flight path.

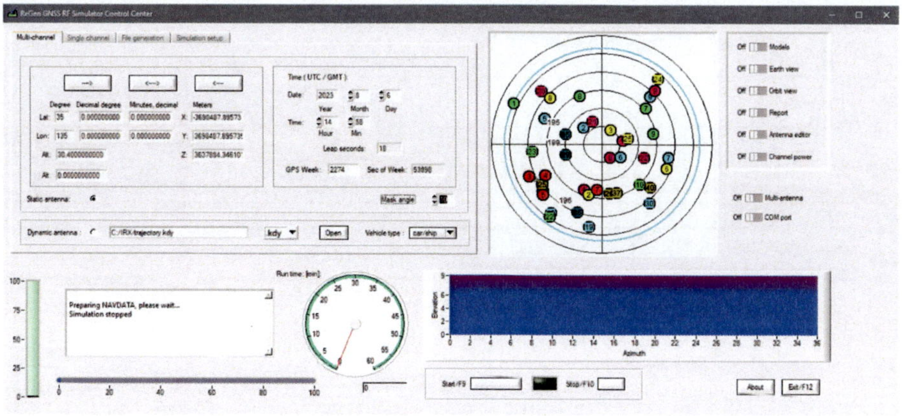

Fig. 9. IP-Solutions ReGen GNSS simulator screenshot with simulated GPS, GLONASS, Galileo, BeiDou, QZSS, NavIC and SBAS satellites in Tokyo region.

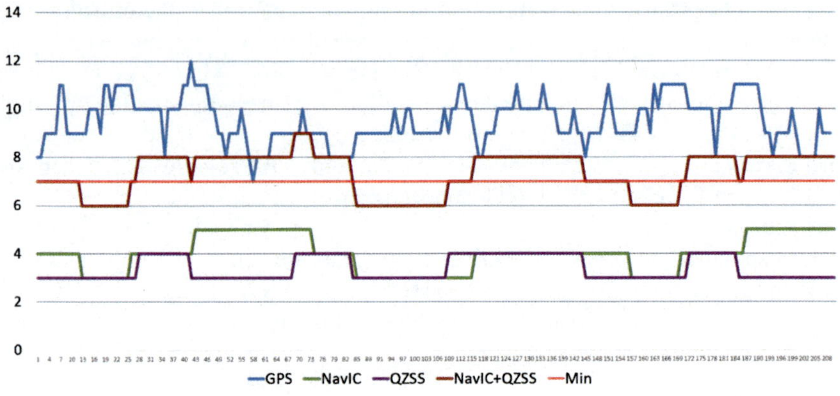

Fig. 10. Satellite visibility for waypoint Sapporo.

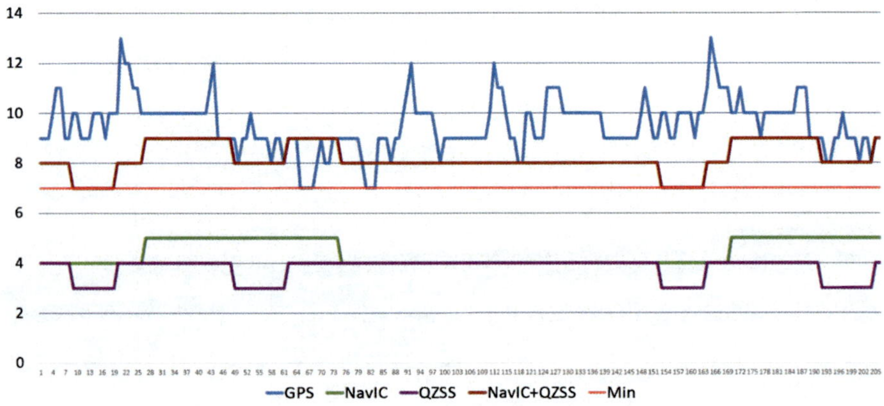

Fig. 11. Satellite visibility for waypoint Tokyo.

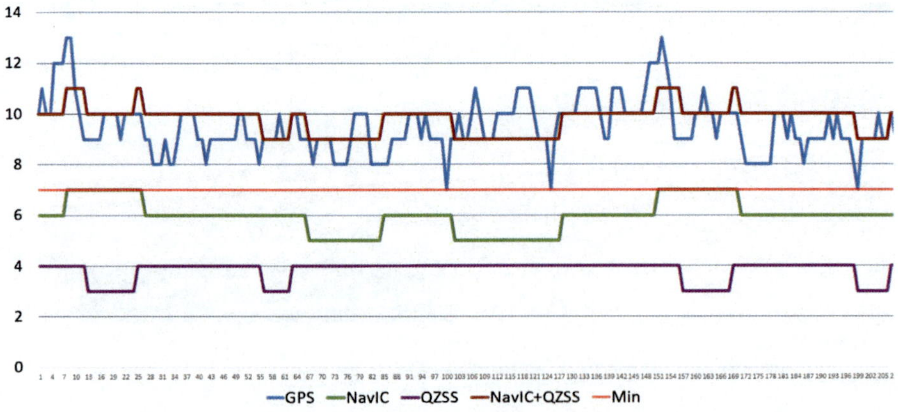

Fig. 12. Satellite visibility for waypoint Xián.

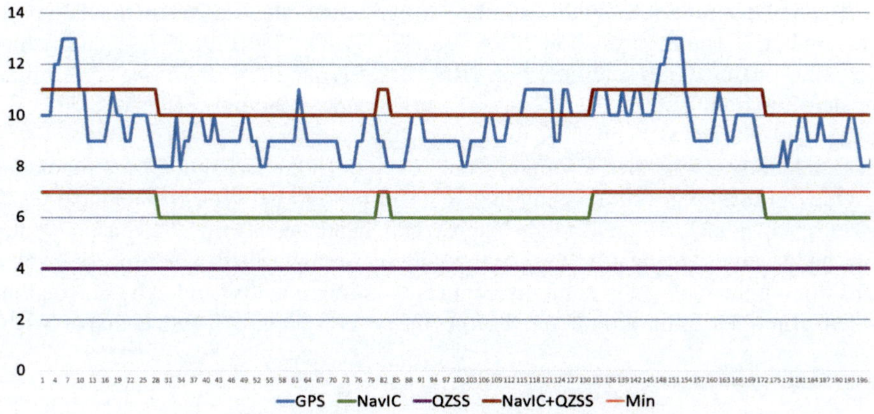

Fig. 13. Satellite visibility for waypoint Delhi.

The NavIC satellites (green line) and QZSS satellites (magenta) (Figs. 10, 11, 12 and 13), when combined (brown line), provide almost enough satellites, except of higher latitude represented by Sapporo. However, the NavIC and QZSS constellation are growing. The ICD provide PRN placeholders for more satellites. The skycharts on the Figs. 14, 15 and 16 show the satellite positions on the sky, GPS (cyan), NavIC (red), QZSS (light green) with additional SBAS satellites (brown).

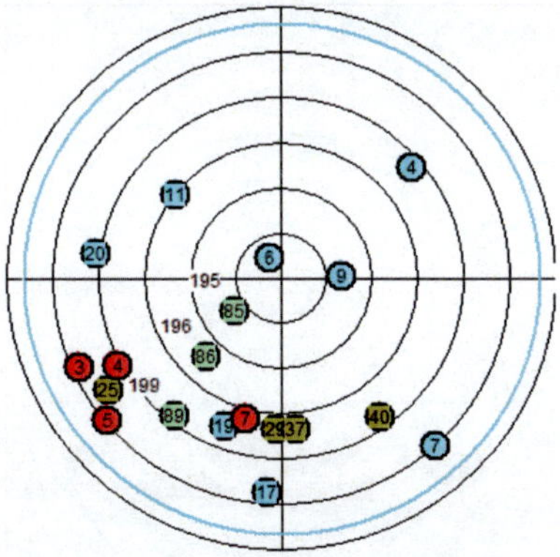

Fig. 14. Skychart for waypoint Sapporo with additional SBAS GEO satellites.

The simulation and processing were done with 5-degree elevation mask angle. The MSAS and GAGAN satellites would provide additional satellites, which will allow to exceed the threshold in case of Sapporo latitude (see Fig. 14) though the satellite geometry would become less favorable, apparently with higher DOPs. This would allow to speculate that in case of civil aviation the RNSS backed navigation should be applied to limited number of routes depending on the latitude.

The SBAS and QZSS satellite numbers inside of the circles are shown with omitted first number 1. The Figures show MSAS satellites with PRN 129, 137, Luch satellites PRN 125, 140, GAGAN satellites with PRN 127(657) and PRN 128(634). The satellites are shown by numbers 29, 37, 25, 40, 27 and 28 correspondingly. The simulation was based on QZSS and GPS almanac of January 2024.

Conducting a simulated test flight from Tokyo to Delhi, utilizing GNSS simulators, provides an ideal platform to study the handover process between two regional GNSS systems: Japan's QZSS and India's NavIC. This handover is critical to ensure continuous and seamless navigation service throughout the flight path. The process involves transitioning the navigation solution from being predominantly guided by one regional system to another, as the aircraft moves from the coverage area of QZSS to that of NavIC.

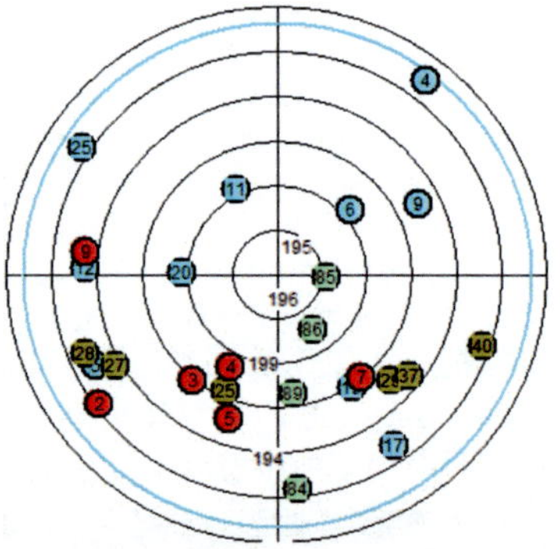

Fig. 15. Skychart for waypoint Xian, China with additional SBAS GEO satellites.

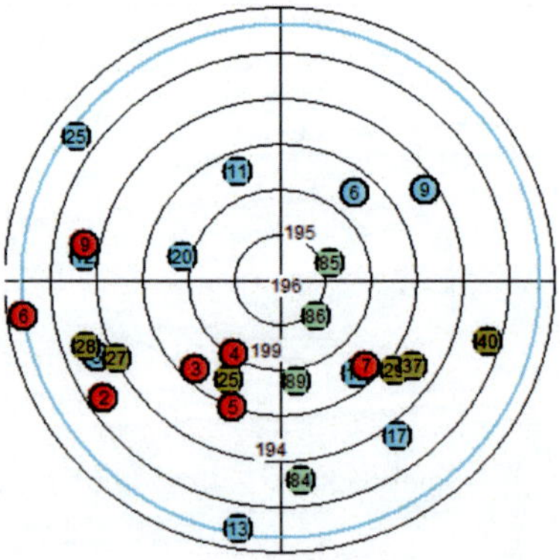

Fig. 16. Skychart for waypoint Dehli with additional SBAS GEO satellites.

In addition, the platform allows us to investigate two important issues. The first on is usage of QZSS L1S signal, which allows RNSS based system to enhance integrity service as it is provided by SBAS. It provides such information as warning of volcanic activity, ash fall, tornado, which is really necessary for civil aviation safety. At least this feature of RNSS should be undoubtedly implemented in civil aviation. The Fig. 17 shows ARAMIS receiver tracking and decoding L1S JAXA emergency signal. The skycharts shows the QZSS satellites with both PRN for each satellite, a number on the white background is L1C/A PRN (PRN 194, 195, 196, 199) and L1S PRN within a circle with first number omitted (PRN 184, 185, 186, 189) The Fig. 18 shows the ARAMIS tracking simulated NavIC L5 and S RF signal during the test. The NavIC S-band placed signal provides additional advantage due to large frequency separation both because better potential for measuring and compensating ionospheric error and better resilience to possible unintentional interference. This advantage is especially important for AI-based algorithms, such as genetic algorithm optimization [9].

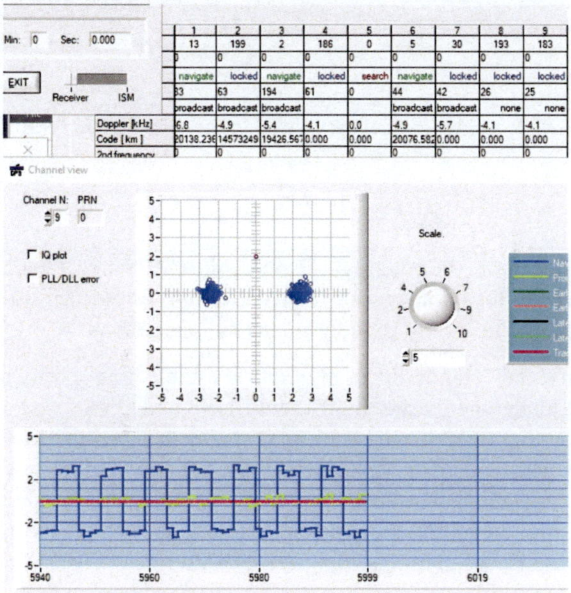

Fig. 17. SDR ARAMIS tracking of L1S QZSS satellite (PRN 183, 193).

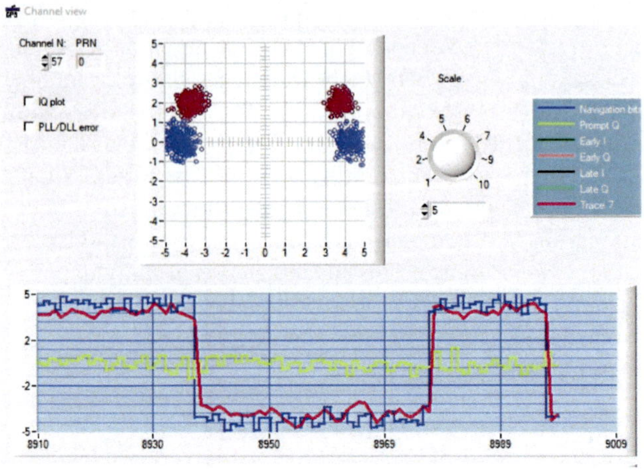

Fig. 18. SDR ARAMIS tracking of ReGen simulated L5 and S RF signals.

4 Conclusions and Future Directions

The initial tests underscore the potential of RNSS, like QZSS and NavIC, in enhancing the redundancy and reliability of navigation in civil aviation. While the results are promising, they also highlight the necessity for continued development as follows:

1. Interoperability and standardization require establishing protocols for seamless integration and handover between different RNSS and GNSS systems.
2. Regulatory compliance and certification are required to ensure that RNSS solutions meet the rigorous standards set by international aviation authorities.
3. Continued testing and validation are important to validate the reliability and accuracy of RNSS solutions under various conditions, the task for which the agile development platform as one discussed in this paper are essential.

References

1. Desai, M.V., Darshna, J., Shweta, S.N.: Implementation and analysis of grid-based ionospheric correction technique and positioning errors of NavIC + GPS ARAMIS SDR receiver. IETE Tech. Rev. (2022). https://doi.org/10.1080/02564602.2022.2152880
2. Petrovski, I.G., II., Hugentobler, U.: Analysis of impact of onboard time scale instrumental error accuracy of Earth satellite ephemerides estimation with one-way measurements. In: Proceedings of the European Navigation Conference ENC-2006. Royal Institute of Navigation, Manchester, UK (2006)
3. Ostroumov, I.: Performance evaluation of positioning methods used in the flight management system. In: Ostroumov, I., Zaliskyi, M. (eds.) Proceedings of the International Workshop on Advances in Civil Aviation Systems Development. ACASD 2023. Lecture Notes in Networks and Systems, vol. 736, pp. 3–16. Springer, Cham (2023)
4. Wertz, J.R.: Mission Geometry; Orbit and Constellation Design and Management, p. 1008. Microcosm Press, USA (2001)

5. Petrovski, I.G., II., Tsujii, T.: Digital Satellite Navigation and Geophysics: A Practical Guide with GNSS Signal Simulator and Receiver Laboratory. Cambridge University Press, New York (2012)
6. Petrovski, I.G., II.: QZSS - Japan's new integrated communication and positioning service for mobile users. GPS World **14**(6), 24–29 (2006)
7. Petrovski, I.G., II.: Ionosphere with GNSS SDR: Constructing and Employing Specialized Software-Defined Radio for In-Depth Ionospheric Research with Global Navigation Satellite System. Forthcoming, p. 350. Springer, Cham (2024)
8. Konin, V., Pogurelskiy, O., Prykhodko, I., Maliutenko, T., Sushych, O., Ishchenko, O.: Multi-GNSS in limited navigation satellite availability. In: Ostroumov, I., Zaliskyi, M. (eds.) Proceedings of the International Workshop on Advances in Civil Aviation Systems Development. Lecture Notes in Networks and Systems, vol. 736. pp. 126–140. Springer, Cham (2023). https://doi.org/10.1007/978-3-031-38082-2_10
9. Petrovski, I.G., II.: RTK-B enhancing real-time kinematic (RTK) positioning with genetic algorithms and long-code GNSS signals for rapid ambiguity resolution and improved accuracy. In: Ostroumov, I., Zaliskyi, M. (eds.) Proceedings of the International Workshop on Advances in Civil Aviation Systems Development. Lecture Notes in Networks and Systems, vol. 736. pp. 116–125. Springer, Cham (2023). https://doi.org/10.1007/978-3-031-38082-2_9

Inertial Navigation System

Autonomous Latitude Determination Using an Inertial Measuring Unit

Vadym Avrutov(✉), Oleksii Hehelskyi, Lev Ryzhkov, Sergii Rupich, and Alexander Zamorsky

National Technical University of Ukraine "Igor Sikorsky Kyiv Polytechnic Institute", Beresteiskyi Avenue, 37, Kyiv 03056, Ukraine
v.avrutov@kpi.ua

Abstract. The novel latitude determination method on the fixed base is presented. The method uses the Inertial Measurement Unit (IMU). The IMU comprises accelerometers, gyroscopes, and a signal-processing circuit. Two new expressions for autonomous latitude determination of a location for a fixed base are obtained. Along with the previously obtained expression, there are only three complete analytical expressions for determining latitude. This makes it possible to determine latitude in three ways. Triple counting undoubtedly improves the reliability of latitude calculations. The constant tilt of the IMU relative to the horizon plane, as well as the deviation of the IMU in azimuth, does not affect the determination of the initial latitude of the location. A full error model of the latitude determination method is developed. It is assumed that the latitude determination error depends on errors of gyroscopes and accelerometers and the difference between geocentric latitude, measured by GNSS, and geodetic latitude. The results of the calculation data are presented to demonstrate the performance of the new latitude determination method. Experiments have confirmed this method. The experiments have been provided on IMU with MEMS gyroscopes and accelerometers and wavelet de-noising of output signals.

Keywords: Inertial Measurement Unit · Latitude · Accelerometers · Gyroscopes

1 Introduction

Nowadays Global Navigation Satellite Systems (GNSS) are the main navigation systems. For example, GPS GNSS receiver provides an error in determining a location of less than 1 m. Equipping satellites with high-speed inter-satellite communication channels will make it possible to determine the user's location with an error of up to 0.15 m [1]. However, unfortunately, GNSS is not autonomous system. During international crises, war, and military conflicts, GNSS is susceptible to attacks by electronic countermeasures. Therefore, the development of autonomous navigation systems has recently been of great importance.

The primary system that provides an autonomous solution to the navigation problem is Inertial Navigation Systems (INS) [2]. For normal operation of INS at the stage of their

I. Ostroumov and M. Zaliskyi (Eds.): ACASD 2024, LNNS 992, pp. 169–178, 2024.
https://doi.org/10.1007/978-3-031-60196-5_14

initial installation, it is necessary to know the initial values of the latitude and longitude of location. Accurate location information, especially latitude, is a critical requirement for existing strap-down INS initial alignment methods [3–6]. However, calibration of gyroscopes, accelerometers, and data acquisition are also important stages in preparing for the operation of INS [7–9].

A well-known autonomous means of determining latitude is the gyro latitude finder (type II Foucault gyroscope), which is a one-degree gyroscope in a gimbal suspension [10]. To construct a gyro latitude finder, you need a vertical gyro reference, in which the plane of the internal frame is oriented, and a gyrocompass, which can set the axis of rotation of the frame in the East-West direction. In addition, the time to determine latitude will depend on the transient time of such a mechanical system.

In the method of determining latitude using two gyroscopes [11], it is necessary to level the base on which the gyroscopes are located. The method of autonomous determination of latitude, based on the use of an Inertial Measurement Unit (IMU) [12], which contains three accelerometers and three gyroscopes, does not have these disadvantages. It is shown that the constant tilt of the IMU relative to the horizon plane, as well as the azimuth deviation of the IMU, do not affect the determination of the initial latitude of a location. Experiments with laser gyroscopes and navigation accelerometers confirmed the validity of this method [13]. The same method was used for IMU with MEMS gyroscopes and accelerometers that had undergone preliminary calibration and filtering of noise components of the output signals [14].

This paper presents a new algorithm for autonomous determination of latitude and also considers a complete mathematical model of the error of autonomous determination of latitude.

The paper is organized as follows. The known and two novel latitude determination methods are explained in Sect. 2. Section 3 presents a mathematical model of the autonomous determination latitude error. The experimental latitude determination using IMU is presented in Sect. 4 and finally, conclusions with future work directions are provided at the end of the paper.

2 Description of Latitude Determination Method Using an Inertial Measurement Unit

We introduce the following coordinate systems into consideration: $O\xi\eta\zeta$ is geographic coordinate system (Fig. 1), with the axis $O\eta$ directed North, axis $O\xi$ is to the East, $O\zeta$ is local vertical, $Oxyz$ is coordinate system associated with the IMU.

Denote by φ the latitude of point, $\vec{g}$ is a gravity acceleration vector, $\vec{\Omega}$ is a vector of the Earth's angular rate. Using the scalar product of two vectors, $\vec{g}$ and $\vec{\Omega}$, we get an expression for determining the latitude of the point [14]:

$$sin\ \varphi = -\frac{1}{\Omega \cdot g}\left(\Omega_x \cdot g_x + \Omega_y \cdot g_y + \Omega_z \cdot g_z\right), \tag{1}$$

where $g = \sqrt{g_x^2 + g_y^2 + g_z^2}$ is a gravity acceleration, $\Omega = \sqrt{\Omega_x^2 + \Omega_y^2 + \Omega_z^2}$ is the Earth's angular rate; $\Omega_x, \Omega_y, \Omega_z$ are the projections of the Earth's angular rate on $Oxyz$, g_x, g_y, g_z are the projections of the gravity acceleration.

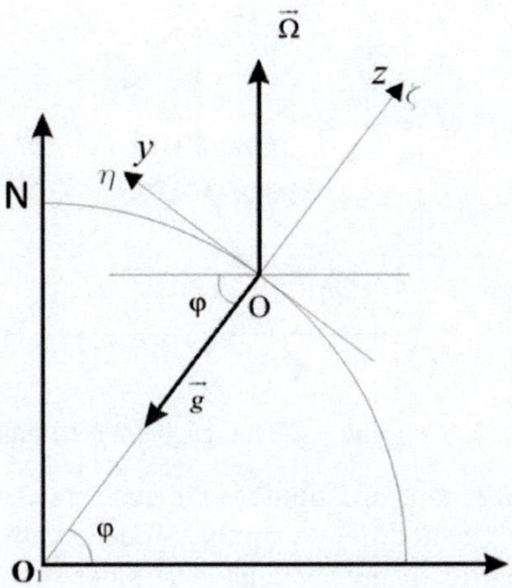

Fig. 1. Position of the IMU relative to the Earth.

Similarly, using the vector product of the gravity acceleration vector and the Earth's angular rate, we can obtain the expression:

$$cos\ \varphi = \frac{1}{\Omega g}\sqrt{(\Omega_y g_z - \Omega_z g_y)^2 + (\Omega_z g_x - \Omega_x g_z)^2 + (\Omega_x g_y - \Omega_y g_x)^2}. \quad (2)$$

Based on the expressions (1) and (2), we could obtain a formula for determining the latitude of IMU location:

$$tg\varphi = -\frac{\Omega_x \cdot g_x + \Omega_y \cdot g_y + \Omega_z \cdot g_z}{\sqrt{(\Omega_y g_z - \Omega_z g_y)^2 + (\Omega_z g_x - \Omega_x g_z)^2 + (\Omega_x g_y - \Omega_y g_x)^2}}. \quad (3)$$

So, to calculate the latitude, it is necessary to know the values of the projections of the angular velocity of the Earth's rotation and the projections of the acceleration of gravity on the axes associated with the IMU.

To check the validity of the last expression, we will assume that the axes of the IMU coincide with the axes of the geographic coordinate system: $Ox = O\xi$, $Oy = O\eta$, $O\zeta = Oz$. Thus:

$$\Omega_x = 0, \quad \Omega_y = \Omega\ cos\ \varphi, \quad \Omega_z = \Omega\ sin\ \varphi, \quad g_x = g_y = 0, \quad g_z = -g. \quad (4)$$

So, both expressions (2) and (3) could be used for latitude determination for a fixed base. There are three complete analytical expressions for determining latitude: by Eqs. (1), (2), and (3) determine geodetic latitude (Fig. 2).

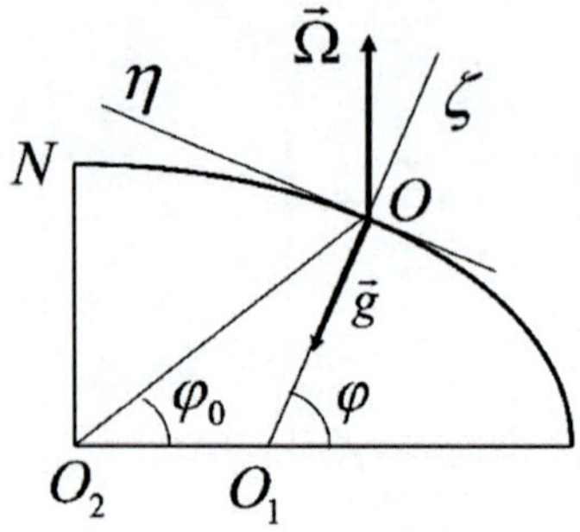

Fig. 2. Geocentric ϕ_0 and geodetic ϕ latitudes.

As a rule, the reference value of latitude is the geocentric value of latitude received from the GNSS receiver, which differs from the geodetic value of latitude. It is known a relation between geocentric φ_0 and geodetic φ latitudes [15]:

$$tg\varphi = \frac{a^2}{b^2} tg\varphi_0. \tag{5}$$

It is possible to obtain the difference (correction) (Fig. 3):

$$\Delta\varphi = \varphi - \varphi_0, \tag{6}$$

$$\Delta\varphi = arctg \frac{\left(\frac{a^2}{b^2} - 1\right) tg\varphi_0}{1 + \frac{a^2}{b^2} \cdot tg^2\varphi_0}, \tag{7}$$

where a, b are the major and minor semi-axes of the Earth's ellipsoid.

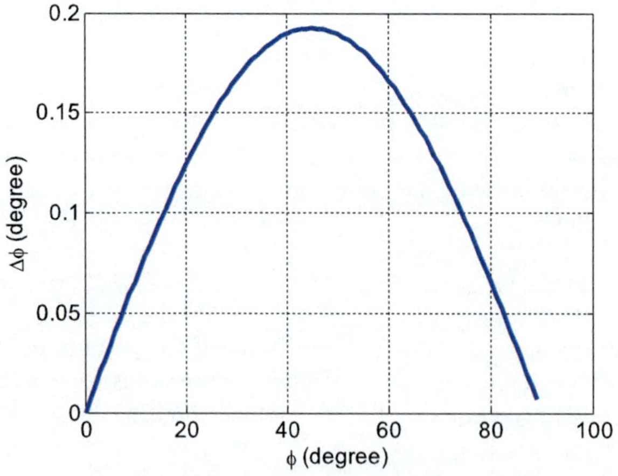

Fig. 3. The difference between geocentric ϕ_0 and geodetic ϕ latitudes.

If the WGS-84 model is taken as the reference ellipsoid, then a=6378.137 km, b = 6356.752 km. Figure 3 presents the differences $\Delta\varphi$. The maximum value $\Delta\varphi$ is observed at latitude 45° (South of Odesa city). For a given latitude, the linear coordinate error $\Delta s = \Delta\varphi R$, , caused by the difference (4) will be equal to 21.140 km.

Thus, it is possible to conclude about the need to take into account the correction $\Delta\varphi$ when obtaining the geodetic latitude according to expressions (1)–(3) for comparing it with the geocentric latitude.

3 Mathematical Model of the Autonomous Determination Latitude Error

We will obtain a mathematical model of the error of the autonomous latitude determination. It follows from expression (3) that:

$$\varphi = -arctg\frac{N_1}{N_2}, \tag{8}$$

where $N_1 = \Omega_x g_x + \Omega_y g_y + \Omega_z g_z$;

$$N_2 = \sqrt{\left(\Omega_y g_z - \Omega_z g_y\right)^2 + (\Omega_z g_x - \Omega_x g_z)^2 + \left(\Omega_x g_y - \Omega_y g_x\right)^2}. \tag{9}$$

Let's use the Taylor series expansion of function (8). Let us represent the latitude as a function F of two variables N_1 and N_2.

We get the latitude error by varying the function $F(N_1, N_2)$:

$$\Delta\varphi = -\sum\nolimits_{i=1}^{2}\frac{\partial F}{\partial N_i}\Delta N_i, \tag{10}$$

$$\Delta N_i = \frac{\partial N_i}{\partial \Omega_x}\Delta\Omega_x + \frac{\partial N_i}{\partial \Omega_y}\Delta\Omega_y + \frac{\partial N_i}{\partial \Omega_z}\Delta\Omega_z + \frac{\partial N_i}{\partial g_x}\Delta g_x + \frac{\partial N_i}{\partial g_y}\Delta g_y + \frac{\partial N_i}{\partial g_z}\Delta g_z, \tag{11}$$

where $\Delta\Omega_x$, $\Delta\Omega_y$, $\Delta\Omega_z$ are gyroscope errors; Δg_x, Δg_y, Δg_z are accelerometer errors.

After cumbersome transformations, a complete mathematical model of the error of autonomous latitude determination was obtained:

$$\Delta\varphi = \frac{1}{\Omega^2 g^2}(N_1 \Delta N_2 - N_2 \Delta N_1), \tag{12}$$

$$\Delta N_1 = g_x\Delta\Omega_x + g_y\Delta\Omega_x + g_z\Delta\Omega_x + \Omega_x\Delta g_x + \Omega_y\Delta g_y + \Omega_z\Delta g_z, \tag{13}$$

$$\begin{aligned}
\Delta N_2 = \frac{1}{N_2}\{&\left[\left(\Omega_x g_y - \Omega_y g_x\right)g_y - (\Omega_z g_x - \Omega_x g_z)g_z\right]\cdot\Delta\Omega_x + \\
&+\left[\left(\Omega_y g_z - \Omega_z g_y\right)g_z - \left(\Omega_x g_y - \Omega_y g_x\right)g_x\right]\cdot\Delta\Omega_y + \\
&+\left[(\Omega_z g_x - \Omega_x g_z)g_x - \left(\Omega_y g_z - \Omega_z g_y\right)g_y\right]\cdot\Delta\Omega_z + \\
&+\left[(\Omega_z g_x - \Omega_x g_z)\Omega_z - \left(\Omega_x g_y - \Omega_y g_x\right)\Omega_y\right]\cdot\Delta g_x + \\
&+\left[\left(\Omega_x g_y - \Omega_y g_x\right)\Omega_x - \left(\Omega_y g_z - \Omega_z g_y\right)\Omega_z\right]\cdot\Delta g_y + \\
&+\left[\left(\Omega_y g_z - \Omega_z g_y\right)\Omega_y - (\Omega_z g_x - \Omega_x g_z)\Omega_x\right]\cdot\Delta g_z.
\end{aligned} \tag{14}$$

For a numerical evaluation of the obtained error model, it is worth considering a separate case when the axes of the IMU coincide with the axes of the geographic coordinate system, i.e. $Ox = O\xi, \quad Oy = O\eta, \quad O\zeta = Oz$. In this case, from the complete mathematical model (12) we can obtain:

$$\Delta\varphi = \frac{1}{\Omega g}\left(\Omega\Delta g_y + g\ sin\ \varphi\ \Delta\ \Omega_y - g\ cos\ \varphi\ \Delta\ \Omega_z\right). \tag{15}$$

Consider the case of systematic errors. Expression (15) can be represented as follows:

$$\Delta\varphi = \frac{1}{\Omega g}\left(\Omega\Delta g_y^{sys} + g\ sin\ \varphi\ \Delta\ \Omega_y^{sys} - g\ cos\ \varphi\ \Delta\ \Omega_z^{sys}\right), \tag{16}$$

where $\Delta\Omega_y^{sys}$ and $\Delta\Omega_z^{sys}$ are systematic errors of the gyroscopes (the constants in this run) and Δg_y^{sys} is accelerometer error respectively.

As an example, consider the IMU with parameters from Table 1.

Table 1. Technical data of IMU.

Item	Value
Temperature	$-40 + 85$ °C
Gyro scale range	±400°/s
Gyro Bias Stability	0.5°/hr
Angle Random Walk (*ARW*)	$0.15°/\sqrt{}$hr
Gyro Frequency Range	262 Hz
Accelerometer scale range	±10*g*
Accelerometer Bias Stability	0.05 *mg*
Velocity Random Walk (*VRW*)	0.07 m/s$^2/\sqrt{}$Hz
Accelerometer Frequency Range	214 Hz

We will accept next data: $\Delta\Omega_y^{sys} = \Delta\Omega_z^{sys} = 0.5\ °/h$; $\Delta g_y^{sys} = 0.05mg$; $\varphi = 50°$. Calculation according to formula (16) gives $\Delta\varphi = 0.235°$, and the main contribution is the errors of the gyroscopes.

Consider the effect of random noise at the IMU output on the latitude determination error. A value of Angle Random Walk (ARW) is usually determined to characterize the noise of gyroscopes. The noise of accelerometers is characterized by the value Velocity Random Walk (VRW). First, move in expression (15) from the absolute values of errors $\Delta\Omega_y$, $\Delta\Omega_z$; Δg_y to the root-mean-square values σ_{Ω_y}, σ_{Ω_z}; σ_{ac}. We separately highlight the influence of gyroscopes on the error in determining latitude, assuming that $\sigma_{\Omega_y} = \sigma_{\Omega_z}$:

$$\sigma_\varphi^{gyro} = \frac{\sigma\Omega_y}{\Omega}(sin\varphi - cos\varphi) \tag{17}$$

Influence of accelerometers on the error in determining latitude:

$$\sigma_{\varphi}^{ac} = \frac{\sigma_{ac}}{g} \tag{18}$$

Standard deviations at the output of gyroscopes and accelerometers:

$$\sigma_{\Omega_y} = ARW\sqrt{f_{gyro}} \tag{19}$$

$$\sigma_{ac} = VRW\sqrt{f_{ac}}. \tag{20}$$

Substituting the data from Table 2, we first obtain the standard deviation values $\sigma_{\Omega_y} = \sigma_{\Omega_z} = 145.68°/\text{hr}$ and $\sigma_{ac} = 1.02\text{m/sec2}$. Now we calculate the root-mean-square (RMS) error in determining latitude for $\phi = 50°$: $\sigma_{\varphi}^{gyro} = 68.48°$ and $= \sigma_{\varphi}^{ac} = 5.73°$.

As we can see, the standard deviation of latitude determination due to the accelerometer error is an order of magnitude smaller than the standard deviation due to the gyroscope error. However, the magnitude of the overall error is significant, which implies filtering of the output signals of gyroscopes and accelerometers. For this, we will use wavelet filtering of gyroscopes' and accelerometers' output signals.

4 Experimental Latitude Determination Using IMU

For experimental latitude determination, an IMU STIM300 was used, consisting of three orthogonally located MEMS gyroscopes and three accelerometers. The signals received from the gyroscopes and accelerometers were averaged over a period of 10 min. After that, the average values were used in expression (3), and the correction $\Delta\varphi$, determined by formula (4), was subtracted from the calculated latitude values. For the measurement latitude, it is $\Delta\varphi = 0.1909$. The results of the calculations showed that with an increase in the number of points in the sample or with an increase in the averaging time, the accuracy of latitude determination increases. To filter the output signals of gyroscopes and accelerometers, a sample of signals was taken with the number of $N = 38281$, a sampling frequency of 25 Hz, and a signal quantization time of 0.04 s.

To select the optimal wavelet, the Wavelet Toolbox of the Matlab application package was used. Wavelet Toolbox contains programs and functions for time-frequency signal analysis and multi-level image analysis. The programs allow to perform time-frequency analysis, signal filtering, or image analysis, and generate Matlab scripts to automate work on wavelets. The Wavelet Signal Denoiser tool was used to select the wavelet transform parameters. This is a program for interactive visualization and filtering of one-dimensional signals using wavelet transformation. After selecting the parameters of the wavelet transform, the Wavelet Signal Denoiser program generated a script for the Matlab program to filter the signals. Various types of wavelets were tried, including Symlet, Coiflet, Haar, and Fejer-Korovkin wavelets, but the best result in terms of reducing the location latitude error was achieved with a Daubechies 10 wavelet transform with 12 levels of decomposition. Figure 4 shows graphs of output signals (left) and signals cleaned of noise (right) of the accelerometer unit.

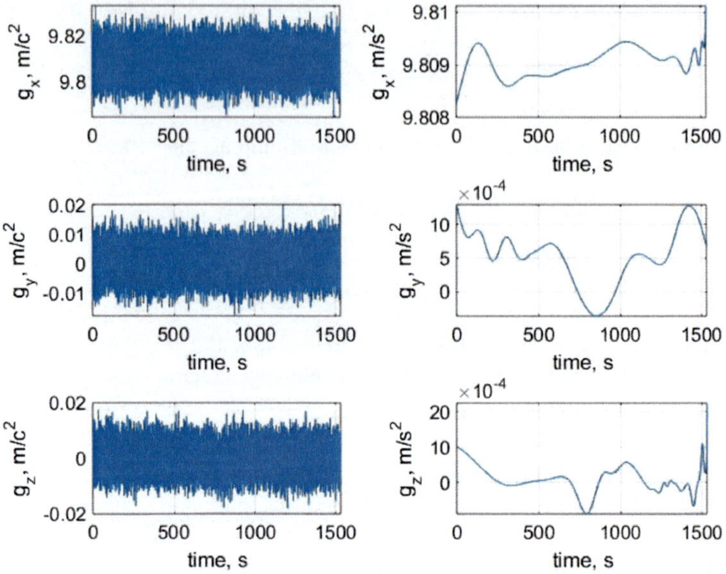

Fig. 4. Graphical dependencies of the initial (left) and cleaned (right) signals of the accelerometer unit.

After wavelet filtering, the gyroscope and accelerometer signals were averaged. Then, according to expression (3), the values of the latitude of the site were calculated. Table 2 shows the average values of the projections of the angular velocity of the Earth's rotation and the average values of the projections of the acceleration of gravity, obtained before and after wavelet decomposition.

Table 2. Results of wavelet filtering of IMU output signals.

Parameter	Mean before de-noising	Mean after de-noising	RMS before de-noising	RMS after de-noising	Units
Ω_x	11.607	11.532	34.457	1.845	°/hr
Ω_y	−10.127	−10.120	33.491	2.265	°/hr
Ω_z	0.131	0.144	33.477	1.293	°/hr
a_x	9.80903	9.80904	0.005927	2.5896·10–4	m/s2
a_y	0.00051	0.00051	0.004418	$4.0962 \cdot 10^{-4}$	m/s2
a_z	$6.99 \cdot 10^{-5}$	$9.97693 \cdot 10^{-5}$	0.004391	$3.8572 \cdot 10^{-4}$	m/s2
$\Delta\varphi_1$, deg	0.26379				
$\Delta\varphi_2$, deg	0.0602				

Here, RMS is a root mean square deviation; $\Delta\varphi_1$ and $\Delta\varphi_2$ represent the latitude determination errors, calculated before and after wavelet decomposition, compared to the absolute value of the geocentric latitude. The values of $\Delta\varphi_1$ and $\Delta\varphi_2$ show that due to taking into account the difference between the geocentric φ_0 and geodetic φ latitudes and the performed wavelet filtering of the noise of gyroscopes and accelerometers, the latitude determination error was reduced by more than four times.

5 Conclusions

Three expressions for autonomous latitude determination for a fixed base are considered. For this, one needs to have an IMU, which contains three accelerometers and three gyroscopes. The constant tilt of the IMU relative to the horizon plane, as well as the deviation of the IMU in azimuth, does not affect the determination of the initial latitude of the location.

A complete mathematical model of the error for determining the latitude of a place has been developed, which has shown that the main reasons for the error in determining the latitude of a place are the errors of gyroscopes and accelerometers. The results of numerical calculations showed that for the practical implementation of this method of determining the latitude of a place, fairly accurate gyroscopes and navigation accelerometers are required. The experiments carried out confirmed the validity of the analyzed method. An experiment carried out with an IMU manufactured using MEMS technology and undergoing preliminary calibration confirmed the results of numerical calculations. Wavelet filtering of the noise components of the output signals of gyroscopes and accelerometers made it possible to increase the accuracy of latitude determination by four times.

Future work involves continuing the search for advanced methods for filtering the output signals of MEMS gyroscopes and accelerometers on board the vehicle. In addition, to improve the accuracy of latitude determination, it is planned to use IMU errors' auto-compensation methods.

References

1. Schmidt, G.T.: GPS Based navigation systems in difficult environments. Gyroscopy Navig. **10**(2), 41–53 (2019)
2. Grewal, M., Weill, L., Andrews, A.: Global Positioning Systems, Inertial Navigation, and Integration. Wiley-Interscience, Hoboken (2007)
3. Titterton, D.H., Weston, J.L.: Strapdown Inertial Navigation Technology. IEE Radar, Sonar, Navigation and Avionics Series, vol. 17 (2004)
4. Silva, F.O., Hemerly, E.M., Filho, W.C.L.: Error Analysis of analytical coarse alignment formulations for stationary SINS. IEEE Trans. Aerosp. Electron. Syst. **52**, 1777–1796 (2016)
5. Yang, H.T., Zhou, B., Wang, L.X., Wei, Q., Zhang, R.A.: Novel method for fast stationary initial alignment based on extended measurement information. IEEE Access **7**, 165873–165883 (2019)
6. Zheng, Z.Y., Zhou, A.J., Tang, J., Xu, X.B.: An initial alignment method of SINS without latitude based on calculation of Earth axis vector. J. Astronaut. **40**, 77–84 (2019)

7. Sushchenko, O.A., Golitsyn, V.O.: Data processing system for altitude navigation sensor. In: 2016 IEEE 4th International Conference on Methods and Systems of Navigation and Motion Control (MSNMC 2016), pp. 84–87. IEEE, Kyiv (2016)
8. Sushchenko, O., Chikovani, V., Tsiruk, H.: Redundant information processing techniques comparison for differential vibratory gyroscope. Eastern-Europ. J. Enterp. Technol. **4**(7–82), 45–52 (2016)
9. Sushchenko, O.A., Bezkorovainyi Y.M., Golitsyn, V.O.: Fault-tolerant inertial measuring instrument with neural network. In: 40th International Conference on Electronics and Nanotechnology (ELNANO), pp. 797–801. IEEE, Kyiv (2020)
10. Pavlovsky, M.A.: Theory of the Gyroscopes (in ru), Kiev (1986)
11. Patent of RF No. 2 572 651 (2014)
12. Avrutov, V.V.: Autonomous determination of initial latitude with an inertial measuring unit. Int. Appl. Mech. **54**(5), 594–599 (2018)
13. Wang, S., Yang, G., Chen, W., Wang, L.: Latitude determination and error analysis for stationary SINS in unknow-position condition. Sensors **20**(9), 2558 (2020)
14. Avrutov, V., et al.: Wavelet de-noising and Kalman filtering of MEMS sensors for autonomous latitude determination. Int. J. Sens., Wirel. Commun. Control **12**(5), 344–351 (2022)
15. Babich, O.A.: Data processing in navigation systems (in ru). Mashinostroenie. Moscow (1991)

Instrument Landing System

Comparison of Instrument and Satellite Landing Systems

Leonid Sibruk(✉), Viktor Sibruk, and Ihor Zakutynskyi

National Aviation University, Liubomyra Huzara Avenue, 1, Kyiv 03058, Ukraine
sibruk@nau.edu.ua

Abstract. The paper deals with engineering and cost specifications of Instrument Landing System (ILS) and GBAS Approach Service Type (GAST), which are used for precision approach in civil aviation. In the process of gradually replacing the ILS system with the GAST system, it is advisable to briefly consider all the factors that affect the decision regarding the scope and timing of the GAST implementation. The paper briefly deals with technical characteristics of systems, destabilizing effects on the systems accuracy, vulnerability to jamming and spoofing, and future benefits. Main advantages and drawbacks of the navigational systems are considered. The use of GAST-D (category II and III) gives an opportunity to change air traffic management procedures in order to increase runway throughput, which is important for busy airports, especially in low visibility conditions. Cost analysis includes only some items: equipment cost, airfield works connected with installation, maintenance, and flight inspection. Some examples from airports "Zhulyany" and "Boryspil" (Ukraine) are given.

Keywords: Instrument Landing System · GBAS Approach Service Type · Critical Zone · Maintenance · Runway Throughput · Jamming

1 Introduction

Normative documents in civil aviation consider several types of instrumental systems for precision approach: Instrument Landing System (ILS), Microwave Landing System (MLS), and Ground Based Augmentation System (GBAS) Approach Service Type (GAST). Against International Civil Aviation Organization (ICAO) former roadmap of complete replacement ILS on MLS till 1985 now MLS is not used. A common explanation is its higher price. But in 90-th the ratio of equipment cost was something about two times. MLS is free from main ILS disadvantages and can ensure a precision approach in conditions of difficult relief configuration. Nowadays, development of technologies makes phase antenna arrays of MLS beacons (the most expensive part of MLS) significantly cheaper.

Demands for landing system operation are expressed in conditions of runway visibility range and altitude decision. Systems ensure precision lateral and vertical guidance according to the demands of categories [1].

I. Ostroumov and M. Zaliskyi (Eds.): ACASD 2024, LNNS 992, pp. 181–189, 2024.
https://doi.org/10.1007/978-3-031-60196-5_15

In accordance with Category 1 (CAT I), the landing system provides information in the vertical plane up to a height of at least 60 m. At this runway visibility must be not less than 800 m in the horizontal plane or a visible edge of the runway should be not less than 550 m.

In accordance with Category II (CAT II), the landing system provides information in the vertical plane up to a height not less than 30 m and a visible edge of the runway should be not less than 300 m in the horizontal plane.

Category III (CAT III) has three subdivisions.

- A. Landing system provides information in the vertical plane up to a height from 30 m to zero height. In the horizontal plane, the visible edge of the runway should be not less than 175 m.
- B. Landing system provides information in the vertical plane up to a height from 15 m to zero height. In the horizontal plane, the visible edge of the runway should be in the range from 175 m to 50 m.
- C. Landing system has no restrictions on visibility in both vertical and horizontal planes.

ILS supports CAT I-IIIA. ILS specifications and information for guidance in the application of the Standards and Recommended Practices are given in [2]. Today landing system GAST-C by CAT I, operates in about 100 airports and this number is growing each year. GAST-D is designed to support CAT II/CAT III [2, 3]. It is no doubt, that in the near future certified GAST-D and corresponding avionics will be available.

GAST implementation is a complex process that includes system approval by Federal Aviation Administration (FAA) and ICAO, facility approval, and service approval [4]. In contrast, the ILS infrastructure exists and functions satisfactorily. The replacement of the landing system should be well-planned. Thus, it could be useful to provide a comparative analysis of both ILS and GBAS systems to ensure the safety of air transportation with consideration of some cost evaluation.

The paper contains eight sections. The first section describes introduction. The second section briefly highlights technical specifications of both landing systems. The third section describes siting restrictions. The fourth section compares systems maintenance. The fifth section presents problems of ILS in low visibility conditions. The sixth section includes information on resilience systems to spoofing and jamming. The seventh section deals with cost evaluations. Finally, the eighth section is the conclusion.

2 Comparison of Systems Technical Specifications

ILS is a measuring system with amplitude modulation of radio frequency carrier by 90 Hz and 150 Hz tones. ILC consists of two main parts: localizer and glide-path beacons. The beacons create electromagnetic fields with zero level of information parameter (difference in the depth of modulation, DDM) along landing course and glide-path lines. Airborne receivers measure angular lateral and vertical displacements of an aircraft from these reference lines.

ILS CAT I uses single-frequency beacons, ILS CAT II/III are dual frequencies. This is explained by necessity to diminish influence of multipath propagation, caused by radio wave reflections from terrain, buildings, and other obstructions. Reflected waves interfere with direct wave at airborne antenna. Relationships between the levels of the modulation signals 90 Hz and 150 Hz are violated, which leads to errors in DDM determining, also in the aircraft guidance during landing. Errors due to the multipath are the main factor of bending course and glide path. Allowable values of bends are normalized depending on the ILS category and the distance from the runway using ILS points.

Dual frequency localizer has course and clearance channels, which carriers are shifted only on 8 kHz. The patterns of course channel is narrow (±10°) in order to reduce influence of multipath errors. The clearance signal is used to find the side of airplane position relatively to course. Its level diminishes along course line. Antennas of both channels radiate the carrier side band (CSB) and the sideband only (SBO) signals which are shown in Fig. 1.

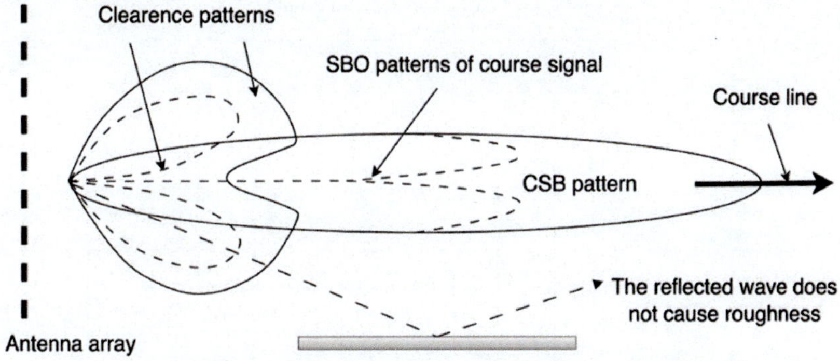

Fig. 1. The antenna patterns of dual frequency localizer.

Besides using the antenna patterns of given forms, another means to restrict bends of course and glide-path lines are critical and sensitive areas [10]. ILS critical areas are areas about beacon antennas, where vehicles, including aircraft, are excluded during ILS operation in order to prevent radio wave interference. Sensitive area is a critical area, where interference is possible [11]. Dimensions of critical zone for localizer depend on used antenna array (60–120 m long), category of operation, and aircraft type. Geometry of localizer critical area is presented in Fig. 2.

Critical zone for glide path beacon (Fig. 3) is important for bend errors and stability of glide path angle. This is connected with dependence on the pattern of horizontal dipole (radiating element of antenna system) from antenna hanging height. Glide path angle corresponds to zero directions between two lobes of SBO channel antenna. When the grass is growing the glide path angle decreases. So, the grass needs to be removed. Also, the same problem is caused by snow during winter times. Therefore, the surface in the direction of radiation must be flat and its level must be stable.

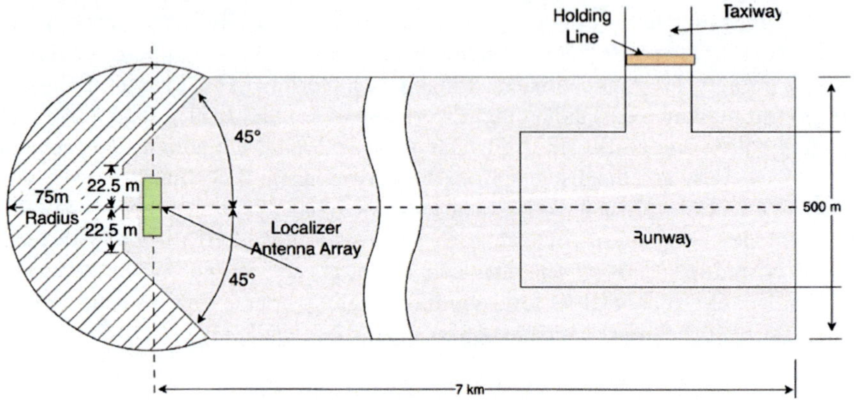

Fig. 2. Localizer critical zone.

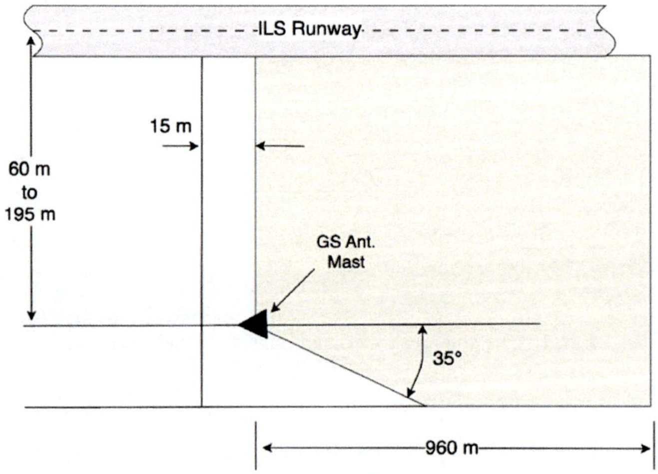

Fig. 3. Critical zone of glide path beacon.

Primary ILS installation includes some building works and smoothing terrain in critical areas and out of it (if possible).

The final problem of ILS operation is connected with construction of high-rise buildings near the airport area. The developer company must coordinate the construction near the airfield with the aviation authorities. For example, in Kyiv airport, Zhulyany operates ILS CAT I. When there was a question about replacement system on ILS CAT II they met difficulties related to new buildings.

GAST structure contains a Global Navigation Satellite System (GNSS), GBAS, and avionics equipment, which are shown in Fig. 4. GAST is a digital rangefinder system with complex signal processing. Existing systems GAST-C and designing GAST-D are single-frequency (L1) systems. Besides GNSS has begun to radiate on other frequency (L5) with greater power level (on 1.5 dB) for applications including civil aviation. At this GAST-C and GAST-D remain to be the single frequency systems. Dual frequency system GAST-F for operation to the CAT-II/III minima is in the study. So, for real applications, it is expedient to consider GAST-C and GAST-D. In order to better understand systems possibilities let's consider the sources of errors influencing accuracy characteristics. Minimum GAST coverage is close to ILS and is shown in Fig. 5 and Fig. 6.

The time division multiple access, which is used GAST, is based on frames and time slots. Each 1-s epoch contains two frame of 0.5 s each. The first frame starts at the beginning of the epoch. The frame is time division multiplexed.

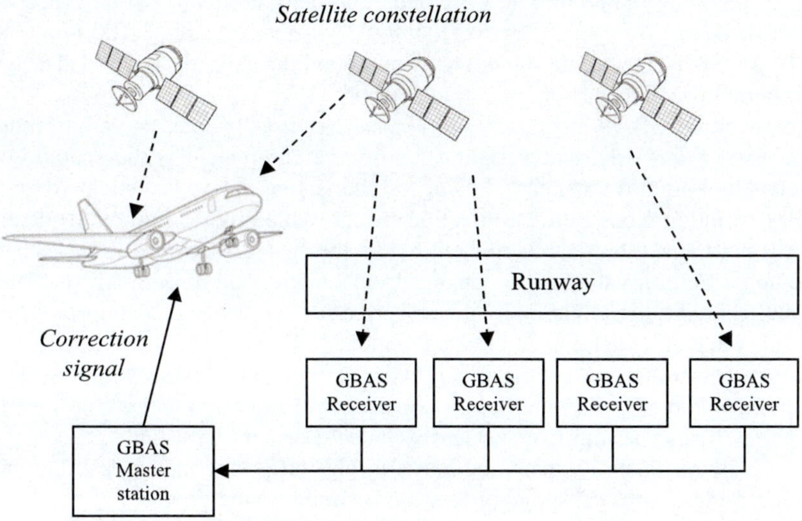

Fig. 4. GAST structure.

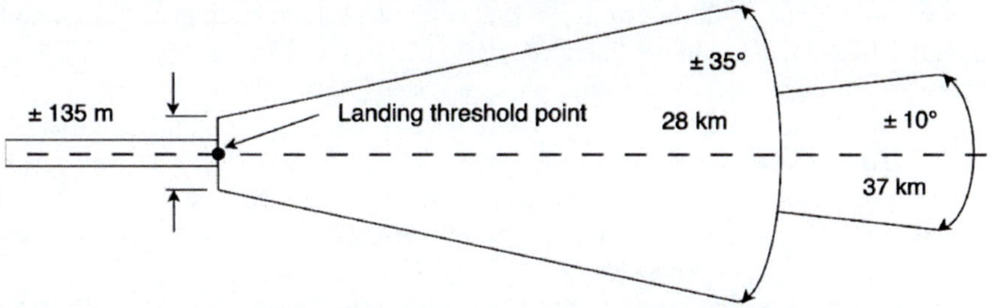

Fig. 5. Minimum GAST coverage in horizontal plane.

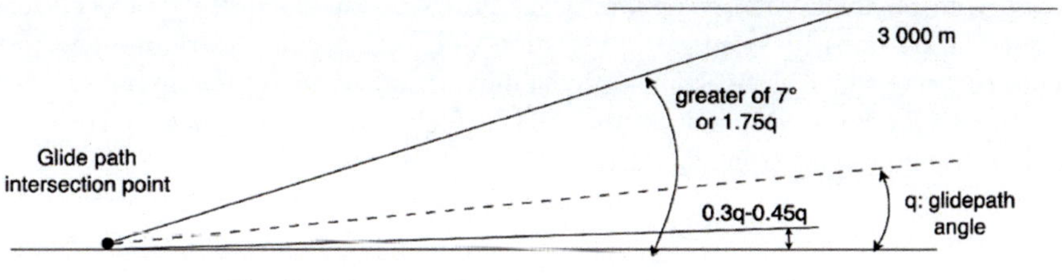

Fig. 6. Minimum GAST coverage in vertical plane.

Many works are devoted to GAST accuracy and quality parameters such as availability and integrity, which is defined by errors [5–10]. Clock errors of pseudoranges appear as GNSS and GBAS clocks are not synchronized. Besides are ephemeris errors, ionospheric and tropospheric errors due to the radio waves propagation in medium, errors caused by multipath, noise and possible interference in ground and airborne antennas. Precision measurements of carrier phase are characterized by corresponding errors.

Common errors for ground equipment and aircraft, such as clock offset, ephemeris, GNSS hardware errors, the ionosphere and troposphere errors are correlated over short distances and thus can be removed. Remain noise and multipath errors, which are not correlated. These errors can be corrected through carrier-smoothing (100 s for GAST-C and 30 s for GAST-D) to little value. The same smoothing algorithm is used in the ground and airborne for compensation of common errors.

The residual GBAS errors include ionosphere, troposphere, noise and multipath components. Main reason of variation ionosphere component is the spatial position of ground and aircraft antennas, causing varying signal delay. The standard deviation of delay variation along vertical direction is represented by σ_{vig} (mm/km) parameter, which is normalized by value sufficient bound the spatial ionosphere decorrelation in the country (region). Values of σ_{vig} change from 2.07 mm/km in Germany to 40 mm/km in Brazil [11]. At this in Germany ionosphere error can have small impact on system availability and essential impact in Brazil region.

Consistency of the corrections is checked by calculating *B*-value, which is compared with the threshold. When *B*-value is less than threshold the corrections can be transmitted. Otherwise, *B*-value permits to detect and exclude the data from the fault receiver. *B*-value is affected by residual multipath and noise, its absolute value increases at law satellite elevation angles. Airborne signal processing is similar to the ground.

Accuracy of GAST-C in lateral and vertical planes are equal to 16 m and 4 m respectively, for GAST-D: 5 m and 2.9 m [2]. Integrity is a measure of trust put in the correctness of the position estimates. It is defined by three components: probability of integrity loss during landing (integrity risk), maximum position errors (risk limits) and times from error occurrence until an appropriate alert or removing the erroneous measurements (time to alert). Integrity risk for GAST-C is 2×10^{-7} and for GAST-D is 1×10^{-9}. Time to alert is 6 s for GAST-C and 2 s for GAST-D. The availability is also important parameter, which is determined by the percentage of time during which the system is ready for use and operational. The values of system integrity and availability are chosen as a result of trade-off.

In order to ensure that integrity and availability are within tolerances, the ground equipment includes monitoring apparatus, which protect the users from any threat, connecting with residual small errors and large errors of low probability of occurrence. Information on integrity is added in radiation slot. As a result of airborne processing the protection levels in vertical and horizontal planes are defined for nominal case without false and for cases with fault conditions [2]. Technical characteristics determine the peculiarities of systems siting and use.

3 Siting Criteria

The minimum permissible distance from the runway threshold to ILC localizer antenna is 330 m and maximum is 600 m. The terrain before antenna to the end of the runway should be flat and without obstructions. The cost of the first-time beacon siting includes the works of installing power lines and control cables, site grading, antenna construction, and the cost of laying roads [11]. Demands to terrain in front of glide-slope antenna is similar. Additional attention is paid to the transverse slope in the first 600 m in front of antenna because this could be the reason for glide path angle changes at the final stage of landing.

Siting criteria for GAST is not so severe as for ILS. The main system's parts in view of siting are four antennas of GNSS receivers, the shelter with equipment, and several antennas of VHF Data Broadcast (VDB) signal. The Local Object Consideration Area is the clear volume near the ground GBAS antennas, which is mainly defined by the probability of multipath reception. Antennas should be placed in the points with a masking effect not greater than 3° [13]. Minimum separation between GNSS antennas is about 100 m. The VDB antenna should be 200 m outside the nearest aircraft taxiway and at distance less than 5.5 km from the runway threshold being served. The last is important for serving all the landing directions of airport with one GAST. Besides GAST includes monitors, such as the ionospheric gradient monitor. Monitors ensure the system integrity.

4 Maintenance and Flight Inspection

The volume and cost of ILS routine maintenance depend on local conditions. For example, Boryspil International Airport is equipped with three Thales ILS CAT I and one CAT III. System maintenance of eight beacons with complex antennas is performed by a group of 14 technicians. Power consumption of one beacon is at the level from 1.5 kW to 2 kW. In the case of implementation GAST-D it will be one or two systems with shelters, receivers, monitors, and digital VDB radio line. Despite the complex processing of the received signals, amount of equipment is less and antennas are simpler. So, it requires less maintenance.

GAST requires lower amount of flight tests in comparison with ILS [14]. Both ILS and GAST have almost the same coverage, which is under inspection. Therefore, the number of flight hours during flight inspection cannot be very different. The annual cost of flight inspection of ILS systems at "Boryspil" airport is about $0.2 million.

5 Runway Throughput

The presence of critical and sensitive zones in the ILS system, where should be no sources of interference (aircraft), limits the capacity of the runway and the airport as a whole. The negative impact increases in conditions of low visibility when due to the low visibility procedures the traffic on the airfield is limited, and spacing tolerances between aircrafts become greater. This is unacceptable for airports with flight intensity, which tends to maximum capacity.

The use GAST-D in low visibility conditions will change air traffic management procedures in order to increase runway throughput. Among them is reducing the final approach spacing airplanes on 1 NM [14].

6 Jamming and Spoofing

There are two common models of ILS spoofing [15], which include overshadow and single-tone attacks. In overshadow model, the ILS fake signals are transmitted with greater power in comparison with legitimate ILS signals. The power difference of attacked and real signals must be about 20.5 dB for the localizer and 27.5 dB for glide-path beacon. So, this model is closed to narrowband jamming and takes into account 15 W output power of localizer CSB channel needs a high level of jamming power.

The single-tone spoofing model is based on radiation only one part of SBO channel signals. That is modulation tones 90 Hz or 150 Hz. This leads to the distortion of the localizer signal spectrum, including phase relationships, and a deviation of the course line from the given position. The radiation power of spoofing signal is less than real signal. Realization of a single-tone model using software-defined radio is rather simple and cheap.

The GNSS receivers operate with a typical level of input signals −125 dBm. This is a very low level in comparison with a minimum level of 5 mV in localizer receiver, which on 50 Ω corresponds to −93 dBm.

7 Cost Evaluation

The cost of ILS equipment including installation depends on whether it is a primary placement or a replacement. For CAT I primary installation costs about $1.6 million and replacement is $0.4 million [16]. For CAT II/III respectively $2.5 million and $0.45 million. The price of GAST-C is equal to $0.5 million and installation costs $0.12 million [17].

ILS annual maintenance cost depends on many factors and can be greater than $200 thousand. Maintenance cost of GAST-C is about four times less. The costs for flight inspection of systems are approximately the same.

The revenue growth associated with increased runway capacity when GAST-D is implemented depends on the airport's planned load and annual weather conditions.

8 Conclusions

The continuous growth of civil aviation in the near future increase airport capacity, especially in low visibility conditions. One of the possible solutions to ensure flight safety could be a wide implementation of GAST-D. This seems to be the main advantage of GAST, which will also provide significant economic benefits in the future.

Contrary to expectations, the equipment costs of two landing systems in the case of ILS equipment replacement, including deployment, are at the same level. System maintenance costs are noticeably lower in GAST.

There is a flight safety risk associated with jamming of GNSS signals. In this case, service to all airport landings is stopped. It should be taken into account that the practice of GNSS jamming has been tested in many cases including military conflicts. In case of justified risks of using jamming, it is advisable to consider the possibility of maintaining at least one ILS system.

References

1. Manual of All-Weather Operations. Doc. 9365. ICAO (2013)
2. Aeronautical Telecommunications. Radio Navigation Aids, vol. 1, an. 10 to the Convention of International Civil Aviation. ICAO (2018)
3. Minimum Operational Performance Standards for GPS Local Area Augmentation Airborne Equipment. DO-235D. Radio Technical Commission for Aeronautics (2017)
4. GBAS Program Update, FAA (2016)
5. Circiu, M.S.: Integrity aspects for dual-frequency dual-constellation ground based augmentation system (GBAS). Doctorate dissertation, RWTH Aachen University (2020). https://doi.org/10.18154/RWTH-2020-07268
6. Felux, M., Lee, J., Holzapfel, F.: GBAS ground monitoring requirements from an airworthiness perspective. GPS Solut. **19**, 393–401 (2015). https://doi.org/10.1007/s10291-014-0398-0
7. Saito, S., Yoshihara, T.: Evaluation of extreme ionospheric total electron content gradient associated with plasma bubbles for GNSS Ground-Based Augmentation System. Radio Sci. **52**(8), 951–962 (2017). https://doi.org/10.1002/2017RS006291
8. Weverton, C., Monico, F.: GBAS: fundamentals and availability analysis according to sigma. J. Geodetic Sci. **12**(1), 22–37 (2022). https://doi.org/10.1515/jogs-2022-0132
9. Machuta, J., Kraus, J.: SBAS avionics compared to GBAS on-board equipment. Mag. Aviat. Dev. **6**(1), 11–16 (2017). https://doi.org/10.14311/MAD.2018.01.02
10. Felux, M.: Total system performance of GBAS-based automatic landings. Doctorate dissertation. Technical University of Munich (2018)
11. Siting Criteria for Instrument Landing Systems. FAA Order 6750.16 (2014)
12. Ground Architecture and Airport Installation. SESAR Project № 15.03.06, D04 GBAS Cat II/III L1 Approach. SESAR Joint Undertaking (2015)
13. Siting Criteria for Ground Based Augmentation System (GBAS). FAA Order 6884.1 (2010)
14. GBAS CAT II-III Functional Description Update Report - Update for V3. Project Number 06.08.05, SESAR, Edition 00.01.01 (2015)
15. Sathaye, H., Schepers, D., Ranganathan, A., Guevara, N.: Wireless attacks on aircraft instrument landing system. In: 28th USENIX Security Symposium, Santa Clara, USA, pp. 357–367 (2019). https://doi.org/10.1145/3317549.3326298
16. Xiasong, X.: ILS replacement. ACI World Safety Seminar. Airport Council International, Beijing (2008). https://docplayer.net/21826474-Ils-replacement-aci-world-safety-seminar-18-19-november-2008-kempinski-hotel-beijing-lufthansa-centre.html
17. Oliveira, P., Silva, J., Soares, P.A.: Comparative study between ILA and GBAS approaches: the case of Viseu airfield. J. Airline Airport Manag. **10**(2), 66–75 (2020). https://doi.org/10.3926/jairm.168

Alert System Notifying the Crew of the Necessity for a Go-Around

Yurii Hryshchenko, Oleksii Chuzha(✉), Oleksii Romanenko, and Oleksandra Bankova

National Aviation University, Liubomyra Huzara Avenue, 1, Kyiv 03058, Ukraine
achuzha@ukr.net

Abstract. This paper examines the peculiarities of aircraft piloting by the crew during the landing phase. Specifically, it focuses on investigating the issues related to the control mode of piloting an aircraft during the landing phase. The feasibility of implementing an onboard system to alert pilots about the advisability of initiating a go-around is substantiated. The paper proposes a functional scheme for such a system and develops a methodology for its operation. The uniqueness of the system lies in its utilization of only standard onboard equipment, serving as the source of primary input information. The flight data processing is carried out using the on-board avionics computer system. Analytical solutions in the system are formulated using a geometric model of probabilistic boundaries for the aircraft's entry into the glide path. The model is represented as a three-axial ellipsoid, where the entry point into the glide path simultaneously serves as the center of the spatial coordinate system and the ellipsoid. The current aircraft position is determined in the coordinate system. Landing is considered hazardous if the aircraft deviates beyond the spatial boundaries of the ellipsoid. This implies that the aircraft's deviation from the specified boundaries may lead to an unsuccessful landing. The primary objective of the developed system is to enhance the safety of flights during the landing phase, alleviate the psycho-physiological stress on pilots, and provide operational assistance to the crew in making the correct decision regarding whether to continue the landing or initiate a go-around.

Keywords: Safety · Landing · Glide Path · Entry Point into the Glide Path

1 Introduction

One of the key objectives of civil aviation is to ensure the safety of aircraft flights. This is achieved through the high reliability of aviation technology, effective operation of aircraft maintenance systems and air traffic control systems, the high professional training of pilots and controllers, specialized onboard and ground technical equipment, and other contributing factors.

The most critical phases of flight are considered to be the landing and takeoff of an aircraft. In particular, special attention is given to the landing phase. This is due to the fact that during the landing approach, the aircraft and its crew are influenced by various factors that increase the risks associated with flight execution. For the aircraft, these factors

I. Ostroumov and M. Zaliskyi (Eds.): ACASD 2024, LNNS 992, pp. 190–202, 2024.
https://doi.org/10.1007/978-3-031-60196-5_16

include low speed and lift, an increased angle of attack, an elevated coefficient of frontal resistance, and a limited margin of flight altitude. For the crew, these factors encompass continuous visual monitoring of external space and instrument equipment, information exchange with ground dispatch services, control of wing mechanization, management of angular and trajectory flight parameters, and time constraints for decision-making and actions.

The complexity for the crew in executing flight during landing phase lies in the necessity to perform all specified actions within a very short timeframe, considering the rapidity of processes. This demands a high level of attention and concentration, coordinated actions, and professional training from the pilots, involving the application of specialized training methodologies.

2 Literature Review and Problem Statement

Considerable attention has been devoted to the investigation of issues related to the landing phase in specialized scientific literature. Particularly, the impact of human factors on flight safety during the landing approach is extensively discussed [1]. This concern is also evident in various circulars recommended for implementation by the International Civil Aviation Organization (ICAO) [2, 3].

The study in [4] focuses on a method for assessing the quality of aircraft piloting during the landing approach. The proposed approach involves evaluating the piloting technique processes through the spectra of non-normalized autocorrelation functions, specifically examining autocorrelation functions for roll and pitch. The authors argue that analyzing the spectra of these functions allows for the detection of piloting process characteristics and drawing general conclusions about its quality.

In [5], a numerical method is developed to assess the quality of the aircraft's entry into the glide path. Research indicates that imprecise entry into the glide path can significantly degrade the quality of the aircraft's landing.

Statistical research on flight parameters under various conditions with consistent pilot training was conducted in [6]. The distribution law of flight parameters under normal conditions followed a normal distribution, while under significantly degraded flight conditions, it conformed to a Weibull distribution.

An analysis of factors provoking pilots to make abrupt control inputs was carried out in [7]. The negative consequences of such actions on flight safety were demonstrated, and a warning system to prevent sudden control movements was developed. The study [7] investigated the impact of faults in indicated airspeed and angle of attack indicators on flight safety. A warning system for pilots regarding faults and an algorithm for actions in such situations were developed.

While many authors emphasize the importance of researching processes occurring on the glide path during the landing approach, some overlook the events immediately preceding the landing. However, works [8–10] highlight the significance of this flight phase and emphasize that the quality of piloting during this phase significantly influences the subsequent success of the aircraft landing.

3 Problem Definition, Research Challenges, and Dataset Description

3.1 Problem Definition

Ensuring the safety of aircraft operation is a prioritized task for all sectors of civil aviation [11–17]. Particular attention is given to the safety of flights during the landing phase. This is attributed to the fact that controlling an aircraft during the landing approach is the most intricate and responsible aspect of flight operations. According to statistical data, more than one-third of aviation incidents occur during the landing phase.

The primary avenues for addressing this issue involve enhancing the quality of automation and refining landing systems, alleviating the high workload of the crew, and improving other factors associated with the human factor.

3.2 Research Challenges

For pilots, the complexity of executing the landing approach is rooted in the following issue. If we consider the pilot as an autonomous control system, it operates as a single-channel system. This means that the pilot, like any operator, can track only the flow of information changes for one parameter at a time. Simultaneously perceiving, assessing, and analyzing information from two or more sources is impossible for the pilot. It is precisely for this reason that pilots experience psychophysiological stress. This stress is significantly heightened when flight conditions become more challenging, a phenomenon known as cognitive overload. The intensified psychophysiological stress on pilots often leads to errors in the control process and an inadequate perception of information, potentially resulting in aviation incidents and accidents.

Thus, one of the tasks for a safe aircraft landing is to alleviate the workload on pilots through the use of technical means and analytical systems.

Therefore, the aim of our research is to assess the feasibility of forming and implementing a system to alert pilots about the necessity of initiating a go-around, utilizing standard onboard equipment based on a geometric model of probabilistic boundaries for the aircraft's entry into the glide path.

3.3 Dataset Description

In our previous research, a method for assessing the accuracy of an aircraft's glide path entry was developed based on autocorrelation functions of flight parameters, considering various levels of flight complexity. Using this method, probabilistic boundaries for the aircraft's entry into the glide path were determined. Geometrically, these boundaries are described by a second-order surface with three-axial ellipsoid [18].

The functional implementation of the system to alert pilots about the necessity of initiating a go-around is built upon the previously developed geometric probabilistic model in the form of a three-axial ellipsoid. This system generates warnings for pilots about the potential for an unsuccessful aircraft landing. The proposed system utilizes information generated by standard onboard equipment as its input data.

For the Instrument Landing System (ILS), the glide path entry point is situated at an altitude of 400 m and a distance of 7.4 km from the runway threshold.

4 Analysis and Description of System

4.1 System Analysis

Initially, we will assess the feasibility of implementing proposed system using standard onboard equipment of aircraft. The key element in the system is the entry point into the glide path. The geodetic coordinates of this point are known for each airport in FMS database. Therefore, each arriving aircraft has access for coordinates of the entry point into the glide path. Another crucial component of the system is the ability to assess the current position of the aircraft concerning the entry point into the glide path. In this regard, two scenarios need to be considered: the aircraft location before entering the glide path and after entering it.

In the first scenario, determining the aircraft location is accomplished using FMS and satellite navigation system. In the second scenario, aircraft position is determined using ground and onboard instrumental landing system (ILS). Additionally, for informational support, altitude of flight is a crucial input parameter, determined by altimeters. The mathematical implementation of the proposed system operation can be achieved with FMS.

4.2 Description of System

Let's consider main components of proposed system (Fig. 1). FMS computer processes all input data and generates a control signal, which is then sent to the Electronic Flight Instrument System (EFIS). This system issues warnings to the pilots about the possibility of an unsuccessful landing. The advisability of initiating a go-around is displayed on the multi-functional display, the master caution indicator, and the voice warning system.

The multifunctional display serves as the primary source of information for the pilots. The voice warning system generates an aural message about the advisability of initiating a go-around.

During the flight, the aircraft maintains communication with two radio beacons along the route. Relative to these beacons, the onboard equipment measures the azimuth and distance to beacons [19, 20]. By these data a precise location of the aircraft on the route could be calculated. The process of determining the coordinates of the location is carried out by FMS computer [21]. The pressure altitude from air data system is used as primary altimeter for measuring current altitude of aircraft before entering the glide path.

Determining aircraft position after entering the glide path is done using ILS and the onboard equipment of the aircraft. ILS includes ground equipment of course, glide path, and marker radio beacons. The glide path is formed by the intersection of service areas of course and glide path radio beacons. ILS onboard equipment receives radio signals from ground radio beacons. Using the course and glide path signals onboard equipment generates deviations from the glide path line, ε_c and ε_g. The distance to runway L is determined using DME. Based on the values of L, ε_c, ε_g, the location of the aircraft on the glide path is determined. The radio altimeter serves as the primary altimeter, measuring the current altitude of the aircraft on the glide path.

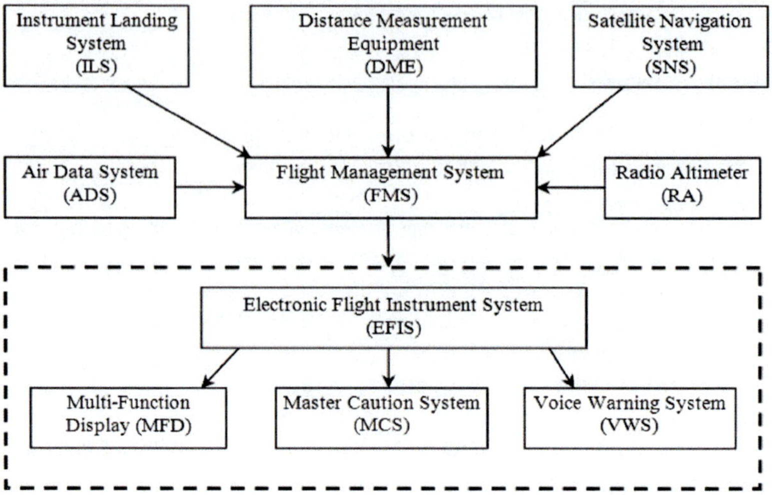

Fig. 1. System for Notifying the Crew of the Necessity to Go-Around.

5 Methodology

Let's consider the methodology that forms the basis of the functional operation of the proposed system. Flight is performed along a predetermined trajectory. However, due to external factors, the actual flight trajectory may deviate from the planned one. The relationship between the deviation of the aircraft flight trajectory from the intended course is characterized by the correlation function between the planned and realized flight trajectories. Over a certain segment of the aircraft route, with a length L, the square of the integral difference Δ between the planned and actual flight trajectories is determined by the formula:

$$\Delta = \int_{x_1}^{x_2} \left[Z_p(x) - Z_a(x)\right]^2 dx, \tag{1}$$

where $Z_p(x)$ is the planned flight trajectory, $Z_a(x)$ is the realized flight trajectory; x_1, x_2 are the initial and final points for tracking the flight trajectories in the horizontal plane.

Utilizing the formula for the square of difference between the planned and actual flight trajectories in an expanded form, we express the relationship of the function Δ to its length L at a certain deviation χ. Let's examine this relationship in various spatial planes. The following autocorrelation dependencies are obtained:

$$\frac{\Delta_i}{L_i} = \left(\frac{1}{3} - \frac{3\chi_i}{L_i} + \frac{2\chi_i^2}{L_i^2} - \frac{\chi_i^3}{3L_i^3}\right) i^2, \tag{2}$$

where $i = [1, 3]$ is x, y, z axes of spatial coordinate system.

The plots of autocorrelation functions (2) is presented in Fig. 2 for $\mathrm{L} = 1200$ m; $\chi = 600$ m; $\mathrm{y} = -300 : 300$ m; $\mathrm{z} = -5 : 5$ m lower hemisphere and Fig. 3 for upper hemisphere.

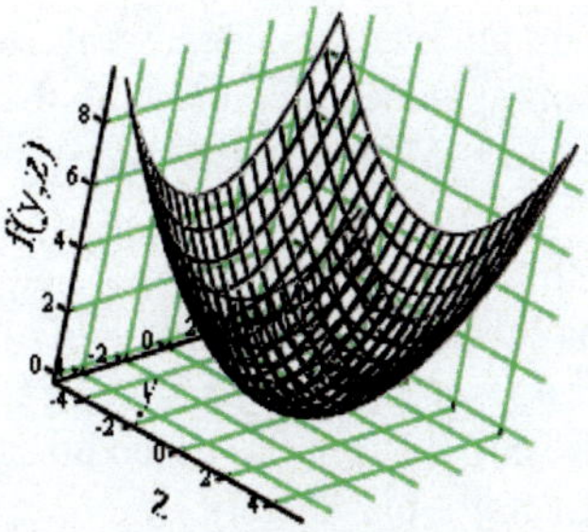

Fig. 2. Autocorrelation function (2) for lower hemisphere.

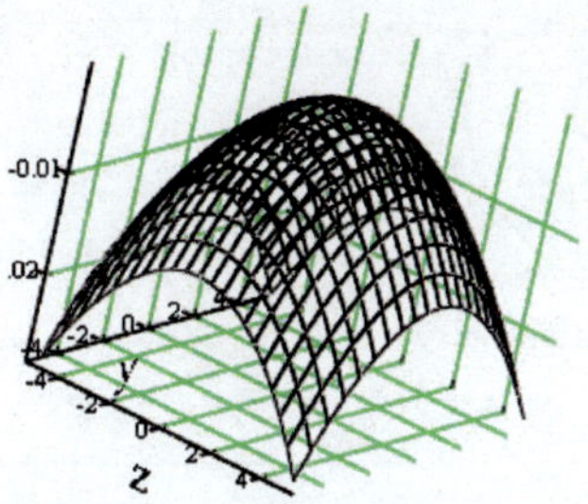

Fig. 3. Autocorrelation function (2) for upper hemisphere.

Taking into account (2) for case of the aircraft deviation from the glide path entry point:

$$\frac{x^2}{a^2} + \frac{y^2}{b^2} + \frac{z^2}{c^2} \leq 1, \tag{3}$$

$$a = \left(\frac{\Delta}{L} \cdot \frac{1}{\frac{1}{3} - \frac{3\chi_1}{L_1} + \frac{2\chi_1^2}{L_1^2} - \frac{\chi_1^3}{3L_1^3}}\right)^{\frac{1}{2}}, \tag{4}$$

$$b = \left(\frac{\Delta}{L} \cdot \frac{1}{\frac{1}{3} - \frac{3\chi_2}{L_2} + \frac{2\chi_2^2}{L_2^2} - \frac{\chi_2^3}{3L_2^3}}\right)^{\frac{1}{2}}, \tag{5}$$

$$c = \left(\frac{\Delta}{L} \cdot \frac{1}{\frac{1}{3} - \frac{3\chi_3}{L_3} + \frac{2\chi_3^2}{L_3^2} - \frac{\chi_3^3}{3L_3^3}}\right)^{\frac{1}{2}}, \tag{6}$$

where a, b, c are semi-axes of the ellipsoid.

Consider the aircraft flight along the path line (Fig. 4), heading towards the glide path entry point. At the glide path entry point, the aircraft changes its horizontal position to the landing one with a certain pitch angle. The aircraft lands along the glide path line, and its projection on the horizontal plane coincides with the axial line of the runway. The center of the three-axial ellipsoid is located at the glide path entry point. The semi-axes a, b, c have boundary values: $a = b = 300$ m, $c = 46$ m.

If the aircraft, both during the approach to the glide path entry point and after passing through it, enters the inner spatial region of the ellipsoid, the probability of a successful landing is high. If the aircraft does not enter this zone, it indicates a high probability of an unsuccessful landing. These conclusions are based on statistical data and the analysis of autocorrelation functions of flight parameters.

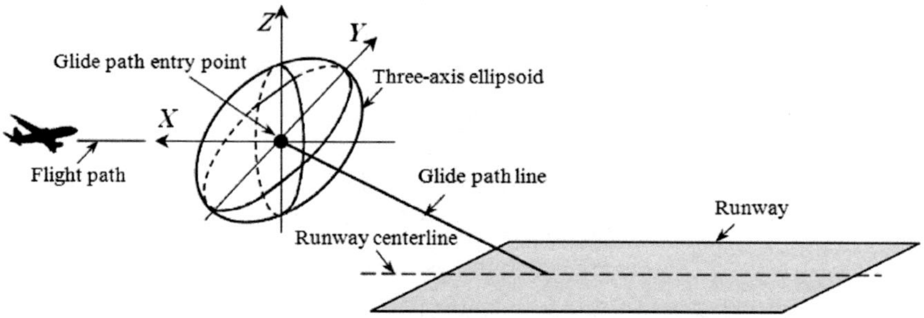

Fig. 4. Aircraft entry into the three-axis ellipsoid space.

Aircraft current location will be determined in a coordinate system with its origin coinciding with the glide path entry point. The aircraft location and the position of the glide path entry point are determined using geographical coordinates longitude (λ) and latitude (φ). Thus, we need to convert WGS 84 geographical coordinates to cartesian coordinates (x, y, z):

$$x = (N + h)cos\varphi cos\lambda, \tag{7}$$

$$y = (N + h)cos\varphi sin\lambda, \tag{8}$$

$$z = \left(\left(\frac{b^2}{a^2}\right)N + h\right)sin\varphi, \tag{9}$$

where h is the height above the Earth's surface; N is the radius of curvature of the first vertical; b is the minor axis of the ellipsoid model of the Earth; a is the major axis of the ellipsoid model of the Earth.

The radius of curvature of the first vertical is calculated by the formula:

$$N = \frac{\mathrm{a}}{\sqrt[2]{1 - (\mathrm{e}\ \sin\varphi)^2}}, \tag{10}$$

where e is the eccentricity of WGS 84 ellipsoid model.

The coordinate system located at the entry point into the glide path has its origin at point O_1. We will assume that the spatial orientation of the axes in the two systems coincides. Take an arbitrary point $A(x, y, z)$ in space, corresponding to aircraft location. Then, to transition from the x, y, z coordinate system to the $x_1y_1z_1$ system, the following formulas are required:

$$x_1 = x - R_z - h, \tag{11}$$

$$y_1 = y - R_z - h, \tag{12}$$

$$z_1 = z - R_z - h, \tag{13}$$

where R_z is the Earth radius ($R_z = N$).

During the execution of the steering flight, the pilot activates the landing approach mode as soon as the aircraft reaches the entry point into the glide path. At this point, the process of determining the coordinates of the aircraft location relative to the navigation beacons is terminated. After entering the glide path point, the spatial orientation of the aircraft is determined exclusively with respect to the glide path, formed by the landing beacons. The aircraft location is determined based on the parameters of the angular deviation from the glide path, ε_C and ε_g, and the distance to the runway, L. Signals ε_C and ε_g are fed to the command arrows of the landing indicator. Pilots use them to navigate how to control the aircraft to maintain it on the glide path.

Thus, the task is to use the data from the landing instruments to determine the spatial coordinates of the aircraft along the glide path. For this purpose, we introduce a runway coordinate system $X_2Y_2Z_2$(Fig. 5).

Let's position runway coordinate system so that the x-axis coincides with the runway's axial line, and the origin is at the end of the descent glide path. The geodetic coordinates of this point are known. To calculate the coordinates of the spatial position of the aircraft $(x_2, y_2, z_2)(x_2, y_2, z_2)$, we will use the spherical coordinate system. It is the closest approximation to the polar coordinate system currently used in aircraft landing systems.

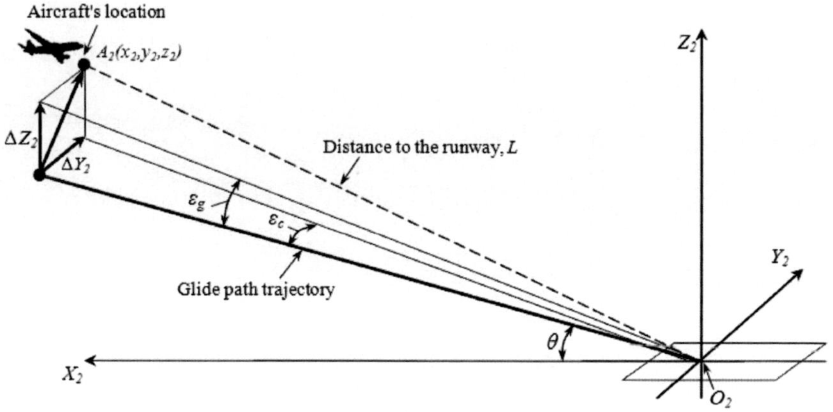

Fig. 5. Runway coordinate system and aircraft position relative to the glide path.

Consider an arbitrary point, which defines aircraft location with a deviation from the glide path (Fig. 5). The values of the horizontal projection of the deviation ΔY_2 and the vertical projection of the deviation ΔZ_2 are determined by the formulas:

$$\Delta \mathrm{Y}_2 = Lsin\,\varepsilon_c, \tag{14}$$

$$\Delta\,\mathrm{Z}_2 = Lsin\,\varepsilon_g. \tag{15}$$

The spatial coordinates of the aircraft's location along the glide path are determined by the following formulas:

$$x_2 = Lcos\,\varepsilon_c\,sin\big(\theta + \varepsilon_g\big), \tag{16}$$

$$\mathrm{y}_2 = Lsin\,\varepsilon_c\,sin\big(\theta + \varepsilon_g\big), \tag{17}$$

$$\mathrm{z}_2 = Lcos\big(\theta + \varepsilon_g\big), \tag{18}$$

where θ is angle of the glide path ($\theta = 3°$).

As coordinate system $X_2Y_2Z_2$ is oriented to the axial line of the runway of a specific airport, it will have spatial angular displacement concerning the XYZXYZ coordinate system. To transition from the $X_2Y_2Z_2$ coordinate system to $X_1Y_1Z_1$ (or XYZXYZ), Euler angles transformation is necessary.

Let's determine the coordinates of point O_2, which is located on the axial line of the runway and serves as the center of the $X_2Y_2Z_2$ coordinate system (Fig. 6):

$$x_{O2} = N_2 cos\varphi_2 cos\lambda_2, \tag{19}$$

$$y_{O2} = N_2 cos\varphi_2 sin\lambda_2, \tag{20}$$

$$z_{O2} = 0, \tag{21}$$

where N_2 is the radius of the first vertical curve for point O_2; and the geographical coordinates are the longitude λ_2 and latitude φ_2 of point O_2.

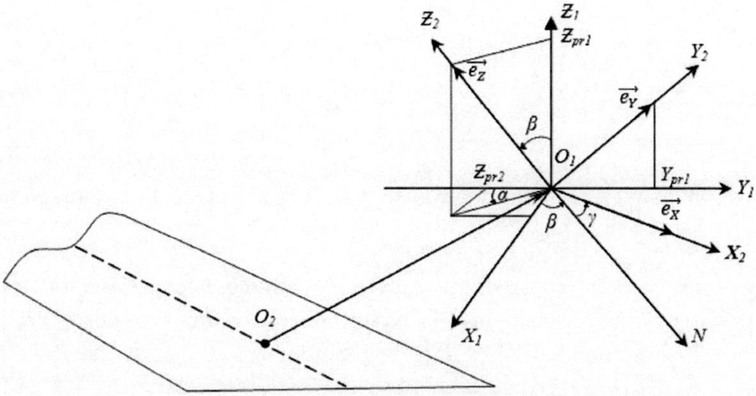

Fig. 6. Transition from the $X_2Y_2Z_2$ coordinate system to the $X_1Y_1Z_1$ system.

Figure 6 presents transition from the $X_2Y_2Z_2$ coordinate system to the $X_1Y_1Z_1$ system using Euler angles:

$$\overrightarrow{O_2O_1} = (x_{O1} - x_{O2}; y_{O1} - y_{O2}; z_{O1} - z_{O2}), \tag{22}$$

where O_1 is the point of entry into the glide path and the center of the coordinate system $X_1Y_1Z_1$.

On the coordinate axes of the $X_2Y_2Z_2$ coordinate system, denote the unit vectors as $\overrightarrow{e_X}, \overrightarrow{e_Y}, \overrightarrow{e_Z}$. Move the coordinate system $X_2Y_2Z_2$ parallel to the vector $\overrightarrow{O_2O_1}$ to the point O_1. During the displacement of the $X_2Y_2Z_2$ coordinate system, its spatial orientation is preserved. Also, during the alignment of the two systems, a new axis N was formed. It is called the nodal line. The nodal line N is the intersection of the plane $X_2Y_2Z_2$ and $X_1Y_1Z_1$. Angles were formed between the axes of the coordinate systems: α is angle between the axis O_1X_1 and O_1N_1; β is angle between the axis O_1Z_1 and O_1Z_2; γ is angle between the axis O_1Z_1 and O_1Z_2; α, β, γ are Euler angles. When projecting the unit vector $\overrightarrow{e_Z}$ onto the O_1Z_1 axis, its projection Z_{pr1} is formed:

$$Z_{pr1} = cos\beta. \tag{23}$$

Projection of the unit vector $\overrightarrow{e_Z}$ onto the O_1Y_1 is Z_{pr2}:

$$Z_{pr2} = cos\alpha sin\beta. \tag{24}$$

When projecting the unit vector $\overrightarrow{e_Y}$ onto the O_1Y_1 axis, the projection Y_{pr1} is determined:

$$Y_{pr1} = cos\gamma sin\beta. \tag{25}$$

Then the Euler angles are determined by the formulas:

$$\alpha = arccos\frac{Z_{pr2}}{(1 - (Z_{pr1})^2)^{1/2}}, \tag{26}$$

$$\beta = \arccos(Z_{pr1}), \tag{27}$$

$$\gamma = arccos\frac{Y_{pr1}}{(1 - (Z_{pr1})^2)^{1/2}}. \tag{28}$$

Transition matrix A from the coordinate system $X_2Y_2Z_2$ to $X_1Y_1Z_1$ through the Euler angles:

$$A = \begin{bmatrix} \cos\alpha\cos\gamma - \sin\alpha\cos\beta\sin\gamma & -\cos\alpha\sin\gamma - \sin\alpha\cos\beta\cos\gamma & \sin\alpha\sin\beta \\ \sin\alpha\cos\gamma + \cos\alpha\cos\beta\sin\gamma & -\sin\alpha\sin\gamma + \cos\alpha\cos\beta\cos\gamma & -\cos\alpha\sin\beta \\ \sin\beta\sin\gamma & \sin\beta\cos\gamma & \cos\beta \end{bmatrix}. \tag{29}$$

Thus, the coordinates of any point in the $X_1Y_1Z_1$ system are calculated by the formula:

$$(X_1Y_1Z_1) = \mathrm{A}(X_2Y_2Z_2). \tag{30}$$

6 Conclusions, Results and Discussions

Thus, the analysis conducted substantiates the feasibility of implementing a system to warn pilots about the advisability of initiating a go-around using the aircraft onboard equipment. The onboard equipment is utilized to obtain flight information as input data for system, and FMS could be used for computation. The functioning of the system is based on a geometric model of probabilistic boundaries for aircraft entry into the glideslope point. The geometric model represents a three-axial ellipsoid, with its center serving as the origin of a rectangular spatial coordinate system coinciding with the entry point into the glideslope. The current position of the aircraft is referenced relative to this system. The criterion for danger during the landing approach is defined by the aircraft deviating beyond the spatial boundaries of the ellipsoid. Any deviation of the aircraft beyond these specified limits signifies a high probability of an unsuccessful landing.

Advantages of the developed system:

- Enhances the safety of aircraft flights during the most critical phase.
- Assists the crew in making decisions regarding whether to continue the landing or initiate a go-around.
- Partially alleviates the psycho-physiological stress on the crew.
- Allows for the evaluation of landing quality before entering the glideslope point.
- Absence of additional hardware in the system that could alter the mass indicators of onboard equipment. This aspect is crucial for aircraft.
- System universality. It can be adapted for any type of aircraft.

The next step in research will involve developing a mathematical model of the system, considering all the peculiarities of aircraft piloting during pre-landing maneuvers. Proposed system evaluates spatial boundaries only around the entry point into the glideslope. This capability enables it to assess landing quality even before entering the glideslope and on the initial segment of the glideslope. Unfortunately, this ability is absent on other segments of the glideslope. Therefore, there are plans to expand the analytical space down along the glideslope to decision-making altitude. This expansion will be supported by a new probabilistic geometric model.

The current system is designed only for manual piloting mode. Therefore, a prospective direction for the development of this system is its transformation for automatic landing mode. This development will consider the functioning features of the automatic control system during the landing approach and external factors influencing normal flight.

References

1. Reva, O., Borsuk, S., Kharchenko, V.: Multiplication of air accidents frequency and hazard desirability coefficients for ICAO safety risk tolerability matrix solution. Logist. Transp. **25**, 63–69 (2015)
2. Annual Safety Report Group. ICAO. Fifth Meeting (ASRG/5) (2023)
3. Human factors digest no. 16 cross-cultural factors in aviation safety. ICAO Circular 302 (2004)
4. Hryshchenko, V., Romanenko, V., Hryshchenko, Y.: Quality of piloting during the approach. In: Ostroumov, I., Zaliskyi, M. (eds.) Proceedings of the International Workshop on Advances in Civil Aviation Systems Development. ACASD 2023. LNNS, vol. 736. pp. 260–270. Springer, Cham (2023). https://doi.org/10.1007/978-3-031-38082-2_20
5. Hryshchenko, Y. Romanenko, V., Chuzha, O., Nych, E.: Estimation of the quality of glide path entrance by the spectra of the roll angle autocorrelation functions. In: 6th International Conference APUAVD, pp. 150–153. IEEE, Kyiv, Ukraine (2020). https://doi.org/10.1109/MSNMC50359.2020.9255566
6. Hryshchenko, Y., Zaliskyi, M., Pavlova, S., Solomentsev, O., Fursenko, T.: Data processing in the pilots' training process on the integrated aircraft simulator. Electr. Control. Commun. Eng. **17**(1), 67–76 (2021)
7. Hryshchenko, Y.V., Pavlova, S.V., Pipa, D.M., Kravets, I.V.: Assesment of the flight quality in vertical plane. Electron. Control Syst. **2**(64), 87–92 (2020)
8. Solomentsev, O., Zaliskyi, M., Kozhokhina, O., Herasymenko, T.: Efficiency of data processing for UAV operation system. In: 4th International Conference on Actual Problems of Unmanned Aerial Vehicles Developments, pp. 27–31. IEEE, Kyiv, Ukraine (2017). https://doi.org/10.1109/APUAVD.2017.8308769
9. Okoro, O.C., Zaliskyi, M., Dmytriiev, S., Solomentsev, O., Sribna, O.: Optimization of maintenance task interval of aircraft systems. Int. J. Comput. Netw. Inf. Secur. **14**(2), 77–89 (2022). https://doi.org/10.5815/ijcnis.2022.02.07
10. Solomentsev, O., Zaliskyi, M.: Correlated failures analysis in navigation system. In: 5th International Conference on Methods and Systems of Navigation and Motion Control, pp. 41–44. IEEE, Kyiv, Ukraine (2018). https://doi.org/10.1109/MSNMC.2018.8576306
11. Cui, A., Zhao, H., Zhang, X., Zhao, B., Li, Z.: Power system real time data encryption system based on DES algorithm. In: 13th International Conference on Measuring Technology and Mechatronics Automation (ICMTMA), pp. 220–228. IEEE, Beihai, China (2021). https://doi.org/10.1109/ICMTMA52658.2021.00056
12. Pakkiraiah, Ch., Satyanarayana, R.: Design and FPGA realization of an energy efficient artificial neural modular exponentiation architecture. In: Panda, S.K., Rout, R.R., Sadam, R.C., Rayanoothala, B.V.S., Li, K.C., Buyya, R. (eds.) Computing, Communication and Learning. CoCoLe 2022. CCIS, vol. 1729, pp. 115–126. Springer, Cham (2023). https://doi.org/10.1007/978-3-031-21750-0_10

13. Rozenberg, R.: Human factors and analysis of aviation education content of military pilots. In: New Trends in Aviation Development, pp. 139–144. IEEE, Chlumec nad Cidlinou, Czech Republic (2019). https://doi.org/10.1109/NTAD.2019.8875561
14. Averyanova, Y., et al: Turbulence detection and classification algorithm using data from AWR. In: 2nd Ukrainian Microwave Week, pp. 518–522. IEEE, Kyiv, Ukraine (2022). https://doi.org/10.1109/UkrMW58013.2022.10037172
15. Sushchenko, O.A., Golitsyn, V.O.: Data processing system for altitude navigation sensor. In: 4th International Conference on Methods and Systems of Navigation and Motion Control (MSNMC 2016), pp. 84–87. IEEE, Kyiv, Ukraine (2013)
16. Sushchenko, O., Chikovani, V., Tsiruk, H.: Redundant information processing techniques comparison for differential vibratory gyroscope. Eastern-Eur. J. Enterp. Technol. **4**(82), 45-52. (2016)
17. Sushchenko, O.A., Bezkorovainyi Y.M., Golitsyn, V.O.: Fault-tolerant inertial measuring instrument with neural network. In: 40th International Conference on Electronics and Nanotechnology (ELNANO), pp. 797–801. IEEE, Kyiv, Ukraine (2020)
18. Hryshchenko, Y., Romanenko, V., Zaliskyi, M.: Quality assessment of aircraft glide path entrance. In: CEUR Workshop Proceedings, vol. 2711, pp. 649–660 (2020)
19. Ostroumov, I.V., Kuzmenko, N.S., Marais, K.: Optimal pair of navigational aids selection. In: 5th International Conference on Methods and Systems of Navigation and Motion Control (MSNMC), pp. 32–35. IEEE, Kiev, Ukraine (2018). https://doi.org/10.1109/MSNMC.2018.8576293
20. Ostroumov, I.V., Kuzmenko, N.S.: Accuracy assessment of aircraft positioning by multiple Radio Navigational aids. Telecommun. Radio Eng. **77**(8), 705–715 (2018). https://doi.org/10.1615/TelecomRadEng.v77.i8.40
21. Ostroumov, I.V., Kuzmenko, N.S.: Compatibility analysis of multi signal processing in APNT with current navigation infrastructure. Telecommun. Radio Eng. **77**(3), 211–223 (2018). https://doi.org/10.1615/TelecomRadEng.v77.i3.30

Artificial Intelligence in Civil Aviation Systems

Integration of Deterministic, Stochastic, and Non-stochastic Models to Obtain an Optimal Collaborative Decision in the Flight Emergency

Tetiana Shmelova[1], Yuliya Sikirda[2](✉), Maxim Yatsko[1], and Vadym Stratonov[2]

[1] National Aviation University, Liubomyra Huzara Avenue 1, Kyiv 03058, Ukraine
[2] Flight Academy of the National Aviation University, Stepana Chobanu Street 1, Kropyvnytskyi 25005, Ukraine
sikirdayuliya@ukr.net

Abstract. From 1953 to 2009, the researchers recorded 129 incidents of aircraft hitting volcanic ash clouds. In nine cases, the damage was accompanied by the failure of the engine(s) during the flight. In such flight emergencies (FE), operational interaction between the pilot, air traffic controller, and search and rescue specialist is highly significant. Nevertheless, the interprofessional interaction of aviation specialists is currently not sufficiently developed for joint development and adoption of optimal decisions. Therefore, the development of effective Collaborative Decision Making (CDM) models by aviation specialists in FE is an actual problem. Multi-stage CDM modeling by the aviation specialists in FE with the gradual involvement of additional operators in decision-making is proposed. The decision tree, network graphics, and decision-making matrices for the aviation specialists are presented. Integration of CDM models allows moving from a complex to a simple CDM process by establishing unambiguity and synchronization of joint actions of operators at key points of multialternative interaction and minimizing potential risk/loss. A training example of CDM in FE "Loss of thrust on both engines due to falling into a cloud of volcanic ash with subsequent ditching" is presented. The optimal CDM based on the objective and subjective factors to follow of aviation specialist's decisions by Wald, Laplace, and Hurwitz criteria is determined. After integrating stochastic and non-stochastic models, simplified deterministic CDM models in FE should be used in artificial or hybrid intelligence systems to obtain optimal collaborative solutions.

Keywords: Certainty · Decision-Making Matrix · Decision Tree · Network Graphic · Risk · Structure-Time Table · Uncertainty

1 Introduction

1.1 Introduction to the Problem

As is well known, civil aviation in terms of its scale and other attributes acquired its modern form in the period after the end of the World War II. Around the same time, scientists began to study cases of the impact of volcanic eruption products on aircraft

I. Ostroumov and M. Zaliskyi (Eds.): ACASD 2024, LNNS 992, pp. 205–223, 2024.
https://doi.org/10.1007/978-3-031-60196-5_17

safety. Paper [1] analyzes the cases of aircraft collisions with volcanic ash that occurred in the period from 1953 to 2009. During the reporting period, the researchers recorded 129 incidents of aircraft hitting volcanic ash clouds. In 94 of these cases, traces of ash were found on the aircraft after the flight, and 79 of them sustained damage of varying severity (airframe or engine). In 26 cases, the degree of damage was determined as "significant or very serious", of which nine were accompanied by the failure (shutdown) of the engine(s) during the flight, which was nevertheless subsequently restarted.

In general, the authors divided all analyzed cases into five classes, according to the degree of damage to the aircraft. Although engine failures (class four) were recorded, the crews managed to resume their work after emerging from the ash cloud. There were no cases corresponding to class five (accompanied by engine failure and other damage that led to an accident/air crash).

The vast majority of cases with severe consequences occurred within 24 h after the start of the eruption and at a distance of less than 1000 km from its epicenter. The authors also identified zones of volcanic activity, each of which resulted in at least five aircraft incidents. In addition to the already mentioned cases, these are the areas of volcanoes – Pacaya (Guatemala), Pinatubo (Philippines), Sakura-jima (Japan), and Soufriere Hills (Lesser Antilles, United Kingdom). The latest incident that led to almost a week of flight restrictions in most of Europe and parts of the North Atlantic was the eruption of the Icelandic volcano Eyjafjallajökull.

It notes that the greatest risk of aircraft collision with ash clouds occurs in the early stages of an eruption, as the aggregate data on the situation at that time is limited and the ways of warning (communication) have not yet been established. An effective risk reduction strategy combines real-time monitoring of volcanoes, tracking ash clouds using satellite technology, modeling methods (predicting the likely direction of movement), and a system for promptly sending warning messages.

Among the promising areas that will help raise awareness of volcanic eruptions were also mentioned [1]:

- sampling from ash clouds for their analysis;
- placement of sensors (atmospheric condition analyzers) on board aircraft.

1.2 Motivation

Generally known that it is prohibited to operate civil aviation aircraft without search and rescue support. This statement is implemented by ICAO through the adoption of Standards and Recommended Practices (SARPs) by Article 38 of the Convention on International Civil Aviation. SARPs are published as Annexes to the Chicago Convention. Annex 12 "Search and Rescue" contains information on the organization, interaction, and other measures aimed at ensuring search and rescue operations during an emergency [2]. The practical implementation of Annex 12 is carried out through the International Air and Maritime Search and Rescue (IAMSAR) Manual (Doc. 9731), which is set out in three volumes [3–5]. These facts motivate to carry out scientific research to improve operational interaction in flight emergencies (FE) between the pilot, air traffic controller (ATCO), and search and rescue specialist (SRS).

1.3 Contribution

This research contributes to the enhancement of the interaction between the aviation specialists in FE based on the integration of deterministic, stochastic, and non-stochastic decision-making models. Using integrated collective models in the Intelligent Decision Support System will allow for obtaining optimal collaborative decisions.

1.4 The Organization of the Paper

The paper consists of seven sections. The first section is an introduction. The second section concentrates on the analysis of related works and problem statements. The third section considers the multi-stage modeling of CDM by the aviation specialists in the FE, and the fourth – describes the illustrative example of CDM in the FE "Loss of thrust on both engines due to falling into a cloud of volcanic ash with subsequent ditching". The fifth section presents results and discussions. The sixth section is a conclusion. The seventh section describes the future research directions.

2 Literature Review and Problem Statement

The beginning of the era of volcanic research coincides with the beginning of the era of maritime exploration by Spain and Portugal in the early sixteenth century, which opened up Latin America and much of the Pacific Ocean to European explorers. The development of territories such as Alaska, Hawaii, and New Zealand also contributed to the study and documentation of volcanic activity [6].

To date, there has been no significant increase in volcanic activity. Analyzing the activity of volcanoes over the past two centuries, it was noted that the number of eruptions has not increased, but only the information about such facts has improved due to the development of communication technologies, as well as the fact that the number of people living around the zones of volcanic activity has increased over time. This observation, therefore, denies the claim that the number of such events increased in the twentieth century compared to the nineteenth or earlier.

Statistics on volcanic activity over the past two centuries also show that peaks and periods of decline in activity are closely related to various historical events, such as global conflicts, economic crises, and others. That is, this does not indicate a change in volcanic activity, but rather that humanity was engaged in other things at a certain time than recording volcanic eruptions. At the same time, volcanic eruptions that were accompanied by a large number of deaths later generated increased interest in the problem (the cases were properly recorded). These statements are also confirmed by the information from the graph of volcanic activity in the period from 1800 to 2000 presented in [6], as the number of powerful eruptions (which generated at least 0.1 km^3 of debris, ash, etc.) remained almost unchanged during the observation period.

The extent of coverage of volcanic eruptions often depended on the location of the volcanoes (near densely populated areas or not), as well as the number of deaths. For example, the coverage of the Kamchatka eruption and the St. Helens volcano in Washington State (USA) were quite different in terms of the degree of resonance in society. Coverage of the respective cases in the period after the end of World War II was facilitated by the launch of a global cataloging of active volcanoes and eruptions (early 1950s – late 1960s). The global spread of the Internet and other methods of processing and transmitting information, satellite monitoring of activity zones, and the installation of infrared sensors for detecting activity since the 1990s have greatly facilitated the detection of eruptions that could have been overlooked in previous periods of observation [6].

Cases of volcanic eruptions that are accompanied by ash emission and can interfere with flights are quite rare. To date, no cases of forced landing of aircraft on the water surface caused by volcanic activity have been recorded [7]. However, there are documented cases when civilian aircraft engines failed due to volcanic ash clouds, forcing crews to make emergency landings outside the destination aerodrome.

One of the incidents occurred in 1982 with a Boeing 747-236B en route from Heathrow Airport to Auckland Airport (New Zealand). While flying over Indonesia, the aircraft hit a cloud of volcanic ash containing particles of stones, glass, and other materials. As a result, all four engines failed. After descending to flight level (FL) 120, the crew was able to successfully restart all four engines. However, one of them had to be shut down before landing due to the impact of another ash cloud. The aircraft was then directed to land at Jakarta Airport. Even though the aircraft sustained significant damage to its engines and airframe, and smoke entered the cabin during the flight, none of the passengers were killed. As it became known after the accident, the crew was not notified in advance of the volcanic activity by ATCO (although the volcanic eruption had occurred two hours earlier). This case served to revise the requirements for informing about flight hazards caused by volcanic activity [8].

Another accident involving an aircraft hitting a volcanic ash cloud occurred over Alaska in late 1989 [9]. Boeing 747-400 en route at FL 250 hit a volcanic ash cloud that resulted from the eruption of Mount Redoubt stratovolcano. The pilots were unable to avoid hitting the cloud because the cloud looked normal and there was no indication of its volcanic origin. Within 10–15 s of entering the danger zone, all four engines and the electrical system failed. After numerous attempts, the crew succeeded in restarting the engines before reaching FL 110. After that, a successful emergency landing was made at the Anchorage airport. In addition to the engines and electrical system, the windshield, some internal systems, and avionics elements of the aircraft were damaged. As in the previous case, information about the active volcano was not transmitted in time to both ground services and the aircraft crew [10].

At the same time, despite the redundancy of the main systems of modern aircraft, in the past, there have been accidents that led to forced landings due to the failure of all engines. For example, in one case, this was due to the fact that during maintenance, an ATR-72 aircraft had a fuel level indicator from an ATR-42 installed. When preparing for the flight, the aircraft contained only 1255 kg of fuel after refueling, instead of the 3800 kg indicated by the indicator. During the flight, the aircraft exhausted the available fuel, which led to engine shutdown and forced landing in the Mediterranean Sea [11]. In another case, the human factor was the cause of the shutdown of all three engines of the Yak-40 aircraft [12]. During the landing approach, the flight engineer mistakenly moved the engine control levers to the "Stop" position instead of the "Low Throttle" position. As a result, the aircraft landed on a wetland in the suburbs of Kyiv.

However, the most resonant was the accident that occurred on January 15, 2009. The Airbus A320 was flying from LaGuardia Airport (New York) to Seattle. Approximately three minutes after takeoff, at an altitude of 2818 ft, the aircraft collided with a flock of geese. The birds got into the working engines, which led to the failure of both power units. The aircraft almost immediately began to lose speed and altitude.

After the onset of the emergency on board, the aircraft commander transmitted a Mayday signal to ATCO and requested an emergency landing at the airport of origin. However, during the subsequent radio exchange with the controller, having further assessed the situation, he refused to land at the airport of departure and the airport in New Jersey. Instead, the pilot decided to divert to the Hudson River. Over the next two minutes, the crew brought the aircraft on a course along the river and approximately two minutes later made the landing. The forced landing on the water surface was successful because although water began to flow into the cabin, it did not significantly impede the evacuation process. After evacuating the cabin, the passengers were placed on the escape ladders (which served as life rafts) and on the wing of the aircraft. Both U.S. Coast Guard vessels and civilian ferries participated in the rescue operation. The last person left the aircraft 25 min after it was ditched. Thanks to the coordinated actions at all stages and the pilot's skill, none of those on board died, only a few people were seriously injured, and most did not even need hospitalization [13].

The analysis shows that there is still a possibility of ditching due to falling into the volcanic activity zone.

According to the concept of Collaborative Decision Making (CDM) [14–17], effective interaction of aviation specialists is a prerequisite for ensuring safety at all stages of aircraft flight in normal and emergency situations. Aviation specialists strictly adhere to the regulatory documents tested in their professional training and professional activities. At the same time, the content of professional training and guidance documents often differs, which makes it impossible to develop a unified algorithm for joint actions, especially in FE. Between the decisions and actions of the CDM-participants the conflicts arise.

The authors' works [18–22] initiated the study of modeling, optimization, intellectualization, and training of CDM by teams of Air Navigation System's operators and auxiliary personnel (pilot – ATCO, UAV operator – ATCO, UAV operator – flight dispatcher, UAV operator – ATCO – engineer, UAV operator – ATCO – pilot, UAV operator – ATCO – telemedical personnel, pilot – ATCO – emergency service specialist, etc.) in different FE.

Nevertheless, the interprofessional interaction of aviation specialists in FE is currently not sufficiently developed for joint development and adoption of optimal decisions.

Therefore, the development of effective CDM models by aviation specialists in FE is an actual problem.

The purposes of this work are:

- Multi-stage CDM modeling based on deterministic, stochastic, and non-stochastic models in conditions of certainty, risk, and uncertainty.
- Building the integrated CDM models of the pilot, ATCO, and SRS in the FE "Loss of thrust on both engines due to falling into a cloud of volcanic ash with subsequent ditching".

3 Multi-stage Modeling of Collaborative Decision-Making (CDM) by the Aviation Specialists in the Flight Emergency

Development of the FE can be presented as a Markov network (Fig. 1).

Multi-stage CDM modeling by the aviation specialists in the FE is proposed. In the first stage, only the pilot is involved in decision-making, in the second stage – joins ATCO, in the third stage – joins SRS, and the result – is the optimal collaborative decision for all aviation specialists.

In Fig. 2 the decision tree (the stochastic model in risk conditions) of the FE development, in Fig. 3 network graphics (the deterministic model in certainty conditions) and decision-making matrices (the non-stochastic model in uncertainty conditions) for the aviation specialists are presented.

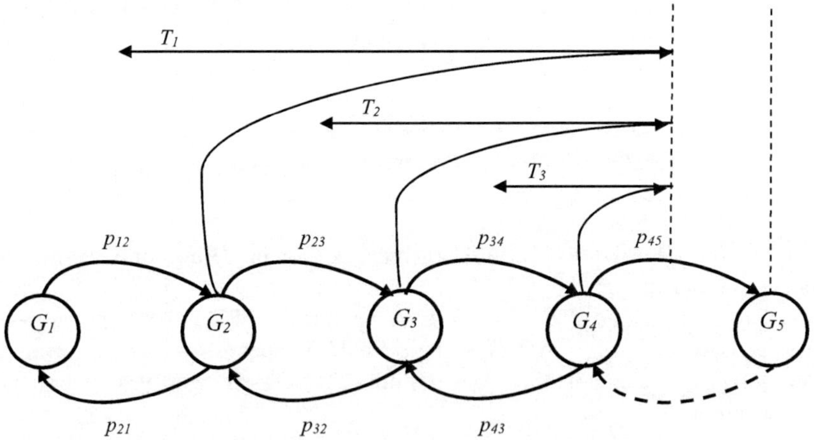

Fig. 1. Markov network of the FE development: G_1 – normal situation; G_2 – complicated situation; G_3 – complex situation; G_4 – emergency situation; G_5 – catastrophic situation; p_{ij} – the probabilities of situation development; T – the time of the decision-making by the operators.

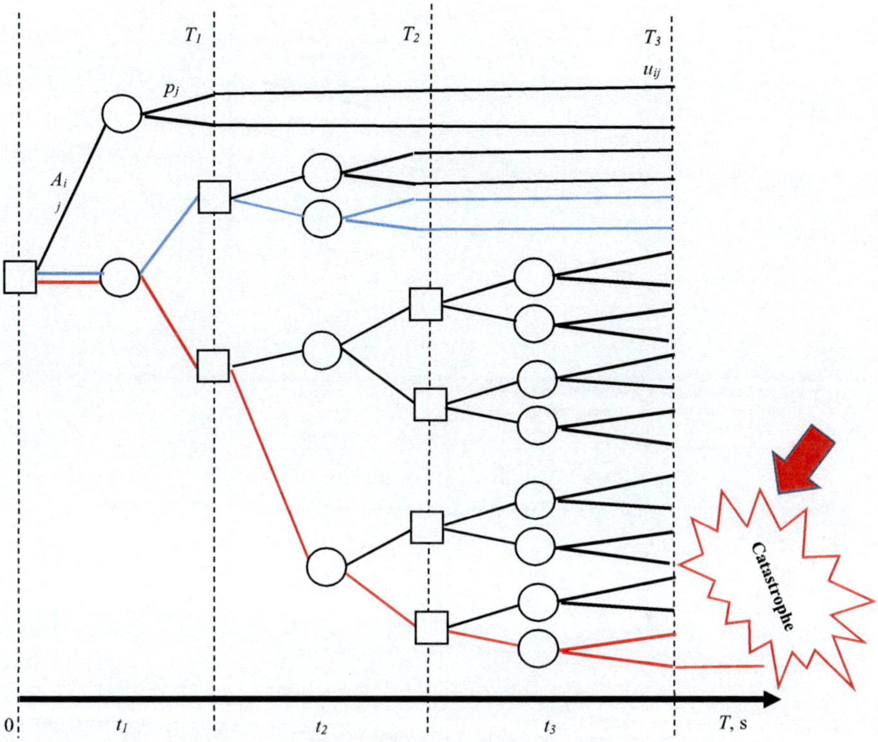

Fig. 2. Decision tree of the FE development: A_i – the alternative decisions; u_{ij} – the expected outcomes; p_j – the probabilities of the expected outcomes; t – the stages of the decision-making; T – the time of the decision-making by the operators; $R_i = \sum_{j=1}^{m} p_j u_{ij}$ – the risk of i-alternative choice.

Integration of non-stochastic, stochastic, and deterministic CDM models by aviation specialists allows moving from a complex to a simple CDM process by establishing unambiguity and synchronization of joint actions of operators at key points of multi-alternative interaction under conditions of complete or incomplete uncertainty and minimizing potential risk/loss.

Simplified deterministic CDM models in FE after integrating in them stochastic and non-stochastic models under conditions of uncertainty should be further used in artificial or hybrid intelligence systems to obtain optimal collaborative solutions [23, 24].

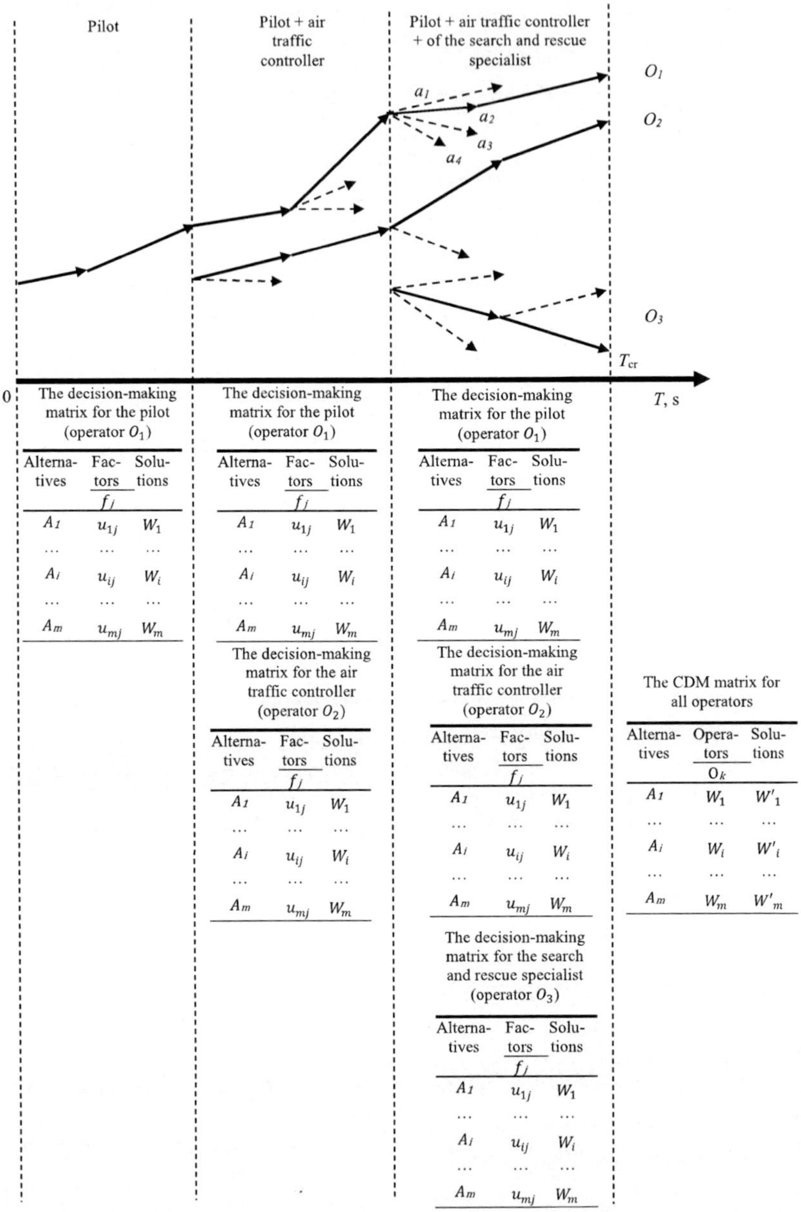

Fig. 3. Network graphics and decision-making matrices for the operators: O_1 – pilot, O_2– ATCO, O_3 – SRS; A_i – the alternative decisions; f_j – the factors influencing decision-making in FE; u_{ij} – the expected outcomes; W_i – the optimal solutions based on the Wald, Laplace, Hurwitz, and Savage criteria.

4 The Illustrative Example of Collaborative Decision-Making (CDM) in the Flight Emergency "Loss of Thrust on Both Engines Due to Falling into a Cloud of Volcanic Ash with Subsequent Ditching"

There is presented a training example of CDM in the FE "Loss of thrust on both engines due to falling into a cloud of volcanic ash with subsequent ditching" [25, 26].

Characterization of the aircraft and flight conditions:

- light aircraft (up to 5700 kg) Cessna 310 with two Continental O-470-B horizontally opposed piston engines [27];
- one crew member;
- four passengers;
- the flight is performed according to the Flight Instrument Rules (FIR);
- absolute flight altitude $H = 1380$ m (recommended altitude for short flights $H = 4500$ ft);
- aircraft aerodynamic quality $K_{\max} = 13$;
- flight airspeed $V = 330$ km/h;
- magnetic flight course at the site Napoli – Palermo $MFC = 200°$ (approximately);
- distance between airports Napoli and Palermo $S_a = 318$ km;
- distance between coastlines of Napoli and Palermo $S_c = 313$ km;
- total flight time $t = 60$ min.

Given the flight conditions and the type of FE [28], the Equal Time Point (ETP) and Point of Safe Return (PSR) were not calculated.

Characterization of the air and meteorological situation:

- the ATCO has one aircraft on the radio;
- the flight is carried out in the absence of wind;
- the presence of a volcanic ash cloud, visibility about 5 km;
- the FE occurred in late March, in the afternoon.
- Characterization of the FE:
- as a result of entering the volcanic ash cloud of the Stromboli volcano (whose eruption intensified after the flight was authorized), both engines lost thrust. Due to deficiencies in preflight preparation, the meteorological hazards en route were not analyzed, so the crew was unable to identify the cloud of volcanic origin in advance. The engine screws automatically switched to autorotation mode, and smoke began to enter the cabin of the aircraft. The pilot planned to touch down the windward side and parallel to waves and swells;
- the pilot sent a Mayday signal;
- loss of thrust on both engines occurred over the Tyrrhenian Sea, 185 km from the coast of Naples (140 km northwest of the eruption epicenter) at the 40th minute of the flight (Fig. 4).

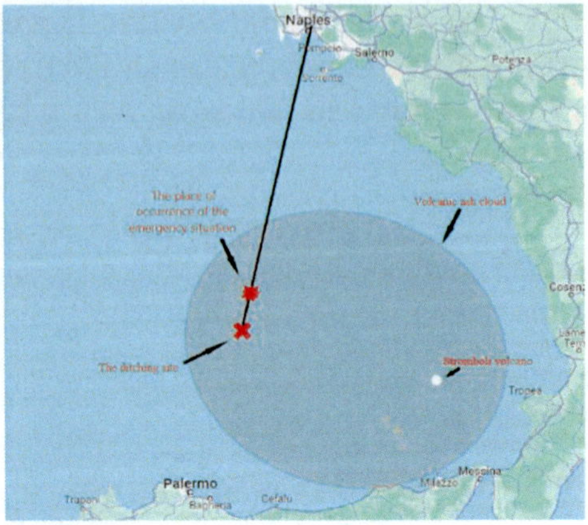

Fig. 4. The place of the FE after loss of thrust on both engines.

Input data:

- the altitude at which the loss of thrust on both engines occurred $H = 1380$ m;
- horizontal planning speed of the aircraft after loss of thrust on both engines $V_{pl} =$ 125 km/h;
- aircraft aerodynamic quality $K_{\max} = 13$;
- minimum planning angle for aircraft with engines that have lost thrust $\theta_{\min} = 4.4°$:

$$tg\theta_{\min} = \frac{1}{K_{\max}};\ \theta_{\min} = arctan\frac{1}{K_{\max}};$$
$$\theta_{\min} = arctan\frac{1}{13} = arctan\ 0.0769 = 4.397 \approx 4.4^{\circ}; \tag{1}$$

- planning distance (the limit of the aircraft's range) in a straight flight (2):

$$\begin{gathered} L_{pl} = H \cdot k; \\ L_{pl} = 1380 \cdot 13 = 17940\ \text{m}; \end{gathered} \tag{2}$$

- time of planning for aircraft with engines that have lost thrust (3):

$$t_{pl} = \frac{L_{pl}}{V_{km/h}/3.6};$$
$$t_{pl} = \frac{17940}{125/3.6} = 516\ \text{s}; \tag{3}$$

- aircraft vertical descent speed (4):

$$V_v = \frac{H}{t_{pl}};$$
$$V_v = \frac{1380}{516} = 2,7\text{m/s}. \tag{4}$$

From the moment both engines lost thrust until the pilot signaled Mayday, 132 s were passed. During this time, the aircraft at $V_{pl} = 125$ km/h flew $S_{132s} = 4583$ m. After the pilot sends the Mayday signal, there are still 384 s.

Given the current conditions, it is logical to consider the options for ditching (Fig. 5):

- a_1 – do not change the flight course heading for the destination airport Palermo;
- a_2 – continue flying southwest to Ustica Island (the island for the emergency landing was selected based on the decision-making matrix in Table 1);
- a_3 – make a turn and plan in the direction of the departure airport of Naples.

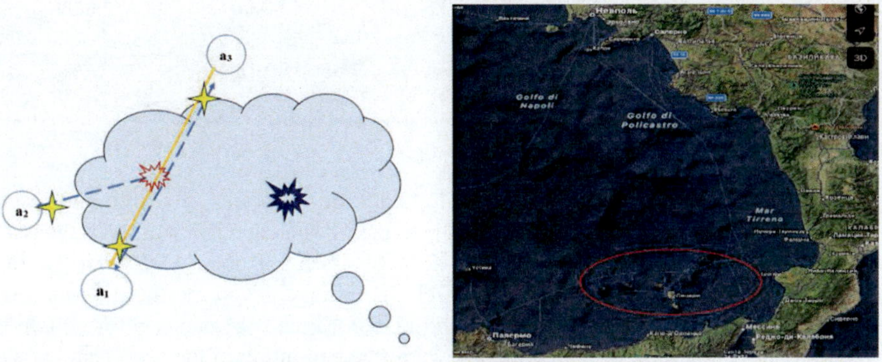

Fig. 5. The options for a ditching.

The decision-making matrix of selected the island for the emergency landing is presented in Table 1. According to calculations based on the objective factors that influence the decision-making with the help of the Expert Judgment Method (EJM) (1 – the worst variant, 10 – the best variant), it is advisable to consider Ustica Island as a place for the emergency landing.

Thus, without changing the direction of flight (distance to the coast of Palermo) (a_1) in the remaining time (384 s), the aircraft will overcome 13 333 m:

$$S_{a1} = 313\,000 - 185\,000 - 4\,583 - 13\,333 = 110084 \text{ m} \approx 110 \text{ km}.$$

Ditching (a_1) will be carried out 110 km off the coast of Palermo.

Accordingly, the landing point for ditching (a_2) will be $S_{a2} = 55$ km northeast of the coast of Ustica Island.

When making a turn towards the departure airport of Naples (a_3):

$$S_{a3} = 185000 + 4583 - 13333 = 176250\text{m} \approx 176\text{km}.$$

Ditching (a_3) will be carried out 176 km from the coast of Naples.

Table 1. The decision-making matrix of selected the island for the emergency landing.

Islands	Objective factors that influence the decision-making			Solutions	
	Remoteness from the emergency place	Remoteness from the volcano	Availability of sea routes nearby	Wald criterion	Laplace criterion
Lipari Island	7	1	9	1.00	5.67
Ustica Island	8	7	3	**3.00**	**6.00**
Ventotene Island	3	7	1	1.00	3.67

Next, let's calculate the time of arrival at each of the points of search and rescue vehicles. According to [29], the closest place to base Italian air search and rescue assets is in Catania. This airbase is home to ATR 42 MP, P 166 DL3, and P 180 (Orca) aircraft, which are only capable of performing search operations. Given the fact that the pilot was able to transmit the Mayday signal and his coordinates by radio before ditching, we will assume that they were already known at the beginning of the search and rescue operations (SRO), so it is logical to involve helicopters from the airbase in Catania.

In addition to the aircraft, helicopters AB 412 CP (Koala) and AW 139 CP (Nemo) are based there. Calculations will be made taking into account the AW 139 CP (Nemo) helicopter, which is capable of evacuating victims from the water surface using special means, and given the distribution of possible injuries that those on board may receive, we will assume that at least one person will need assistance in a specialized medical facility.

Based on this, let's calculate the time for evacuation from each of the places of arrival (a_1, a_2, a_3) to a medical facility in Palermo:

- a_1 – the total length of the route will be 326 km. The AW 139 CP helicopter, which has a cruising speed of 306 km/h, will cover this distance in approximately 64 min;
- a_2 – the total length of the route will be 331 km. The distance will be covered in approximately 65 min;
- a_3 – the total length of the route will be 372 km. The distance will be covered in approximately 73 min.

The Italian Coast Guard has more than 600 rescue ships and boats located in 113 ports [30]. The Coast Guard also has a representative office on Ustica Island. However, no marine search and rescue vessel can provide evacuation to a medical facility in Palermo in a shorter time than a helicopter.

The analysis of the pilot and ATCO actions in case of engines failure and emergency landing is carried out in [22]. Structure-time table of the SRS's actions during ditching is presented in Table 2.

Table 2. Structure-time table of the SRS's actions during ditching – an example.

The technological operation, s	The sequence of the technological operations	Previous operation	Time of operation, s
s_1	The moment when information about the FE on board the aircraft is received by the coordination center	–	–
s_2	Appointment of the SRO coordinator by the governing body, organization of the headquarters	s_1	30
s_3	Appeal to the ATCO to create conditions for the use of airspace by search and rescue aircraft	s_2	30
s_4	Studying and evaluating all available information (on the search object, hydrometeorological conditions, and means involved in SRO) on FE	s_2	300
s_5	Notification of aircraft and vessels located in possible search areas about their involvement as search and rescue assets	s_2	30 (if available)
s_6	Starting to organize SRO: marking the search area on the navigation map; making decisions on the methods and search and rescue means to be used; calculating the parameters of the search area	s_4	180
s_7	Allocation of aviation and maritime forces and means of constant readiness for search and rescue operations	s_6	5

(continued)

Table 2. (*continued*)

The technological operation, *s*	The sequence of the technological operations	Previous operation	Time of operation, *s*
s_8	Familiarizing captains of marine and pilots of air search and rescue assets with the relevant search and rescue plans	s_7	60
s_9	Informing on safety measures during SRO and assignment of primary and secondary communication channels	s_8	40
s_{10}	Search for aircraft, rescue of persons at the ditching site, and their evacuation to a medical facility	s_9	a_1– 64 min a_2 – 65 min a_3– 73 min
In total			a_1– 75 min 15 s a_2 – 76 min 15 s a_3 – 84 min 15 s

According to the technological operations of the pilot and ATCO [22], and SRS (Table 2), the network graphic of the synchronized actions of the operators during ditching is built (Fig. 6).

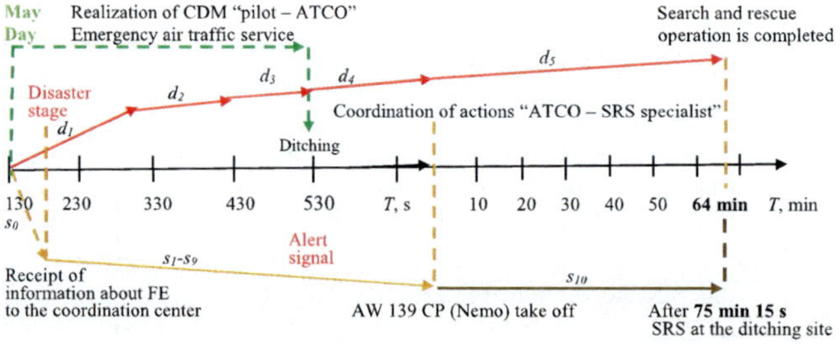

Fig. 6. The network graphic of the synchronized actions of the ATCO and SRS during ditching: *d* – technological operations of the ATCO; *s* – technological operations of the SRS.

The individual decision-making matrices for the aviation specialists in FE "Loss of thrust on both engines due to falling into a cloud of volcanic ash with subsequent ditching" are in Table 3. Optimal solutions are based on the Wald (W), Laplace (L), and Hurwitz (H) criteria. Expected outcomes considered by the pilot, ATCO, and SRS are obtained with the help of the EJM (1 – the worst variant, 10 – the best variant). Objective factors that influence the decision-making: f_1 – fuel reserve on board (not produced so $f_1 = \text{const}$); f_2 – remoteness from the land; f_3 – remoteness from the SRS; f – meteorological conditions (direction and velocity of wind, pressure, temperature, dangerous weather phenomena, etc.); f_5 – aircraft performance (time of planning); f_6 – air traffic intensity; f_7 – search and rescue category; f_8 – alternative detection methods; f_9 – remoteness from the volcano.

Table 3. The individual decision-making matrices for the aviation specialists (pilot, ATCO, and SRS).

Alternatives	Objective factors that influence the decision-making									Solutions		
	f_1	f_2	f_3	f_4	f_5	f_6	f_7	f_8	f_9	W	L	$H, \alpha = 0.5$
	Pilot (O_1)											
a_1	const	6	7	7	7	const	10	8	7	**6.00**	**7.43**	**8.00**
a_2	const	4	4	5	3	const	10	8	4	3.00	5.43	6.50
a_3	const	7	8	6	9	const	4	5	7	4.00	6.57	6.50
	ATCO (O_2)											
a_1	const	7	7	6	7	7	10	8	7	**6.00**	**7.38**	**8.00**
a_2	const	4	4	5	3	7	10	8	4	3.00	5.63	6.50
a_3	const	7	7	6	9	2	4	5	7	2.00	5.88	5.50
	SRS (O_3)											
a_1	const	9	9	6	7	7	10	8	7	**6.00**	**7.88**	**8.00**
a_2	const	6	6	5	3	7	10	8	4	3.00	6.13	6.50
a_3	const	7	7	6	9	2	4	7	7	2.00	6.13	5.50

The optimal option for a ditching follows the pilot's, ATCO's, and SRS's individual decisions is continuing flying towards the destination airport Palermo (a_1).

To determine the coherence of aviation specialists, the collective decision-making matrix is constructed (see Table 4), in which the objective factors for the operators (pilot (O_1), ATCO (O_2), SRS (O_3)) are similar, and the solutions of the operators are taken from the individual matrices. In the CDM matrix, the operators' attitudes (subjective factors) are used. The optimal CDM for operators in the FE "Loss of thrust on both engines due to falling into a cloud of volcanic ash with subsequent ditching" based on the Wald (W), Laplace (L), and Hurwitz (H) criteria is shown in Table 4.

Table 4. The collective matrix for all aviation specialists (pilot, ATCO, and SRS).

Alternatives	Operators/Solutions											
	O_1	O_2	O_3	W	O_1	O_2	O_3	L	O_1	O_2	O_3	H, $\beta = 0.5$
a_1	6.00	6.00	6.00	**6.00**	7.43	7.38	7.88	**6.00**	8.00	8.00	8.00	**6.00**
a_2	3.00	3.00	3.00	3.00	5.43	5.63	6.13	3.00	6.50	6.50	6.50	3.00
a_3	4.00	2.00	2.00	2.00	6.57	5.88	6.13	2.67	6.50	5.50	5.50	3.00

The optimal CDM based on the objective and subjective factors to follow of pilot's, ATCO's, and SRS's decisions by Wald (W), Laplace (L), and Hurwitz (H) criteria is continuing flying towards the destination airport Palermo (a_1) and ditching at 110 km of its coast.

In [31] an example of CDM by the aviation specialists in the emergency "Cargo fire during the climb" with taking into account objective factors' priority is considered.

5 Results and Discussions

Multi-stage CDM modeling by the aviation specialists in the FE is proposed. At the first stage, only one operator is involved in decision-making, at each subsequent stage – join next operators, and the result – is the optimal collaborative decision for all aviation specialists. The decision tree (the stochastic model in risk conditions), network graphics (the deterministic model in certainty conditions), and decision-making matrices (the non-stochastic model in uncertainty conditions) for the aviation specialists are presented.

A training example of CDM in the FE "Loss of thrust on both engines due to falling into a cloud of volcanic ash with subsequent ditching" during the Cessna 310 flight from Napoli to Palermo is presented. The options for ditching are considered: a_1 – do not change the flight course heading for the destination airport Palermo; a_2 – continue flying southwest to Ustica Island; a_3 – make a turn and plan in the direction of the departure airport of Naples. Structure-time table of the SRS's technological operations and the network graphic of the synchronized actions of the ATCO and SRS during ditching are built. The calculations are shown that the total time of SRO is: a_1 – 75 min 15 s; a_2 – 76 min 15 s; a_3 – 84 min 15 s.

Nine main objective factors influencing decision-making in the case of falling into a cloud of volcanic ash are determined: fuel reserve on board; remoteness from the land; remoteness from the SRS; meteorological conditions (direction and velocity of wind, pressure, temperature, dangerous weather phenomena, etc.); aircraft performance (time of planning); air traffic intensity; search and rescue category; alternative detection methods; remoteness from the volcano. The optimal option for a ditching follows the pilot's, ATCO's, and SRS's individual decisions is continuing flying towards the destination airport Palermo (a_1). The optimal CDM based on the objective and subjective factors to follow of pilot's, ATCO's, and SRS's decisions by Wald, Laplace, and Hurwitz criteria is continuing flying towards the destination airport Palermo (a_1) and ditching at 110 km of its coast. The AW 139 CP helicopter will cover this distance in approximately 64 min; SRS will arrive at ditching site after 75 min 15 s.

6 Conclusion

From 1953 to 2009, the researchers recorded 129 incidents of aircraft hitting volcanic ash clouds. In 94 of these cases, traces of ash were found on the aircraft after the flight, and 79 of them sustained damage of varying severity (airframe or engine). In 26 cases, the degree of damage was determined as "significant or very serious", of which nine were accompanied by the failure (shutdown) of the engine(s) during the flight. In such FE operational interaction between the pilot, ATCO, and SRS is highly significant. Nevertheless, the interprofessional interaction of aviation specialists is currently not sufficiently developed for joint development and adoption of optimal decisions. Therefore, the development of effective CDM models by aviation specialists in FE is an actual problem.

Multi-stage CDM modeling by the aviation specialists in the FE with the gradual involvement of additional operators in decision-making is proposed. The decision tree (the stochastic model under risk conditions), network graphics (the deterministic model under certainty conditions), and decision-making matrices (the non-stochastic model under uncertainty conditions) for the aviation specialists are presented. Integration of non-stochastic, stochastic, and deterministic CDM models by aviation specialists allows moving from a complex to a simple CDM process by establishing unambiguity and synchronization of joint actions of operators at key points of multi-alternative interaction under conditions of complete or incomplete uncertainty and minimizing potential risk/loss.

A training example of CDM in the FE "Loss of thrust on both engines due to falling into a cloud of volcanic ash with subsequent ditching" during the Cessna 310 flight from Napoli to Palermo is presented. Structure-time table of the SRS's technological operations and the network graphic of the synchronized actions of the ATCO and SRS during ditching are built. The optimal CDM based on the objective and subjective factors to follow of pilot's, ATCO's, and SRS's decisions by Wald, Laplace, and Hurwitz criteria is determined.

After integrating stochastic and non-stochastic models under uncertainty conditions, simplified deterministic CDM models in FE should be used in artificial or hybrid intelligence systems to obtain optimal collaborative solutions.

7 Future Scope

Further research will be aimed at creating an Intelligent System of Collaborative Decision-Making during the Training (IS CDM-T) for the collective training of aviation professionals (pilots, UAV operators, air traffic controllers, flight dispatchers, ground handling agents, technicians, etc.) based on big data analysis and machine learning. To manage artificial intelligence solutions by aviation professionals, it is necessary to use hybrid intelligent systems that combine machine and human competencies.

References

1. Lechner, P., Tupper, A., Guffanti, M., Loughlin, S., Casadevall, T.: Volcanic Ash and Aviation—The Challenges of Real-Time, Global Communication of a Natural Hazard. In: Fearnley, C.J., Bird, D.K., Haynes, K., McGuire, W.J., Jolly, G. (eds.) Observing the Volcano World. AV, pp. 51–64. Springer, Cham (2017). https://doi.org/10.1007/11157_2016_49
2. Dempsey, P.S., Jakhu, R.S. (eds.): Routledge Handbook of Public Aviation Law, 1st edn. Routledge, London (2016). https://doi.org/10.4324/9781315297774
3. International Aeronautical and Maritime Search and Rescue Manual. Volume I: Organization & Management. Doc. 9731-1, 12th edn. International Civil Aviation Organization, Montreal, Canada (2022)
4. International Aeronautical and Maritime Search and Rescue Manual. Volume II: Mission Co-ordination. Doc. 9731-2, 9th edn. International Civil Aviation Organization, Montreal, Canada (2022)
5. International Aeronautical and Maritime Search and Rescue Manual. Volume III: Mobile Facilities. Doc. 9731-3, 12th edn. International Civil Aviation Organization, Montreal, Canada (2022)
6. Smithsonian Institution. Global Volcanism Program. Has volcanic activity been increasing? https://volcano.si.edu/faq/index.cfm?question=historicalactivity. Accessed 11 Jan 2024
7. Aviation Safety Network. Aviation events. Cause: Ash cloud. https://aviation-safety.net/database/event/WXA/list. Accessed 04 Jan 2024
8. Accidents and incidents. Aircraft accident Boeing 747-236B. https://skybrary.aero/accidents-and-incidents/b742-en-route-south-southeast-jakarta-indonesia-1982. Accessed 04 Jan 2024
9. Takebayashi, M., Onishi, M., Iguchi, M.: Large volcanic eruptions and their influence on air transport: the case of Japan. J. Air Transp. Manag. **97**, 102136 (2021)
10. Aviation Safety Network. Aircraft accident Boeing 747-406M. https://aviation-safety.net/database/record.php?id=19891215-1. Accessed 04 Jan 2024
11. Aviation Safety Network. Aircraft accident ATR 72-202. https://aviation-safety.net/database/record.php?id=20050806-0. Accessed 04 Jan 2024
12. Aviation Safety Network. Aircraft accident Yak-40. https://aviation-safety.net/database/record.php?id=19760602-0. Accessed 04 Jan 2024
13. Aircraft Accident Report. Loss of thrust in both after encountering a flock of birds and subsequent ditching on the Hudson River US Airways Flight 1549 Airbus A320-214. https://www.ntsb.gov/investigations/AccidentReports/Reports/AAR1003.pdf. Accessed 04 Jan 2024
14. Manual on Collaborative Air Traffic Flow Management (ATFM). Doc. 9971, 3rd edn. International Civil Aviation Organization, Montreal, Canada (2018)
15. A-CDM Concept of Operations. Airservices Australia, Canberra, ACT, Australia (2014)
16. Airport CDM Implementation: Manual. EUROCONTROL, Brussels, Belgium (2017)
17. Airport-Collaborative Decision Making: IATA Recommendations. International Air Transport Association, Montreal, Canada (2018)
18. Shmelova, T., Sikirda, Yu., Rizun, N., Salem, A.M., Kovalyov Yu. (eds.): Socio-technical Decision Support in Air Navigation System: Emerging Research and Opportunities. IGI Global Publishing, Hershey (2018). https://doi.org/10.4018/978-1-5225-3108-1
19. Shmelova, T., Chialastri, A., Sikirda, Yu., Yatsko, M.: Decision making models by the pilot in flight emergency "engine failure during take-off". In: CEUR Workshop Proceedings, vol. 3101, pp. 347–365 (2021)
20. Shmelova, T., Sikirda, Yu., Yatsko, M., Kasatkin, M.: Collective models of the aviation human-operators in emergency for intelligent decision support system. In: CEUR Workshop Proceedings, vol. 3156, pp. 160–174 (2022)

21. Shmelova, T, Yatsko, M, Sikirda, Y.: Collaborative-factor models of decision making by operators of the air navigation system in conflict or emergency situations. Commun. Comput. Inf. Sci. (CCIS) **1635**, 391–409 (2022)
22. Shmelova, T., Sikirda, Yu., Yatsko, M., Kolotusha, V.: Intelligent integrated training system for the aviation specialists "collaborative decision-making – education" (CDM-E). In: CEUR Workshop Proceedings, vol. 3538, pp. 168–180 (2023)
23. Alharbi, B., Prince, M.: A hybrid artificial intelligence approach to predict flight delay. Int. J. Eng. Res. Technol. **13**(4), 814–822 (2020). https://doi.org/10.37624/IJERT/13.4.2020.814-822
24. Artificial Intelligence Roadmap: A human-centric approach to AI in aviation. European Union Aviation Safety Agency, Cologne, Germany (2020)
25. Engine failure in cruise. SKYbrary. https://skybrary.aero/articles/engine-failure-cruise. Accessed 08 Jan 2024
26. Low altitude engine failure scenarios and engine failure during flight. Flight Study. https://www.flight-study.com/2019/12/engine-failure-after-lift-off.html. Accessed 08 Jan 2024
27. Cessna 310. C310. https://skybrary.aero/aircraft/c310. Accessed 09 Jan 2024
28. Ren, L., Castillo-Effen, M.: Air Traffic Management (ATM) operations: a review. Report number 017GRC0222. GE Global Research, Niskayuna, New York, United States (2017)
29. Airbases, Guardia Costiera. https://www.guardiacostiera.gov.it/en/Pages/airbases.aspx. Accessed 10 Jan 2024
30. Vessels, Guardia Costiera. https://www.guardiacostiera.gov.it/en/Pages/archive-vessels.aspx. Accessed 10 Jan 2024
31. Shmelova, T., Sikirda, Y., Yatsko, M.: Nondeterministic collaborative decision-making models in flight emergencies based on the factors' priority. In: Ostroumov, I., Zaliskyi, M. (eds.) Proceedings of the International Workshop on Advances in Civil Aviation Systems Development. ACASD 2023. Lecture Notes in Networks and Systems, vol. 736. Springer, Cham (2023).https://doi.org/10.1007/978-3-031-38082-2_22

Comparison of Neural Network and Statistical Approaches to the Problem of Signal Detection

Ihor Prokopenko, Kostiantyn Prokopenko, and Anastasiia Dmytruk(✉)

National Aviation University, Liubomyra Huzara Avenue, 1, Kyiv 03058, Ukraine
gyhyre@gmail.com

Abstract. With the growing interest in deep learning technologies and empirical data confirming the effectiveness of neural networks in solving tasks, including signal detection problems, the effectiveness of neural networks under the influence of non-Gaussian interferences, a factor that significantly complicates the detection process, remains understudied. Thus, the paper considers the possibility of using neural network technologies to detect signals in scenarios characterized by complex non-Gaussian disturbances with inherent uncertainty in the probability distribution. The research carried out a synthesis and comparative analysis of two algorithms: a statistical adaptive algorithm and a neural network with Fourier transform, designed to ascertain a harmonic signal characterized by an undetermined phase against the background of interference, simulated using an autoregressive model, caused by non-Gaussian interference. According to the simulation results, implementing of neural network technologies for signal detection in the conditions of complex non-Gaussian interferences of the impulse type characterized by a priori uncertainty of probability distributions is characterized by high resistance to the effects of disturbances. However, the application of statistical procedures enables the synthesis of optimal algorithms, robust to intricate interference.

Keywords: Signal Detection · Neural Network · Adaptive Algorithm · Non-Gaussian Interference

1 Introduction

Detection of signals is one of the leading tasks, for the solution of which it becomes promising to use the possibilities of artificial intelligence (AI) and deep learning technologies (DL).

At its core, the resolution of the signal detection problem relies on the synthesis of optimal algorithms, originating from foundational works established in the mid-twentieth century. These works predominantly address scenarios characterized by Gaussian probability distributions of both signals and interference.

Nevertheless, given that the predominant majority of signals and interference encountered in diverse applications, for instance, processing speech signals, images, reception and processing of radar signals from moving targets and meteorological formations, as well as in the utilization of information and measurement systems (IMS) within

I. Ostroumov and M. Zaliskyi (Eds.): ACASD 2024, LNNS 992, pp. 224–241, 2024.
https://doi.org/10.1007/978-3-031-60196-5_18

radio defense conditions, exhibit non-Gaussian distribution, addressing this challenge necessitates the synthesis of novel approaches.

A key strategy in addressing non-Gaussian interference involves the synthesis of robust procedures [1, 2], incorporating statistical signal processing methods that maintain performance metrics across a broad spectrum of signal-interference scenarios [3, 4]. This encompasses situations influenced by disturbances with varying probability distributions in both time and space, particularly instances of organized impulse interference.

However, recently, attention has increasingly begun to be paid to the use of the potential of neural networks for the problem of signal detection in various fields of science [5–7] since the results of the work of neural networks demonstrate that they can be a powerful tool for solving this kind of problems.

Neural networks have extensive applications to detect sound signals in speech-processing tasks [8]. They can recognize speech, identify sounds, and perform audio data classification tasks. In the works [9–11], a meaningful analysis of the use of deep learning for processing and detection of radar signals is given, and limiting factors manifested by computational complexity, limitations of the size of the training sample, and the peculiarity of the deployment and architecture of networks are defined [12–14]. In turn, the articles [15, 16] investigated the effectiveness of using neural network technologies and conducted an experimental comparison with existing radar detection methods, the results of which demonstrate the accuracy of the proposed algorithms and the possibility of increasing the detection characteristics at a sufficiently low signal/interference ratio.

In addition, there is a surge of interest in the application of neural network technologies for signal detection in the presence of intricate non-Gaussian impulse disturbances, accompanied by a priori uncertainty in probability distributions [17, 18]. This interest is motivated by the observed heightened resilience of neural networks to the impact of disturbances in such conditions.

Taking into account the fact that the use of neural network algorithms in the problem of signal detection has not yet been widely used, and their effectiveness has not been sufficiently investigated, the question arises about the feasibility and possibility of using AI and DL technologies for this kind of problem.

Thus, the paper focuses on the synthesis and comparative analysis of the effectiveness of harmonic signal detection algorithms. The first approach will be based on the statistical description of the mixture of signal and interference and the procedure of statistical synthesis of the adaptive signal detection algorithm. In turn, the other one will be based on the application of a neural network in combination with the Fourier transform.

2 Formulation of the Problem

The synthesis of the adaptive algorithm [19] and deployment of the neural network involves mathematical modeling of the signal and interference. In this context, the harmonic signal can be formalized by the following expression

$$S(t) = U\cos(\omega_0 t + \varphi). \tag{1}$$

At the same time, the autoregressive random process of the k-th order (AR-k) is used as a mathematical model [20, 21] of the correlated interference, which allows describing random processes with a complex frequency spectrum

$$x_i = \sum_{j=1}^{k} a_j x_{i-j} + \eta_j, \tag{2}$$

where $a_1, ..., a_j$ are autoregressive coefficients, η_j are counts of a generating random process with a non-Gaussian probability distribution.

Samples of the generating random process η_j are formed as a mixture of uncorrelated Gaussian interference v_i and an impulse process z_i

$$\eta_j = v_i + z_i, i = \overline{1, n} . \tag{3}$$

In turn, the distribution of the impulse process will be described by:

$$f(z) = (1-p)\delta(z) + p\frac{1}{2\lambda} exp\left(-\frac{|z_i|}{\lambda}\right), \tag{4}$$

where p is the probability of occurrence of a pulse with a Laplace distribution with a scale parameter λ.

Considering the above, the density distribution of the mixture will be expressed as

$$f_\eta(x) = (1-p)\frac{1}{\sqrt{2\pi\sigma_v^2}} exp\left(-\frac{x^2}{2\sigma_v^2}\right) + p\frac{1}{2\lambda} exp\left(-\frac{|x|}{\lambda}\right). \tag{5}$$

Taking into account the peculiarities of the formation of the detection algorithm, it is appropriate to describe the AR process (2) through the multidimensional density of the probability distribution, which can be represented by the product:

$$f(x_1, .., x_n, t_1, .., t_n) = f(x_1, t_1) \times f(x_2, ..., x_k) \times f(x_{k+1}, ..., x_n), \tag{6}$$

where $f(x_1, t_1)$ is unconditional probabilities density (pdf) of the first samples x_1; $f(x_2, ..., x_k) = \prod_{i=2}^{k} f(x_i, t_i | x_{i-1}, t_{i-1}, ..., x_{i-min(k,i)}, t_{i-min(k,i)})$ is the second term of conditional pdf of the AR-k process; $f(x_{k+1}, ..., x_n) = \prod_{i=k+1}^{n} f(x_i, t_i | x_{i-1}, t_{i-1}, ..., x_{i-k}, t_{i-k})$ is the second term of conditional pdf.

3 Synthesis of Harmonic Signal Detection Algorithms

3.1 Statistical Synthesis of the Adaptive Detection Algorithm

The resolution to the challenge of signal detection in the sample (2) is derived through an examination of the conditional probability distributions associated with the hypotheses. Specifically, hypothesis H_0 posits that the sample exclusively comprises interferences, while hypothesis H_1 posits that the sample contains both interferences and a signal.

Therefore, in the context of hypothesis H_0, the conditional multivariate density of the probability distribution associated with the generated autoregressive process (2), taking into account the second term of conditional pdf of (6) will be described as:

$$f(x_{k+1}, .., x_n|x_1, .., x_k, H_0)$$
$$= \prod_{i=k+1}^{n}\left[\frac{(1-p)}{\sqrt{2\pi\sigma_v^2}}exp\left(-\frac{\left(x_i - \sum_{j=1}^{k} a_j x_{i-j}\right)^2}{2\sigma_v^2}\right)\right.$$
$$\left.+ \frac{p}{2\lambda}exp\left(-\frac{\left|x_i - \sum_{j=1}^{k} a_j x_{i-j}\right|}{\lambda}\right)\right]. \tag{7}$$

In turn, hypothesis H_1 postulates the presence of a signal in the sample, and the corresponding conditional multivariate probability distribution is expressed in the form

$$f(x_{k+1}, .., x_n|x_1, .., x_k, b \neq 0, H_1) =$$

$$f(x_{k+1}, .., x_n|x_1, .., x_k, b \neq 0, H_1)$$
$$= \prod_{i=k+1}^{n}\left[\frac{(1-p)}{\sqrt{2\pi\sigma_v^2}}exp\left(-\frac{\left(\sum_{i=k+1}^{n}(x_i - bS_i) - \sum_{j=1}^{k} a_j x_{i-j} - \sum_{j=1}^{k} a_j bS_{i-j}\right)^2}{2\sigma_v^2}\right)\right.$$
$$\left.+ \frac{p}{2\lambda}exp\left(-\frac{\sum_{i=2}^{n}\left|x_i - \sum_{j=1}^{k} a_j x_{i-j}\right|}{\lambda}\right)\right], \tag{8}$$

where S_i, $i = \overline{1, n}$ represents the values of a signal at the i-th moment of time; b denotes the signal parameter.

The derived multivariate probability distributions for hypotheses (7) and (8) can be employed to formulate the optimal signal detection algorithm

$$L(x_1, ., x_n, b) = \frac{f(x_{k+1}, .., x_n|x_1, .., x_k, b, H_1)}{f(x_{k+1}, .., x_n|x_1, .., x_k, b = 0|H_0)}. \tag{9}$$

Considering the signal model (1), the distribution (8) can be formulated as presented in Eq. (10):

$$f(x_{k+1}, .., x_n|x_1, .., x_k, b \neq 0, H_1)$$
$$= \prod_{i=k+1}^{n}\left[\frac{(1-p)}{\sqrt{2\pi\sigma_v^2}}exp\left(-\frac{\left(\sum_{i=k+1}^{n}(x_i - b\cos(\omega_0 t_i + \varphi)) - \sum_{j=1}^{k} a_j x_{i-j} - \sum_{j=1}^{k} a_j b\cos(\omega_0 t_{i-j} + \varphi)\right)^2}{2\sigma_v^2}\right)\right.$$
$$\left.+ \frac{p}{2\lambda}exp\left(-\frac{\sum_{i=2}^{n}\left|x_i - \sum_{j=1}^{k} a_j x_{i-j}\right|}{\lambda}\right)\right]. \tag{10}$$

Taking into account the above-mentioned, an optimal signal detection algorithm, robust to the pulse interference, can be determined by

$$l(x_1, .., x_n) = \prod_{i=1}^{n} \frac{\vartheta(x_i, \overline{\theta}) + Q(x_i)}{1 + Q(x_i)} = \prod_{i=1}^{n} l(x_i, \overline{\theta}) > V(p, \alpha), \tag{11}$$

where

$$l\left(x_i, \overline{\theta}\right) = \frac{f\left(x_i|x_{i-1}, .., x_{i-k}, \overline{\theta}, b\right)}{f\left(x_i|x_{i-1}, .., x_{i-k}, \overline{\theta}, b=0\right)} = \frac{exp\left(b\left(x_i - \sum_{j=1}^{k} a_j x_{i-j}\right)\left(S_i - \sum_{j=1}^{k} a_j S_{i-j}\right)/2\sigma^2\right) + Q(x_i)}{1 + Q(x_i)};$$

$$Q(x_i) = \frac{\sqrt{2\pi}\sigma p exp\left(-\frac{\left|x_i - \sum_{j=1}^{k} a_j x_{i-j}\right|}{\lambda}\right)/\lambda}{2(1-p)exp\left(-\left(x_i - \sum_{j=1}^{k} a_j x_{i-j}\right)^2/2\sigma^2\right)}; \overline{\theta} = \{a_1, .., a_k, \sigma, \lambda\}.$$

In this case, $\overline{\theta}$ is parameter vector.

As the paper addresses the challenge of detecting a harmonic signal characterized by a known frequency but an unknown phase, it becomes essential to devise a phase-invariant algorithm. Following this approach, we shall reformulate Eq. (11) by decomposing the signal into quadrature components. This decomposition enables us to express the algorithm in terms of both amplitude and phase.

The resulting algorithmic representation allows integration of the obtained statistics over the phase φ

$$l(x_i, \overline{\theta}) = \frac{exp\left(b(A(x_i, \omega)\cos(\varphi) + C(x_i, \omega)\sin(\varphi))/2\sigma^2\right) + Q(x_i)}{1 + Q(x_i)}, \tag{12}$$

where $(x_i, \omega) = \left(x_i - \sum_{j=1}^{k} a_j x_{i-j}\right)\left(\cos(\omega t_i) - \sum_{j=1}^{k} a_j \cos\left(\omega t_{i-j}\right)\right)$; $C(x_i, \omega) = \left(x_i - \sum_{j=1}^{k} a_j x_{i-j}\right)\left(\sin(\omega t_i) - \sum_{j=1}^{k} a_j \sin\left(\omega t_{i-j}\right)\right)$.

This enables us to present $l(x_i)$ in the following form:

$$l(x_i, \overline{\theta}) = \frac{exp\left((b(R(x_i, \omega)\cos(G(x_i, \omega) + \varphi)))/2\sigma^2\right) + Q(x_i)}{1 + Q(x_i)}, \tag{13}$$

where $R(x_i, \omega) = \sqrt{A(x_i, \omega)^2 + C(x_i, \omega)^2}$; $G(x_i, \omega) = arctg[C(x_i, \omega)/A(x_i, \omega)]$.

To obtain a phase-invariant algorithm, it is necessary to integrate the acquired statistics over the variable φ

$$\lambda(x_i) = \int_0^{2\pi} \frac{exp\left(bR(x_i, \omega)/\sigma^2\right)\cos(\varphi - G(\omega)) + Q(x_i)}{1 + Q(x_i)} d\varphi = \frac{I_0(bR(x_i, \omega)/\sigma^2) + Q(x_i)}{1 + Q(x_i)}, \tag{14}$$

where $I_0(z)$ is a modified zero-order Bessel function.

Utilizing the approximation of the Bessel function within the region $z \in [0, 10]$:

$$I_0(z) \approx \frac{exp(z/1.15)}{\sqrt{2\pi}}.$$

The decision rule can be reformulated as follows:

$$l(x_1, ..., x_n, \overline{\theta}) = \prod_{i=1}^{n} \frac{exp\left(b(R(x_i, \omega)/1.15)/\sqrt{2\pi} + Q(x_i)\right)}{1 + Q(x_i)}. \tag{15}$$

Local-optimal decision rule can be obtained as derivate of (15) over signal parameter in the point $b = 0$. Therefore, the detection algorithm can be simplified to the form:

$$\frac{d}{db}\ln(l(x_1, \ldots, x_n)) = \sum_{i=1}^{n} \frac{d}{db}\ln\left(\frac{bR(x_i) + Q_i}{1 + Q_i}\right)\Bigg|_{b=0} = \sum_{i=1}^{n} \Phi(R_i, Q_i) > V(p, \alpha) \tag{16}$$

As a result, the structure of the algorithm (16) is formed, as illustrated in Fig. 1.

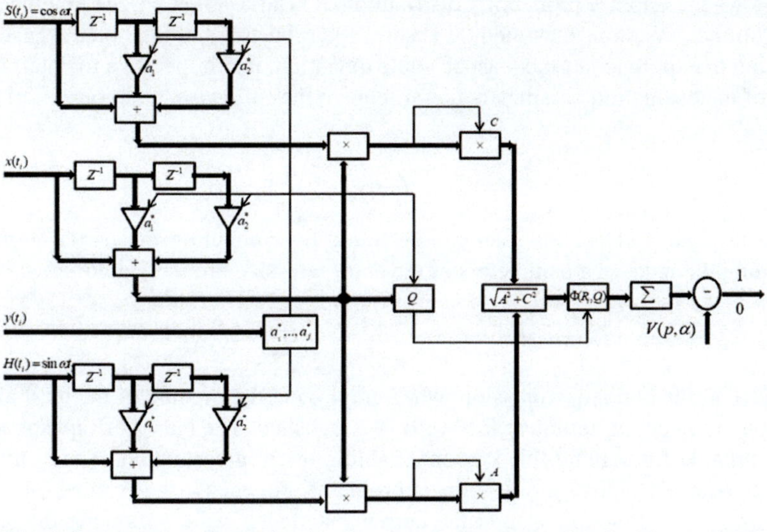

Fig. 1. Block diagram of a robust adaptive detector of a harmonic signal with an unknown phase against a mixture of correlated autoregressive and impulse interference.

The detector comprises three identical Finite Impulse Response (FIR) filters, each including two delay lines Z^{-1}, two weighting factors a_1^* and a_2^*, the values of which correspond to the estimated autoregression coefficients, which are determined in the parameter estimation block a_1^*, ..., a_j^* and adder (+). The upper and lower filters receive the quadrature components of the signal $S(t_i)$ and $H(t_i)$. The middle filter processes a mixture of the signal and interference, denoted as $x(t)$, which after filtering is fed to two multipliers (×). The outcomes of these multipliers are directed to the input blocks where the squares of the components are computed. Subsequently, these squares are connected to the input of the statistics calculation block, denoted as R. Following this, the derived statistics are forwarded to the nonlinear converter block $\Phi(R_i, Q_i)$, generating a value that undergoes comparison with the decision threshold, represented by $V(p, \alpha)$.

In this case, upon surpassing the threshold, a_1 is generated at the output of the detector, in the opposite case, if the threshold is not exceeded, a_0 is generated.

3.2 Synthesis of the Algorithm for Estimating Parameters of Autoregressive Interference

The parameters of autoregressive interference are estimated by utilizing a training sample comprising solely the interference $y_1, \ldots, y_m$.

The formulation of the evaluation algorithm is constructed through the utilization of an empirical Bayesian approach, wherein the estimation of unknown parameters is determined by the method of maximum likelihood

$$\frac{d}{da_l} lnf\,(y_{k+1}, \ldots, y_m | y, \ldots, y_k) = 0, l = \overline{1, k}\ . \tag{17}$$

Deriving from the maximum likelihood equation, assuming the description of obstacles follows the Laplace probability distribution, it is advisable to use the method of the least absolute deviation. This method ensures the robustness of the obtained estimates. Following the approach of the least absolute deviation, it is imperative to determine the values of the evaluation parameters that minimize the following function

$$L = \sum\nolimits_{i=1}^{n} \left| y_i - \sum\nolimits_{j=1}^{k} a_j y_{i-j} \right| \rightarrow min. \tag{18}$$

The minimum of the function can be found by computing the partial derivatives concerning the unknown parameters of the autoregressive process, yielding:

$$\frac{d}{da_l} \sum\nolimits_{i=k+1}^{n} lnf\,(y_{k+1}, \ldots, y_m | y_1, \ldots, y_k, \overline{a}) = 0 \rightarrow \varphi_l = \sum\nolimits_{i=k+1}^{n} sgn\left(y_i - \sum\nolimits_{j=1}^{k} a_j y_{i-j} \right) y_{i-l} = 0, l = \overline{1, k}\ . \tag{19}$$

Based on the derived expression, when employing the method of the least absolute deviation, a system of nonlinear Eqs. (19) is established. The Newton-Raphson method proves suitable for solving this system, enabling the transformation of the solution of nonlinear equations into a series of linear problems. For each n-th iteration, this method approximates the function around the current parameter value, allowing the determination of the value that yields the sought minimum. Given that the function sgn is non-differentiable, to enable the application of the Newton-Raphson method, it can be approximated by a logistic function referred to as a sigmoid:

$$sgn(q) \approx S(q) = \frac{1}{1 + exp(-\alpha q)} - 0.5. \tag{20}$$

Considering this, expression (19) can be reformulated as follows (21):

$$\varphi_{l'} = \sum\nolimits_{i=k+1}^{n} S\left(y_i - \sum\nolimits_{j=1}^{k} a_j y_{i-j} \right) y_{i-l} = 0, l = \overline{1, k}\ . \tag{21}$$

As a result of the substitution, we get the calculated values by solving Eq. (21) using the modified Newton-Raphson method:

$$\overline{a_{j+1}^*} = \overline{a_j^*} - c_j H^{-1}\left(\overline{a_j^*}\right)\nabla\varphi\left(\overline{a_j^*}\right), \tag{22}$$

where $H = \begin{bmatrix} \frac{d}{da_l}\varphi_l{}', \ldots, \frac{d}{da_l}\varphi_l{}' \\ \ldots \quad \ldots \quad \ldots \\ \frac{d}{da_l}\varphi_k{}', \ldots, \frac{d}{da_l}\varphi_k{}' \end{bmatrix}$ is Hessian matrix; $c_j = \frac{1}{1+0.1j}$ is damping coefficients; $\nabla\varphi\left(\overline{a_j^*}\right)$ is the gradient of the function $\varphi\left(\overline{a_j^*}\right)$.

3.3 Neural Network Detection Algorithm

The neural network architecture employed for the detection of a harmonic signal with an unknown phase is depicted in Fig. 2.

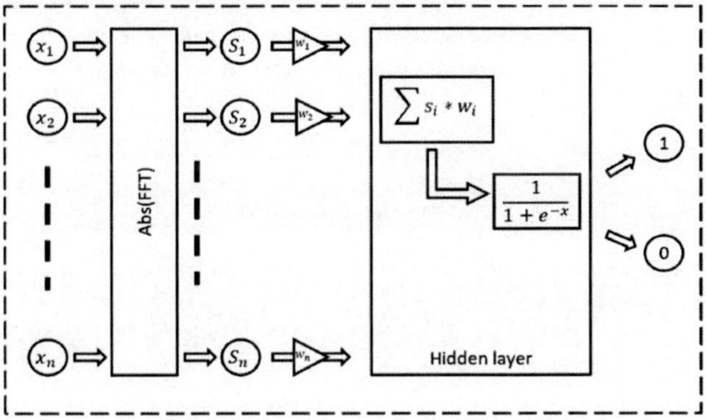

Fig. 2. The architecture of the neural network detector.

The detector comprises a preprocessor that executes the Fourier transform of the input sample$x_1, \ldots, x_n$, representing a combination of signal and noise. The transformation coefficients $S_1, \ldots, S_n$ constitute the input layer of the neural network, feeding into a single hidden layer with a single logistic regression node.

A neural network is defined by a specific set of parameters. It is classified as a standard classification neural network, implemented using the Python Multi-layer Perceptron classifier (Sklearn MLPClassifier). The network architecture includes an input layer with 128 elements, one hidden layer containing a single logistic regression node, and an output layer consisting of 2 classes.

The training dataset is composed of a matrix of Fourier transform coefficients and a target vector comprising zeros and ones. In this context, zero corresponds to a noise-only sample, while one corresponds to a signal-only sample. Throughout the training process, the coefficients $w_1, \ldots, w_n$ are iteratively adjusted using the backpropagation of the error method to ensure optimal performance.

The training dataset is composed of 10,000 examples of interference and 10,000 examples of the signal without interference. Perturbation parameters involve an autoregressive process with coefficients ($a_1 = 1.95, a_2 = -0.99$) characterized by a Gaussian distribution with a variance of 1. The implementation includes a count of $n = 128$.

For signal parameters, a harmonic signal with a known frequency and random phase is considered, with the implementation count also set to $n = 128$.

Further processing of the dataset involves applying the Fourier transform, determining the modulus, and subsequent normalization of the resulting values. This normalization is accomplished by dividing each value by the sum of all sample values.

4 Computer Simulation

4.1 Simulation of Adaptive Algorithm

The results of modeling the work of the adaptive algorithm are shown in Fig. 4, Fig. 5, and Fig. 6, where the spectra of the signal and the mixture of interferences before filtering and after filtering and the detection characteristics of the proposed algorithm (Fig. 6) which in turn allow to analyze its effectiveness. In turn, Fig. 3 shows the implementation of the autoregressive process without impulse interference and under the action of this type of interference.

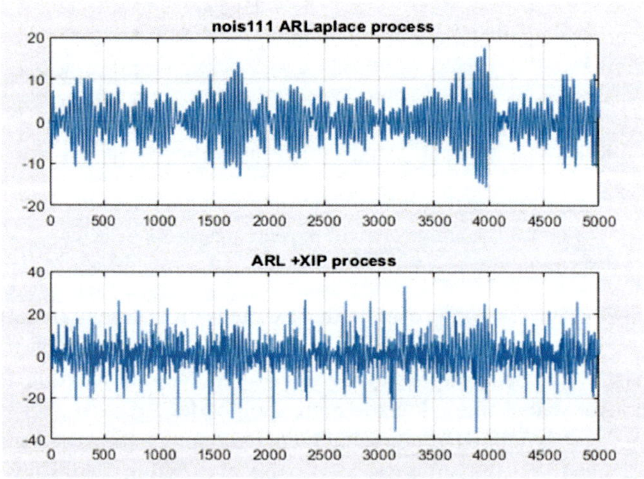

Fig. 3. Time diagram of the autoregressive process without impulse interference and under the action of impulse interference.

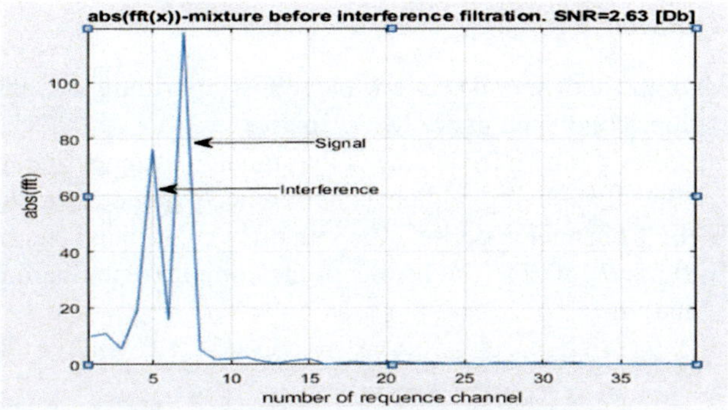

Fig. 4. Spectrum of mixture before filtration.

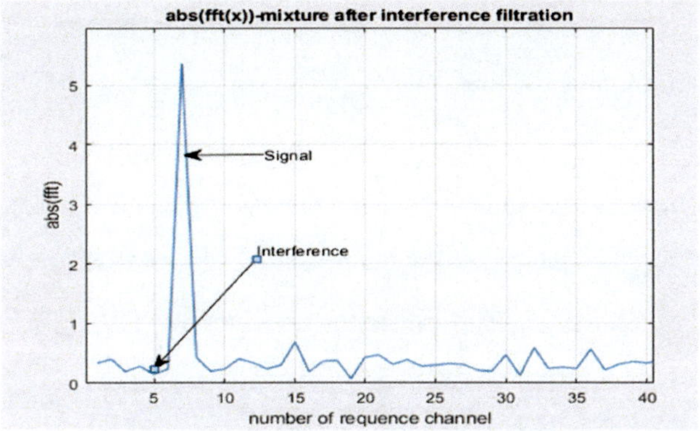

Fig. 5. Spectrum of mixture after filtration.

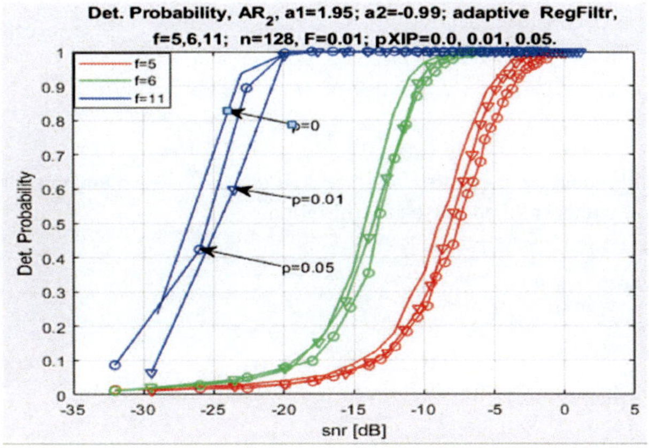

Fig. 6. Detection characteristics of algorithm (16) at different signal frequencies $f = [5; 6; 11]$.

4.2 Testing Scenarios

The effectiveness of the both neural network and robust algorithms was analyzed through computer simulations conducted under two scenarios.

Scenario 1 involves a known frequency of the signal (5 channel FFT, period = 25.6), with the phase of the signal being random. Interference parameters include $a_1 = 1.95$, and $a_2 = -0.99$ (FFT channel 4, period 32), with different impulse interference parameter $p = 0, 0.01, 0.05, 0.1, 0.25$. The results of the computer simulation are illustrated in Fig. 7, Fig. 8, and Fig. 9.

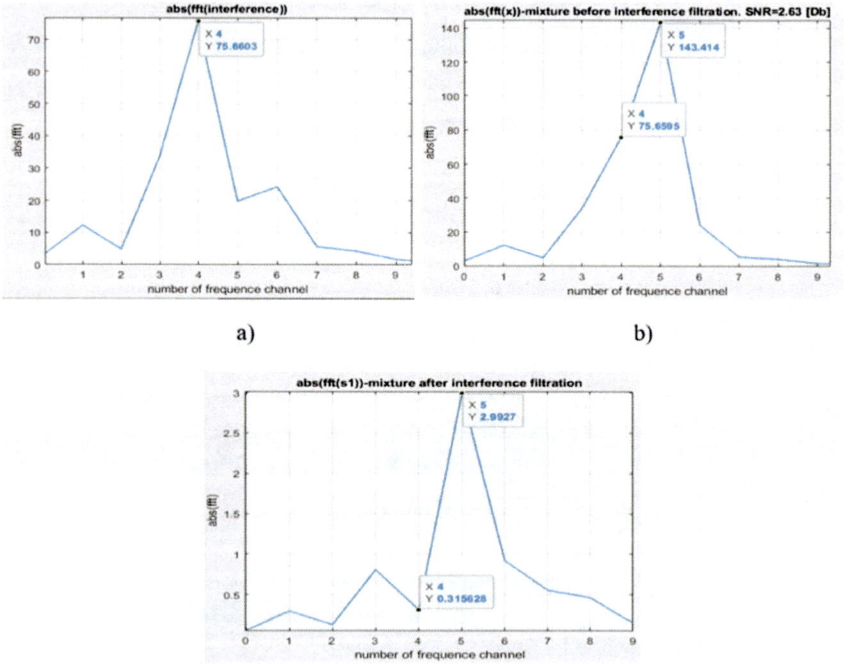

a) b)

c)

Fig. 7. Signal and interference spectra: a) correlated noise; b) signal and correlated noise before filtration; c) signal and correlated noise after filtration.

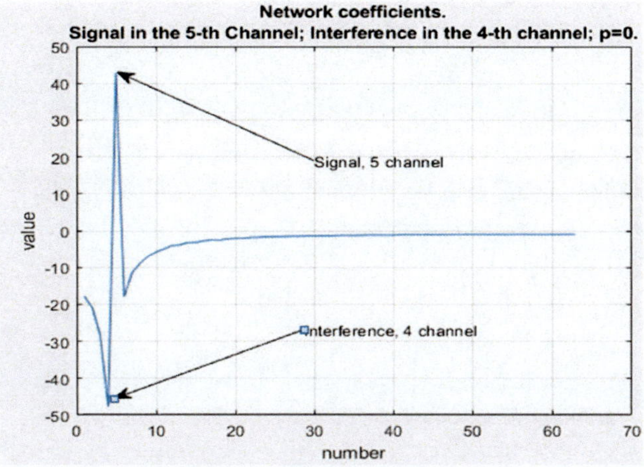

Fig. 8. Network coefficients.

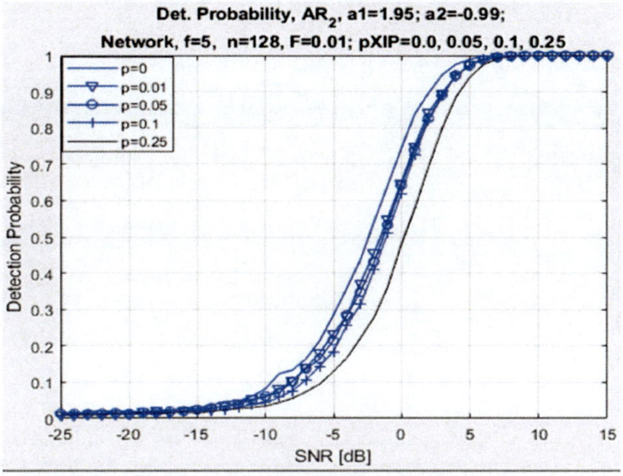

Fig. 9. Detection characteristics of the neural network algorithm at scenario 1 in the fifth frequency channel under the influence of impulse interference. The probability of appearance impulse interference is $p = 0, 0.01, 0.05, 0.1, 0.25$.

In Scenario 2, the signal frequency is known (FFT channel 6, period = 21.3), and the signal phase is random. Interference parameters include $a_1 = 1.95$, $a_2 = -0.99$ (FFT channel 4, period 32), with the presence of impulse interference ($p = 0, 0.01, 0.05, 0.1, 0.25$). The simulation results are depicted in Fig. 10, 11, and 12.

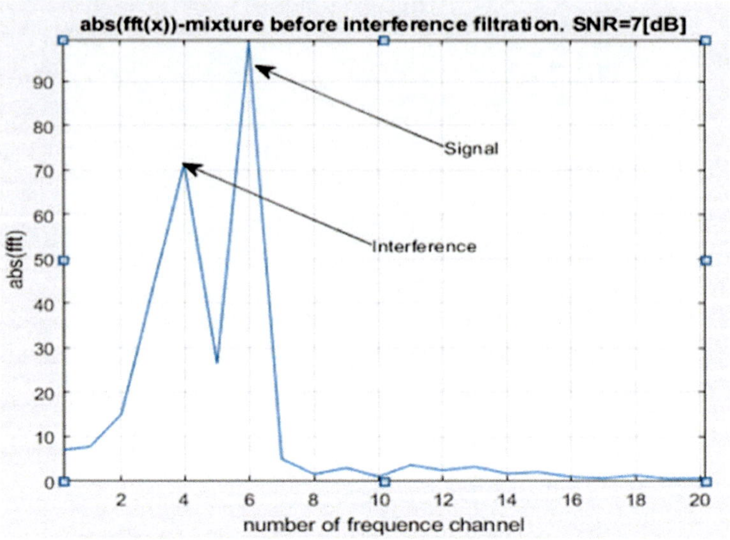

a)

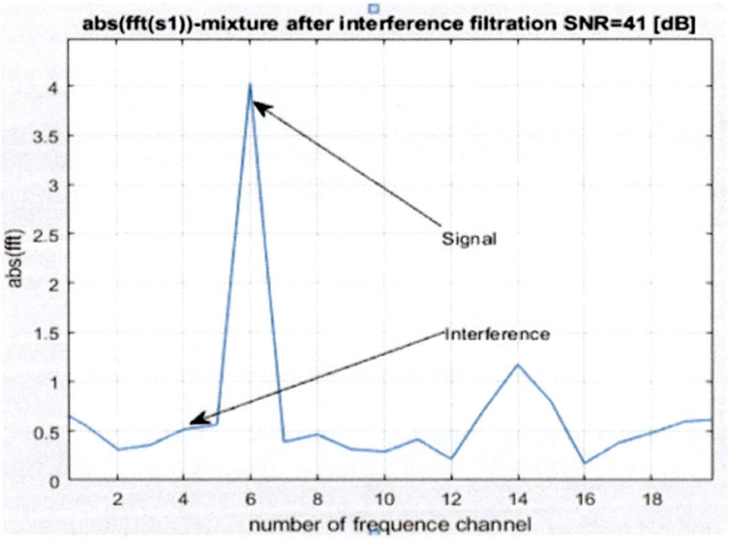

b)

Fig. 10. Signal and interference spectra: a) signal and correlated noise before filtration; b) signal and correlated noise after filtration

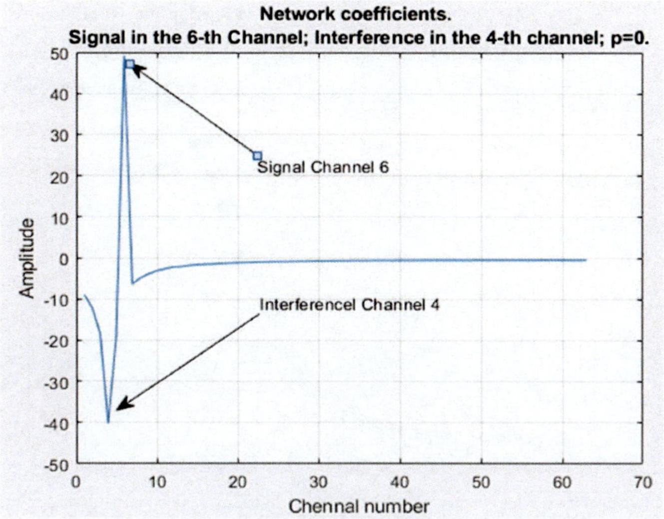

Fig. 11. Network coefficients

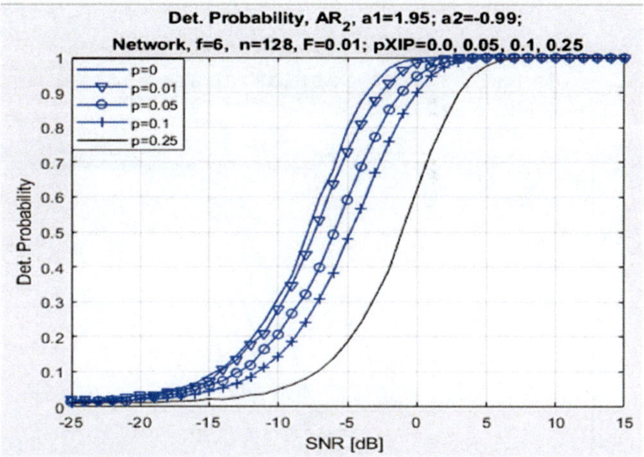

Fig. 12. Detection characteristics of the neural network algorithm at scenario 2 in the 6-th frequency channel under the influence of impulse interference. The probability of appearance impulse interference is $p = 0, 0.01, 0.05, 0.1, 0.25$.

4.3 Comparative Analysis

Integrating the simulation outcomes of both algorithms and visualizing the detection characteristics enables the analysis of the performance of the synthesized algorithms (Fig. 13, Fig. 14, and Fig. 15).

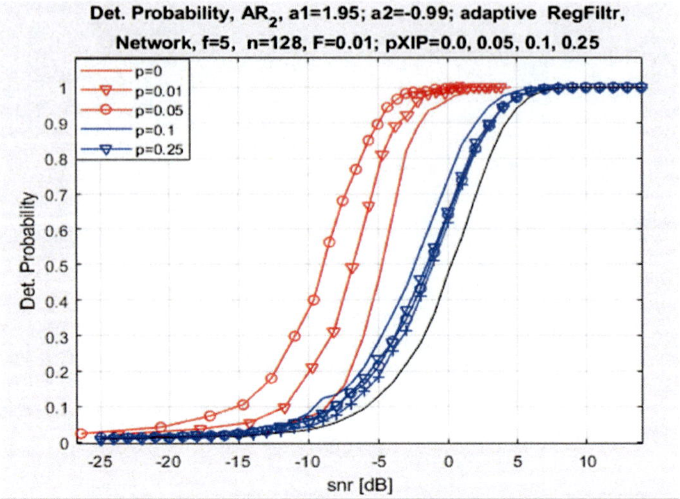

Fig. 13. Detection characteristics of the adaptive rejected filter (Fig. 1) and neural network algorithm (Fig. 5) with different probabilities of impulse interference $p = 0, 0.01, 0.05, 0.1, 0.25$. Harmonic Signal is located in the 5-th frequency channel ($f = 5$).

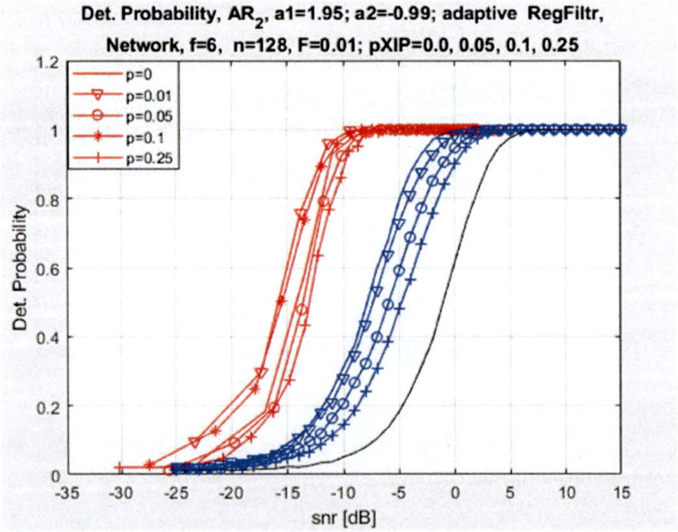

Fig. 14. Detection characteristics of the adaptive rejected filter (Fig. 1) and neural network algorithm (Fig. 5) with different probabilities of impulse interference $p = 0, 0.01, 0.05, 0.1, 0.25$. Harmonic Signal is located in the 6-th frequency channel ($f = 6$).

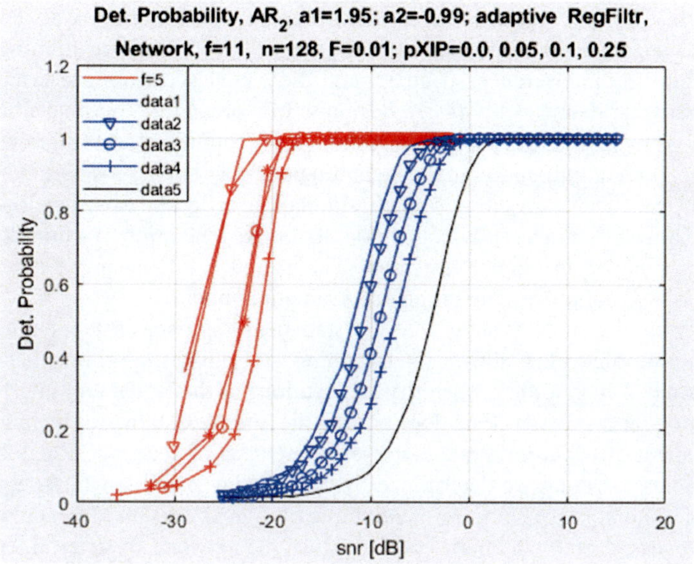

Fig. 15. Detection characteristics of the adaptive rejected filter (Fig. 1) and neural network algorithm (Fig. 5) with different probabilities of impulse interference. Harmonic Signal is located in the 11-th frequency channel ($f = 11$).

According to the simulation results and comparative analysis of both algorithms, it is worth noting that the effectiveness of the network algorithm is lower than that of the optimal one. The difference in the threshold signal at the level of ensuring the detection probability of 0.9 varies from 5 dB, when the signal is in the channel adjacent to the interference (the 5th channel), to 15 dB when the signal frequency (the 11th channel) is significantly higher than the central frequency of the narrowband interference (4-th channel).

At the same time, both algorithms ensure the robustness of detection characteristics to the action of intense impulse interference. The range of changes in the threshold signal when the probability of impulse interference changes from 0 to 0.25 varies within from 5 to 6 dB.

5 Conclusions

The paper addresses the synthesis of two algorithms designed for detecting a harmonic signal amidst correlated interference.

The first algorithm (Fig. 1) is synthesized through a statistical approach, incorporating a statistical interference model and accounting for a priori uncertainty in the parameters of both the interference model and the signal. The outcome of this synthesis yields a quadrature structure where the phase is estimated, resulting in a phase-invariant detector.

The synthesis of the second algorithm (Fig. 2) is grounded in a neural network approach, coupled with preprocessor processing of a signal and interference mixture utilizing a fast Fourier transform algorithm. A simulation program was developed in the Python language for implementing this algorithm. During the training phase of the algorithm, the interference spectrum and the signal spectrum are evaluated. The coefficients of the neural network are designed to replicate the optimal filter's shape, akin to the Wiener filter. This structure results in the suppression of interference in the clutter region and amplification in the signal region.

For performance analysis, a Matlab simulation program was developed, which includes a harmonic signal module, the module for the generation of a mixture of correlated autoregressive interference and impulse interference, a signal processing algorithm module, and a signal processing efficiency evaluation module.

The results of the study show that the statistical approach in the synthesis of the signal detection algorithm allows for obtaining an optimal algorithm that uses all a priori information about the probability distributions of the signal and interference. At the same time, network synthesis does not require complex statistical techniques and a priori statistical distribution models, it uses a universal structure of a neural network, the coefficients of which are determined by the learning process, which requires large volumes of simulation of samples of mixtures of signals and interference, and a long learning process, during which the coefficients of the network are determined.

Based on the simulation results (Fig. 13, Fig. 14, and Fig. 15) and the comparative analysis of the two algorithms, it is important to emphasize that the efficiency of the network algorithm is inferior to the efficiency of the optimal one.

Nevertheless, it is worth noting that both algorithms demonstrate robustness in maintaining detection characteristics under the influence of intense impulse interference.

References

1. Prokopenko, I.: Robust methods and algorithms of signal processing. In: Microwaves, Radar and Remote Sensing Symposium (MRRS), Kiev, Ukraine, pp. 71–74. IEEE (2017). https://doi.org/10.1109/MRRS.2017.8075029
2. Prokopenko, I., Dmytruk, A.: Robust algorithm for detecting moving targets against the background of non-Gaussian clutter. In: Workshop on Microwave Theory and Techniques in Wireless Communications (MTTW), Riga, Latvia, pp. 129–133. IEEE (2022). https://doi.org/10.1109/MTTW56973.2022.9942645
3. Prokopenko, I., Yanovsky, F., Ligthart, L.: Adaptive algorithms for weather radar. In: First European Radar Conference, Amsterdam, Netherlands, pp. 329–332. IEEE (2004)
4. Solomentsev, O., Zaliskyi, M., Nemyrovets Y., Asanov, M.: Signal processing in case of radio equipment technical state deterioration. In: Signal Processing Symposium (SPSympo), Debe, Poland, pp. 1–5. IEEE (2015). https://doi.org/10.1109/SPS.2015.7168312
5. Shevitski, B., Watkins, Y., Man, N., Girard, M.: Digital Signal Processing Using Deep Neural Networks. U.S. Department of Energy, USA (2023). https://doi.org/10.2172/1984848
6. Tian, C., Hong, M., Li, D., Yuan, D.: Deep recurrent neural network for ground-penetrating radar signal denoising. In: 4th International Conference on Intelligent Information Processing, Guangzhou, China, pp. 85–88. IEEE (2022). https://doi.org/10.1109/IIP57348.2022.00024
7. Acharya, U., Fujita, H., Oh, S., Hagiwara, Y., Tan, J., Adam, M.: Application of deep convolutional neural network for automated detection of myocardial infarction using ECG signals. Inf. Sci. **415**, 190–198 (2017). https://doi.org/10.1016/j.ins.2017.06.027

8. Purwins, H., Li, B., Virtanen, T., Schlüter, J., Chang, S., Sainath, T.: Deep learning for audio signal processing. IEEE J. Sel. Top. Signal Process. **13**(2), 206–219 (2019). https://doi.org/10.1109/JSTSP.2019.2908700
9. Geng, Z., Yan, H., Zhang, J., Zhu, D.: Deep-learning for radar: a survey. IEEE Access **9**, 141800–141818 (2021).https://doi.org/10.1109/ACCESS.2021.3119561
10. Nanduri, A., Sherry, L.: Anomaly detection in aircraft data using Recurrent Neural Networks (RNN). In: Integrated Communications Navigation and Surveillance (ICNS), Herndon, VA, USA, pp. 5C2-1–5C2-8. IEEE (2016). https://doi.org/10.1109/ICNSURV.2016.7486356
11. Jiang, W., Ren, Y., Liu, Y., Leng, J.: Artificial neural networks and deep learning techniques applied to radar target detection: a review. Electronics **11**(1), 156 (2022). https://doi.org/10.3390/electronics11010156
12. Delamou, M., Bazzi, A., Chafii, M., Amhoud, E.: Deep learning-based estimation for multitarget radar detection. In: 97th Vehicular Technology Conference (VTC2023-Spring), Florence, Italy, pp. 1–5. IEEE (2023)
13. Wang, L., Tang, J., Liao, Q.: A study on radar target detection based on deep neural networks. IEEE Sens. Lett. **3**(3), 1–4 (2019)
14. Mason, E., Yonel, B., Yazici, B.: Deep learning for radar. In: IEEE Radar Conference (RadarConf), Seattle, WA, USA, pp. 1703–1708. IEEE (2017)
15. Carrera, E., Lara, F., Ortiz, M., Tinoco, A., Leon, R.: Target detection using radar processors based on machine learning. In: 2020 IEEE ANDESCON, Quito, Ecuador, pp. 1–5. IEEE (2020). https://doi.org/10.1109/ANDESCON50619.2020.9272173
16. Chen, S., Wang, H., Xu, F., Jin, Y.: Target classification using the deep convolutional networks for SAR images. IEEE Trans. Geosci. Remote Sens. **54**(8), 4806–4817 (2016)
17. Garth, L., Poor, H.: Detection of non-Gaussian signals: a paradigm for modern statistical signal processing. Proc. IEEE **82**(7), 1061–1095 (1994). https://doi.org/10.1109/5.293163
18. Hummels, D., Ahmed, W., Mousavi, M.: Adaptive detection of small sinusoidal signals in non-Gaussian noise using an RBF neural network. IEEE Trans. Neural Netw. **6**(1), 214–219 (1995)
19. Kelly, E.: An adaptive detection algorithm. IEEE Trans. Aerosp. Electron. Syst. **AES-22**(2), 115–127 (1986)
20. Prokopenko, I., Omelchuk, I., Chirka, J., Yanovsky, F.: Detection of Markovian signals on the background of Markovian interferences: prior uncertainty case. In: Proceedings of SPIE 7502, Photonics Applications in Astronomy, Communications, Industry, and High-Energy Physics Experiments (2009)
21. Box, G., Jenkins, G.: Time Series Analysis: Forecasting and Control, Revised Edition. Holden-Day, Oakland (1976)

Optimal Pathfinding Based on Artificial Intelligence Tools

Serhii Migel, Maryna Maloied, Maksym Zaliskyi(✉), Anzhela Lelechenko, Alina Osipchuk, and Oleksandr Solomentsev

National Aviation University, Liubomyra Huzara Avenue, 1, Kyiv 03058, Ukraine
mzaliskyi@nau.edu.ua

Abstract. Nowadays, artificial intelligence tools are developing at a fast pace, improving their efficiency and productivity. Large volumes of data generated in various spheres of human activity have not become a problem for modern information technologies. The artificial intelligence tools use statistical theory, methods of machine and deep learning. These tools are the basis for obtaining new knowledge, decision-making, optimization, and are the main element of the well-known Industry 4.0 approach. Transport and logistics problems can also be solved with the help of artificial intelligence. Therefore, this paper focuses on researching the use of reinforcement learning to find the best vehicle route. The developed method consists of three elements: a nonparametric clustering procedure, Dijkstra's algorithm and a method of accelerating calculations. In general, the proposed method is reduced to: 1) transformation of the original image of the map into a maze with the designation of permitted and prohibited passages; 2) determination of the fastest path from the starting point to the exit from the maze. The paper also contains the program realization for implementing the proposed method.

Keywords: Machine Learning · Reinforcement Learning · Pathfinding · Nonparametric Clustering · Dijkstra's Algorithm · DBSCAN

1 Introduction

1.1 Introduction to the Problem

In accordance with the trends of sustainable development, the complication and integration of spheres of human activity requires the use of new approaches, methods and technologies [1, 2]. The modern industry is considered in the Big Data framework [3], therefore there is an urgent need to improve the methods of intelligent data processing and decision-making based on them [4, 5]. Nowadays, artificial intelligence (AI) tools are developing at a fast pace, improving their efficiency and productivity [6, 7]. Those trends are related to use of intelligent technologies and robots [8]. Intellectualization is usually based on statistical theory, methods of machine learning (ML) and deep learning (DL) [9, 10]. The further development of computer information technologies will contribute to the expansion of ML technologies to different fields of modern daily human life [11].

I. Ostroumov and M. Zaliskyi (Eds.): ACASD 2024, LNNS 992, pp. 242–261, 2024.
https://doi.org/10.1007/978-3-031-60196-5_19

Nevertheless, the neural networks are constantly spreads to modern applications, the important role inside the AI remains under ML control [12]. The most promising ML branch that cumulates advantages of supervised and unsupervised learning is the reinforcement learning (Fig. 1). It includes the environment representation in form of a maze and pathfinding algorithms usage for task solving.

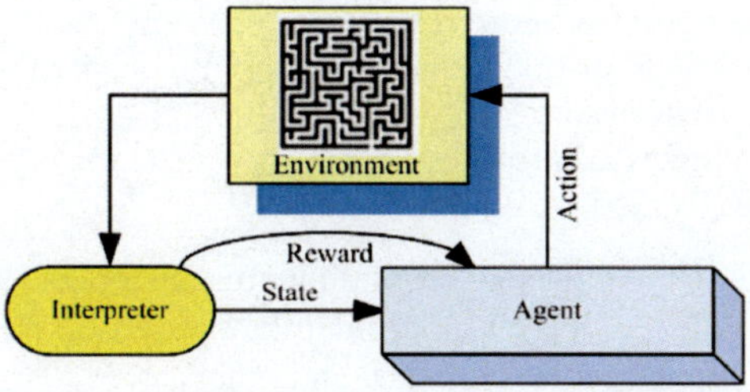

Fig. 1. The reinforcement learning approach.

The main advantages of reinforcement learning based on pathfinding algorithms are:

1. The possibility of correction and adaptation of results according to real-time data.
2. The possibility of taking into account each limitation and changeable information.
3. Data-driven decision-making [13, 14].

The widespread form for the environment representation inside ML applications, especially equipped by computer vision systems is a form of an image. Those images can be shown to interpreter (Fig. 1) for assigning a new task. If task conditions are entered like a pair of starting and finishing points and the existing ways between them are clear then the image turns into a classical maze that can be successfully solved by any ML pathfinding method. But in the case when system cannot clearly distinct passages from walls during the image analysis, and other ways for task clarification is unavailable, it needs to evolve additional techniques for the available passages exploration [15].

This paper considers the DBSCAN clustering algorithm for the conversion of an image into the maze by passages exploration. Then maze can be solved by any pathfinding method but the Dijkstra's algorithm has the largest compatibility with DBSCAN comparing with other pathfinding methods. This paper demonstrates the main advantages of Dijkstra's algorithms together with DBSCAN.

1.2 Motivation

Machine learning is a modern trend in the field of AI and data science [16, 17]. ML methods are mainly aimed at solving regression and classification problems based on mathematical tools of statistics, clustering, regression analysis, optimization theory, and others [18, 19].

The AI tools based on DL give the possibility to deal with sophisticated tasks in conditions of multivariate data distributions with soft requirements to mathematical skills of researcher [20]. In most cases these methods need to be trained on big dataset and have a weakness in learning time and computing complexity [21]. These tools are the basis for obtaining new knowledge, decision-making, optimization, and are the main element of the well-known Industry 4.0 approach [22]. Transport and logistics problems can also be solved with the help of artificial intelligence [23].

For transport infrastructure management, the AI-based tools provide cost-effective solving the following problems:

- determining the flight routes of aircraft;
- finding the optimal route for taxi services taking into account traffic jam;
- optimization of delivery system;
- unmanned aerial vehicles control;
- robots' automation [24, 25].

The are many research of optimal pathfinding presented in the modern literature [12, 26]. The main approaches to solve this problem are based on mathematical statistics, theory of navigation, graph theory, logistics, and others.

Alternative way is to use ML and DL, among which clustering methods, genetic algorithms, recurrent neural networks, methods of regression, iterative optimization procedures have obtained the widespread utilization [16, 23].

Therefore, the methods of finding the optimal route are still evolving, so this motivated us to synthesize a new method based on artificial intelligence with reinforcement learning in order to minimize the duration of the route and simplify the computer calculations.

1.3 Contribution

The scientific novelty of this paper is the further improvement of the methods of finding the optimal path based on the application of reinforcement learning technologies. The main contribution concerns the development of a new approach to solving the problem of finding a way out of a maze based on machine learning methods. This can be considered an extension of the algorithmic support of AI tools for solving industry problems.

1.4 The Organization of the Paper

The paper consists of four sections. The first section presents the background information and highlights the necessity of considered problem solving. The second section explains the steps of proposed ML method for pathfinding in the maze. The third section contains the program code for implementing the proposed method. The fourth section is conclusion.

2 Materials and Methods

2.1 Classical and Grid-Based DBSCAN

The classical DBSCAN (density-based spatial clustering of applications with noise) algorithm includes following three types of classifiers (Fig. 2, 3): C_1 – points classifier into core and non-core points (Fig. 3, left side); C_2 – core points classifier into cluster groups (Fig. 3, right side, black markers); C_3 – non-core points classifier into cluster groups or noise (Fig. 3, right side, gray and white markers).

During cluster formation procedure the important role is also devoted to queue with a list of core points already attached to the common cluster group.

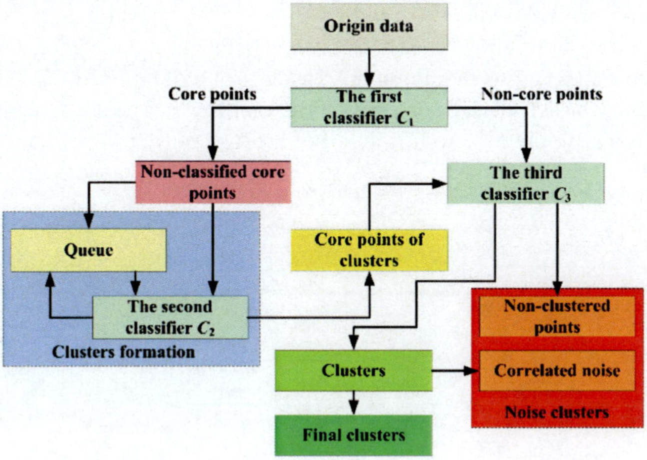

Fig. 2. The interconnection of three types of classifiers.

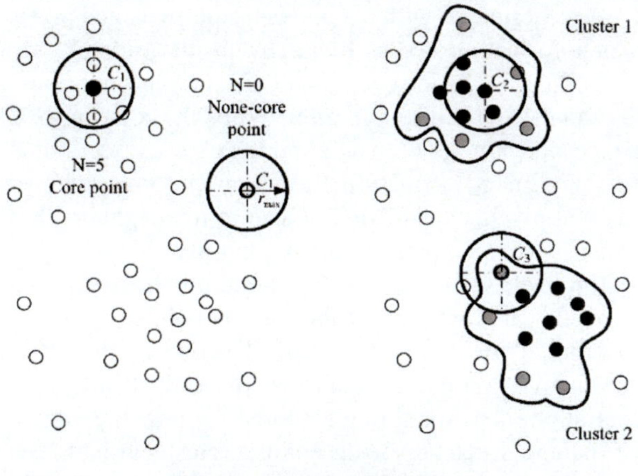

Fig. 3. Clusters formation.

The main method of class membership recognition inside C_1, C_2, C_3 classifiers (Fig. 2, Fig. 3) are the distance measurements between neighboring points and comparing it with a given limit range r_{max}. This method remains actual in the case of an image with only the difference, that the image clustering has deal with 5D space containing an information about the color in the RGB coding system (3 colors) and the position (2 coordinates).

Let's omit well known details of classical DBSCAN clustering algorithm realization and go deeper to the Grid-based version of DBSCAN because it is more capable for an image clustering [27, 28]. The simplest way how Grid-based DBSCAN extracts clusters from an image is following:

1. Assume that all 5D points of an image are core points. It eliminates the usage of two of three mentioned classifiers (C_1 and C_3 from Fig. 2). Then remained $C_2(X_{cur}, X_{px})$ classification function (Fig. 4) should check, whether or not the 5D distance between the currently attached to the common cluster group 5D point (X_{cur}) and yet not a clustered (X_{px}) based on usage of a given limit range r_{max}. The positive result indicates that X_{px} is close to the X_{cur} so it is a part of the X_{cur} cluster.

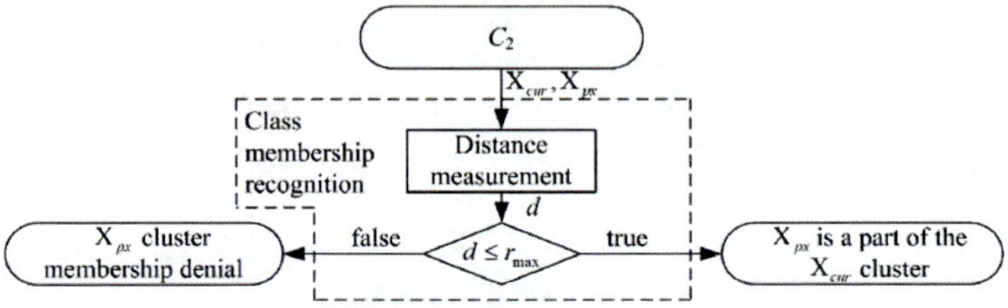

Fig. 4. The second classifier.

2. The grid cell spaces of images with 5D points inside are called pixels. They represent the variation of information access hierarchy inside grid-split space by a unique indexed address of its cell. Inside every pixel the color information (3 colors) and the position (2 coordinates) have different roles. While the color from 5D point is a main 5D space distance indicator between neighboring pixels, its coordinates are indicators of pixel address. For image navigation, the address information opens the access not only to the internal pixel information but also to its neighbors, because the known address difference between a pair of neighboring pixels is widely programmatically used for neighbor pixels searching acceleration through the grid (Fig. 5).
3. The neighbor pixels can contain neighboring 5D points or not. But pixels that are not neighbors definitely do not contain them. That is why the Grid-based DBSCAN introduces the window – a concise list or a map of relatively neighbor pixels addresses around the central one that should be explored for possible neighboring 5D points detection for its potential attaching to the common cluster group. The simplest window form is shown on Fig. 6. Its space exploration sequence expects four $C_2(X_{cur}, X_{px})$ classifications between neighbor pixel pairs placed in same row or column. The window's currently central pixel (X_{cur}) is already attached to the common cluster group and neighbor pixels (X_{nb1}, …, X_{nb4}) if they are not a part of some cluster, they should be explored to their closeness to X_{cur} for further attaching to its cluster group. According to neighbor numeration, the neighbor with the lowest index (X_{nb1}) is supposed to be explored in the first turn (by $C_2(X_{cur}, X_{nb1})$ function). Next window explores the neighbor pixel with the second index (X_{nb2}), and so on. The complete window space exploration algorithm is shown on Fig. 7.

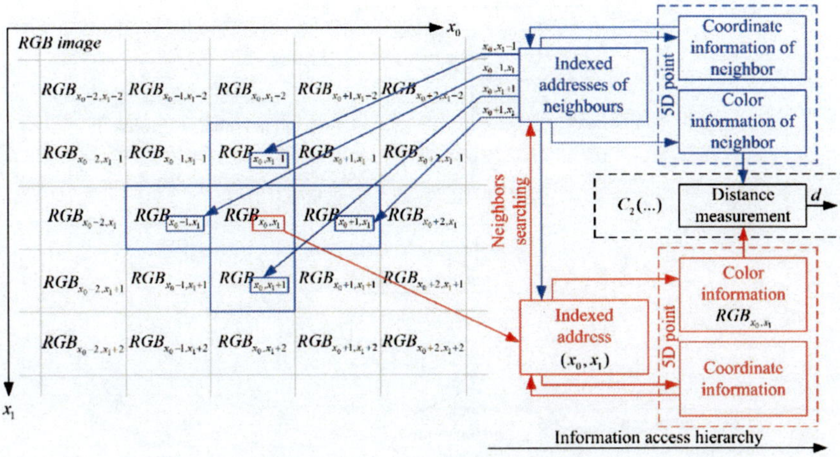

Fig. 5. The grid cell spaces of images.

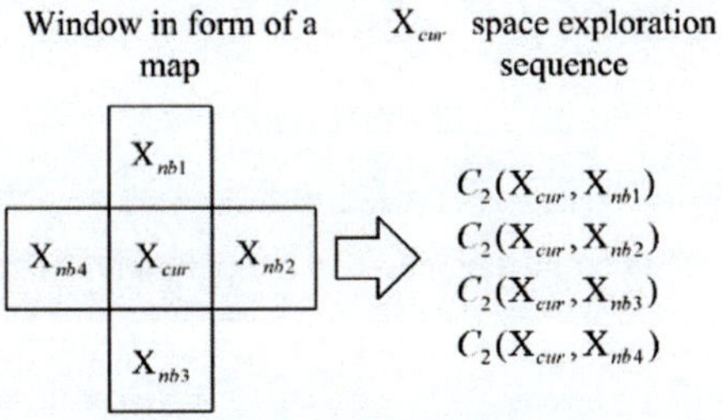

Fig. 6. The simplest window form.

4. The cluster formation procedure involves all previously mentioned tools in one common system. The aim of this procedure is gathering contiguous fragments of similar pixels to the same cluster. It is realized by the window's space exploration (Fig. 7) through the queue (Fig. 8). During cluster formation, the list of pixels already attached to the common cluster group is placed to the queue with some queue positions. Then window sequentially uses each queue position like its central one (X_{cur}) for the exploration of neighboring pixels for the cluster membership. As a result, pixels ($X_{px0}, X_{px1}, \ldots$) inside the queue with lower queue position than the currently central pixel of the window (X_{cur}) are considered already explored while the rest pixels in the queue are considered like yet not explored by the window. The newly attached clusters members after window's space exploration will be added to the end of the queue with higher queue position than the queue position of window's currently central pixel (X_{cur}). So, the spaces around them gradually become explored on further steps of queue processing.

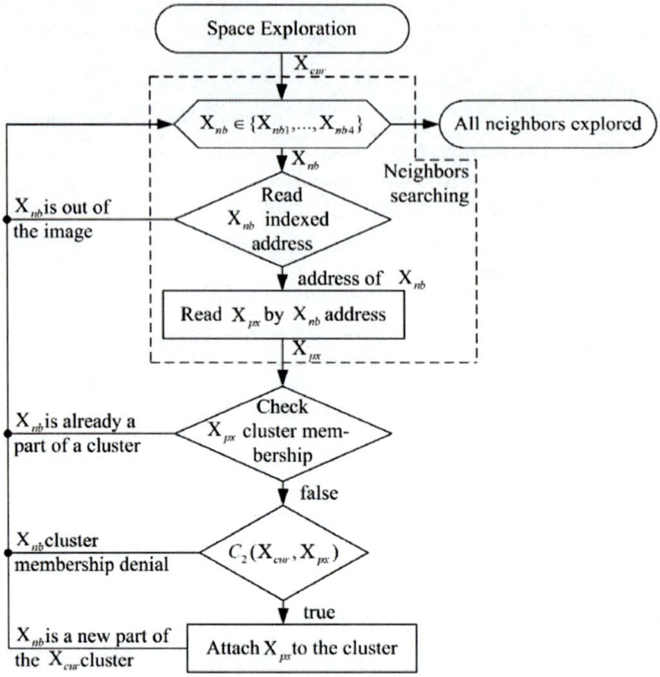

Fig. 7. The complete window space exploration algorithm.

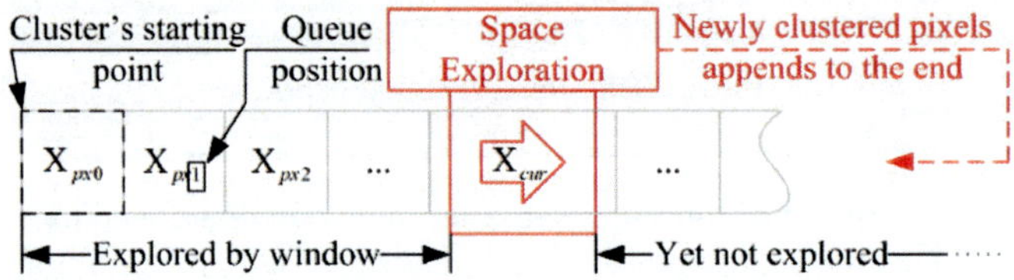

Fig. 8. The queue formation.

To conclude Grid-based DBSCAN operation with window acceleration let's consider an example. Let's say we have an image (Fig. 9), where contiguous yellow pixel group should be turned to the common cluster. For this we use the cluster formation procedure with window from Fig. 6 and with the queue (Fig. 8). After all similar pixels from Fig. 9 will be attached to the common cluster group their queue positions $(X_{px0}, X_{px1}, \ldots)$ become in the sequence shown on the figure.

The first 3 steps of cluster formation are shown separately on Fig. 10 and considered on Fig. 11.

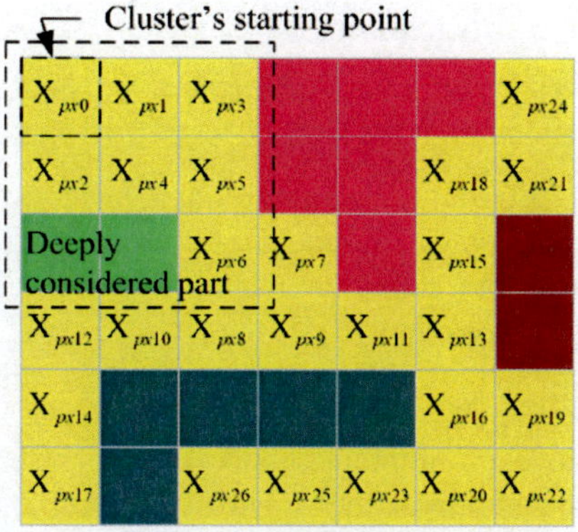

Fig. 9. Example of image.

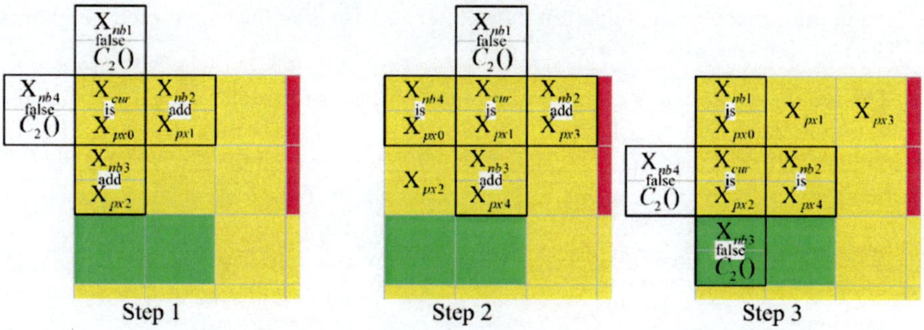

Fig. 10. The cluster formation for the image part.

Exploration step			Space exploration sequence	Space exploration results description	Cluster changes
Preparation			From left top corner	Starting point is added to cluster	X_{px0} is added
Step 1	X_{cur} is X_{px0}	Fig. 8, left side	$C_2(X_{cur}, X_{nb1}) = false$	X_{nb1} is out of the image	
			$C_2(X_{cur}, X_{nb2}) = true$	X_{nb2} is a new part of the X_{cur} cluster	X_{px1} is added
			$C_2(X_{cur}, X_{nb3}) = true$	X_{nb3} is a new part of the X_{cur} cluster	X_{px2} is added
			$C_2(X_{cur}, X_{nb4}) = false$	X_{nb4} is out of the image	
				Move X_{cur} to next position inside queue	X_{px0} is explored
Step 2	X_{cur} is X_{px1}	Fig. 8, center side	$C_2(X_{cur}, X_{nb1}) = false$	X_{nb1} is out of the image	
			$C_2(X_{cur}, X_{nb2}) = true$	X_{nb2} is a new part of the X_{cur} cluster	X_{px3} is added
			$C_2(X_{cur}, X_{nb3}) = true$	X_{nb3} is a new part of the X_{cur} cluster	X_{px4} is added
			$C_2(X_{cur}, X_{nb4}) = true$	X_{nb4} is already a part of a cluster	
				Move X_{cur} to next position inside queue	X_{px1} is explored
Step 3	X_{cur} is X_{px2}	Fig. 8, right side	$C_2(X_{cur}, X_{nb1}) = true$	X_{nb1} is already a part of a cluster	
			$C_2(X_{cur}, X_{nb2}) = true$	X_{nb2} is already a part of a cluster	
			$C_2(X_{cur}, X_{nb3}) = false$	X_{nb3} cluster membership denial	
			$C_2(X_{cur}, X_{nb4}) = false$	X_{nb4} is out of the image	
				Move X_{cur} to next position inside queue	X_{px2} is explored
....	....	....	....	....	....

Fig. 11. The iterative procedure of cluster formation.

2.2 Dijkstra's Algorithm Integration

Dijkstra's algorithm is a basic pathfinding algorithm. It operates according to following principle:

1. First it explores the shortest routes to neighboring points from starting that are attached with them immediately. Then it explores the shortest routes from previously explored points to the little further its neighbors. Next, algorithm continues the exploration until the finish point is achieved.
2. While the exploration procedure of the algorithm builds a table with distance measurements results to different points along different routes and sort its points depending on their range. Each its exploration step is performed around the points that simultaneously satisfies two conditions: they have the least distance from the start and yet not explored neighbors.
3. Depending on the distance it divides routes into two types: the shortest routes to the points and routs that are not the shortest. On each its step it tries to prolong to current one point only the routes that are the shortest for its previously explored neighbors. The rest routs become removed.

The result of Dijkstra's algorithm application has a form of a table of achievable points of the maze with the shortest possible ways required for their achieving. For the example from Fig. 9 the Dijkstra's algorithm table is shown on the Fig. 12. But because of the Dijkstra's algorithm needs a finish point we introduced it at the right bottom corner of the image from Fig. 9 (X_{px22}). That is why the table from Fig. 12 has its last row with the description of pixel X_{px22}. For the future examples in this paper with Fig. 9, exactly this finish point (X_{px22}) is used for Dijkstra's pathfinding algorithm conception explanation.

Current point	Previous neighbors	Shortest routes
X_{px0} (Start)	-	-
X_{px1}	X_{px0}	X_{px0}
X_{px2}	X_{px0}	X_{px0}
X_{px3}	X_{px1}	X_{px0}, X_{px1}
X_{px4}	X_{px1} X_{px2}	X_{px0}, X_{px1} X_{px0}, X_{px2}
X_{px5}	X_{px3} X_{px4}	$X_{px0}, X_{px1}, X_{px3}$ $X_{px0}, X_{px1}, X_{px4}$ $X_{px0}, X_{px2}, X_{px4}$
....	....	....
X_{px22} (Finish)	X_{px19} X_{px20}	$X_{px0}, X_{px1}, X_{px3}, X_{px5}, \ldots, X_{px19}$ $X_{px0}, X_{px2}, X_{px4}, X_{px5}, \ldots, X_{px20}$

Fig. 12. The Dijkstra's algorithm table.

But in this case the Dijkstra's algorithm has following important features:

1. The distance estimation inside an image is a step number that need be passed between the closest start and current pixel through the available passages between neighbor pixel pairs placed in the same row or column (Fig. 13).

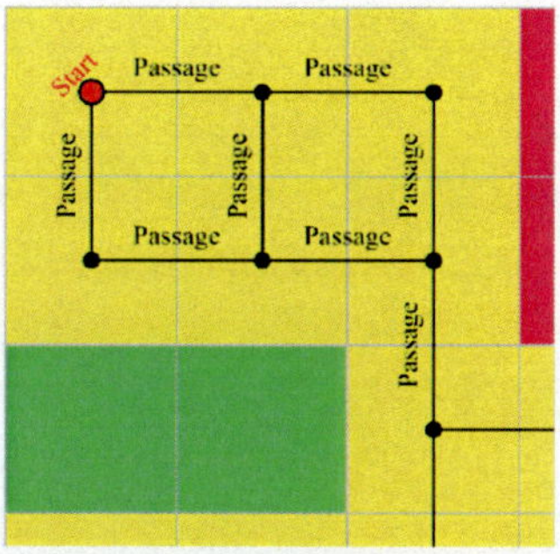

Fig. 13. The passage through the maze.

2. The large maze images can have a lot of equidistant branches that is often inappropriate to combine them into a common table. For those situations the Dijkstra's algorithm expects table usage with other concise format: where third column (the column that is most filled) of the table (Fig. 12) is completely removed, but its restoration is achievable by its second column by means of specified for those purposes the Dijkstra's reverse transition algorithm.
3. The concise Dijkstra's algorithm table creation is performed by Dijkstra's direct transition algorithm. But in the case of an image, it has concrete sequence for points attaching to table. According this sequence the sorted by distance criteria table without any points permutations is generated.

These concrete points attaching sequence has a significant impact for DBSCAN integration to Dijkstra's algorithm. It turns out that the points attaching sequence during the cluster formation procedure by DBSCAN is almost entirely the same as for the Dijkstra's direct transition algorithm. This statement is briefly demonstrated by table from Fig. 12 where it is not a coincidence that the first column of this table is completely the same as the sequence of pixels placed to queue during the DBSCAN's cluster formation procedure with window from Fig. 6. This interesting fact gives us the reason to assert that the Dijkstra's direct transition algorithm is can be practically completely replaced by the DBSCAN while its Dijkstra's algorithm table can be practically completely replaced to the queue of DBSCAN.

The result of the Dijkstra's direct transition algorithm on base of DBSCAN (Fig. 14) is a queue that can be transformed into the complete Dijkstra's algorithm table inside the Dijkstra's reverse transition algorithm. The principles for that queue to table transformations are following:

1. The rule of the special pixel positions. It declares the queue positions for starting and finishing pixels. According this rule, the queue positions of maze's starting pixels are the first while the queue positions of maze finish pixel is the last. It leads to introduction of urgent interruption of cluster formation after the closest finish pixel reaching (Fig. 14). This rule is only the one rule from the list that cannot be completely serviced by means of DBSCAN. That is why for this rule servicing in part of finish pixels the urgent interruption of cluster formation after the closest finish pixel reaching is provided.

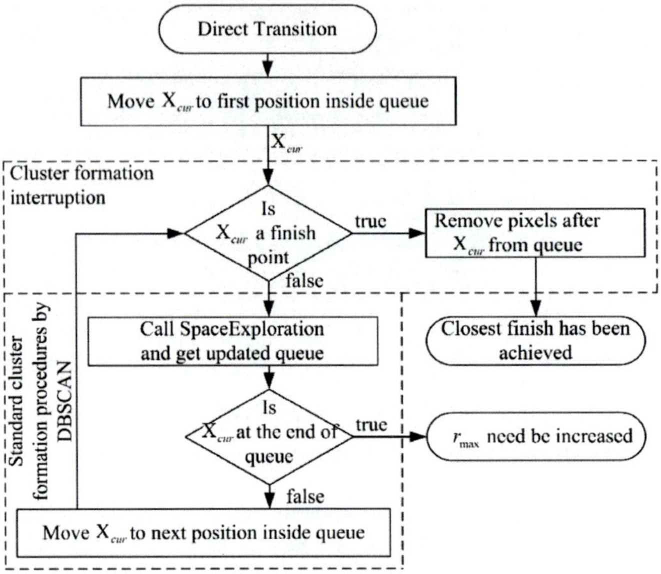

Fig. 14. The Dijkstra's direct transition algorithm on base of DBSCAN.

Although unlike the standard DBSCAN, Dijkstra's algorithm allows several starts so several first pixels in the queue can be marked like starting, but their presence needs any additional commands inside the Dijkstra's direct transition algorithm for their service. Considering the example from Fig. 9, in the case when the finish pixel is placed on the right bottom corner of the image (X_{px22}), it becomes obvious that the last window space exploration is hold around the window's currently central pixel (X_{cur}) with queue position $X_{cur} = X_{px21}$, because this is the step before X_{px22} that is detected like finish. So, unlike the standard DBSCAN clustering algorithm, pixels with further queue positions than X_{px22} (that are $X_{px23}, \ldots, X_{px26}$) no longer be added to the cluster during the Dijkstra's direct transition algorithm or will be immediately removed from it.

2. The rule of passage availability. It declares the distinction between passages and walls by their pixels presence inside the queue. According this rule, while a pixel pair is not included to the queue, the transition between them is considered inadmissible. It leads to introduction for cluster formation the r_{max} value adjusting procedure.

The DBSCAN clustering is used for the distinction of passages and walls inside an image between each pair of neighboring pixels (X_{cur} and X_{nb}). The algorithm decides, are neighboring pixels ($X_{nb1}, \ldots, X_{nb4}$) should be placed to the queue or not. For the correct recognition between the situations, have we transition or no, the r_{max} value should be chosen minimal but enough for turning all passable pixels into a common cluster separately from walls. But when the cluster formation by the algorithm completes without reaching finish pixel (Fig. 14) it definitely indicates that the r_{max} value need to be increased.

The example from Fig. 9 is briefly demonstrates that each pair of neighboring and even not neighboring pixels from the queue ($X_{px0}, X_{px1}, \ldots$) are connected by common transition. This rule has an exception only when a pair of pixels from the queue is clustered from different starts. But this situation is successfully solved by Dijkstra's reverse transition algorithm.

3. The rule of the distance. It declares the queue position as an implicit distance indicator from the closest starting pixel. According this rule, the queue positions of pixels placed nearby starts are low while the queue positions of pixels placed far away from starts are high. It leads to the conclusion that the closest finish pixel is always be found the first because it got the lowest queue position.

As it shown from the example in Fig. 9, inside the queue, the equidistant pixel groups become clustered to sequential queue positions so they go one after the other depending on their range from the closest starting pixel: $\{X_{px1}, X_{px2}\}$, $\{X_{px3}, X_{px4}\}$, $\{X_{px7}, X_{px8}\}$, etc. As higher the pixels queue positions as far it away from its closeness start. This also concerns to the finish pixel (X_{px22}) – the closest finish pixel is always be found the first together with its equidistant pixels group as soon as it far from the closest starting pixel.

4. The rule of the direction. It declares the navigation rule for finding the closest starting pixel from any other pixel already placed to properly formatted queue. According this rule each step inside the window (Fig. 6) through the sequence of pixels with lower and lower queue positions finally leads to the closest starting point along one of the shortest possible ways. But if currently central pixel (X_{cur}) is already a part of the shortest route than each neighboring pixel (X_{nb}) with lower queue position becomes its continuation.

This principle is a main navigational basis for shortest route tracing realization for Dijkstra's reverse transition algorithm. As we can see from Fig. 9 and from table in Fig. 12, having a currently central pixel of the window (X_{cur}) on any clustered pixel of the image, each neighbor ($X_{nb1}, \ldots, X_{nb4}$) with a lower queue position is placed one step closer to the starting than X_{cur}, while each neighbor ($X_{nb1}, \ldots, X_{nb4}$) with a higher queue position is placed one step away from the starting than X_{cur}. The sequential passage to neighbors with exclusively lower queue position follows each next step closer and closer to starting pixel, providing the start reaching in one of the shortest of possible ways. The Dijkstra's reverse transition algorithm on base of DBSCAN (Fig. 14) is a practical realization of "The rule of the direction". It carries out immediately after the direct transition reaches the closest finish pixel assigning to it the highest queue position and mark that it is a part of the shortest route. The processing procedure of Dijkstra's reverse transition algorithm starts from this finish in reversed order comparing to queue processing direction from Dijkstra's direct transition algorithm. But from the programming point of view the easiest way for the distinction of pixels with lover queue positions from others is the removing each pixel from the queue after its usage as a window's currently central pixel (X_{cur}). It leads to result that we have only pixels with lower queue position inside the queue and the positive answer for question in the form "Is this pixel at the queue?" is simultaneously gives us the positive answer to question in the form "Is this pixel at the queue has lower queue position then window's currently central pixel". For this processing way providing, each window processing step expects marks putting to each pixel inside the window around the pixel that is a part of the shortest route that they are also its continuations (Fig. 15). The pixels from the queue that are not received such a label at the time of their application as current (X_{cur}) are considered like pixels that do not belong to the shortest route.

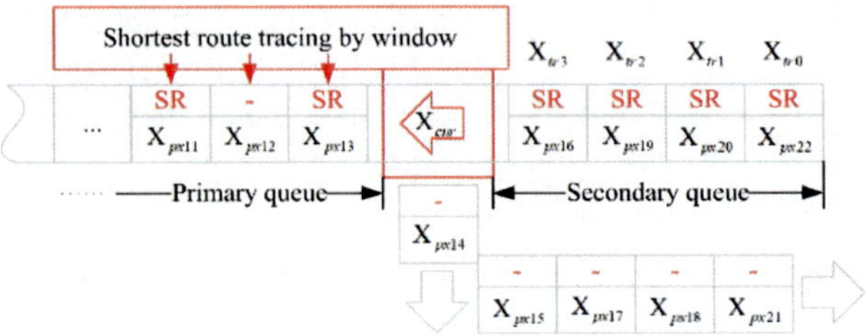

Fig. 15. The shortest route tracing.

Other components of Dijkstra's reverse transition algorithm (Fig. 16) have straight logic by following principles:

1. Because of not every pixel in primary queue belongs to the shortest route we need to support two procedures for its tracing: the secondary queue for the accumulation of pixels already detected like shortest route parts, and the window for tracing those pixels from primary queue. So, inside the Dijkstra's reverse algorithm transition realization the window is used completely differently than during the direct transition – like a filter-decoder of the shortest route branches to secondary queue from the cluster's pixels saved in primary queue.
2. According to "The rule of the direction", if currently central pixel (X_{cur}) is already a part of the shortest route than each neighboring pixel (X_{nb}) with lower queue position is its continuation. That is why the main windows task is the exploration around the currently placed to the secondary queue pixel (X_{cur}) the neighboring pixel (X_{nb}) with lower queue position for placing them to secondary queue for further steps executions. It is done in the special indirect way, described above (Fig. 15). But the first step of the Dijkstra's reverse algorithm transition is acceptable only with not empty markers of the shortest route belonging. For this, the founded closest finish pixel should be marked like the first pixel of the desired shortest route the next parts of each will be further traced by the window.
3. The reverse primary queue transition is realized by secondary queue that during the reverse transition becomes filled in reverse direction by pixels from the primary queue. The secondary queue processing by the window continues until it reaches to the starting pixel group trough the primary queue. It means that secondary queue is already gathered all pixels of shortest route and its equidistant branches like a sequence of its pixels.

Let's consider a one important remark concerned the Dijkstra's direct transition algorithm. In the case with Dijkstra's direct transition algorithm on base of DBSCAN the first cluster formation will be immediately interrupted after the finish pixel (X_{px22}) achieving without attaching to cluster all its components (Fig. 17).

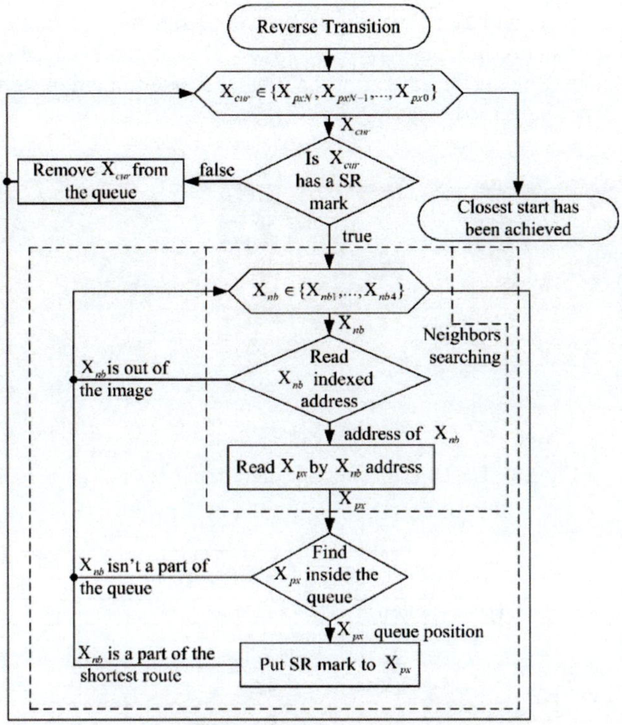

Fig. 16. The Dijkstra's reverse transition algorithm.

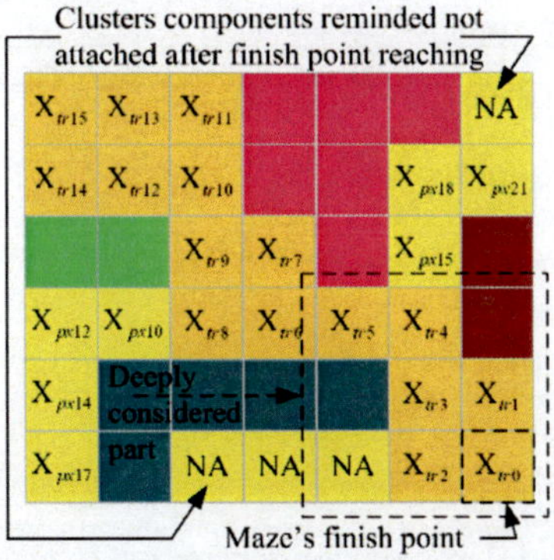

Fig. 17. The finish pixel achieving without attaching to cluster all its components.

After this, the finish pixel from secondary queue X_{px22} is duplicated to primary queue like X_{tr0} for further usage like currently central pixel of the window (X_{cur}) for the first Dijkstra's reverse algorithm transition step. This step and also two others are shown separately on Fig. 18 and considered on Fig. 19. The numbers on Fig. 17 ($X_{tr0}, X_{tr1}, \ldots$) are the pixels of equidistant branches of the traced shortest route and indicate pixels positions inside the secondary queue after the reverse transition realization completed. The queue positions with indexes $X_{px23}, \ldots, X_{px26}$ are missed because of they are not added to the cluster. The queue state after the first 4 reverse transition steps is shown on Fig. 15.

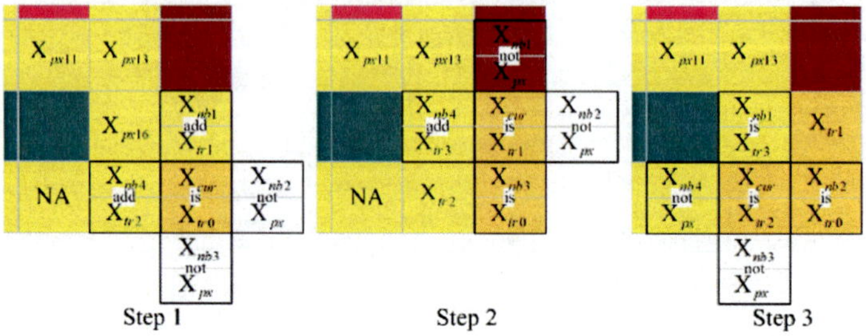

Fig. 18. The Dijkstra's reverse algorithm transition step.

Exploration step			Shortest route tracing results description	Shortest route changes
Preparation			Finish point is added to secondary queue	$X_{px22} \leftarrow SR$
Step 1	X_{cur} is X_{tr0}	Fig. 16, left side	X_{nb1} is a part of the shortest route	$X_{px19} \leftarrow SR$
			X_{nb2} is out of the image	
			X_{nb3} is out of the image	
			X_{nb4} is a part of the shortest route	$X_{px20} \leftarrow SR$
			Move X_{cur} to next position inside secondary queue	X_{tr0} is added
Step 2	X_{cur} is X_{tr1}	Fig. 16, center side	X_{nb1} isn't a part of the queue	
			X_{nb2} is out of the image	
			X_{nb3} is a part of the shortest route	$X_{px16} \leftarrow SR$
			X_{nb4} is a part of the shortest route	
			Move X_{cur} to next position inside secondary queue	X_{tr1} is added
Step 3	X_{cur} is X_{tr2}	Fig. 16, right side	X_{nb1} is a part of the shortest route	
			X_{nb2} is a part of the shortest route	
			X_{nb3} is out of the image	
			X_{nb4} isn't a part of the queue	
			Move X_{cur} to next position inside secondary queue	X_{tr2} is added
....	....	....	....	....

Fig. 19. The iterative procedure of shortest pathfinding.

2.3 Hybrid Algorithm Implementation for Image Processing

The complete Dijkstra's algorithm realization on the base of the DBSCAN is shown on the Fig. 20.

The aim of the direct transition is the exploration of available passages from an image and finding the closest one finish pixel (the end of the shortest route). The aim of the reverse transition is gathering each pixel of the shortest route to second new queue. Because of the shortest route can have the equidistant branches, all those pixels should also be placed to this second queue. In the simplest case, the optimal route formation algorithm only expects choosing from the shortest route branches. It gives us following advantages in group with DBSCAN:

1. Dijkstra's algorithm allows us solving mazes with several inputs and several outputs for the shortest route tracing inside mazes with race tasks.
2. It assumes the entering starting and finishing points to the maze in form of pixels with special colors for marking.
3. The r_{max} presence automates the differentiation between the situations where is a passage and not.
4. The equality of passages weights between neighboring pixels leads to the similarity of the direct transition of the maze by Dijkstra's algorithm with points attaching sequence during the cluster formation procedure by DBSCAN. It makes available the integration of Grid-based DBSCAN to Dijkstra's algorithm.
5. After the integration of Grid-based DBSCAN to Dijkstra's algorithm, it becomes realized like a part of the first cluster formation by DBSCAN. But even this first cluster formation should not be necessary finished after the closest finish pixel finding.

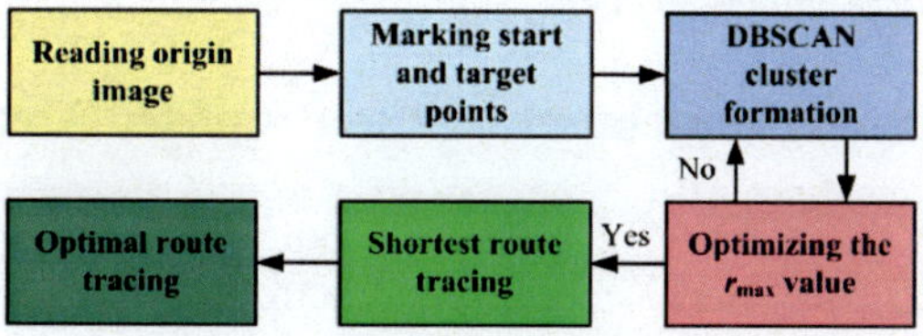

Fig. 20. The complete Dijkstra's algorithm.

The following examples of this paper consider the usage of Dijkstra's algorithm realization on the base of DBSCAN for pathfinding inside the properly formatted images.

The proper image format for those images are:

1. Each starting point is marked by red RGB (255, 0, 0).
2. Each finish point is marked by blue RGB (0, 0, 255).
3. Each starting and finishing point groups are placed inside the passages without any direct contact with walls.
4. Any other pixel at the image is not red RGB (255, 0, 0) except starting and any other pixel at the image is not blue RGB (0, 0, 255) except finishing. If we initially have the pixels of those colors, we need to convert them into the closest one red RGB (254, 0, 0) or blue RGB (0, 0, 254) that is visually do not changes image quality.

To provide the service of listed image format we need also introduce to Dijkstra's algorithm realization on the base of the DBSCAN some modifications:

1. We add for each pixel a 6-th dimension (additionally to 3 colors and 2 coordinates) with following numeral information, both of them have two different meanings: "–1" – pixel is start and also pixel is a part of the first queue (and also of a cluster); "–2" – pixel is finish and also pixel is a part of the shortest route. That binary mining of value "–1" eliminates us from several "if()" questions inside the Dijkstra's direct

transition algorithm programming realization (Fig. 14) while the binary meaning of value "–2" eliminates us from adding the first achieved finish point to secondary queue.
2. We add algorithm that extends the image space of start pixels not only to each red pixel but also to its immediate neighbors and that extends the image space of finish pixels not only to each blue pixel but also to its immediate neighbors. Because of the mazes passages can have a color different from red and blue but after this preparation procedure some starts and finishes become a part of the same cluster. Without those procedures the starting and finishing pixels cannot become a part of the same cluster.
3. Algorithm in Fig. 7 performs processing of large arrays. If it is made by a separate subprogram, then according to the rules of MathCAD, copies of the original arrays for processing will be created in the middle of this subprogram, which is very expensive from the point of view of resource consumption. Therefore, the algorithm in Fig. 14 is implemented immediately with the algorithm (Fig. 7) built into it.
4. Dijkstra's reverse transition algorithm (Fig. 16) for the purpose of simplification and, accordingly, the current visualization task is not fully executed and marks "–2" on the impassable parts of the maze, but visually, the shortest path between the specified points is correctly provided by it.

3 The Program for Implementing the Proposed Method

This section shows the example of proposed method realization in MathCAD. The developed program is quite simple and does not require powerful computation resources.

The software implementation of the described algorithm within the MathCAD environment can be considered in the following step framework:

Step 1. Typing the name of the file with the maze to read and displaying its contents.
Step 2. Formation of the RGB array with the combined color component of each pixel.
Step 3. Formation of the "Q" array with an additional information component to each pixel (6D format).
Step 4. Visualization of the "Q" array in order to make sure that the image formatting is correct (that all start and finish points are found, that no extraneous image points are recognized as start or finish points):
Step 5. Enter the starting points in the queue.
Step 6. Form a window (Fig. 6):
Step 7. Enter the limit range and function $C_2(X_{cur}, X_{px})$ (Fig. 4):
Step 8. Dijkstra's direct transition algorithm (Fig. 14).
Step 9. Display the form of the obtained cluster. Make sure that the r_{max} value is chosen enough for finish point achieving and for turning all passable pixels into a common cluster separately from walls.
Step 10. Dijkstra's reverse transition algorithm (Fig. 16).
Step 11. The shortest routes visualization.

The program realization in MathCAD is shown in Fig. 21 for common labyrinth illustration.

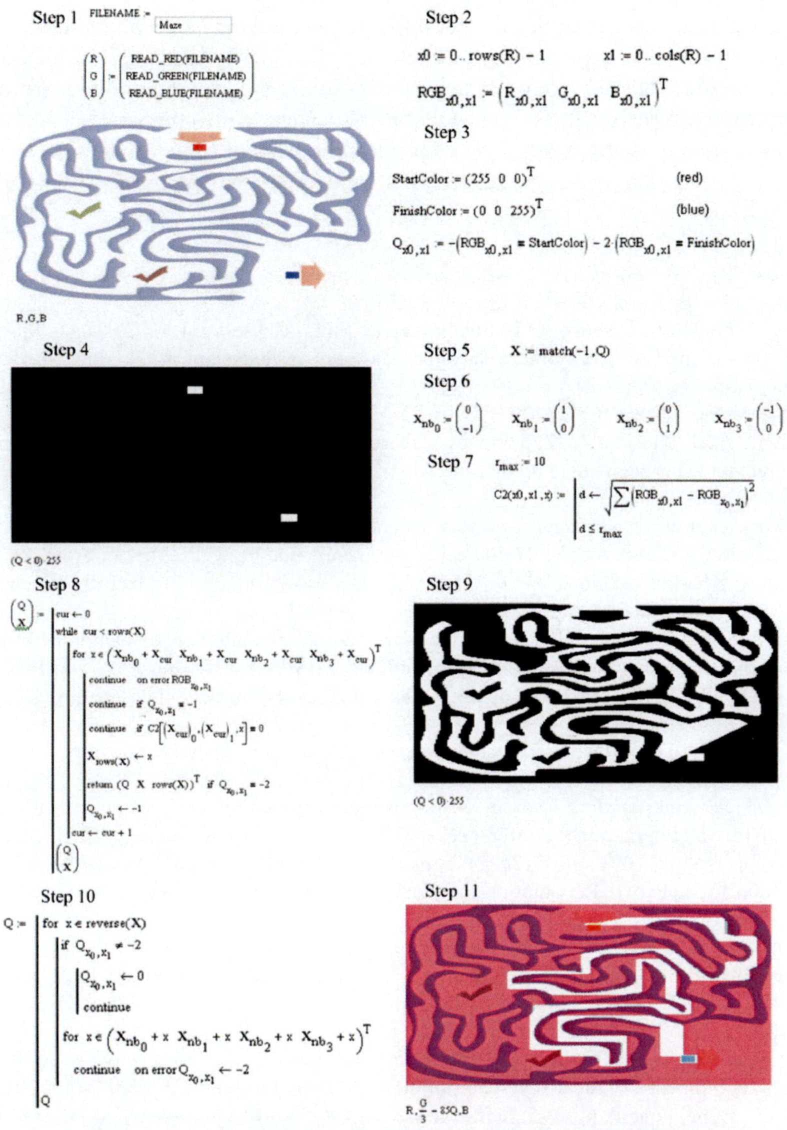

Fig. 21. The program example for implementing the proposed method.

4 Conclusions

The paper presents the artificial intelligence tool to solve the problem of optimal pathfinding. The proposed method is based on DBSCAN nonparametric clustering procedure and Dijkstra's algorithm. The combination of these methods made it possible to realize the features of reinforcement learning, which are associated with advantages in terms of duration and simplicity of calculations.

The paper also concentrates on step-by-step methodology for optimal routes determining with detailed explanations of each step. For clarity of the given method, the authors provided the example of its implementation in the programming environment.

The future research directions will be associated with comparative analysis of developed methods with existing ones for different cases of route calculation.

References

1. Poole, D.L., Mackworth, A.K.: Artificial Intelligence: Foundations of Computational Agents. Cambridge University Press, Cambridge (2017)
2. Tachinina, O.M., Lysenko, O.I., Ponomarenko, S.O., Alekseeva, I.V.: Conceptual proposals for the creation of a fully reusable light-class aerospace system in Ukraine. In: IEEE 6th International Conference on Methods and Systems of Navigation and Motion Control, Kyiv, Ukraine, pp. 85–88. IEEE (2020). https://doi.org/10.1109/MSNMC50359.2020.9255504
3. Al-Azzeh, J., Mesleh, A., Zaliskyi, M., Odarchenko, R., Kuzmin, V.: A method of accuracy increment using segmented regression. Algorithms **15**(10), 378, 1–24 (2022). https://doi.org/10.3390/a15100378
4. Larin, V., et al.: Prediction of the final discharge of the UAV battery based on fuzzy logic estimation of information and influencing parameters. In: 3rd KhPI Week on Advanced Technology, Kharkiv, Ukraine, pp. 1–6. IEEE (2022). https://doi.org/10.1109/KhPIWeek57572.2022.9916490
5. Ostroumov, I.V., Kuzmenko, N.S.: Accuracy estimation of alternative positioning in navigation. In: 4th International Conference on Methods and Systems of Navigation and Motion Control (MSNMC), Kiev, Ukraine, pp. 291–294. IEEE (2016). https://doi.org/10.1109/MSNMC.2016.7783164
6. Kubat, M.: An Introduction to Machine Learning. Springer, Cham (2017). https://doi.org/10.1007/978-3-319-63913-0
7. Sushchenko, O., et al.: Airborne sensor for measuring components of terrestrial magnetic field. In: 41st International Conference on Electronics and Nanotechnology (ELNANO), Kyiv, Ukraine, pp. 687–691. IEEE (2022). https://doi.org/10.1109/ELNANO54667.2022.9926760
8. Bishop, C.M.: Pattern Recognition and Machine Learning. Springer, Singapore (2006). https://doi.org/10.1007/978-0-387-45528-0
9. Yanovsky, F.J., Ivashchuk, V.E., Prokhorenko, V.P.: Through-the-wall surveillance technologies. In: 6th International Conference on Ultrawideband and Ultrashort Impulse Signals, Sevastopol, Ukraine, pp. 30–33. IEEE (2012). https://doi.org/10.1109/UWBUSIS.2012.6379723
10. Solomentsev, O., Zaliskyi, M., Nemyrovets, Yu., Asanov, M.: Signal processing in case of radio equipment technical state deterioration. In: Signal Processing Symposium 2015 (SPS 2015), Debe, Poland, pp. 1–5. IEEE (2015). https://doi.org/10.1109/SPS.2015.7168312
11. Kharchenko, V.P., Kuzmenko, N.S., Ostroumov, I.V.: Identification of unmanned aerial vehicle flight situation. In: 5th International Conference of Actual Problems of Unmanned Aerial Vehicles Development (APUAVD), Kyiv, Ukraine, pp. 116–120. IEEE (2017). https://doi.org/10.1109/APUAVD.2017.8308789
12. Ivashchuk, O., et al.: A configuration analysis of Ukrainian flight routes network. In: 16th International Conference on the Experience of Designing and Application of CAD Systems, Lviv, Ukraine, pp. 6–10. IEEE (2021). https://doi.org/10.1109/CADSM52681.2021.9385263
13. Sushchenko, O., et al.: Integration of MEMS inertial and magnetic field sensors for tracking power lines. In: XVIII International Conference on the Perspective Technologies and Methods in MEMS Design, Polyana (Zakarpattya), Ukraine, pp. 33–36. IEEE (2022). https://doi.org/10.1109/MEMSTECH55132.2022.10002907

14. Zaliskyi, M., Solomentsev, O., Kozhokhina, O., Herasymenko, T.: Reliability parameters estimation for radioelectronic equipment in case of change-point. In: Signal Processing Symposium 2017 (SPSympo 2017), Jachranka Village, Poland, pp. 1–4. IEEE (2017). https://doi.org/10.1109/SPS.2017.8053676
15. Solomentsev, O., Zaliskyi, M., Herasymenko, T., Kozhokhina, O., Petrova, Y.: Efficiency of operational data processing for radio electronic equipment. Aviation **23**(3), 71–77 (2019). https://doi.org/10.3846/aviation.2019.11849
16. Gopal, M.: Applied Machine Learning. McGraw Hill Education, Noida (2018)
17. Okoro, O.C., Zaliskyi, M., Dmytriiev, S., Solomentsev, O., Sribna, O.: Optimization of maintenance task interval of aircraft systems. Int. J. Comput. Netw. Inf. Secur. **14**(2), 77–89 (2022). https://doi.org/10.5815/ijcnis.2022.02.07
18. Deisenroth, M.P., Faisal, A.A., Ong, C.S.: Mathematics for Machine Learning, Cambridge University Press, Cambridge (2020)
19. Averyanova, Yu., et al.: Turbulence detection and classification algorithm using data from AWR. In: IEEE 2nd Ukrainian Microwave Week (UkrMW), Ukraine, pp. 518–522. IEEE (2022). https://doi.org/10.1109/UkrMW58013.2022.10037172
20. Müller, A.C., Guido, S.: Introduction to Machine Learning with Python. O'Reilly, Sebastopol (2017)
21. Solomentsev, O., Zaliskyi, M., Zuiev, O.: Estimation of quality parameters in the radio flight support operational system. Aviation **20**(3), 123–128 (2016). https://doi.org/10.3846/16487788.2016.1227541
22. Bokal, Z.M., Sinitsyn, R.B., Yanovsky, F.J.: Generalized copula ambiguity function application for radar signal processing. In: Microwaves, Radar and Remote Sensing Symposium (MRRS), Kiev, Ukraine, pp. 313–316. IEEE (2011). https://doi.org/10.1109/MRRS.2011.6053663
23. Bonaccorso, G.: Machine Learning Algorithms. Packt Publishing, Birmingham (2018)
24. Ostroumov, I.V., Kuzmenko, N.S., Marais, K.: Optimal pair of navigational aids selection. In: 5th International Conference on Methods and Systems of Navigation and Motion Control (MSNMC), Kiev, Ukraine, pp. 32–35. IEEE (2018). https://doi.org/10.1109/MSNMC.2018.8576293
25. Solomentsev, O., Zaliskyi, M., Kozhokhina, O., Herasymenko, T.: Efficiency of data processing for UAV operation system. In: 4th International Conference Actual Problems of Unmanned Aerial Vehicles Developments (APUAVD), Kyiv, Ukraine, pp. 27–31. IEEE (2017). https://doi.org/10.1109/APUAVD.2017.8308769
26. Ostroumov, I.V., Kuzmenko, N.S.: Accuracy improvement of VOR/VOR navigation with angle extrapolation by linear regression. Telecommun. Radio Eng. **78**(15), 1399–1412 (2019). https://doi.org/10.1615/TelecomRadEng.v78.i15.90
27. Boonchoo, T., Ao, X., Liu, Y., Zhao, W., Zhuang, F.: Grid-based DBSCAN: indexing and inference. Pattern Recogn. **90**, 271–284 (2019). https://doi.org/10.1016/j.patcog.2019.01.034
28. Wang, Y., Gu, Y., Shun, J.: Theoretically-efficient and practical parallel DBSCAN. In: International Conference on Management of Data (SIGMOD), Portland, USA, pp. 1–20 (2020)

Self-organization Technique with a Norm Transformation Based Filtering for Sustainable Infocommunications Within CNS/ATM Systems

Oleksii Holubnychyi[1] (✉), Maksym Zaliskyi[1], Ivan Ostroumov[1], Olha Sushchenko[1], Oleksandr Solomentsev[1], Yuliya Averyanova[1], Yurii Bezkorovainyi[1], Olena Sokolova[1], Roman Voliansky[2], Ihor Bovdui[3], Viktoriia Ivannikova[4], Kostiantyn Cherednichenko[1], Tatyana Nikitina[5], and Borys Kuznetsov[3]

[1] National Aviation University, Liubomyra Huzara Avenue, 1, Kyiv 03058, Ukraine
oleksii.holubnychyi@npp.nau.edu.ua

[2] National Technical University of Ukraine "Igor Sikorsky Kyiv Polytechnic Institute", Beresteiskyi Avenue, 37, Kyiv 03056, Ukraine

[3] Anatolii Pidhornyi Institute of Mechanical Engineering Problems of the National Academy of Sciences of Ukraine, Pozhars′koho Street, 2/10, Kharkiv 61046, Ukraine

[4] Dublin City University, Collins Ave. Ext, Dublin, Ireland

[5] Educational Scientific Professional Pedagogical Institute Ukrainian Engineering Pedagogical Academy, Universitetska Str, 16, Kharkiv 61003, Ukraine

Abstract. A self-organization machine learning technique for sustainable infocommunications within communications, navigation, and surveillance / air traffic management (CNS/ATM) systems is proposed in the paper. The proposed technique is based on the modification of the expectation-maximization algorithm with adding of components of Gaussian mixture model. The proposed technique allows for an unsupervised self-organization of system parameters into ranges (e.g., frequency bands and any other groups of homogenous parameters), which simplifies a general tuning of infocommunications for aeronautical purposes in dynamically changing conditions. The proposed technique uses a norm transformation filtering to restrict possible influence of outliers and anomalies in input system parameters. The feature that only observed input system parameters are required for all stages of data processing characterizes the proposed technique. Setting of initial parameters, stopping criteria for internal and external iterative machine learning processes, robustness and computational cost within the proposed technique are described and analyzed. An example of simulation of the proposed technique, which presents an unsupervised automatic clustering of the available radio spectrum recourse, is also shown in the paper.

Keywords: CNS/ATM Systems · Machine Learning · Sustainable Infocommunications

I. Ostroumov and M. Zaliskyi (Eds.): ACASD 2024, LNNS 992, pp. 262–278, 2024.
https://doi.org/10.1007/978-3-031-60196-5_20

1 Introduction

1.1 Introduction to the Problem

Communications, navigation, and surveillance/air traffic management (CNS/ATM) systems implementation and related activities take into account different issues, including environmental [1]. Currently, these issues can also be considered within a framework of the long-term global aspirational goal (LTAG) for international aviation of net-zero carbon emissions by 2050, which was adopted at the 41st ICAO Assembly [2]. In this regard, the relevant information and communication components of CNS/ATM systems can be presented and analyzed as sustainable information and communications technologies [3, 4] or sustainable infocommunications.

Sustainable infocommunications may be perceived as:

- Information and communication systems, which are directly characterized by the sustainable and environment-friendly structure or parameters, e.g., resource-oriented blockchain systems [5], low-power [6] and green communications [7].
- Information and communication systems, which aim to create more sustainable and environment-friendly infrastructure, systems or their features, e.g., climate change strategies [8, 9] and sustainable cities [10].

1.2 Motivation

Relevance of the above-mentioned aspects in sustainable infocommunications for aeronautical purposes within CNS/ATM systems in dynamically changing conditions can be caused by both the technical development of information and communication technology and the economic, social environmental development and protection. These features and trends in general were implemented in the Sustainable development organizing principle as part of it along with other problems and challenges in different areas of development of human society [11–15]. However, at this point possible influence of outliers and anomalies in sustainable infocommunications must be correctly taken into account. This motivates to design a technique for an unsupervised self-organization of system parameters (e.g., frequency bands and any other groups of homogenous parameters) for sustainable infocommunications within CNS/ATM systems, which simplifies their tuning and adaptation in dynamically changing conditions, and restricts such influence of outliers and anomalies.

1.3 Contribution

This research contributes a new self-organization machine learning (ML) technique for sustainable infocommunications within CNS/ATM systems, which also provides restriction of possible influence of outliers and anomalies in input system parameters due to a norm transformation filtering. The proposed technique is based on the modification of the expectation-maximization (EM) algorithm with adding of components of Gaussian mixture model (GMM) [16–18].

1.4 The Organization of the Paper

The paper consists of six sections. The introduction is presented by the first section. Second section includes a literature review and problem statement. A proposed self-organization technique for sustainable infocommunications within CNS/ATM systems is described in the third section. Results and discussions, and conclusions are presented in the fourth and fifth sections, relatively. Finally, a future research scope is given in the sixth section.

2 Literature Review and Problem Statement

The Sustainable development organizing principle in turn unfolds through the 17 United Nations Sustainable Development Goals (SDGs), which are also being dealt with their structural parts in the context of infocommunication systems and applications [11, 13–15]. Infocommunications can be considered within the both SDG 9 "Industry, innovation and infrastructure", which takes into account information and communication technology, and SDG 11 "Sustainable cities and communities", which takes into account a safe housing under the rapid growth of cities.

Sustainable, in particular, cognitive environment-friendly infocommunications, taking into account that the Dynamic Spectrum Access (DSA) and Signal to Noise Ratio (SNR) scenarios can vary widely in different cities depending on applications [19, 20], are based on the following systemic components (Fig. 1):

- Organizational component, which is responsible for interrelations with external infrastructure that aims to be sustainable and environment-friendly.
- Engineering component, which is responsible for a structure, technical parameters, features, efficiency of infocommunication system and serves to achieve the goals of the organizational component.

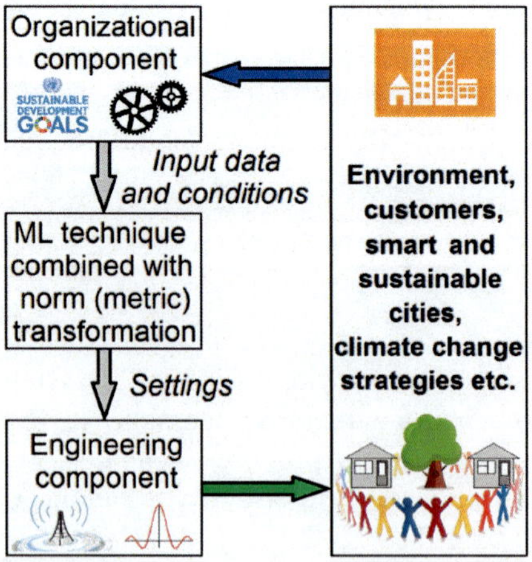

Fig. 1. Sustainable cognitive environment-friendly infocommunication system.

The paper is focused on the engineering component, which can additionally be regulated by the organizational component using the norm (metric) transformation approach [21, 22]. This approach can robustly limit the influence of outliers and anomalies that are observations in infocommunication system, which deviate so much from other observations as to arouse suspicions that they were generated by a different mechanism or appear to be inconsistent with some set of data [22–25].

The engineering component, taking into account DSA and SNR scenarios in environment-friendly cognitive radio, for modern infocommunications includes various features such as low SNR physical access, e.g., Direct Sequence Spread Spectrum and relevant spreading sequences, spatial DSA, e.g., Multiple Input Multiple Output (MIMO) and polarization antenna systems, their relevant electromagnetic field parameters etc.

Both organizational and engineering components of infocommunication system are important for sustainable development of environment-friendly air transport infrastructural parts, e.g., aircraft positioning, automatic dependent surveillance-broadcast systems, and European navigational aids network [26].

Properties and capabilities such as self-organization, self-adaptation, fault tolerance, and self-healing are promising for effective interaction between engineering and organizational components in cognitive environment-friendly infocommunications due to the possibility of autonomic precisely controlling system parameters under conditions of uncertainty [27]. These properties and capabilities can be realized effectively by means of ML and Artificial Intelligence (AI) [28] techniques, in particular neural networks, which are becoming more widespread in modern infocommunications.

The scientific problem with which the paper deals is a designing self-organization, self-adaptation and fault tolerance organizational and engineering components of sustainable environment-friendly infocommunications for aeronautical purposes. In this regard, the aim of the paper is development of self-organization technique for sustainable environment-friendly infocommunications within CNS/ATM systems, which is based on the ML approach combined with a norm (metric) transformation filtering.

3 Development of Self-organization Technique for Sustainable Infocommunications Within CNS/ATM Systems

3.1 ML Technique

Among many theoretical and practical ML approaches and techniques to the self-organization feature of a system we highlight in the paper the EM algorithm [16–18] for data clustering. The EM algorithm deals with GMM and aims to estimate their parameters and perform cluster analysis of elements within the GMM framework [18] for different purposes [15, 29].

Let $\mathbf{d} = (d_1, d_2, \ldots, d_N)^{\mathrm{T}}$ be the vector of homogenous data in an infocommunication system, where N is the amount of data. These data can be associated with parameters of infocommunication system, which are formed by the organizational component. They are also the input data for the ML self-organization technique.

The goals of cognitive part, which is based on the ML self-organization technique, are:

- Clustering of homogenous data $\mathbf{d}$ into clusters, which are subsets of parameters; this has to be done for each kind of homogenous data (e.g., the first kind of homogenous data $\mathbf{d}_1$ corresponds to spectral characteristics in DSA scenario, the second one $\mathbf{d}_2$ corresponds to energy characteristics in SNR scenario etc.).
- Clustering of heterogeneous data, which are previously clustered homogenous data sets $\mathbf{d}_k$, $k = \overline{1, K}$, where K is the number of different kinds of homogenous data in sustainable infocommunication system within CNS/ATM systems.

In the paper, we consider the achievement of the first goal, i.e. the clustering of homogenous data $\mathbf{d}$.

When using the EM algorithm, a decision on the clustering for each n-th element of homogenous data $\mathbf{d}$ to be clustered is made according to the criterion of maximum probability among posterior probabilities $\boldsymbol{\alpha} = (\alpha_{n,m})$, $n = \overline{1, N}$, $m = \overline{1, M}$, where m is the current cluster number, M is the total number of clusters. Probabilities $\boldsymbol{\alpha}$ can be estimated at the expectation step (E-step) of the EM algorithm:

$$\alpha_{n,m} = \frac{\frac{\gamma_m}{\sqrt{2\pi\sigma_m^2}}\exp\left[-\frac{(d_n-\mu_m)^2}{2\sigma_m^2}\right]}{\sum_{i=1}^{M}\frac{\gamma_i}{\sqrt{2\pi\sigma_i^2}}\exp\left[-\frac{(d_n-\mu_i)^2}{2\sigma_i^2}\right]},$$

where $\boldsymbol{\gamma} = (\gamma_1, \gamma_2, \ldots, \gamma_M)^{\mathrm{T}}$ is the vector of GMM weighting coefficients; $\boldsymbol{\mu} = (\mu_1, \mu_2, \ldots, \mu_M)^{\mathrm{T}}$ is the vector of mean values of Gaussians, which are GMM components; $\boldsymbol{\sigma} = (\sigma_1, \sigma_2, \ldots, \sigma_M)^{\mathrm{T}}$ is the vector of standard deviations of Gaussians, which are GMM components.

The vectors of GMM parameters $\boldsymbol{\theta} = \{\boldsymbol{\gamma}, \boldsymbol{\mu}, \boldsymbol{\sigma}\}$ can be estimated at the maximization step (M-step) of the EM algorithm:

$$\gamma_m = \frac{1}{N}\sum_{n=1}^{N}\alpha_{n,m};$$

$$\mu_m = \frac{\sum_{n=1}^{N}\alpha_{n,m}d_n}{\sum_{n=1}^{N}\alpha_{n,m}};$$

$$\sigma_m = \sqrt{\frac{\sum_{n=1}^{N}\alpha_{n,m}(d_n - \mu_m)^2}{\sum_{n=1}^{N}\alpha_{n,m}}}$$

The implementation of the EM algorithm is an iterative process, consisting of alternating repeats of expectation and maximization steps. This process starts from initial GMM parameters $\boldsymbol{\theta}^{(0)} = \{\boldsymbol{\gamma}^{(0)}, \boldsymbol{\mu}^{(0)}, \boldsymbol{\sigma}^{(0)}\}$ and ends with a convergence of the log-likelihood function $L(\boldsymbol{\theta}|\mathbf{d})$ while maximizing its value at the observed data $\mathbf{d}$:

$$L(\theta|\mathbf{d}) = \ln\prod_{n=1}^{N}p(d_n|\theta) = \sum_{n=1}^{N}\ln\sum_{m=1}^{M}\frac{\sigma_m}{\sqrt{2\pi\sigma_m^2}}\exp\left[-\frac{(d_n-\mu_m)^2}{2\sigma_m^2}\right];$$

$$L(\boldsymbol{\theta}|\mathbf{d}) \to \max,$$

where $p(d_n|\boldsymbol{\theta})$ is the value of the GMM probability density function $p(d)$, which is expressed in (1), for the n-th element of data $\mathbf{d}$ at the GMM parameters $\boldsymbol{\theta}$.

$$p(d) = \sum\nolimits_{m=1}^{M} \gamma_{\mathrm{m}} \mathcal{N}\left(\mathrm{d};\, \mu_{\mathrm{m}}, \sigma_{\mathrm{m}}^2\right) = \sum\nolimits_{\mathrm{m}=1}^{\mathrm{M}} \frac{\gamma_{\mathrm{m}}}{\sqrt{2\pi\sigma_{\mathrm{m}}^2}} \exp\left[-\frac{(\mathrm{d} - \mu_{\mathrm{m}})^2}{2\sigma_{\mathrm{m}}^2}\right], \quad (1)$$

where $\mathcal{N}\left(d;\, \mu_m, \sigma_m^2\right)$ is the m-th Gaussian of GMM.

A scheme for the EM algorithm, as for the iterative process, is shown in Fig. 2.

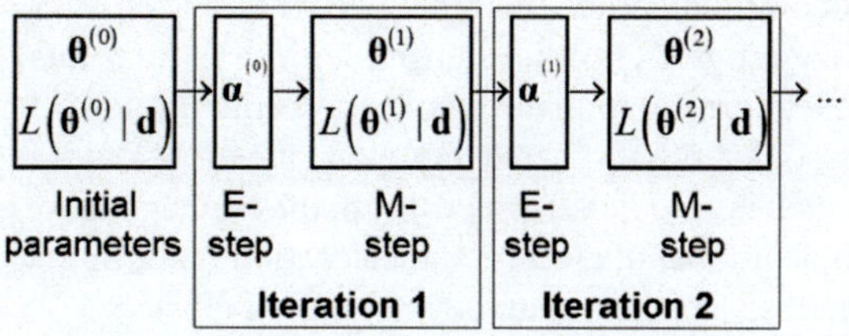

Fig. 2. Scheme for the EM algorithm, as for the iterative process.

In Fig. 2 and elsewhere in the paper the upper index in the notations $\boldsymbol{\theta}^{(\beta)} = \left\{\boldsymbol{\gamma}^{(\beta)}, \boldsymbol{\mu}^{(\beta)}, \boldsymbol{\sigma}^{(\beta)}\right\}$ and $\boldsymbol{\alpha}^{(\beta-1)}$ denotes the iteration number β within the EM algorithm.

There are two criteria for determining the convergence of the EM algorithm, i.e. stopping criteria for its iterative process, can be used:

- $L\left(\boldsymbol{\theta}^{(\beta)}|\mathbf{d}\right) - L\left(\boldsymbol{\theta}^{(\beta-1)}|\mathbf{d}\right) < \varepsilon$ where ε is a small positive value for the first criterion, which is sufficient to make a decision on convergence.
- $\left(\alpha_{n,m}^{(\beta-1)} \leq \lambda\right) \vee \left(\alpha_{n,m}^{(\beta-1)} \geq 1 - \lambda\right)$, $n = \overline{1, N}$, $m = \overline{1, M}$, where λ is a small positive value for the second criterion, which is sufficient to make a decision on convergence.

When using either first or second criterion, $\boldsymbol{\alpha}^{(\beta-1)}$ is used to make a decision on clustering of homogenous data $\mathbf{d}$ into clusters.

The first criterion is better for estimation of GMM parameters $\boldsymbol{\theta}$, whereas the second one is better for clustering tasks. The second criterion is closely associated with the robustness of clustering because for $\lambda \to 0$ the values $\alpha_{n,m}^{(\beta-1)} \to 0$ or $\alpha_{n,m}^{(\beta-1)} \to 1$ are being obtained, which characterizes strong clustering behavior and a high degree of certainty of belonging of the n-th element d_n to the m-th cluster that is also the m-th Gaussian of GMM. However, it is not always possible to achieve such robust clustering, and it significantly depends on both the observed input data $\mathbf{d}$, initial GMM parameters $\boldsymbol{\theta}^{(0)}$, and the total number of clusters M, which can be taken for an internal parameter of the EM algorithm. In this regard, it is prudent to combine these two criteria in the stopping criteria for the EM algorithm in the following way:

$$\left\{L\left(\boldsymbol{\theta}^{(\beta)}|\mathbf{d}\right) - L\left(\boldsymbol{\theta}^{(\beta-1)}|\mathbf{d}\right) < \delta\right\} \vee \left\{\left(\alpha_{n,m}^{(\beta-1)} \leq \lambda\right) \vee \left(\alpha_{n,m}^{(\beta-1)} \geq 1 - \lambda\right)\right\} \forall n, m, \quad (2)$$

where δ is a small positive value, and $\delta \ll \varepsilon$ for the first criterion within the combined criterion (2).

This approach may require an increase in the total amount of iterations, but improves the robustness of stopping criterion for the EM algorithm by the preferred second criterion within the combined criterion (2).

The clustering problem is complicated by the fact that the total number of clusters M can be unknown a priori. It also forms a background to features of application of the combined criterion (2).

There are known different modifications of the EM algorithm, in particular, the modification with adding of clusters [18] and the modification with removing of clusters [15, 30] during clustering. In the paper, we propose to use in a self-organization technique for sustainable infocommunications within CNS/ATM systems the modification of the EM algorithm with adding of clusters, starting with the initial total number of clusters $M = 2$. Such approach forms within the whole clustering framework an external iterative process, which is associated with increasing the value of M, with respect to an internal iterative process, which is associated directly with the EM algorithm and shown in Fig. 2.

Let M_ξ be the total number of clusters at the iteration ξ, starting with $\xi = 1$, within the external iterative process, i.e. the initial total number of clusters $M_1 = 2$ and $M_\xi = \xi + 1$, taking into account the above-mentioned modification of the EM algorithm with adding of clusters.

As a stopping criterion for the external iterative process, the criterion of minimum posterior estimated value of λ in the combined criterion (2) can be used. In this case, adding of clusters continues until the condition

$$\left\{\widetilde{\Lambda}_{\xi+1} \geq \widetilde{\lambda}_\xi\right\} \vee \left\{\widetilde{\lambda}_\xi \leq \lambda\right\} \tag{3}$$

is met, where $\widetilde{\lambda}_\xi$ is the posterior estimated value of λ at the ξ-th iteration of external iterative process. The value of $\widetilde{\lambda}_\xi$ has a meaning that

$$\left(\alpha_{n,m}^{(\beta-1)} \leq \widetilde{\lambda}_\xi\right) \vee \left(\alpha_{n,m}^{(\beta-1)} \geq 1 - \widetilde{\lambda}_\xi\right) \forall n, m$$

at the last iteration β within the ξ-th iteration of external iterative process.

Results of clustering at the total number of clusters M_ξ are being taken into account for the self-organization of sustainable infocommunication system within CNS/ATM systems.

The initial GMM parameters $\boldsymbol{\theta}^{(0)}$ can be determined using for initial approximation the principle of equidistance between clusters and the three-sigma rule of thumb [31] for a width of clusters $m = \overline{1, M}$, which are Gaussians of GMM:

$$\gamma_m^{(0)} = 1/M;$$

$$\mu_m^{(0)} = \min(d) + \frac{\max(d) - \min(d)}{M}\left(m - \frac{1}{2}\right); \tag{4}$$

$$\sigma_m^{(0)} = \frac{\max(d) - \min(d)}{6M}.$$

The location of Gaussians of GMM $\mathcal{N}\left(d; \mu_m, \sigma_m^2\right)$ (i.e., clusters) at the initial GMM parameters $\boldsymbol{\theta}^{(0)}$, which are determined by (4), are shown in Fig. 3.

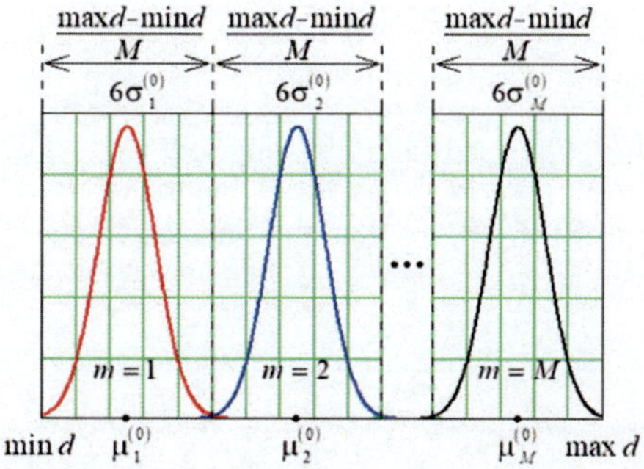

Fig. 3. Location of Gaussians (clusters) at the initial GMM parameters.

3.2 Norm Transformation Based Filtering

The norm (metric) transformation based filtering approach serves to limit possible influence of outliers and anomalies in the observed input data **d**. This filtering is carried out through the transformation of **d** using a non-Euclidian norm:

$$\mathbf{d}^* = \mathcal{F}_2^{-1}[\mathcal{F}_{\mathrm{NE}}(\mathbf{d})], \tag{5}$$

where $\mathbf{d}^*$ is the filtered observed input data **d**; $\mathcal{F}_{\mathrm{NE}}(\cdot)$ denotes the transformation in a non-Euclidian norm; $\mathcal{F}_2^{-1}(\cdot)$ denotes the inverse transformation in the Euclidian ℓ^2 norm.

A scheme for the norm transformation based filtering approach under consideration is shown in Fig. 4.

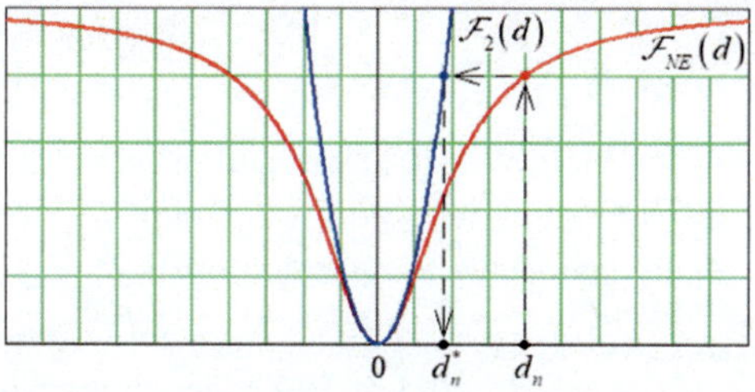

Fig. 4. Scheme for the norm transformation based filtering approach.

The transformation in a non-Euclidian norm $\mathcal{F}_{\mathrm{NE}}(d)$ can be based on different functions, which are bounded from above or, at least, characterized by their moderate values for large absolute values of argument d in comparison with a quadratic function in case of ℓ^2 norm (the example is shown in Fig. 4).

The most common kinds of non-Euclidian norms, which are used in statistical signal and data processing, at their core use functions such as [22, 32]:

- Magnitude function $\mathcal{F}_1(d) = |d|$, which is instrumental in describing Manhattan ℓ^1 norm.

- Hyperbolic function $\mathcal{F}_{21}(d) = \sqrt{s^2 + d^2} - s$.
- Geman-McClure function $\mathcal{F}_{20}(d) = d^2/(s^2 + d^2)$.

The functions $\mathcal{F}_{21}(d)$ and $\mathcal{F}_{20}(d)$ contain parameter s ($s \geq 0$), which determines their behavior. For example, for $|d| \ll s$ the behavior of $\mathcal{F}_{21}(d)$ and $\mathcal{F}_{20}(d)$ is close to the quadratic function $\mathcal{F}_2(d) = d^2$ that can be seen in Fig. 5, as well as $\mathcal{F}_{21}(0) = \mathcal{F}_{20}(0) = \mathcal{F}_2(0) = 0$.

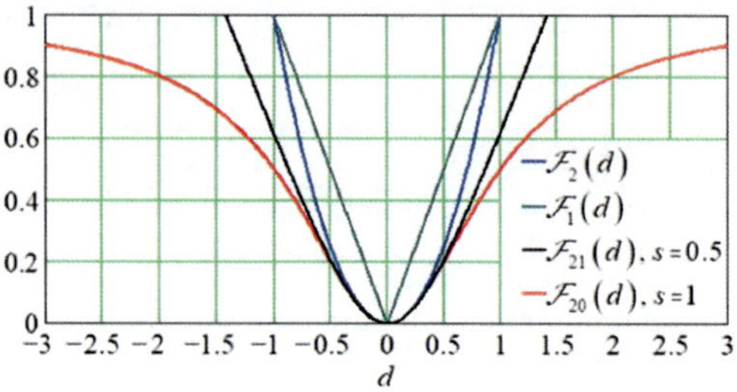

Fig. 5. Functions $\mathcal{F}_2(d)$, $\mathcal{F}_1(d)$, $\mathcal{F}_{21}(d)$, and $\mathcal{F}_{20}(d)$.

In the case of $s = 0$, $\mathcal{F}_{21}(d) = \mathcal{F}_1(d)$ (i.e. $\mathcal{F}_{21}(d)$ represents d in ℓ^1 norm), and $\mathcal{F}_{20}(d|d \neq 0) = 1$ (i.e. $\mathcal{F}_{20}(d)$ represents d in ℓ^0 "norm" as the number of non-zero elements d).

Among the considered examples, the Geman-McClure function [33] is of particular interest for the norm transformation based filtering because of its bounded from above behavior, i.e. $0 \leq \mathcal{F}_{20}(d) < 1$ for finite values of d.

The transformation (5) when using the Geman-McClure function and $d_n \geq 0$, $n = \overline{1, N}$, is expressed as:

$$\mathbf{d}^* = \mathcal{F}_2^{-1}[\mathcal{F}_{20}(\mathbf{d})] \iff d_n^* = d_n/\sqrt{s^2 + d_n^2}. \tag{6}$$

In the case of using the Geman-McClure function in norm transformation based filtering, an unsupervised automatic tuning of the parameter s is required as a part of self-organization ML technique. In the paper we propose to determine the parameter s using the following features of the Geman-McClure function:

- $\mathcal{F}_{20}'(d) = \frac{2ds^2}{(d^2+s^2)^2} > 0$ when $d > 0$, i.e. $\mathcal{F}_{20}(d)$ increases when $d > 0$.
- $\mathcal{F}_{20}''(d) = \frac{2s^2(s^2-3d^2)}{(d^2+s^2)^3}$ at $d = \sqrt{3}s/3$ or when $s = \sqrt{3}d$, $\mathcal{F}_{20}''(d) > 0$ on $d \in \left[0, \sqrt{3}s/3\right)$, and $\mathcal{F}_{20}''(d) < 0$ on $d \in \left(\sqrt{3}s/3, +\infty\right)$, i.e. the increasing of $\mathcal{F}_{20}(d)$ begins to decrease after the value of d where $s = \sqrt{3}d$.

In view of the above, it is advisable to tune the Geman-McClure function using the parameter s as the normalized arithmetic or geometric mean of the observed input data $\mathbf{d}$ (when $d_n \geq 0$, $n = \overline{1, N}$), s_A and s_G respectively:

$$s_A = \frac{\sqrt{3}\sum_{n=1}^{N} d_n}{N}; \tag{7}$$

$$s_G = \sqrt{3}\left(\prod_{n=1}^{N} d_n\right)^{\frac{1}{N}} = \exp\left(\frac{1}{2}\ln 3 + \frac{1}{N}\sum_{n=1}^{N} \ln d_n\right) \quad (8)$$

The use of $s = s_G$ is appropriate when $\min(d) \ll \max(d)$ in the observed input data **d**, e.g., for the case when the self-organization technique for sustainable infocommunications within CNS/ATM systems will use the wide band spectrum monitoring system from 30 MHz to 1800 MHz [34], i.e. $\min(d) = 3 \cdot 10^7$ Hz and $\max(d) = 1.8 \cdot 10^9$ Hz in possible DSA scenario.

Filtered data **d*** are used as the input data in the proposed ML self-organization technique.

4 Results and Discussions

4.1 Simulation of the Proposed Technique

Let a sustainable infocommunication system within CNS/ATM systems uses DSA scenario, and within the self-organization framework approach, an engineering goal boils down to an unsupervised automatic clustering of the available radio spectrum recourse into subsets of frequencies.

For this case, suppose that the observed input homogenous data **d** are the following $N = 21$ frequencies in MHz:

$$\mathbf{d} = (2150, 2110, 728, 740, 3315, 3905, 3720, 660, 2650, 2170, 2490, 17125, 2510, 710, 6275, 6015, 6200, 5925, 5050, 780, 1910)^{\mathrm{T}}. \quad (9)$$

As can be seen from (9), the element $d_{12} = 17125$ in the observed data **d** might be perceived as an outlier or anomaly if used definitions and criteria for them from [23].

Determine the parameter s using (7) and (8): $s_A = 6362$; $s_G = 4422$.

Taking into account that $\max(d)/\min(d) = 26$ and $\min(d) \ll \max(d)$ in the observed input data **d**, the parameter $s = s_G$ in (6) is being used, and as a result the following data after norm transformation based filtering are obtained:

$$\mathbf{d}^* = (0.437, 0.431, 0.162, 0.165, 0.600, 0.662, 0.644, 0.148, 0.514, 0.441, 0.491, 0.968, 0.494, 0.159, 0.817, 0.806, 0.814, 0.801, 0.752, 0.174, 0.396)^{\mathrm{T}}$$

Let the criterion (2) is used as the stopping criterion for the internal ML iterative process with parameters $\delta = 0.1$ and $\lambda = 0.01$.

Adding of clusters continues until the stopping criterion (3) for the external ML iterative process is met. Simulation results for this case are obtained using developed by authors program and shown as a listing in Fig. 6.

β	$\gamma_1^{(\beta)}$	$\gamma_2^{(\beta)}$	$\gamma_3^{(\beta)}$	$\gamma_4^{(\beta)}$	$\gamma_5^{(\beta)}$	$\mu_1^{(\beta)}$	$\mu_2^{(\beta)}$	$\mu_3^{(\beta)}$	$\mu_4^{(\beta)}$	$\mu_5^{(\beta)}$	$\sigma_1^{(\beta)}$	$\sigma_2^{(\beta)}$	$\sigma_3^{(\beta)}$	$\sigma_4^{(\beta)}$	$\sigma_5^{(\beta)}$	$L(\theta^{(\beta)} \mid d^*)$
Iteration $\xi = 1$ (external ML iterative process), total number of clusters (Gaussians of GMM) $M_1 = 2$																
0	0.500	0.500				0.353	0.763				0.068	0.068				−17.366
1	0.571	0.429				0.334	0.762				0.149	0.107				0.018
	Checking the criterion (2): $L(\theta^{(1)} \mid d^*) - L(\theta^{(0)} \mid d^*) = 17.384 > \delta$; e.g., $\alpha_{5,1}^{(0)} = 0.025$, $\lambda < \alpha_{5,1}^{(0)} < 1-\lambda \Rightarrow L(\theta^{(1)} \mid d^*)$, $\alpha^{(0)}$ do not meet (2).															
2	0.591	0.409				0.348	0.764				0.161	0.114				0.268
	Checking the criterion (2): $L(\theta^{(2)} \mid d^*) - L(\theta^{(1)} \mid d^*) = 0.250 > \delta$; e.g., $\alpha_{5,1}^{(1)} = 0.384$, $\lambda < \alpha_{5,1}^{(1)} < 1-\lambda \Rightarrow L(\theta^{(2)} \mid d^*)$, $\alpha^{(1)}$ do not meet (2).															
3	0.603	0.397				0.355	0.766				0.167	0.115				0.330
	Checking the criterion (2): $L(\theta^{(3)} \mid d^*) - L(\theta^{(2)} \mid d^*) = 0.062 < \delta$; e.g., $\alpha_{5,1}^{(2)} = 0.459$, $\lambda < \alpha_{5,1}^{(2)} < 1-\lambda \Rightarrow L(\theta^{(3)} \mid d^*)$ meets (2).															
Checking the criterion (3): $\tilde{\lambda}_1 = 0.459$ ($\alpha_{5,1}^{(2)} = \tilde{\lambda}_1$), $\tilde{\lambda}_1 > \lambda \Rightarrow \tilde{\lambda}_1$ does not meet (3); the condition $\tilde{\lambda}_2 \ge \tilde{\lambda}_1$ in (3) will be checked at $\xi = 2$.																
Iteration $\xi = 2$ (external ML iterative process), total number of clusters (Gaussians of GMM) $M_2 = 3$																
0	0.333	0.333	0.333			0.284	0.558	0.831			0.046	0.046	0.046			−22.875
1	0.303	0.410	0.286			0.214	0.527	0.826			0.102	0.084	0.067			2.244
	Checking the criterion (2): $L(\theta^{(1)} \mid d^*) - L(\theta^{(0)} \mid d^*) = 25.119 > \delta$; e.g., $\alpha_{2,1}^{(0)} = 0.223$, $\lambda < \alpha_{2,1}^{(0)} < 1-\lambda \Rightarrow L(\theta^{(1)} \mid d^*)$, $\alpha^{(0)}$ do not meet (2).															
2	0.267	0.441	0.292			0.190	0.516	0.822			0.082	0.089	0.073			3.632
	Checking the criterion (2): $L(\theta^{(2)} \mid d^*) - L(\theta^{(1)} \mid d^*) = 1.388 > \delta$; e.g., $\alpha_{21,2}^{(1)} = 0.711$, $\lambda < \alpha_{21,2}^{(1)} < 1-\lambda \Rightarrow L(\theta^{(2)} \mid d^*)$, $\alpha^{(1)}$ do not meet (2).															
3	0.242	0.460	0.298			0.166	0.509	0.818			0.035	0.089	0.076			8.056
	Checking the criterion (2): $L(\theta^{(3)} \mid d^*) - L(\theta^{(2)} \mid d^*) = 4.424 > \delta$; e.g., $\alpha_{6,3}^{(2)} = 0.215$, $\lambda < \alpha_{6,3}^{(2)} < 1-\lambda \Rightarrow L(\theta^{(3)} \mid d^*)$, $\alpha^{(2)}$ do not meet (2).															
4	0.238	0.458	0.304			0.161	0.506	0.815			0.009	0.088	0.079			12.824
	Checking the criterion (2): $L(\theta^{(4)} \mid d^*) - L(\theta^{(3)} \mid d^*) = 4.768 > \delta$; e.g., $\alpha_{6,2}^{(3)} = 0.716$, $\lambda < \alpha_{6,2}^{(3)} < 1-\lambda \Rightarrow L(\theta^{(4)} \mid d^*)$, $\alpha^{(3)}$ do not meet (2).															
5	0.238	0.451	0.311			0.161	0.504	0.812			0.009	0.086	0.081			12.867
	Checking the criterion (2): $L(\theta^{(5)} \mid d^*) - L(\theta^{(4)} \mid d^*) = 0.043 < \delta$; e.g., $\alpha_{6,3}^{(4)} = 0.351$, $\lambda < \alpha_{6,3}^{(4)} < 1-\lambda \Rightarrow L(\theta^{(5)} \mid d^*)$ meets (2).															
Checking the criterion (3): $\tilde{\lambda}_2 = 0.351$ ($\alpha_{6,3}^{(4)} = \tilde{\lambda}_2$), $\tilde{\lambda}_2 < \tilde{\lambda}_1$, $\tilde{\lambda}_2 > \lambda \Rightarrow \tilde{\lambda}_2$ does not meet (3); the condition $\tilde{\lambda}_3 \ge \tilde{\lambda}_2$ will be checked at $\xi = 3$.																
Iteration $\xi = 3$ (external ML iterative process), total number of clusters (Gaussians of GMM) $M_3 = 4$																
0	0.250	0.250	0.250	0.250		0.250	0.455	0.660	0.866		0.034	0.034	0.034	0.034		−14.211
1	0.238	0.333	0.184	0.244		0.161	0.458	0.662	0.839		0.009	0.039	0.054	0.064		14.560
	Checking the criterion (2): $L(\theta^{(1)} \mid d^*) - L(\theta^{(0)} \mid d^*) = 28.771 > \delta$; e.g., $\alpha_{19,4}^{(0)} = 0.131$, $\lambda < \alpha_{19,4}^{(0)} < 1-\lambda \Rightarrow L(\theta^{(1)} \mid d^*)$, $\alpha^{(0)}$ do not meet (2).															
2	0.238	0.332	0.164	0.266		0.161	0.457	0.651	0.830		0.009	0.039	0.053	0.068		14.883
	Checking the criterion (2): $L(\theta^{(2)} \mid d^*) - L(\theta^{(1)} \mid d^*) = 0.323 > \delta$; e.g., $\alpha_{19,3}^{(1)} = 0.353$, $\lambda < \alpha_{19,3}^{(1)} < 1-\lambda \Rightarrow L(\theta^{(2)} \mid d^*)$, $\alpha^{(1)}$ do not meet (2).															
3	0.238	0.331	0.151	0.279		0.161	0.457	0.642	0.826		0.009	0.039	0.045	0.070		15.196
	Checking the criterion (2): $L(\theta^{(3)} \mid d^*) - L(\theta^{(2)} \mid d^*) = 0.313 > \delta$; e.g., $\alpha_{19,3}^{(2)} = 0.191$, $\lambda < \alpha_{19,3}^{(2)} < 1-\lambda \Rightarrow L(\theta^{(3)} \mid d^*)$, $\alpha^{(2)}$ do not meet (2).															
4	0.238	0.333	0.141	0.288		0.161	0.458	0.636	0.824		0.009	0.039	0.033	0.071		15.615
	Checking the criterion (2): $L(\theta^{(4)} \mid d^*) - L(\theta^{(3)} \mid d^*) = 0.419 > \delta$; e.g., $\alpha_{6,4}^{(3)} = 0.076$, $\lambda < \alpha_{6,4}^{(3)} < 1-\lambda \Rightarrow L(\theta^{(4)} \mid d^*)$, $\alpha^{(3)}$ do not meet (2).															
5	0.238	0.333	0.136	0.292		0.161	0.458	0.635	0.823		0.009	0.039	0.027	0.071		15.764
	Checking the criterion (2): $L(\theta^{(5)} \mid d^*) - L(\theta^{(4)} \mid d^*) = 0.149 > \delta$; e.g., $\alpha_{6,3}^{(4)} = 0.914$, $\lambda < \alpha_{6,3}^{(4)} < 1-\lambda \Rightarrow L(\theta^{(5)} \mid d^*)$, $\alpha^{(4)}$ do not meet (2).															
6	0.238	0.334	0.136	0.293		0.161	0.458	0.634	0.822		0.009	0.039	0.026	0.071		15.765
	Checking the criterion (2): $L(\theta^{(6)} \mid d^*) - L(\theta^{(5)} \mid d^*) = 0.001 < \delta$; e.g., $\alpha_{6,4}^{(5)} = 0.095$, $\lambda < \alpha_{6,4}^{(5)} < 1-\lambda \Rightarrow L(\theta^{(6)} \mid d^*)$ meets (2).															
Checking the criterion (3): $\tilde{\lambda}_3 = 0.095$ ($\alpha_{6,4}^{(4)} = \tilde{\lambda}_3$), $\tilde{\lambda}_3 < \tilde{\lambda}_2$, $\tilde{\lambda}_2 > \lambda \Rightarrow \tilde{\lambda}_3$ does not meet (3); the condition $\tilde{\lambda}_4 \ge \tilde{\lambda}_3$ will be checked at $\xi = 4$.																
Iteration $\xi = 4$ (external ML iterative process), total number of clusters (Gaussians of GMM) $M_4 = 5$																
0	0.200	0.200	0.200	0.200	0.200	0.230	0.394	0.558	0.722	0.886	0.027	0.027	0.027	0.027	0.027	−30.419
1	0.238	0.193	0.203	0.186	0.180	0.161	0.427	0.534	0.726	0.853	0.009	0.019	0.053	0.066	0.069	13.851
	Checking the criterion (2): $L(\theta^{(1)} \mid d^*) - L(\theta^{(0)} \mid d^*) = 44.270 > \delta$; e.g., $\alpha_{16,4}^{(0)} = 0.415$, $\lambda < \alpha_{16,4}^{(0)} < 1-\lambda \Rightarrow L(\theta^{(1)} \mid d^*)$, $\alpha^{(0)}$ do not meet (2).															
2	0.238	0.179	0.209	0.193	0.181	0.161	0.426	0.525	0.726	0.846	0.009	0.018	0.058	0.076	0.076	14.252
	Checking the criterion (2): $L(\theta^{(2)} \mid d^*) - L(\theta^{(1)} \mid d^*) = 0.401 > \delta$; e.g., $\alpha_{18,4}^{(1)} = 0.427$, $\lambda < \alpha_{18,4}^{(1)} < 1-\lambda \Rightarrow L(\theta^{(2)} \mid d^*)$, $\alpha^{(1)}$ do not meet (2).															
3	0.238	0.169	0.213	0.198	0.182	0.161	0.426	0.519	0.726	0.841	0.009	0.018	0.060	0.081	0.080	14.373
	Checking the criterion (2): $L(\theta^{(3)} \mid d^*) - L(\theta^{(2)} \mid d^*) = 0.121 > \delta$; e.g., $\alpha_{18,4}^{(2)} = 0.436$, $\lambda < \alpha_{18,4}^{(2)} < 1-\lambda \Rightarrow L(\theta^{(3)} \mid d^*)$, $\alpha^{(2)}$ do not meet (2).															
4	0.238	0.162	0.214	0.203	0.182	0.161	0.426	0.514	0.725	0.838	0.009	0.018	0.060	0.084	0.083	14.435
	Checking the criterion (2): $L(\theta^{(4)} \mid d^*) - L(\theta^{(3)} \mid d^*) = 0.062 < \delta$; e.g., $\alpha_{18,4}^{(3)} = 0.441$, $\lambda < \alpha_{18,4}^{(3)} < 1-\lambda \Rightarrow L(\theta^{(4)} \mid d^*)$ meets (2).															
Checking the criterion (3): $\tilde{\lambda}_4 = 0.441$ ($\alpha_{18,4}^{(3)} = \tilde{\lambda}_4$), $\tilde{\lambda}_4 > \tilde{\lambda}_3$, $\tilde{\lambda}_4 > \lambda \Rightarrow \tilde{\lambda}_4$ meets (3). Results of clustering at $M_3 = 4$ are taking into account.																

Fig. 6. Simulation results (listing).

As shown in Fig. 6, at the iteration $\xi = 4$ the condition for the external ML iterative process $\tilde{\lambda}_4 > \tilde{\lambda}_3$, which is a part of the stopping criterion (3), is met. Results of clustering at the iteration $\xi = 3$ are taken into account. This means that the filtered data $\mathbf{d}^*$ and the related input data $\mathbf{d}$ are organized into $M_3 = 4$ clusters with the following probabilities

$\alpha_{n,m}$ that the n-th element of input data $\mathbf{d}$ belongs to the m-th cluster:

$$\alpha^{(5)} = \begin{pmatrix} 0.000 & 1.000 & 0.000 & 0.000 \\ 0.000 & 1.000 & 0.000 & 0.000 \\ 1.000 & 0.000 & 0.000 & 0.000 \\ 1.000 & 0.000 & 0.000 & 0.000 \\ 0.000 & 0.006 & 0.981 & 0.013 \\ 0.000 & 0.000 & 0.905 & 0.095 \\ 0.000 & 0.000 & 0.965 & 0.035 \\ 1.000 & 0.000 & 0.000 & 0.000 \\ 0.000 & 1.000 & 0.000 & 0.000 \\ 0.000 & 1.000 & 0.000 & 0.000 \\ 0.000 & 1.000 & 0.000 & 0.000 \\ 0.000 & 0.000 & 0.000 & 1.000 \\ 0.000 & 1.000 & 0.000 & 0.000 \\ 1.000 & 0.000 & 0.000 & 0.000 \\ 0.000 & 0.000 & 0.000 & 1.000 \\ 0.000 & 0.000 & 0.000 & 1.000 \\ 0.000 & 0.000 & 0.000 & 1.000 \\ 0.000 & 0.000 & 0.000 & 1.000 \\ 0.000 & 0.000 & 0.000 & 1.000 \\ 1.000 & 0.000 & 0.000 & 0.000 \\ 0.000 & 1.000 & 0.000 & 0.000 \end{pmatrix} \quad (10)$$

Using the criterion of maximum probability among posterior probabilities, the observed input data $\mathbf{d}$ are distributed over 4 clusters $\mathbf{d}^{(1)}$, $\mathbf{d}^{(2)}$, $\mathbf{d}^{(3)}$, and $\mathbf{d}^{(4)}$ as follows:

$$\begin{aligned} \mathbf{d}^{(1)} &= \{d_3, d_4, d_8, d_{14}, d_{20}\} = (728, 740, 660, 710, 780)^{\mathrm{T}}; \\ \mathbf{d}^{(2)} &= \{d_1, d_2, d_9, d_{10}, d_{11}, d_{13}, d_{21}\} \\ &= (2150, 2110, 2650, 2170, 2490, 2510, 1910)^{\mathrm{T}}; \\ \mathbf{d}^{(3)} &= \{d_5, d_6, d_7\} = (3315, 3905, 3720)^{\mathrm{T}}; \\ \mathbf{d}^{(4)} &= \{d_{12}, d_{15}, d_{16}, d_{17}, d_{18}, d_{19}\} \\ &= (17125, 6275, 6015, 6200, 5925, 5050)^{\mathrm{T}}. \end{aligned}$$

4.2 Discussions

Thus, the proposed ML self-organization technique for sustainable infocommunications within CNS/ATM systems boils down to an unsupervised self-organization of system parameters into ranges, which simplifies a general tuning of such infocommunications in dynamically changing conditions. These ranges can be frequency bands in DSA scenario, SNR ranges in SNR scenario, ranges of time durations and any other groups of parameters, which can be for the engineering component (see Fig. 1).

The proposed technique is based on the modification of EM algorithm with adding of clusters within the GMM framework, which is combined with the norm transformation based filtering approach using, in particular, the Geman-McClure function for a core of the norm.

The following features characterize the proposed technique:

- Only observed input data are required for a preliminary tuning (initial GMM parameters) and further processing, which provides a self-adaptation organizational component.
- Fault tolerance is provided by the norm transformation based filtering, which is carried out through the transformation of observed input data using a non-Euclidian norm and limits possible influence of outliers and anomalies in these data, e.g., the influence of element d_{12} in the data (9).
- Clustering of system parameters into ranges is implemented using the criterion of maximum probability among posterior probabilities $\boldsymbol{\alpha}$ with a high certainty, e.g., only 3 of 21 elements in the data (9), namely d_5, d_6, d_7, are organized using (10) into ranges with accuracy $\Delta\alpha > 0.001$ in the sense of the criterion of ideal clustering ($\alpha_{n,m} \to 0$ or $\alpha_{n,m} \to 1$ for $\forall n, m$).
- The total number of ranges is determined automatically using the stopping criterion (3) for the external iterative process within the proposed ML technique.
- The robustness of clustering can be evaluated by the value of $\widetilde{\lambda}$, which shows a maximum deviation from the ideal clustering ($\alpha_{n,m} \to 0$ or $\alpha_{n,m} \to 1$ for $\forall n, m$), e.g., results of clustering of the data (9) are characterized by the value $\widetilde{\lambda} = \widetilde{\lambda}_3 = 0.095$ (see (10) and Fig. 6); in this case the value $\widetilde{\lambda} = 0$ corresponds to the best robustness of clustering (there is no any uncertainty of clustering), and the value $\widetilde{\lambda} = \widetilde{\lambda}_\xi = 1/M_\xi = 1/(\xi + 1)$ corresponds to the worst robustness of clustering (there is no any certainty of clustering), taking into account that $\sum_{n=1}^{M_\xi} \alpha_{n,m} = 1$ for $\forall n, m$.
- Preliminary determination of the number of ranges, their bounds and other thresholds are not required.
- Observed input data may not necessarily be sorted in ascending order, or anything else.
- The total number of iterations is relatively small in comparison with the total number of elements in analyzed data, e.g., 18 internal iterations within 4 external iterations (see Fig. 6) was required for clustering of 21 elements, which are the data (9).

5 Conclusions

Modern infocommunications, in particular for aeronautical purposes, take into account sustainable and environment-friendly aspects. These aspects of infocommunications are mainly concentrated in SDG 9 "Industry, innovation and infrastructure" and SDG 11 "Sustainable cities and communities" within the Sustainable development framework. Self-organization, self-adaptation, fault tolerance and self-healing properties, which are promising for sustainable infocommunications, can be effectively realized in such systems due to ML and AI based cognitive approaches.

The paper has introduced the self-organization ML technique for sustainable infocommunications within CNS/ATM systems. The proposed technique is based on a modification of the EM algorithm with adding of clusters within the GMM framework. The proposed technique allows for an unsupervised self-organization of system parameters into ranges (e.g., frequency bands, SNR ranges, ranges of time durations and any other groups of homogenous parameters), which simplifies a general tuning of infocommunications in dynamically changing conditions in aviation.

A feature of the proposed technique is the use of norm transformation filtering to restrict possible influence of outliers and anomalies in input system parameters. Such norm transformation filtering, which uses the Geman-McClure function based norm, is analyzed in detail. The main advantage of the proposed technique is the fact that only observed input system parameters are required for all stages of data processing. Setting of initial parameters, stopping criteria for internal and external iterative ML processes within the proposed technique are substantiated and analyzed.

The described evaluation of robustness for the proposed technique allows showing in terms of uncertainty for a distribution of probabilities how far an obtained clustering from ideal one. An example of simulation of the proposed technique for DSA scenario, which presents an unsupervised automatic clustering of the available radio spectrum recourse (21 frequencies including 1 possible anomalous value) into 4 frequency bands, is shown in the paper. The example also shows that the proposed technique has an acceptable computational cost (18 internal ML iterations within 4 external ML iterations are required for the clustering of 21 frequencies).

6 Future Scope

Prospects for applying the proposed ML self-organization technique for sustainable infocommunications within CNS/ATM systems are also in the field of statistical data processing for modern green communications [35–37], sustainable decision-making techniques [38], sustainable communications for healthcare [39], self-sustainable UAV-assisted wireless networks [40], sustainable communications in 5G/6G wireless sensor networks [41] and other technologies, which are connected with sustainable and environment-friendly applications.

Acknowledgments. This research is partially supported by the Ministry of Education and Science of Ukraine under the projects "Development of an integrated flight control system" (№ 0121U109490) and "Methods of improving the operational efficiency of telecommunication and radio electronic systems in Ukraine" (№ 0122U001804).

References

1. APANPIRG/13-IP/2. International Civil Aviation Organization. Thirteenth meeting of the Asia/Pacific Air Navigation Planning and Implementation Regional Group (APANPIRG/13, Bangkok, Thailand, 9 to 13 September 2002). https://www.icao.int/Meetings/AMC/MA/2002/APANPIRG13/ip02.pdf. Accessed 09 Jan 2024
2. Long term global aspirational goal (LTAG) for international aviation. International Civil Aviation Organization. https://www.icao.int/environmental-protection/Pages/LTAG.aspx. Accessed 09 Jan 2024
3. IEEE Sustainable ICT. https://sustainableict.ieee.org. Accessed 09 Jan 2024
4. Wu, J., Guo, S., Huang, H., Liu, W., Xiang, Y.: Information and communications technologies for sustainable development goals: state-of-the-art, needs and perspectives. IEEE Commun. Surv. Tutorials **20**(3), 2389–2406 (2018). https://doi.org/10.1109/COMST.2018.2812301
5. Qian, K, Liu, Y, Han, Y, Wang, K.: BCadvisor: enabling green blockchain systems through resource-oriented benchmarking. In: International Conference on Communications (ICC), pp. 4031–4036. IEEE, Seoul (2022). https://doi.org/10.1109/ICC45855.2022.9838249
6. Brockmann, C., et al.: Modular ultra-low-power IoT-core – bridging the gap between power electronics and distributed sensor networks. In: International Exhibition and Conference for Power Electronics, Intelligent Motion, Renewable Energy and Energy Management (PCIM Europe), pp. 1–6. VDE VERLAG, Nuremberg (2022). https://doi.org/10.30420/565822008
7. Ogawa, H.: The research and development of advanced optical communication technology towards a beyond 5G and green society by the Japanese government. In: 27th OptoElectronics and Communications Conference (OECC) and International Conference on Photonics in Switching and Computing (PSC), pp. 1–2. IEEE, Toyama (2022). https://doi.org/10.23919/OECC/PSC53152.2022.9850015
8. Kiss, E., Zichar, M., Fazekas, I., Karancsi, G., Balla, D.: Categorization and geovisualization of climate change strategies using an open-access WebGIS tool. Infocommun. J. **12**(1), 32–37 (2020). https://doi.org/10.36244/ICJ.2020.1.5
9. Averyanova, Y., Ostroumov, I., Sushchenko, O., Larin, V., Zaliskyi, M., Bezkorovainyi, Y.: Frequency characteristics of the channel for atmospheric turbulence intensity estimate with polarimetric radar. In: Ostroumov, I., Zaliskyi, M. (eds.) ACASD 2023. LNCS, vol. 736, pp. 166–175. Springer, Cham (2023). https://doi.org/10.1007/978-3-031-38082-2_13
10. Shehab, M.J., Kassem, I., Kutty, A.A., Kucukvar, M., Onat, N., Khattab, T.: 5G networks towards smart and sustainable cities: a review of recent developments, applications and future perspectives. IEEE Access **10**, 2987–3006 (2021). https://doi.org/10.1109/ACCESS.2021.3139436
11. Calero, C., Moraga, M.A., Garcia, F.: Software, sustainability, and UN sustainable development goals. IT Prof. **24**(1), 41–48 (2022). https://doi.org/10.1109/MITP.2021.3117344
12. Rani, S., et al.: Amalgamation of advanced technologies for sustainable development of smart city environment: a review. IEEE Access **9**, 150060–150087 (2021). https://doi.org/10.1109/ACCESS.2021.3125527
13. Lopez-Vargas, A., Ledezma, A., Bott, J., Sanchis, A.: IoT for global development to achieve the United Nations sustainable development goals: the new scenario after the COVID-19 pandemic. IEEE Access **9**, 124711–124726 (2021). https://doi.org/10.1109/ACCESS.2021.3109338
14. Astobiza, A.M., Toboso, M., Aparicio, M., Lopez, D.: AI ethics for sustainable development goals. IEEE Technol. Soc. Mag. **40**(2), 66–71 (2021). https://doi.org/10.1109/MTS.2021.3056294
15. Holubnychyi, O., Gabrousenko, Y., Taranenko, A., Slobodian, O., Zharova, O.: Machine learning technique based on Gaussian mixture model for environment friendly communications. In: Faure, E., Danchenko, O., Bondarenko, M., Tryus, Y., Bazilo, C., Zaspa, G. (eds.)

ITEST 2022. LNCS, vol. 178, pp. 13–29. Springer, Cham (2023). https://doi.org/10.1007/978-3-031-35467-0_2
16. Dempster, A.P., Laird, N.M., Rubin, D.B.: Maximum likelihood from incomplete data via the EM algorithm. J. R. Stat. Soc. Ser. B **39**(1), 1–38 (1977)
17. Gupta, M.R., Chen, Y.: Theory and use of the EM algorithm. Found. Trends Sig. Process. **4**(3), 223–296 (2011). https://doi.org/10.1561/2000000034
18. Vlassis, N., Likas, A.: A greedy EM algorithm for Gaussian mixture learning. Neural. Process. Lett. **15**, 77–87 (2002). https://doi.org/10.1023/A:1013844811137
19. Li, W., Shentu, G., Gao, X.: Estimation of distribution of primary channel periods based on imperfect spectrum sensing in cognitive radio. IEEE Access **10**, 82025–82035 (2022). https://doi.org/10.1109/ACCESS.2022.3196681
20. Ostroumov, I., Sushchenko, O., Averyanova, Y., Zaliskyi, M., Solomentsev, O., Holubnychyi O.: Performance analysis of compact position report for geodata storing and transfering. In: 13th International Conference on Electronics and Information Technologies (ELIT), pp. 227–231. IEEE, Lviv (2023). https://doi.org/10.1109/ELIT61488.2023.10310829
21. Popescu, M.C., Grama, L., Rusu, C.: Conformal transformation of the metric for k-nearest neighbors classification. In: 16th International Conference on Intelligent Computer Communication and Processing (ICCP), pp. 229–234. IEEE, Cluj-Napoca (2020). https://doi.org/10.1109/ICCP51029.2020.9266240
22. Holubnychyi, O., Lavrynenko, O., Bakhtiiarov, D.: Well-adapted to bounded norms predictive model for aviation sensor systems. In: Ostroumov, I., Zaliskyi, M. (eds.) ACASD 2023. LNCS, vol. 736, pp. 179–193. Springer, Cham (2023). https://doi.org/10.1007/978-3-031-38082-2_14
23. Zhang, Y., Meratnia, N., Havinga, P.: Outlier detection techniques for wireless sensor networks: a survey. IEEE Commun. Surv. Tutorials **12**(2), 159–170 (2010). https://doi.org/10.1109/SURV.2010.021510.00088
24. Sushchenko, O., Averyanova, Y., Ostroumov, I., Zaliskyi, M., Solomentsev, O., Holubnychyi, O.: Robust structural-parametric synthesis of stabilization systems for ground moving vehicle equipment. In: 7th International Conference on Methods and Systems of Navigation and Motion Control (MSNMC), pp. 74–79. IEEE, Kyiv (2023). https://doi.org/10.1109/MSNMC61017.2023.10329013
25. Zaliskyi, M., Solomentsev, O., Holubnychyi, O., Ostroumov, I., Averyanova, Y., Sushchenko, O.: Robust method of data processing during radio equipment operation. In: 18th International Conference on Computer Science and Information Technologies (CSIT), pp. 1–4. IEEE, Lviv (2023). https://doi.org/10.1109/CSIT61576.2023.10323604
26. Ostroumov, I., Kuzmenko, N.: Configuration analysis of European navigational aids network. In: Integrated Communications Navigation and Surveillance Conference (ICNS), pp. 1–9. IEEE, Dulles (2021). https://doi.org/10.1109/ICNS52807.2021.9441576
27. Maksuti, S., Zsilak, M., Tauber, M., Delsing, J.: Security and autonomic management in system of systems. Infocommun. J. **13**(3), 66–75 (2021). https://doi.org/10.36244/ICJ.2021.3.7
28. Frankó, A., Varga, P.: A survey on machine learning based smart maintenance and quality control solutions. Infocommun. J. **13**(4), 28–35 (2021). https://doi.org/10.36244/ICJ.2021.4.4
29. Holubnychyi, O., Zaliskyi, M., Solomentsev, O., Ostroumov, I., Averyanova, Y., Sushchenko, O.: Gaussian mixture model based machine learning approach for detection of threat types in communication networks. In: 13th International Conference on Electronics and Information Technologies (ELIT), pp. 142–147. IEEE, Lviv (2023). https://doi.org/10.1109/ELIT61488.2023.10310785
30. Huang, T., Peng, H., Zhang, K.: Model selection for Gaussian mixture models. Stat. Sin. **27**(1), 147–169 (2017). https://doi.org/10.5705/ss.2014.105

31. Pukelsheim, F.: The three sigma rule. Am. Stat. **48**(2), 88–91 (1994). https://doi.org/10.2307/2684253
32. Carfantan, H., Idier, J.: Statistical linear destriping of satellite-based pushbroom-type images. IEEE Trans. Geosci. Remote Sens. **48**(4), 1860–1871 (2010). https://doi.org/10.1109/TGRS.2009.2033587
33. Geman, S., McClure, D.E.: Statistical methods for tomographic image reconstruction. Bull. Int. Stat. Inst. **52**(4), 5–21 (1987)
34. Humad, Y.A.I., Dudás, L.: Wide band spectrum monitoring system from 30 MHz to 1800 MHz with limited size, weight and power consumption by MRC-100 satellite. Infocommun. J. **14**(2), 56–63 (2022). https://doi.org/10.36244/ICJ.2022.2.6
35. Aljebreen, M., Obayya, M., Mahgoub, H., Alotaibi, S.S., Mohamed, A., Hamza, M.A.: Chaotic equilibrium optimizer-based green communication with deep learning enabled load prediction in Internet of Things environment. IEEE Access **12**, 258–267 (2024). https://doi.org/10.1109/ACCESS.2023.3345803
36. Lin, J., Ma, M., Wang, X.: Joint design of long-term base station activation and short-term beamforming for green wireless networks. IEEE Trans. Wireless Commun. **23**(1), 185–200 (2024). https://doi.org/10.1109/TWC.2023.3276639
37. Yang, T., Cui, Z., Peng, C., Wu, J., Liu, F., Yang, Y.: Integrated communication and computing maritime networks design for green Metaverse. IEEE Wirel. Commun. **30**(5), 120–126 (2023). https://doi.org/10.1109/MWC.017.2300119
38. Boskovic, S., Švadlenka, L., Jovcic, S., Dobrodolac, M., Simic, V., Bacanin, N.: A new FullEX decision-making technique for criteria importance assessment: an application to the sustainable last-mile delivery courier selection. IEEE Access **11**, 137426–137436 (2023). https://doi.org/10.1109/ACCESS.2023.3339580
39. Famitafreshi, G., Afaqui, M.S., Melia-Seguí, J.: Introducing reinforcement learning in the Wi-Fi MAC layer to support sustainable communications in e-Health scenarios. IEEE Access **11**, 126705–126723 (2023). https://doi.org/10.1109/ACCESS.2023.3331950
40. Kurmvanshi, R.S., Prasad, G., Arif, W.: EBi^2 ER protocol for self-sustainable UAV-assisted wireless networks. IEEE Commun. Lett. **27**(12), 3195–3199 (2023). https://doi.org/10.1109/LCOMM.2023.3330749
41. Gururaj, H.L., Natarajan, R., Almujally, N.A., Flammini, F., Krishna, S., Gupta, S.K.: Collaborative energy-efficient routing protocol for sustainable communication in 5G/6G wireless sensor networks. IEEE Open J. Commun. Soc. **4**, 2050–2061 (2023). https://doi.org/10.1109/OJCOMS.2023.3312155

Methods of Operational Efficiency Improvement for Civil Aviation Systems

Efficiency Analysis of Current Repair Procedures for Aviation Radio Equipment

Oleksandr Solomentsev[1], Maksym Zaliskyi[1](✉), Oleksii Holubnychyi[1], Ivan Ostroumov[1], Olha Sushchenko[1], Yurii Bezkorovainyi[1], Yuliya Averyanova[1], Viktoriia Ivannikova[2], Borys Kuznetsov[3], Ihor Bovdui[3], Tatyana Nikitina[4], Roman Voliansky[5], Kostiantyn Cherednichenko[1], and Olena Sokolova[1]

[1] National Aviation University, Liubomyra Huzara Avenue, 1, Kyiv 03058, Ukraine
mzaliskyi@nau.edu.ua

[2] Dublin City University, Collins Avenue Ext, Dublin, Ireland

[3] Anatolii Pidhornyi Institute of Mechanical Engineering Problems of the National Academy of Sciences of Ukraine, Pozhars′koho Street, 2/10, Kharkiv 61046, Ukraine

[4] Educational Scientific Professional Pedagogical Institute Ukrainian Engineering Pedagogical Academy, University Street 16, Kharkiv 61003, Ukraine

[5] National Technical University of Ukraine "Igor Sikorsky Kyiv Polytechnic Institute", Beresteiskyi Avenue, 37, Kyiv 03056, Ukraine

Abstract. One of the key tasks of ensuring the level of flight safety is providing a high reliability of radio electronic equipment. This equipment is designed to detect aircraft, determine their coordinates, and provide radio communication between the aircraft and the air traffic controller. The reliability of aviation radio equipment is determined at the stages of design and operation. At the same time, the main procedures include the following measures: the use of a highly reliable element base, the use of various types of redundancy of the structural components of the equipment, ensuring the efficiency of the main operational processes, and others. The processes that have a direct impact on the reliability of aviation radio equipment are monitoring and control of the technical condition, repair, making decisions on corrective and preventive actions based on collected statistical data, and others. This paper considers the problem of the repair procedures efficiency analysis, which includes the diagnostics and detection of failed structural units of equipment, restoration of serviceability, and measurement of diagnostic variables. Efficiency is considered from the point of view of minimizing the duration of repair procedures, their cost and labor intensity. In general, the method of calculating numerical values of efficiency indicators and their statistical characteristics is presented in the form of step-by-step instructions.

Keywords: Operation · Repair · Data Processing · Aviation Radio Equipment · Reliability · Efficiency

I. Ostroumov and M. Zaliskyi (Eds.): ACASD 2024, LNNS 992, pp. 281–295, 2024.
https://doi.org/10.1007/978-3-031-60196-5_21

1 Introduction

1.1 Introduction to the Problem

One of the key tasks of ensuring the level of flight safety is providing a high level of reliability of the functioning of radio electronic equipment [1, 2]. This equipment is designed to detect aircraft, determine their coordinates, and provide radio communication between the aircraft and the air traffic controller [3].

There are three stages of aviation radio equipment lifecycle: design, production and operation [4]. During the design and operation, the issues of increasing the reliability of equipment are usually solved. The main ways to increase the reliability are:

- the use of a highly reliable element base;
- the use of various types of redundancy of the structural components of the equipment;
- optimization of storage and logistics of spare parts and equipment blocks;
- ensuring the efficiency of the main operational processes;
- monitoring the technical condition of the equipment by measuring the determining parameters and reliability indicators;
- use of intelligent statistical data processing algorithms and decision-making based data processing results;
- prediction of failures and faults, implementation of optimal preventive actions, and others [5, 6].

These tasks are solved in project organizations and equipment operation systems [7]. The operation system (OS) is a complex structural unit containing equipment, means of operation, operating enterprises, technological processes, resource support, managerial influences in the form of control actions and regulatory documentation, personnel, and others [8, 9].

Information signals flow between the structural elements of the OS, which are mostly stochastic in nature [10, 11]. This reason is explained by the influence of many factors, including the environment, the possibility of random failures and faults, and the human factor in the decision-making process. The specified circumstances are a prerequisite for collecting statistical data, storing them in special warehouses and using them in the decision-making process.

1.2 Motivation

Information about quantitative indicators of the quality of operational processes performance is important from the point of view of analyzing their influence on the reliability of equipment functioning. In particular, the determination of the statistical characteristics of the duration of the current repair process gives the possibility for a timely assessment of the equipment steady-state availability and, in case of unsatisfactory numerical estimates, to perform certain preventive actions.

Determining the statistical characteristics of the efficiency of the current repair process provides an opportunity for the further implementation of advanced data-driven decision-making approaches in the practice of operation of aviation radio equipment.

1.3 Contribution

This paper contributes to the theory and practice of operation of aviation radio equipment in terms of statistical analysis of the current repair process. The practical significance of this research is associated with building a step-by-step instruction for numerical assessment of the duration, cost and labor intensity of current repair process.

1.4 The Organization of the Paper

The paper contains six sections. The first section describes introduction to the problem, motivation and contribution. The second section concentrates on literature review and problem statement. The third section considers the step-by-step procedure for numerical evaluation of current repair process efficiency. Fourth section presents results and discussions. Fifth and sixth sections are conclusions and future scope.

2 Literature Review and Problem Statement

The OS of aviation radio equipment contains the processes associated with:

- equipment usage for its functional purpose;
- monitoring and control of technical condition;
- maintenance;
- current repair;
- personnel training;
- decision-making on implementation of corrective and preventive actions;
- statistical data processing, and others [12, 13].

The current repair process is implemented after detecting failure or fault of aviation radio equipment [14]. Firstly, current repair process launches the search of failed element, unit or block of equipment based on different diagnostic programs implementation [15, 16]. The next procedure is the recovery of serviceability. This procedure is carried out based on replacement of failed element, performing recovery operations, regulation of various parameters, and others [17]. The last procedure is parameters monitoring that is implemented for decision-making on possibility of further equipment usage and correctness of previous procedures [18].

The scientific results presented in modern literature do not pay enough attention to the problem of efficiency analysis of current repair process. In general case, scientists usually use only expected value of efficiency index [19, 20]. But due to stochastic nature of studied process such consideration does not provide the required information. It is necessary to estimate the probability density function [21]. However, such problem is very sophisticated and, in some cases, requires a big number of experiments. The simpler approach is to make assumption about gaussian nature of studied process and its stationary property. It gives possibility to obtain only estimates of mean and standard deviation that completely describe this distribution [22].

The efficiency indexes of current repair process are:

- the expected value $E(t_r)$ of duration for one repair of equipment;

- the standard deviation $s(t_r)$ of duration for one repair of equipment;
- the threshold value $t_r^{\text{th}}\left(\theta_t^{(0)}\right)$ of duration for one repair of equipment for significance level $\theta_t^{(0)}$ (significance level is the real probability of repair duration not exceeding the predetermined value);
- the expected value $E(C_r)$ of cost for one repair of equipment;
- the standard deviation $s(C_r)$ of cost for one repair of equipment;
- the threshold value $C_r^{\text{th}}\left(\theta_c^{(0)}\right)$ of cost for one repair of equipment for significance level $\theta_c^{(0)}$ (the real probability of repair cost not exceeding the predetermined value);
- the expected value $E(L_r)$ of labor intensity for one repair of equipment;
- the standard deviation $s(L_r)$ of labor intensity for one repair of equipment;
- the threshold value $L_r^{\text{th}}\left(\theta_L^{(0)}\right)$ of labor intensity for one repair of equipment for significance level $\theta_L^{(0)}$ (the real probability of repair labor intensity not exceeding the predetermined value);
- the probability density function of probability of correct diagnostics $f(P_{\text{cd}})$ [23].

The initial data for efficiency indexes calculation include the organizational structure of current repair process, operational data of equipment, statistical data on different procedures characteristics [24, 25].

Consider a mathematical formulation of the research problem. In general case, the vector of efficiency indexes $\overrightarrow{Efficiency}$ depends on following parameters:

- reliability properties of aviation radio equipment $\vec{R}$;
- diagnostics method for searching failed element $\overrightarrow{DM}$;
- the characteristics of current repair process $\overrightarrow{CRP}$ for one procedure of inspection, control and serviceability recovery from the point of view of duration, cost and labor intensity;
- the veracity of decision-making $\vec{V}$ in case of possibility of the errors of the first and second kind;
- the parameters that take into account human factor $\overrightarrow{HF}$;
- the characteristics of measuring equipment $\overrightarrow{ME}$.

Then, certain efficiency index is some function φ_i that establishes the correlation. $\overrightarrow{Efficiency} = \varphi_i\left(\overrightarrow{DM}, \overrightarrow{CRP} | \vec{R}, \vec{V}, \overrightarrow{HF}, \overrightarrow{ME}\right)$.

The aim of this paper is to determine this correlation and show the step-by-step procedure for numerical evaluation of current repair process efficiency.

3 Materials and Methods

3.1 General Information

The calculation of the efficiency indicators of the current repair process includes the following stages:

- analysis of aviation radio equipment and determination of the object of current repair;

- analysis of the technological subsystem of the current repair of aviation radio equipment;
- calculation of numerical values of conventional indicators of the efficiency of the current repair of aviation radio equipment;
- calculation of numerical values of efficiency indicators of single current repair of all aviation radio equipment.

3.2 Analysis of Aviation Radio Equipment and Determination of the Object of Current Repair

During the analysis of aviation radio equipment, it is necessary: to describe how and with the help of which measuring means it is possible to determine the serviceability of the equipment; to provide data on equipment configuration and perform an analysis of the schematic and structural features of its construction, on the basis of which to distinguish hierarchical levels (availability of cabinets, racks, blocks, etc.); to provide data on spare parts for certain element of given hierarchical level.

The stage of determining the object of current repair is aimed at obtaining the data on the level of controllability of the specified hierarchical levels of equipment, which will enable the subsequent construction of a diagnostic model.

To systematize the received data, there is a need to use the following system of schematic and structural elements of aviation radio equipment. All structural elements (SE) are divided into six groups:

1. Structural elements at the level of cabinets.
2. Structural elements at the level of racks.
3. Structural elements at the block level.
4. Structural elements at the device level.
5. Structural elements at the electric board level.
6. Structural elements at the level of non-recovery elements (radio component base, mechanical units, etc.).

This paper studies the objects of current repair in the form of SE of the fourth and fifth levels. Elements of these levels can be divided into two sets. The first set has spare parts stored in the OS. In case of equipment failure, the failed SE is replaced with a serviceable one from the number of spare parts. The second set does not have spare parts, so restoration of serviceability is carried out in the repair units of the OS organizational structure.

In addition to the scheme of the hierarchical construction of the equipment, the following data must be collected: information about the defining parameters characterizing the serviceability of each SE, the measurement devices for each parameter, the operational tolerances for each parameter.

3.3 Analysis of the Technological Subsystem of the Current Repair of Aviation Radio Equipment

During the analysis of the technological subsystem of the current repair, it is necessary: to determine the organizational structure of this subsystem, to conduct an analysis of technological equipment, to determine current repair technologies and required personnel.

Elements of organizational structure can be:

- automated workplaces for current repair of equipment and its parts;
- spare parts for a certain hierarchical level of equipment;
- repair and operation workshops of air traffic service enterprises;
- manufacturing enterprises or repair enterprises of aviation radio equipment.

The analysis of the organizational structure ends with the development of a suitable scheme with the definition of the input and output flows of requests for the co-responding element.

An automated workplace should include bench equipment, oscilloscopes, voltmeters, frequency meters, ammeters, power sources, generators of special signals, soldering stations, etc. When choosing these tools, it is advisable to determine their approximate cost and terms of operation.

Repair technology will differ for elements with and without spare parts. For the elements of the first group, a search for the failed element, its replacement and a check of the serviceability of the entire SE will be carried out. For the elements of the second group, repair procedures will be performed, including soldering, adjustment, and regulations of the existing parts of SE. Regardless of the SE group, the first stage of the repair technology is the process of developing the diagnostic model and diagnostic program. During the repair of SE without spare parts, the following procedures are performed: input inspection of the block, node, board in order to confirm the fact of failure, external inspection of the SE to detect the failed element, search for the failed element, restoration of serviceability, output inspection of the serviceability of the block, node, board. Studying the issue of the justification of current repair technologies, it is necessary to build structural diagrams of the interconnection of technological procedures for each automated workplace. The process of localization of the failed element must be carried out in stages, successively moving from a higher hierarchical level to a lower one based on the diagnostic program. From the multi-stage nature of this search, it follows that it is necessary to have several diagnostic programs and use several failure search methods for this.

To estimate the cost of resources for current repairs, it is also necessary to determine the quantity and qualifications of the personnel who will work at automated workplaces.

3.4 Calculation of Numerical Values of Conventional Indicators of the Efficiency of the Current Repair of Aviation Radio Equipment

Conditional indicators of the efficiency of the current repair of aviation radio equipment include:

- indicators characterizing the expenditure of resources for the current repair of the j-th basic object, which is part of the K equipment samples, for which there are no spare parts. We assume that there are M samples of aviation radio equipment in total, then $U = M - K$ samples will have spare parts – $\overrightarrow{Efficiency}(j|K)$;
- indicators characterizing the expenditure of resources for the current repair of the remaining $K - 1$ objects for which there are no spare parts – $\overrightarrow{Efficiency}(K - 1|K)$;

- indicators characterizing the expenditure of resources for the ongoing repair of U objects for which there are spare parts – $\overrightarrow{Efficiency}(U)$.

While calculating the numerical values of conditional indicators of the efficiency $\overrightarrow{Efficiency}(j|K)$, the following origin data must be obtained:

- diagnostic model of the j-th basic object of equipment;
- a graphic view of the current repair program, which includes diagnostic programs for the j-th object, as well as technological operations of replacing the failed element and output serviceability check after the repair;
- failure probabilities $\overrightarrow{q_n}$ of elements of the object of current repair, where n is the number of elements of the diagnostic model;
- average values of the duration of technological operations of parameter control t_{ci} $(i = \overline{1, n})$ at the outputs of elements of the diagnostic model, replacement of failed elements $t_{\text{Rep}i}$, and output serviceability check t_{OSC};
- average values of the cost of performing technological operations of parameter control C_{ci} at the outputs of diagnostic model elements, replacement of failed elements $C_{\text{Rep}i}$, and output serviceability check C_{OSC};
- average values of labor intensity of performing technological operations of parameter control L_{ci} at the outputs of elements of the diagnostic model, replacement of failed elements $L_{\text{Rep}i}$, and output serviceability check L_{OSC};
- probabilities of errors of the first and second kind (α and β), characterizing the uncertainty in decision-making regarding the actual values of the controlled parameters.

The diagnostic model of the equipment as the object of current repair is not necessarily a repetition of its structural or functional scheme. This model must be shown in the form of a graph, during the construction of which a number of assumptions must be used. After making a decision on detection of the failed element, it is necessary to replace it with a serviceable one and check the functionality of the entire SE. If the SE is in serviceable condition, the current repair process ends. Otherwise, the follow-up strategy should be implemented for the cases of the first and second type errors presence.

Let us consider analytical correlations in case of absence of errors of the first and second kind. While calculating the failure probabilities $\overrightarrow{q_n}$ of elements of the j-th object of current repair and probabilities $\overrightarrow{Q_{K-1}}$ for others $K - 1$ objects, we can use the method of approximate reliability assessment. For this case, the failure rate λ_j for the j-th object is determined first. After that, the failure rates of other objects are calculated with the help of correction factors ρ_h. Then

$$\Lambda_h = \rho_h \lambda_j, h = \overline{1, K - 1}. \tag{1}$$

This approach makes it possible to simplify the analysis of the efficiency of the current repair process. The coefficient $\rho_h > 1$ if, based on an expert assessment, it is considered that the h-th object is more complex in terms of schematic design than the basic j-th object. Otherwise $\rho_h < 1$.

During $\overrightarrow{Efficiency}(j|K)$ estimation, we introduce the set of technological procedures (STP). The STP contain all required operations necessary to recover equipment serviceability. According to probability theory separate STPs make a complete group of events. The number of STP is equal to the number of elements of the diagnostic model. Each STP is characterized by conditional duration, cost and labor intensity $t(STP_i|j, K)$, $C(STP_i|j, K)$ and $L(STP_i|j, K)$. In this case STP_i denotes the STP for the failure of the i-th SE in the object of current repair.

After averaging for values $t(STP_i|j, K)$, $C(STP_i|j, K)$ and $L(STP_i|j, K)$ and taking into account failure probabilities $\overrightarrow{q_n}$ of elements of the object of current repair, we can the equations for $\overrightarrow{Efficiency}(j|K)$ calculation

$$E(t_r|j, K) = \sum\nolimits_{i=1}^{n} t(STP_i|j, K)q_i, \quad (2)$$

$$E(C_r|j, K) = \sum\nolimits_{i=1}^{n} C(STP_i|j, K)q_i, \quad (3)$$

$$E(L_r|j, K) = \sum\nolimits_{i=1}^{n} L(STP_i|j, K)q_i, \quad (4)$$

$$s(t_r|j, K) = \sqrt{\sum\nolimits_{i=1}^{n} (t(STP_i|j, K) - E(t_r|j, K))^2 q_i}, \quad (5)$$

$$s(C_r|j, K) = \sqrt{\sum\nolimits_{i=1}^{n} (C(STP_i|j, K) - E(C_r|j, K))^2 q_i}, \quad (6)$$

$$s(L_r|j, K) = \sqrt{\sum\nolimits_{i=1}^{n} (t(STP_i|j, K) - E(L_r|j, K))^2 q_i}. \quad (7)$$

The probability of correct diagnostics and decision-making on failed element is characterized by probability mass function with following statistical estimates of expected value and standard deviation

$$E(P_{\text{cd}}|j, K) = \sum\nolimits_{i=1}^{n} D(STP_i|j, K)q_i, \quad (8)$$

$$s(P_{\text{cd}}|j, K) = \sqrt{\sum\nolimits_{i=1}^{n} (D(STP_i|j, K) - E(P_{\text{cd}}|j, K))^2 q_i}, \quad (9)$$

where $D(STP_i|j, K)$ is conditional probability of correct diagnostics in case of the i-th SE failure.

The values of efficiency indicators $\overrightarrow{Efficiency}(j|K)$ are basic values for others $K-1$ objects. The calculation of efficiency indicators $\overrightarrow{Efficiency}(K-1|K)$ can be carried out based on following equations

$$E(t_r|h) = E(t_r|j, K)\rho_{h,1}, \quad (10)$$

$$s(t_r|h) = s(t_r|j, K)\rho_{h,1}, \quad (11)$$

$$E(C_r|h) = E(C_r|j, K)\rho_{h,2}, \quad (12)$$

$$s(C_r|h) = s(C_r|j, K)\rho_{h,2}, \quad (13)$$

$$E(L_r|h) = E(L_r|j, K)\rho_{h,3}, \tag{14}$$

$$s(L_r|h) = s(L_r|j, K)\rho_{h,3}, \tag{15}$$

where $\rho_{h,1}$, $\rho_{h,2}$ and $\rho_{h,3}$ are correction coefficients for evaluating the numerical values of the efficiency indicators of the repair of the h-th object from the studied group of equipment in terms of duration, cost, and labor intensity, respectively.

The next step is approximate estimation of efficiency indicators $\overrightarrow{Efficiency}(U)$ for equipment with spare parts. The equations for calculation the efficiency for the ψ-th object are following

$$\begin{aligned} E(t_r|\psi, U) &= A(\psi)\left[E\left(t_{\text{Rep}}(\psi)\right) + E(t_{\text{OSC}}(\psi))\right] \\ &+ (1 - A(\psi))\left[E\left(t_{Buy}(\psi)\right) + E(t_{Del}(\psi)) + E\left(t_{\text{Rep}}(\psi)\right) + E(t_{\text{OSC}}(\psi))\right], \end{aligned} \tag{16}$$

$$s(t_r|\psi, U) = a_t E(t_r|\psi, U), \tag{17}$$

$$\begin{aligned} E(C_r|\psi, U) &= A(\psi)\left[E\left(C_{\text{Rep}}(\psi)\right) + E(C_{\text{OSC}}(\psi))\right] \\ &+ (1 - A(\psi))\left[E\left(C_{Buy}(\psi)\right) + E(C_{Del}(\psi)) + E\left(C_{\text{Rep}}(\psi)\right) + E(C_{\text{OSC}}(\psi))\right], \end{aligned} \tag{18}$$

$$s(C_r|\psi, U) = a_C E(C_r|\psi, U), \tag{19}$$

$$\begin{aligned} E(L_r|\psi, U) &= A(\psi)\left[E\left(L_{\text{Rep}}(\psi)\right) + E(L_{\text{OSC}}(\psi))\right] + \\ &+ (1 - A(\psi))\left[E\left(L_{Buy}(\psi)\right) + E(L_{Del}(\psi)) + E\left(L_{\text{Rep}}(\psi)\right) + E(L_{\text{OSC}}(\psi))\right], \end{aligned} \tag{20}$$

$$s(L_r|\psi, U) = a_L E(L_r|\psi, U), \tag{21}$$

where $A(\psi)$ is steady-state availability for the system of spare part support for each ψ-th object and considered type of SE;$E\left(t_{\text{Rep}}(\psi)\right)$, $E\left(C_{\text{Rep}}(\psi)\right)$ and $E\left(L_{\text{Rep}}(\psi)\right)$ are expected time, cost and labor intensity for replacement of ψ-th failed object;$E(t_{\text{OSC}}(\psi))$, $E(C_{\text{OSC}}(\psi))$ and $E(L_{\text{OSC}}(\psi))$ are expected time, cost and labor intensity for output serviceability check of ψ-th object;$E\left(t_{Buy}(\psi)\right)$, $E\left(C_{Buy}(\psi)\right)$ and $E\left(L_{Buy}(\psi)\right)$ are expected time, cost and labor intensity for buying ψ-th failed object;$E(t_{Del}(\psi))$, $E(C_{Del}(\psi))$ and $E(L_{Del}(\psi))$ are expected time, cost and labor intensity to deliver ψ-th failed object; a_t. , a_C and a_L are coefficient for estimating standard deviation for time, cost and labor intensity of ψ-th failed object.

Analysis showed that the most suitable coefficients are $A(\psi) = 0.95$, $a_t = 0.1$. , $a_C = 0.15$ and $a_L = 0.15$. .

3.5 Calculation of Numerical Values of Efficiency Indicators of Single Current Repair Procedure

The initial data for calculating numerical values of efficiency indicators of single current repair procedure for all aviation radio equipment contain:

- the probabilities of failure $\overrightarrow{Q_M}$. f all aviation radio equipment;

- numerical values of conditional indicators of the efficiency of the current repair of aviatioradio equipment $\overrightarrow{Efficiency}(j|K)$, $\overrightarrow{Efficiency}(K-1|K)$ and $\overrightarrow{Efficiency}(U)$;
- expected values and standard deviations for time, cost and labor intensity for searching failed SE in the object of current repair: $E(t_{ro})$, $s(t_{ro})$, $E(C_{ro})$, $s(C_{ro})$, $E(L_{ro})$, and $s(L_{ro})$.

In the general case, during the search for a failed element of a certain type of aviation radio equipment, a priori determined number of diagnostic programs must be implemented. Therefore, without losing generality, we will assume that the elements of all M. objects of current repair are detected as a result of the execution of the same number of diagnostic programs.

The formulas for calculating efficiency in the form of the duration of the current repair process will take the following form:

$$E(t_r) = E(t_{ro}) + E\left(t_r|\overrightarrow{Q_M}\right), \tag{22}$$

$$E\left(t_r|\overrightarrow{Q_M}\right) = \sum\nolimits_{\psi=1}^{U} E(t_r|\psi, U)Q_\psi + \sum\nolimits_{h=1}^{K-1} E(t_r|h)Q_h + E(t_r|j, K)Q_j, \tag{23}$$

$$s(t_r) = \sqrt{s^2(t_{ro}) + s^2\left(t_r|\overrightarrow{Q_M}\right)}, \tag{24}$$

$$s^2\left(t_r|\overrightarrow{Q_M}\right) = \sum\nolimits_{\psi=1}^{U} s^2(t_r|\psi, U)Q_\psi + \sum\nolimits_{h=1}^{K-1} s^2(t_r|h)Q_h + s^2(t_r|j, K)Q_j. \tag{25}$$

The formulas for calculating efficiency in the form of the cost of the current repair process will take the following form:

$$E(C_r) = E(C_{ro}) + E\left(C_r|\overrightarrow{Q_M}\right), \tag{26}$$

$$E\left(C_r|\overrightarrow{Q_M}\right) = \sum\nolimits_{\psi=1}^{U} E(C_r|\psi, U)Q_\psi + \sum\nolimits_{h=1}^{K-1} E(C_r|h)Q_h + E(C_r|j, K)Q_j, \tag{27}$$

$$s(C_r) = \sqrt{s^2(C_{ro}) + s^2\left(C_r|\overrightarrow{Q_M}\right)}, \tag{28}$$

$$s^2\left(C_r|\overrightarrow{Q_M}\right) = \sum\nolimits_{\psi=1}^{U} s^2(C_r|\psi, U)Q_\psi + \sum\nolimits_{h=1}^{K-1} s^2(C_r|h)Q_h + s^2(C_r|j, K)Q_j. \tag{29}$$

The formulas for calculating efficiency in the form of the labor intensity of the current repair process will take the following form:

$$E(L_r) = E(L_{ro}) + E\left(L_r|\overrightarrow{Q_M}\right), \tag{30}$$

$$E\left(L_r|\overrightarrow{Q_M}\right) = \sum\nolimits_{\psi=1}^{U} E(L_r|\psi, U)Q_\psi + \sum\nolimits_{h=1}^{K-1} E(L_r|h)Q_h + E(L_r|j, K)Q_j, \tag{31}$$

$$s(L_r) = \sqrt{s^2(L_{ro}) + s^2\left(L_r|\overrightarrow{Q_M}\right)}, \tag{32}$$

$$s^2\left(L_r|\overrightarrow{Q_M}\right) = \sum\nolimits_{\psi=1}^{U} s^2(L_r|\psi, U)Q_\psi + \sum\nolimits_{h=1}^{K-1} s^2(L_r|h)Q_h + s^2(L_r|j, K)Q_j. \quad (33)$$

Numerical values of parameters according to formulas (25), (29) and (33) are approximate. The numerical values of the parameters $E(t_{ro})$, $E(C_{ro})$., $E(L_{ro})$ are determined on the basis of information on the schematic and structural features of the aviation radio equipment, the number of diagnostic programs, as well as the time, costs, and labor intensity during repair.

Then we define the parameters $s(t_{ro})$, $s(C_{ro})$, $s(L_{ro})$ as follows.

$$s(t_{ro}) = a_t E(t_{ro}), s(C_{ro}) = a_c E(C_{ro}), s(L_{ro}) = a_L E(L_{ro}). \quad (34)$$

Further, we believe that the numerical values of each of the parameters t_r, C_r, L_r are characterized by Gaussian (normal) probability distributionTaking this into account, threshold values for duration, costs and labor intensity can be determined that correspond to levels of significance $\theta_t^{(0)}$, $\theta_C^{(0)}$ and $\theta_L^{(0)}$. The threshold values are calculated by solving the equations:

$$P\left(t_r \geq t_r^{\text{th}}\left(\theta_t^{(0)}\right)|E(t_r), s(t_r)\right) \leq \theta_t^{(0)}, \quad (35)$$

$$P\left(C_r \geq C_r^{\text{th}}\left(\theta_C^{(0)}\right)|E(C_r), s(C_r)\right) \leq \theta_C^{(0)}, \quad (36)$$

$$P\left(L_r \geq L_r^{\text{th}}\left(\theta_L^{(0)}\right)|E(L_r), s(L_r)\right) \leq \theta_L^{(0)}. \quad (37)$$

Then

$$t_r^{\text{th}}\left(\theta_t^{(0)}\right) = \gamma_t s(t_r) + E(t_r), \quad (38)$$

$$C_r^{\text{th}}\left(\theta_C^{(0)}\right) = \gamma_C s(C_r) + E(C_r), \quad (39)$$

$$L_r^{\text{th}}\left(\theta_L^{(0)}\right) = \gamma_L s(L_r) + E(L_r), \quad (40)$$

where γ_t, γ_C and γ_L are the arguments of the probability integrals in Laplace form, which are respectively equal to $\left(1-\theta_t^{(0)}\right)$, $\left(1-\theta_C^{(0)}\right)$and $\left(1-\theta_L^{(0)}\right)$.

4 Results and Discussions

This section presents the numerical example of calculation of efficiency of current repair procedures for the case when object of current repair is receiver of radio communication equipment. This device can be presented in form of diagnostic model that twelve SE. We assume that information about failure probabilities $\overrightarrow{q_{12}}$ of elements of the object of current repair is known and the errors of the first and second kind are possible. Let

$$q_1 = 0.023, q_2 = 0.047, q_3 = 0.052, q_4 = 0.059, q_5 = 0.055,$$
$$q_6 = 0.058, q_7 = 0.124, q_8 = 0.082, q_9 = 0.231, q_{10} = 0.111,$$
$$q_{11} = 0.12, q_{12} = 0.028, \alpha = 0.01, \beta = 0.02$$

The search of failed SE was realized using engineering approach. The corresponding diagnostic program is source for obtaining the probability mass function of the probability of correct decision-making. The calculated probability mass function is shown in Fig. 1.

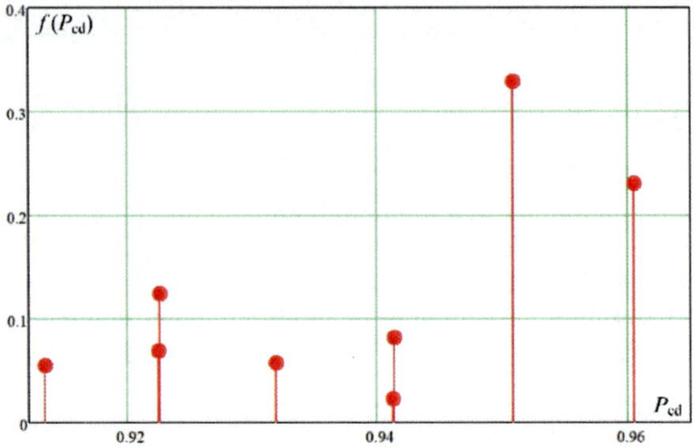

Fig. 1. Visual presentation of probability mass function of the probability of correct decision-making.

The presented probability mass function contains complete information about considered efficiency indicator, including expected value and standard deviation. Figure 2 and Fig. 3 show the dependencies of expected value and standard deviation of the probability of correct decision-making on the probabilities of the errors of the first and second kind. We can see that for $\alpha = \beta = 0$, we get $E(P_{cd}) = 1$ that confirms the correctness of calculation.

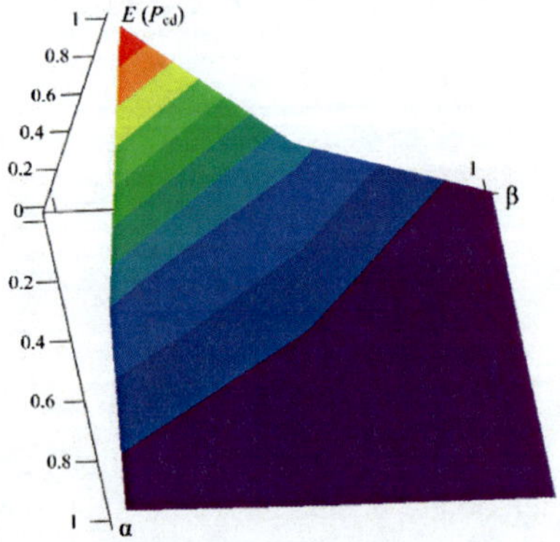

Fig. 2. The dependence of expected value of the probability of correct decision-making on the probabilities of the errors of the first and second kind.

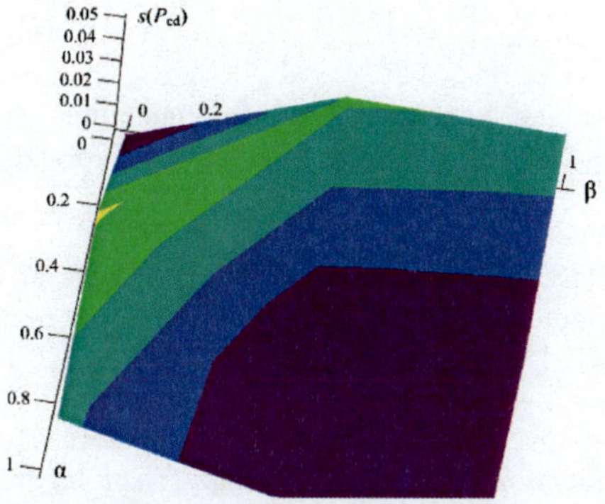

Fig. 3. The dependence of standard deviation of the probability of correct decision-making on the probabilities of the errors of the first and second kind.

5 Conclusions

The paper considers the issue of developing the methodology for calculating the numerical values of the efficiency indicators of the current repair for aviation radio equipment. The analytical ratios are given for the expected values and standard deviations of the duration, costs, and labor intensity of the repair process. The step-by-step instructions for evaluating efficiency are considered. The obtained calculation results confirmed the complex stochastic nature of the evaluated indicators and the necessity of their usage for timely and correct preventive actions.

The presented results can be used while solving the task of creating a subsystem for the current repair of aviation radio equipment that is the part of OS, during the improvement of these subsystems, and during the analysis of certain OS in case of repair procedures implementation.

6 Future Scope

The future research directions will be associated with: the substantiation of statistical characteristics of diagnostics process; the calculation of optimal repair personnel for aviation radio equipment; improvement of operation processes; building the optimal structure of repair enterprises for the level of country.

Acknowledgments. This research is partially supported by the Ministry of Education and Science of Ukraine under the projects "Development of an integrated flight control system" (№ 0121U109490) and "Methods of improving the operational efficiency of telecommunication and radio electronic systems in Ukraine" (№ 0122U001804).

References

1. Goncharenko, A.: Aircraft operation depending upon the uncertainty of maintenance alternatives. Aviation **21**(4), 126–131 (2017). https://doi.org/10.3846/16487788.2017.1415227
2. Solomentsev, O., Zaliskyi, M., Sushchenko, O., Bezkorovainyi, Y., Averyanova, Y., Ostroumov, I., et al.: Data processing through the lifecycle of aviation radio equipment. In: IEEE 17th International Conference on Computer Sciences and Information Technologies, pp. 146–151. IEEE, Lviv, Ukraine (2022). https://doi.org/10.1109/CSIT56902.2022.10000844
3. Ostroumov, I., Kuzmenko, N., Bezkorovainyi, Y., Averyanova, Y., Larin, V., Sushchenko, O., et al.: Relative navigation for vehicle formation movement. In: 3rd KhPI Week on Advanced Technology, pp. 1–4. IEEE, Kharkiv, Ukraine (2022). https://doi.org/10.1109/KhPIWeek57572.2022.9916414
4. Galar, D., Sandborn, P., Kumar, U.: Maintenance Costs and Life Cycle Cost Analysis. CRC Press, Boca Raton (2017)
5. Dhillon, B.: Reliability, quality, and safety for engineers. CRC Press, Boca Raton (2005)
6. Okoro, O.C., Zaliskyi, M., Dmytriiev, S., Solomentsev, O., Sribna, O.: Optimization of maintenance task interval of aircraft systems. Int. J. Comput. Netw. Inf. Secur. **14**(2), 77–89 (2022). https://doi.org/10.5815/ijcnis.2022.02.07
7. Zaliskyi, M., Solomentsev, O., Larin, V., Averyanova, Y., Kuzmenko, N., Ostroumov, I., et al.: Model building for diagnostic variables during aviation equipment maintenance. In: IEEE 17th International Conference on Computer Sciences and Information Technologies, pp. 160–164. IEEE, Lviv, Ukraine (2022). https://doi.org/10.1109/CSIT56902.2022.10000556
8. Davies, A.: Visual inspection systems. In: Davies, A. (eds) Handbook of Condition Monitoring, pp. 57–77. Springer, Dordrecht (1998). doi: https://doi.org/10.1007/978-94-011-4924-2_3
9. Hryshchenko, V., Romanenko, V., Hryshchenko, Y.: Quality of piloting during the approach. In: Ostroumov, I., Zaliskyi, M. (eds.) ACASD 2023. LNNS, vol. 736, pp. 260–270. Springer, Cham (2023). https://doi.org/10.1007/978-3-031-38082-2_20
10. Kuzmenko, N., Ostroumov, I., Bezkorovainyi, Y., Averyanova, Y., Larin, V., Sushchenko, O., et al.: Airplane flight phase identification using maximum posterior probability method. In: 3rd International Conference on System Analysis & Intelligent Computing, pp. 1–5. IEEE, Kyiv, Ukraine (2022). https://doi.org/10.1109/SAIC57818.2022.9922913
11. Larin, V., Solomentsev, O., Zaliskyi, M., Shcherban, A., Averyanova, Y., Ostroumov, I., et al.: Prediction of the final discharge of the UAV battery based on fuzzy logic estimation of information and influencing parameters. In: 3rd KhPI Week on Advanced Technology, pp. 1–6. IEEE, Kharkiv, Ukraine (2022). https://doi.org/10.1109/KhPIWeek57572.2022.9916490
12. Anand, A., Ram, M.: System Reliability Management: Solutions and Techniques. CRC Press, Boca Raton (2021)
13. Raza, A., Ulansky, V.: Optimization of condition monitoring decision making by the criterion of minimum entropy. Entropy (Basel) **21**(12), 1193 (2019). https://doi.org/10.3390/e21121193
14. Modarres, M., Groth, K.: Reliability and Risk Analysis. CRC Press, Boca Raton (2023)
15. Gertsbakh, I.: Reliability theory: with Applications to Preventive Maintenance. Springer, New York (2005)
16. Jardine, A.K.S., Tsang, A.H.C.: Maintenance, Replacement, and Reliability: Theory and Applications, 2nd edn. CRC Press, Boca Raton (2017)

17. Yan, H., Zuo, H., Tang, J., Wang, R., Ma, X.: Predictive maintenance framework of the aircraft system based on PHM information. In: Asia-Pacific International Symposium on Advanced Reliability and Maintenance Modeling, pp. 1–6. IEEE, Vancouver, BC, Canada (2020). https://doi.org/10.1109/APARM49247.2020.9209454
18. Odarchenko, R., Gnatyuk, S., Zhmurko, T., Tkalich, O.: Improved method of routing in UAV network. In: 3rd International Conference on Actual Problems of Unmanned Aerial Vehicles Developments, pp. 294–297. IEEE, Kyiv, Ukraine (2015). https://doi.org/10.1109/APUAVD.2015.7346624
19. Zuiev, O., Petrova, J.: Efficiency analysis of control and regulation processes in modern surveillance systems. In: 9th International Conference on Advanced Computer Information Technologies, pp. 9–12. IEEE, Ceske Budejovice, Czech Republic (2019). https://doi.org/10.1109/ACITT.2019.8780091
20. Al-Azzeh, J., Mesleh, A., Zaliskyi, M., Odarchenko, R., Kuzmin, V.: A method of accuracy increment using segmented regression. Algorithms **15**(10): 378 (2022). 1–24, https://doi.org/10.3390/a15100378
21. Tachinina, O.M., Lysenko, O.I., Ponomarenko, S.O., Alekseeva, I.V.: Conceptual proposals for the creation of a fully reusable light-class aerospace system in Ukraine. In: IEEE 6th International Conference on Methods and Systems of Navigation and Motion Control, pp. 85–88. IEEE, Kyiv, Ukraine (2020). https://doi.org/10.1109/MSNMC50359.2020.9255504
22. Hryshchenko, Y., Skripets, A., Tronko, V.: Mathematical description of phenomenon of amplification of integral-differential motive dynamic stereotype. In: 3rd International Conference on Methods and Systems of Navigation and Motion Control (MSNMC), pp. 71–74. IEEE, Kyiv, Ukraine (2014). https://doi.org/10.1109/MSNMC.2014.6979733
23. Sushchenko, O., Bezkorovainyi, Y., Golitsyn, V., Kuzmenko, N., Averyanova, Y., Zaliskyi, M., et al.: Integration of MEMS inertial and magnetic field sensors for tracking power lines. In: XVIII International Conference on the Perspective Technologies and Methods in MEMS Design, pp. 33–36. IEEE, Polyana (Zakarpattya), Ukraine (2022). https://doi.org/10.1109/MEMSTECH55132.2022.10002907
24. Sushchenko, O., Bezkorovainyi, Y., Solomentsev, O., Kuzmenko, N., Averyanova, Y., Zaliskyi, M., et al.: Airborne sensor for measuring components of terrestrial magnetic field. In: 41st International Conference on Electronics and Nanotechnology (ELNANO), pp. 687–691. IEEE, Kyiv, Ukraine (2022). https://doi.org/10.1109/ELNANO54667.2022.9926760
25. Averyanova, Y., Larin, V., Kuzmenko, N., Ostroumov, I., Zaliskyi, M., Solomentsev, O., et al.: Turbulence detection and classification algorithm using data from AWR. In: IEEE 2nd Ukrainian Microwave Week (UkrMW), pp. 518–522. IEEE, Ukraine (2022). https://doi.org/10.1109/UkrMW58013.2022.10037172

Check for updates

Method of Reliability Increasing Based on Spare Parts Optimization for Telecommunication Equipment

Georgiy Konakhovych, Maksym Zaliskyi(✉), Serhii Tarasiuk, Bohdan Chumachenko, Oleksandr Lavrynenko, and Veniamin Antonov

National Aviation University, Liubomyra Huzara Avenue, 1, Kyiv 03058, Ukraine
mzaliskyi@nau.edu.ua

Abstract. Telecommunication equipment in civil aviation is used for timely transmission of messages and useful information. This equipment performs the functions of establishing communication between the air traffic controller and the aircraft, ensuring the production activities of the airport or operating companies. Ensuring the reliability of the functioning of telecommunication equipment is a complex and sophisticated task, as it directly affects the risks of air navigation services, including the risks of catastrophic situation occurrence. Since modern equipment is developed according to the block and module principle, one of the ways to solve this problem is to plan the optimal quantity of spare parts for certain blocks, modules, nodes, and others. This optimization is usually multi-criteria and is performed on the basis of data on equipment failures and faults, available operating costs, requirements for recovery time. Sometimes the task is also complicated by the need for logistical analysis of spare part transportation. Taking into account the above, this paper is devoted to the development of a model of the procedure's performance of serviceability recovery for the aviation telecommunication equipment in cases of available spare parts and the analysis of the efficiency of the proposed model. The paper contains both analytical calculations and the results of statistical modeling.

Keywords: Operation · Repair · Data Processing · Spare Parts · Control Veracity · Reliability · Optimization

1 Introduction

1.1 Introduction to the Problem

Telecommunication equipment in civil aviation is used for timely transmission of messages and useful information. This equipment performs the functions of establishing communication between the air traffic controller and the aircraft, ensuring the production activities of the airport or operating companies [1, 2].

Telecommunication equipment together with radio technical devices is very important in the process of using airspace by civil aviation [3, 4]. The efficiency of this

I. Ostroumov and M. Zaliskyi (Eds.): ACASD 2024, LNNS 992, pp. 296–309, 2024.
https://doi.org/10.1007/978-3-031-60196-5_22

equipment directly depends on a significant number of factors and processes that are carried out in the operation system (OS) [5].

The main functions of the OS include:

- execution of technological procedures for the use of equipment according to its functional purpose;
- determination of signals from which the useful information about the condition of the equipment and the condition of the structural elements of the operation system can be obtained;
- measured signals and statistical data processing;
- decision-making regarding the implementation of preventive procedures in order to increase operational efficiency;
- implementation of maintenance and repair procedures;
- optimization of resource provision and others [6, 7].

The structural diagram of the OS content is shown in Fig. 1.

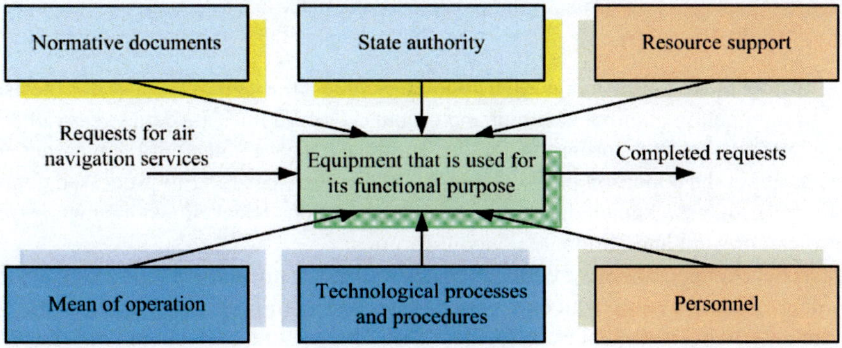

Fig. 1. The OS content.

Ensuring the reliability of the functioning of telecommunication equipment is a complex and sophisticated task, as it directly affects the risks of air navigation services, including the risks of catastrophic situation occurrence [8, 9]. Since modern equipment is developed according to the block and module principle, one of the ways to solve this problem is to plan the optimal quantity of spare parts for certain blocks, modules, nodes, and others [10, 11]. This optimization is usually multi-criteria and is performed on the basis of data on equipment failures and faults, available operating costs, requirements for recovery time [12]. Sometimes the task is also complicated by the need for logistical analysis of spare part transportation [13, 14].

In general, the problems of increasing reliability can be considered from the point of view of improving the processes of statistical data processing [15]. The stochastic occurrence of equipment's structural elements failure, the human factor with possible delays in decision-making or wrong decisions, random fluctuations in the prices of spare parts are the main reasons for the need to use the methods of probability theory, mathematical statistics, methods of machine learning and artificial intelligence to solve such problems [16, 17]. The process of operation from statistical data processing point of view can be explained using Fig. 2.

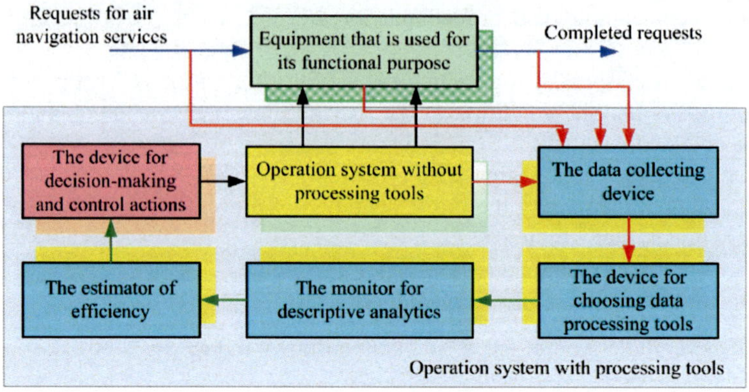

Fig. 2. The data processing in the OS.

Consider the notation in Fig. 2. The blue lines reflect the flow of requests for the use of telecommunication equipment (input and output). The red lines are data streams that are measured and contain information about requests, datasets about equipment parameters, datasets about the condition of OS elements. The green lines refer to processed datasets in the form of new knowledge and useful information. Black lines are management controls to prevent and eliminate nonconformities.

The data collecting device contains nodes for measuring and obtaining all the necessary data during the equipment operation and performs pre-processing in order to reduce the dimensionality of the datasets, select informative features, detect outliers, and others [18, 19]. The device for choosing data processing tools determines the structure and content of algorithms for solving specific operational tasks. Processing results in the form of graphs, parameter estimates, and possible inconsistencies are displayed on the appropriate monitor. The results of dataset processing are also used to calculate the aggregated efficiency index. If the efficiency of operation processes is low or insufficient, decisions are made on preventive and corrective actions.

Evaluation of operational efficiency can be considered within the framework of two approaches:

- assessment and minimization of operating costs;
- ensuring a given level of a certain reliability indicator [20, 21].

Usually, these two approaches are interrelated, that is, the main goal is to maximize reliability with minimal operating costs [22, 23].

Steady-state availability, availability function, probability of failure-free operation, and others can be selected as a reliability indicator [24]. The evaluation of these indicators is sufficiently discussed in the literature [25, 26], however, the analysis of the statistical characteristics of these estimates is still relevant.

1.2 Motivation

According to the principles of system and process approaches, all operational processes must take place in controlled and managed conditions [27]. The implementation of these principles is possible due to the improvement of the process structure, measurement of its characteristics and statistical data on operation for the purpose of further data processing and obtaining more complete information.

The new challenges of the aviation industry, related to the spread of the coronavirus and the possibility of the occurrence of new types of terrorist threats, are the reasons for the decrease in the profits of aviation enterprises. This is associated with the need to implement new security measures and at the same time with a decrease in air passenger traffic and transportation [28, 29]. Such practical circumstances motivate us to improve the structure of resource support for the operation of aviation equipment for the need to reduce the operating costs for maintenance and repair with ensuring guaranteed levels of reliability.

Taking into account the above, this paper is devoted to the development of a model of the procedure's performance of serviceability recovery for the aviation telecommunication equipment in cases of available spare parts and the analysis of the efficiency of the proposed model. The paper contains both analytical calculations and the results of statistical modeling.

1.3 Contribution

This paper contributes to theory of telecommunication equipment maintenance in the framework of optimization tasks of spare parts support and improvement of data processing tools for the OS. The practical value is associated with the possibility of reducing operating costs for guaranteed levels of reliability indicators of telecommunication equipment of aviation enterprises.

1.4 The Organization of the Paper

The paper is organized as follows. The first section describes basic information in the field of telecommunication equipment operation and analyzes the main problems in this domain. The second section presents the optimization of structure for spare part support during aviation telecommunication equipment operation. The third section deals with the estimation of steady-state availability. The last section presents conclusions and future research directions.

2 Materials and Methods

Let's assume that aviation telecommunication equipment consists of one or more easily removable units with different degree of redundancy. An easily removable unit consists of a set of structural and removable units, which are modules or boards. Such boards (modules) are typical for replacement in the unit and are easily accessible in the unit.

Structural removable units include varying amounts of non-recoverable elements. In accordance with the construction of the aviation telecommunication equipment, three possible levels of procedures to maintain and restore their performance are determined. The first-level procedures allow to restore functionality by replacing them with functional ones. At the second level, the unit's operability is restored by replacing a failed structural-removable unit with a functional one. At the third level, the restoration of structurally removable units is carried out by replacing a failed non-recoverable element or group of non-recoverable elements in it.

Figure 3 shows a model of the system for providing spare parts for telecommunication equipment. This model is characterized by the fact that all decisions regarding the use of spare parts are made based on the results of monitoring the functionality of the equipment using built-in control systems. Based on the results of monitoring, the following decisions are made:

- allow the system (unit) to be used for its functional purpose if it is determined to be operational and not apply to the spare parts fund;
- dismantle the failed system (unit) and send it to the manufacturer for restoration;
- request to the spare parts fund for a unit (system) that is known to be operational, install and monitor its performance, and then begin using the system for its functional purpose.

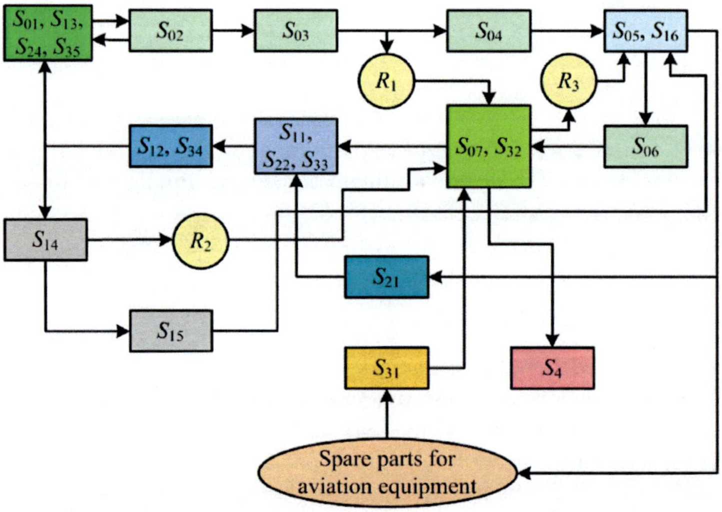

Fig. 3. The model of the system for providing spare parts for telecommunications equipment.

The following notations are introduced in Fig. 3:

- S_{01} is a state of intended use of telecommunications units (systems) installed during manufacture;
- S_{02} is a state of the units (systems) serviceability control using built-in control systems;
- S_{03} is a dismantling the unit in the event of a true positive determination (TP) of inoperability or a false positive (FP) determination of inoperability using the built-in control systems;

- S_{04} is a delivery of the unit to the manufacturer to restore functionality;
- S_{05} is a restoration of serviceability by the manufacturer;
- S_{06} is a delivery of the restored unit from manufacturer;
- S_{07} is a storage of the restored unit in the spare parts fund;
- S_{11} is an installation of unit to the telecommunication equipment;
- S_{12} is performance monitoring using the built-in control system of the unit supplied from the spare parts fund;
- S_{13} is a use of the unit for its functional purpose after correct installation;
- S_{14} is a dismantling the unit detected as inoperative by the built-in control systems;
- S_{15} is delivery of the unit to the manufacturer with TP or FP state;
- S_{16} is a restoration of serviceability by the manufacturer;
- S_{21} is a fast delivery of the unit by the manufacturer in the case of absence of the required unit in the spare parts fund;
- S_{22} is an installation of the delivered unit;
- S_{23} is a monitoring the performance of the installed unit using the built-in control systems;
- S_{24} is an intended use of the unit after delivery and installation;
- S_{31} is a supply of units (systems) by the spare parts fund in accordance with the adopted inventory replenishment strategy;
- S_{32} is a storage of supplied units;
- S_{33} is an installation of the unit received from the spare parts fund;
- S_{34} is monitoring the performance of the unit received from the spare parts fund;
- S_{35} is a use of unit for its functional purpose after restoration serviceability;
- S_4 is an end of operation of the system (unit) after its service life has been declared beyond the determined time;
- R_1 is a request for spare parts of the system (unit);
- R_2 is a request for fast delivery of the unit if it is not included in the spare parts fund;
- R_3 is a request for the restoration of unit serviceability.

The model of the spare parts support system was built for the case of an exponential distribution of the time to failure for units in the telecommunication equipment. Therefore, to simplify the consideration of the model for the formation of the flow of dismantling the unit, the states S_{01}, S_{13}, S_{24}, and S_{35} can be replaced by the state S_{01}.

Thus, state S_{01}, S_{13}, S_{24}, and S_{35} analysis allows to keep track of the individual operating time to failure for the unit or whole. The transition from state S_{01} to S_{03} depends on the probabilities of false and correct block removal.

The goal of creating the system for providing an optimal spare parts fund for the operation process is to create a mechanism for determining the number and composition of reserves that ensure the continuity of equipment use for its functional purpose at minimum specific average costs.

The function for optimizing is the reduced operating costs during the year for a set of similar units of aviation telecommunication equipment

$$C\left(n_{sp}\right) = C_{in}\left(n_{sp}\right) + E_c K\left(n_{sp}\right), \tag{1}$$

where $C_{in}\left(n_{sp}\right)$ is an initial cost for the spare parts of similar units throughout the year, E_c is a standard capital investment efficiency coefficient, $K(n_{sp})$ is a capital investment in the base airport for given unit of equipment, n_{sp} is a number of spare parts for given unit for telecommunications equipment.

Therefore, the optimal number of spare parts in the base airport is determined from the solution of the problem

$$C_{opt}\left(n_{sp}\right) = \min_{n_{sp}} C\left(n_{sp}\right). \tag{2}$$

Taking into account Fig. 3 the operating costs (1) can be presented as follows

$$C\left(n_{sp}\right) = \frac{mn_{sp}T\left(C_1t_1 + C_2t_2 + C_3 + C_4\Delta t(n_{sp})\right)}{MS_0} + (E_0 + E_c)C_5n_{sp}, \tag{3}$$

where m is a quantity of similar units of equipment, T is an average time of TP detection and decision-making; C_1 is an average cost per time for unit dismantling, C_2 is an average cost per time for unit installation, C_3 is the delivery costs, C_4 is the costs associated with aircraft delays, C_5 is cost of one spare part, E_0 is deduction rate for the renovation of spare parts, $\Delta t(n_{sp})$ is average time of aircraft delay at the base airport due to failure and restoration of unit using the spare parts, MS_0 is an average time between two sequential removals of units, t_1 is the time for unit dismantling, t_2 is the time for unit installation.

The average time of aircraft delay at the base airport due to failure and restoration of unit using the spare parts

$$\Delta t\left(n_{sp}\right) = \gamma\left(t_1 + t_2 + t_3 + \Delta t_{sp}\left(n_{sp}\right) - t_4\right), \tag{4}$$

where t_3 is an average costs per time for serviceability monitoring using the built-in control system, t_4 is an average duration of aircraft parking at the base airport, $\Delta t_{sp}\left(n_{sp}\right)$ is an average delay time in satisfying the request for the spare parts at the base airport, γ is indicator defined by expression

$$\gamma = \begin{cases} 0, if t_4 \geq t_1 + t_2 + t_3 + \Delta t_{sp}\left(n_{sp}\right), \\ 1, if t_4 < t_1 + t_2 + t_3 + \Delta t_{sp}\left(n_{sp}\right). \end{cases}$$

It should be noted that $\Delta t\left(n_{sp}\right)$ is an indicator of the sufficiency of the spare parts fund. This indicator is included as a parameter in the objective functions of the spare parts fund optimization for all strategies for constructing the maintenance and repair system. To calculate the sufficiency indicator, we use the theory of Markov processes. For this purpose, let us denote by the number of spare parts of given type. Let the states of equipment are:

- S_1 is state when m units are serviceable and there are n_{sp} spare parts;
- S_2 is state when m units are serviceable and there are $n_{sp} - 1$ spare parts;
- $S_{n_{sp}}$ is state when m units are serviceable and there is only one spare part;
- $S_{n_{sp}+1}$ is state when m units are serviceable and there are no spare parts;
- $S_{n_{sp}+2}$ is state when $m - 1$ units are serviceable and there are no spare parts;
- $S_{n_{sp}+m+1}$ is state when all units with failure and there are no spare parts.

The system state graph is shown in Fig. 4.

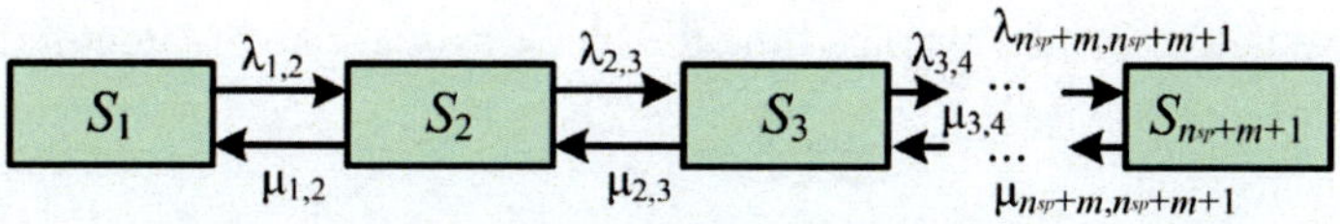

Fig. 4. The system state graph.

The following notations are introduced in Fig. 4: $\lambda_{i,i+1}$ is a rate of block removal that transfers the system from the i-th to the $(i+1)$-th state, $\mu_{i,i+1}$ is a rate of block restoration that transfers the system from the $(i+1)$-th to the i-th state.

The limiting probabilities P_i of states satisfy the system of linear algebraic equations:

$$\begin{cases} \lambda_{1,2}P_1 - \mu_{2,1}P_2 = 0; \\ \lambda_{2,3}P_2 - \mu_{3,2}P_3 = 0; \\ \ldots \\ \lambda_{n_{sp}+m,n_{sp}+m+1}P_{n_{sp}+m} - \mu_{n_{sp}+m+1,n_{sp}+m}P_{n_{sp}+m+1} = 0; \\ \sum_{i=1}^{n_{sp}+m+1} P_i = 1. \end{cases} \tag{5}$$

The solution to the system of Eqs. (5) has the following form:

$$P_1 = \left(1 + \sum_{j=1}^{n_{sp}+m} \frac{\prod_{i=1}^{j} \lambda_{i,i+1}}{\prod_{i=1}^{j} \mu_{i,i+1}}\right)^{-1}. \tag{6}$$

$$P_k = P_1 \frac{\prod_{i=1}^{k} \lambda_{i,i+1}}{\prod_{i=1}^{k} \mu_{i,i+1}}, k \epsilon \left[2; n_{sp} + m + 1\right]. \tag{7}$$

The rates of transitions $\lambda_{i,i+1}$ and $\mu_{i,i+1}$ of the Markov chain for simplest case (only one unit of given type is restored at one time moment) is determined as follows:

$$\lambda_{i,i+1} = \begin{cases} \frac{m}{MS_0}, i = \overline{1, n_{sp} + 1}; \\ \frac{m+n_{sp}+1-i}{MS_0}, i = \overline{n_{sp} + 2, m + n_{sp}}; \end{cases} \tag{8}$$

$$\mu_{i,i+1} = \frac{1}{t_r}. \tag{9}$$

where t_r is mean time to restore.

If the confidence probability $P\left(n_{sp}\right)$ is given, then the procedure for determining the optimal number n_{spopt} of spare parts is as follows. At the first step, we determine the average time the unit is in serviceable condition for the average regeneration cycle using the formula

$$MS_1 = \frac{1 - e^{-\lambda\tau}}{\lambda\left(1 - (1 - \alpha)e^{-\lambda\tau}\right)}, \tag{10}$$

where α is the probability of the first type error, λ is the failure rate, τ is a time.

Then we determine the average time the block is in a state of latent failure during the average regeneration cycle

$$MS_2 = \frac{1}{1-(1-\alpha)e^{-\lambda\tau}}\left(\frac{\tau(1-\beta e^{-\lambda\tau})}{1-\beta} - \frac{1-e^{-\lambda\tau}}{\lambda}\right), \tag{11}$$

where β is the probability of the second type error.

At the next step, we determine the average time between successive removals of units (average regeneration cycle) using the formula

$$MS_0 = MS_1 + MS_2. \tag{12}$$

Then we can calculate the rate of unit removal using the formula

$$\lambda_r = \frac{1}{mMS_0}, \tag{13}$$

The optimal quantity of spare parts is determined from the condition

$$1 - P(n_{sp}) > \frac{(qt_e\lambda_r)^{n_{sp}+1}}{(1+n_{sp})!}e^{-qt_e\lambda_r}, \tag{14}$$

where t_e is the time TP unit removal from telecommunication equipment.

3 Results and Discussions

This section presents example of statistical characteristics of steady-state availability (SSA) determining. Such approach can be used to assess reliability of equipment with spare parts.

The statistical estimates of steady-state availability independently on distribution of time between failures and recovery can be found as

$$A = \frac{MTBF}{MTBF + MTBR}, \tag{15}$$

where *MTBF* is estimate of mean time between failure, *MTBR* is estimate of mean time between restore.

Let us assume that the restoration time is a constant and equal to t_r. Then, $MTBR = t_r$. The operating time between failures can be described by probability density function $f(t_F)$. This function is used to determine $f(MTBF)$. To find the characteristics of SSA we can use theory of functional transformations.

According to normalization condition we can write

$$f(A)dA = f(MTBF)d(MTBF).$$

Then

$$f(A) = f(MTBF)\left|\frac{d(MTBF)}{dA}\right|.$$

The Jacobian of transformation

$$J = \left| \frac{d(MTBF)}{dA} \right|.$$

According to (15) we can get

$$MTBF = MTBR \frac{A}{1 - A}.$$

Therefore,

$$J = \frac{MTBR}{(1 - A)^2}.$$

Summarizing the above, we get

$$f(A) = MTBR \frac{A}{1 - A} f(MTBF) \Big|_{MTBF = MTBR \frac{A}{1-A}}. \tag{16}$$

The statistical estimate of *MTBF* is generated as follows

$$MTBF = \frac{\sum_{i=1}^{S} t_{Fi}}{S}, \tag{17}$$

where S is a quantity of observed failures.

According to the central limit theorem of probability theory for the cases $S \gg 1$ independently on distribution of operating time between failures, the distribution of the sum approximately tends to normal distribution with $1/\lambda$ mean value and $1/\lambda\sqrt{S}$ standard deviation.

The asymptotical distribution of SSA can be determined as follows

$$f(A) = \frac{MTBR}{(1 - A)^2} \frac{\lambda\sqrt{S}}{\sqrt{2\pi}} e^{-0.5\lambda^2 S \left(MTBR \frac{A}{1-A} - \frac{1}{\lambda}\right)^2}. \tag{18}$$

The obtained Eq. (14) can be used to estimate the numerical characteristics of SSA: expected value, variance, standard deviation, skewness, kurtosis, and others.

To analyze the correctness of obtained results, we carried out the statistical simulations. The simulation program generates 1000 datasets. Each dataset contains 1000 samples of operating time between failures. The initial value of failure rate is equal to 0.001 failures per hour. The restoration time is 25 h. The example of the trend of operating time between failures for one procedure of iteration is shown in Fig. 5.

The next step of simulation is estimation of SSA for each observation and iteration. For this, we used Eq. (15). The example of the SSA estimates for one procedure of iteration is shown in Fig. 6.

Figure 7 shows the results of simulation in form of histogram of SSA estimates. The blue line corresponds to Eq. (18). The good coincidence of theoretical distribution (18) and simulation results proved the correctness of considered assumptions.

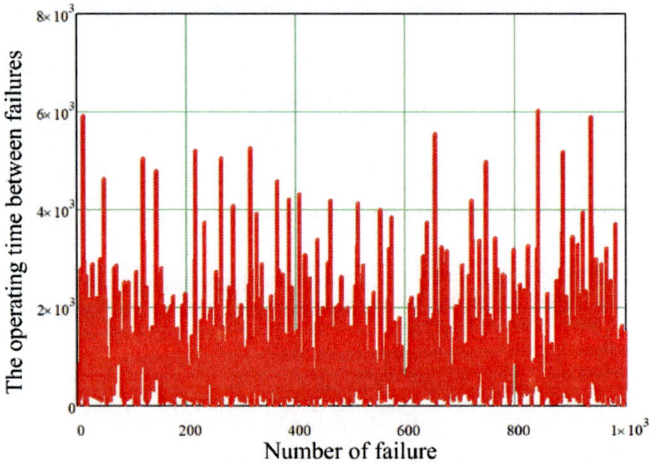

Fig. 5. The dataset of operating time between failures.

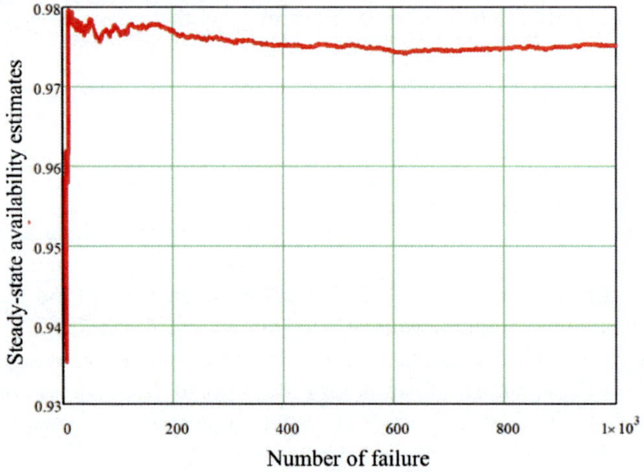

Fig. 6. The estimates of SSA.

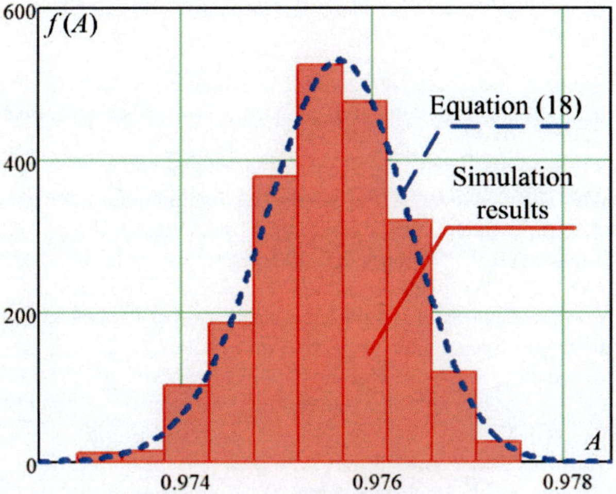

Fig. 7. The histogram of SSA estimates.

4 Conclusions

The paper is devoted to the solution of the actual problem of the theory of operation, connected with increasing the reliability of aviation telecommunication equipment. The main approach to solving this problem was to determine the optimal quantity of spare parts of the equipment.

During the optimization of the spare parts quantity, the main attention was paid to the operating costs of the aviation enterprise and the reliability indicator in the form of the steady-state availability. Operating costs and the optimal quantity of spare parts were calculated on the basis of the proposed model of operation for telecommunication equipment. For the steady-state availability, the probability density function of its estimates was found based on the application of the methods of functional transformations of random processes. To confirm the correctness of the obtained results, statistical simulation was carried out.

The obtained in this paper results can be considered as an improvement of the operation system in the framework of data processing and decision-making.

The future research directions are:

- development of data hub structure and content to improve operational processes;
- synthesis of machine learning methods and other artificial intelligence methods for operational data processing;
- calculation of the optimal quantity of spare parts of equipment for various models of maintenance and repair;
- development of methods of increasing reliability in cases of deterioration of the technical condition of the equipment.

References

1. Kuzmenko, N., Ostroumov, I., Bezkorovainyi, Y., Averyanova, Y., Larin, V., Sushchenko, O., et al.: Airplane flight phase identification using maximum posterior probability method. In: 3rd International Conference on System Analysis & Intelligent Computing, pp. 1–5. IEEE, Kyiv, Ukraine (2022). https://doi.org/10.1109/SAIC57818.2022.9922913
2. Kharchenko, V.P., Kuzmenko, N.S., Ostroumov, I.V.: Identification of unmanned aerial vehicle flight situation. In: 5th International Conference of Actual problems of unmanned aerial vehicles development (APUAVD), pp. 116–120. IEEE, Kyiv, Ukraine (2017). https://doi.org/10.1109/APUAVD.2017.8308789
3. Sineglazov, V.: Main features of terrain-aided navigation systems. In: International Conference on Methods and Systems of Navigation and Motion Control (MSNMC), pp. 49–52. IEEE, Kiev, Ukraine (2014). https://doi.org/10.1109/MSNMC.2014.6979728
4. Ostroumov, I.V., Kuzmenko, N.S.: Accuracy improvement of VOR/VOR navigation with angle extrapolation by linear regression. Telecommun. Radio Eng. **78**(15), 1399–1412 (2019). https://doi.org/10.1615/TelecomRadEng.v78.i15.90
5. Zaliskyi, M., Solomentsev, O., Kozhokhina, O., Herasymenko, T.: Reliability parameters estimation for radioelectronic equipment in case of change-point. In: Signal Processing Symposium 2017 (SPSympo 2017), pp. 1–4. IEEE, Jachranka Village, Poland (2017). https://doi.org/10.1109/SPS.2017.8053676
6. Jardine, A.K.S., Tsang, A.H.C.: Maintenance, Replacement, and Reliability: Theory and Applications, 2nd edn. CRC Press, Boca Raton (2017)
7. Solomentsev, O., Zaliskyi, M., Sushchenko, O., Bezkorovainyi, Y., Averyanova, Y., Ostroumov, I., et al.: Data processing through the lifecycle of aviation radio equipment. In: IEEE 17th International Conference on Computer Sciences and Information Technologies, pp. 146–151. IEEE, Lviv, Ukraine (2022). https://doi.org/10.1109/CSIT56902.2022.10000844
8. Ivashchuk, O., Ostroumov, I., Kuzmenko, N., Sushchenko, O., Averyanova, Y., Solomentsev, O., et al.: A configuration analysis of Ukrainian flight routes network. In: 16th International Conference on the Experience of Designing and Application of CAD Systems, pp. 6–10. IEEE, Lviv, Ukraine (2021). https://doi.org/10.1109/CADSM52681.2021.9385263
9. Ostroumov, I., Kuzmenko, N., Bezkorovainyi, Y., Averyanova, Y., Larin, V., Sushchenko, O., et al.: Relative navigation for vehicle formation movement. In: 3rd KhPI Week on Advanced Technology, pp. 1–4. IEEE, Kharkiv, Ukraine (2022). https://doi.org/10.1109/KhPIWeek57572.2022.9916414
10. Galar, D., Sandborn, P., Kumar, U.: Maintenance Costs and Life Cycle Cost Analysis. CRC Press, Boca Raton (2017)
11. Dhillon, B.: Reliability, Quality, and Safety for Engineers. CRC Press, Boca Raton (2005)
12. Okoro, O.C., Zaliskyi, M., Dmytriiev, S., Solomentsev, O., Sribna, O.: Optimization of maintenance task interval of aircraft systems. Int. J. Comput. Netw. Inf. Secur. **14**(2), 77–89 (2022). https://doi.org/10.5815/ijcnis.2022.02.07
13. Pan, F., Yan, X., Wang, N., Xiao, B.: A prediction model of repairable spare parts utilization rate based on probabilistic method. In: 3rd International Conference on System Reliability and Safety, pp. 187–191. IEEE, Barcelona, Spain (2018). https://doi.org/10.1109/ICSRS.2018.8688876
14. Wang, N., Xia, G., He, Z., Su, L., Zhang, Y., Wang, X.: Research on spare parts requirement quantity forecasting for communication equipment. In: International Conference on Communications, Information System and Computer Engineering, pp. 373–377. IEEE, Haikou, China (2019). https://doi.org/10.1109/CISCE.2019.00088
15. Modarres, M., Groth, K.: Reliability and Risk Analysis. CRC Press, Boca Raton (2023)

16. Anand, A., Ram, M.: System Reliability Management: Solutions and Techniques. CRC Press, Boca Raton (2021)
17. Sineglazov, V.M., Ischenko, V.P.: Integrated navigation complex of UAV on basis of flight controller. In: International Conference Actual Problems of Unmanned Aerial Vehicles Developments, pp. 20–25. IEEE, Kyiv, Ukraine (2015). https://doi.org/10.1109/APUAVD.2015.7346547
18. Al-Azzeh, J., Mesleh, A., Zaliskyi, M., Odarchenko, R., Kuzmin, V.: A method of accuracy increment using segmented regression. Algorithms **15**(10): 378 (2022). 1–24, https://doi.org/10.3390/a15100378
19. Tachinina, O.M., Lysenko, O.I., Ponomarenko, S.O., Alekseeva, I.V.: Conceptual proposals for the creation of a fully reusable light-class aerospace system in Ukraine. In: IEEE 6th International Conference on Methods and Systems of Navigation and Motion Control, pp. 85–88. IEEE, Kyiv, Ukraine (2020). https://doi.org/10.1109/MSNMC50359.2020.9255504
20. Solomentsev, O., Zaliskyi, M., Kozhokhina, O., Herasymenko, T.: Efficiency of data processing for UAV operation system. In: 4th International Conference Actual Problems of Unmanned Aerial Vehicles Developments (APUAVD), pp. 27–31. IEEE, Kyiv, Ukraine (2017). https://doi.org/10.1109/APUAVD.2017.8308769
21. Zaliskyi, M., Solomentsev, O., Larin, V., Averyanova, Y., Kuzmenko, N., Ostroumov, I., et al.: Model building for diagnostic variables during aviation equipment maintenance. In: IEEE 17th International Conference on Computer Sciences and Information Technologies, pp. 160–164. IEEE, Lviv, Ukraine (2022). https://doi.org/10.1109/CSIT56902.2022.10000556
22. Boylan, J.E., Syntetos, A.A.: Spare parts management: a review of forecasting research and extensions. J. Manag. Math. **21**(3), 227–237 (2010). https://doi.org/10.1093/imaman/dpp016
23. Kontrec, N.Z., Milovanovic, G.V., Panic, S.R., Milosevic, H.: A reliability-based approach to nonrepairable spare part forecasting in aircraft maintenance system. Math. Probl. Eng. **2015**, 1–7 (2015). https://doi.org/10.1155/2015/731437
24. Averyanova, Y., Larin, V., Kuzmenko, N., Ostroumov, I., Zaliskyi, M., Solomentsev, O., et al.: Turbulence detection and classification algorithm using data from AWR. In: IEEE 2nd Ukrainian Microwave Week (UkrMW), pp. 518–522. IEEE, Ukraine (2022). https://doi.org/10.1109/UkrMW58013.2022.10037172
25. Ostroumov, I.V., Kuzmenko, N.S.: Accuracy estimation of alternative positioning in navigation. In: 4th International Conference on Methods and Systems of Navigation and Motion Control (MSNMC), pp. 291–294. IEEE, Kiev, Ukraine (2016). https://doi.org/10.1109/MSNMC.2016.7783164
26. Ostroumov, I.V., Kuzmenko, N.S., Marais, K.: Optimal pair of navigational aids selection. In: 5th International Conference on Methods and Systems of Navigation and Motion Control (MSNMC), pp. 32–35. IEEE, Kiev, Ukraine (2018). https://doi.org/10.1109/MSNMC.2018.8576293
27. Sineglazov, V.: Landmarks navigation system software. In: International Conference on Methods and Systems of Navigation and Motion Control (MSNMC), pp. 62–65. IEEE, Kiev, Ukraine (2014). https://doi.org/10.1109/MSNMC.2014.6979731
28. Wang, Y., Ma, L., Zhang, Z., Huang, Z.: Optimization method for spare parts deployment efficiency in development phase. In: 9th International Conference on Reliability, Maintainability and Safety, pp. 1218–1223. IEEE, Guiyang, China (2011). https://doi.org/10.1109/ICRMS.2011.5979455
29. Feng, G., Jingjing, W., Qingbin, W., Suqin, Z.: Research on the demand distribution of aviation equipment maintenance spare parts. In: 6th International Conference on Big Data Analytics, pp. 301–304. IEEE, Xiamen, China (2021). https://doi.org/10.1109/ICBDA51983.2021.9403053

Reliability Assessment of Highly Reliable Samples Using the Tolerance Limits and the Weibull's Law

Valentyn Dyptan[1], Petro Yablonsky[1], Oleksandr Avramenko[1], Volodymyr Klymchuk[2](✉), Pavlo Openko[1], and Vasyl Polishchuk[1]

[1] National Defence University of Ukraine named after Ivan Cherniakhovskyi, Povitroflotskyi Avenue, 28, Kyiv 03049, Ukraine

[2] National Aviation University, Liubomyra Huzara Avenue, 1, Kyiv 03058, Ukraine

klimchuk1591@ukr.net

Abstract. Among the most relevant issues of reliability is the assessment of the reliability of complex technical systems with a small number of observations. In practice, a situation often arises when an estimate of satisfactory accuracy can be obtained only at the end of tests or operation, when it loses its relevance. Then it is better to use methods of processing a small amount of data collected during preliminary tests, to obtain some estimates of reliability indicators, than to postpone the acceptance of more accurate estimates until a time when more reliable statistical material has been accumulated. The paper proposes an approach to the solution of the problem of constructing an estimate of the time of uptime using the theory of β-expected limits. The paper proposes to apply the theory of tolerance limits to determine the reliability of highly reliable samples of complex technical systems when using a failure model in the form of the Weibull distribution law. The choice of the Weibull distribution is due to its universality of use in the analysis of the reliability of mechanical, electronic and other systems. The paper provides examples of the use of the proposed methodology for calculating reliability indicators with a small number of failures. The lower tolerance limits of earnings per failure that correspond to the given level of probability are found. The obtained results are sufficient for making managerial decisions regarding the determination of reliability in the conditions discussed above.

Keywords: Tolerance Limits · Reliability · High-Reliability Engineering Samples · Failure Model Distributed · Weibull Failure Distribution Law · Product Uptime

1 Introduction

Modern samples of weapons and military equipment as samples of complex technical systems (CTSs) have a long-life cycle. According to the results of the research conducted on CTS, which are in service in many countries of the world, it was established that the duration of their life cycle is more than 30 years. Such CTSs are characterized

I. Ostroumov and M. Zaliskyi (Eds.): ACASD 2024, LNNS 992, pp. 310–321, 2024.
https://doi.org/10.1007/978-3-031-60196-5_23

by the gradual exhaustion of the resource, deterioration of the technical condition, and a decrease in their level of reliability, which significantly affects their efficiency. Solving the tasks of technical equipment of the armed forces with CTSs requires the implementation of the tasks of their development, procurement, use, maintenance in a combat-ready state, etc. by creating highly reliable systems. At the same time, as a rule, it is necessary to solve a set of tasks related to the assessment of the influence of the reliability of the components of CTS on the efficiency of the system as a whole. It should be taken into account that the complexity and responsibility of tasks solved by modern systems places high demands on their reliability. In this regard, research and solving the problem of reliable and timely determination of the reliability of CTS and their constituent parts is relevant. Among the most urgent issues of reliability is the assessment of reliability with a small number of observations [1], which is typical for highly reliable systems.

2 Problem Formulation

To maintain CTS in a combat-ready state at the stage of the "use" life cycle, measures are taken that correspond to the stage of the life cycle "support" by conducting inspections of the technical condition, restoring performance, maintenance, etc. They are combined into a maintenance and repair system. Depending on what is implemented by the developer in CTS to maintain the required level of their combat readiness, CTSs are distinguished that are controlled or not controlled, recoverable or not recoverable, serviced or not serviced, etc. In accordance with these features of the operation of CTS, reliability models of objects are used: which are not controlled, restored and maintained; which are monitored, restored and not maintained; which are monitored, restored and maintained, etc.

Thus, in order to solve the actual problem in [2], the failure-free model is considered, which is represented by the exponential distribution of earnings before failure, the Weibull distribution, and the normal distribution; the maintainability model, which is represented by the exponential distribution of the duration of restoration to working condition, and the reliability model of the tractor, as an object that is monitored, restored and not maintained, in the form of a steady-state availability coefficient. In [3], a model of the reliability of CTS (using the example of the oil pipeline system) that is not controlled, restored, and maintained is considered. The process of development of failures of such a complex technical system is divided into two stages: the stage of the occurrence of a defect and the stage of the development of a defect. At the same time, the second stage begins after the first and is characterized by a certain duration. In [4], a simulation model of the reliability of the gas transportation network was developed. In this model, the object of research is such that it is not maintained or restored. At the same time, the influence of the failure of individual nodes and the simultaneous failure of two or more nodes on the reliability and safety of the network operation is investigated. In [5], a reliability model of a wireless sensor network was developed to analyze the impact of hardware failures located on the terrain on its probabilistic characteristics. At the same time, the process of network functioning is modeled taking into account failures of its elements and topology. The simulation object is recoverable and non-maintained. In [6], the model of the reliability of electrical, electronic and programmable electronic systems, which are designed to prevent and mitigate the consequences of dangerous events at various high-risk facilities in the nuclear, processing, oil and gas, chemical, and automotive industries, is considered. The reliability model is based on Markov processes for non-renewable and non-maintained objects.

When developing reliability models of CTS, which are considered above, the features of typical cyclograms of their operation are not considered, which does not allow taking into account all the main factors that significantly affect their reliability.

The issues of developing reliability models of such CTS, which take into account the typical cycle of intended use, are considered in [7, 8]. At the same time, reliability characteristics are considered in relation to a fixed calendar duration of operation, usually at the stage of normal operation, which is characterized by a constant intensity of failures [9–11]. In [12, 13], the influence of the reliability of CTS mobility means on the estimation of its efficiency preservation coefficient is considered. In [11], a model of the effectiveness of a complex technical system is considered, in which it is assumed that the reliability model, which describes the processes of failures and recovery of CTS at the stage of waiting for the intended use, is known. The influence of periodic maintenance and checks of the functioning of CTS on reliability is not taken into account.

The problem of determining minimum sample sizes for setting Weibull tolerance limits from trimmed data is discussed [14] when the numbers or proportions of the left and right trimmed observations are fixed. Step-by-step procedures for determining optimal sampling plans are also presented.

In [15], a general loss function which includes the sampling cost, the time-consuming cost, the salvage value, and the after-sales cost was proposed to determine the Bayes risk and the corresponding optimal sampling plan. Explicit expressions for the Bayes risks for both single and double sampling plans were derived, respectively.

In addition, paper [16] deals with the theoretical aspects of determining the reliability of means of ground support for flights using the theory of tolerance limits. During the description of the logical dependence, only the lower tolerance limit is considered, the value of which is found using the concept of central statistics. At the same time, the distribution function belongs to the class of distribution with shift and scale parameters and is the first marginal distribution (for the minimum) or a double exponential distribution.

Thus, from the review of the literature, it follows that the problematic issues of developing reliability models for CTS that are controlled, restored and maintained are not considered sufficiently. At the same time, the determination of the reliability indicators for CTS in accordance with modern requirements is based on the use of large amounts of information and a significant duration of tests. This especially applies to the development of reliability models of CTS, which take into account its typical cyclogram of intended use and the dynamics of influence on the reliability indicators of control inspections, periodic maintenance and repairs.

Therefore, in practice, a situation often arises when a satisfactory accuracy assessment can be obtained only at the end of the tests, when it loses its relevance. Then it is better to use methods of processing a small amount of data collected during preliminary tests, to obtain some estimates of reliability indicators, than to postpone the acceptance of more accurate estimates until a time when more reliable statistical material is accumulated.

3 Description of the Method and Basic Mathematical Equations

In practice, it is often necessary to know the value of the uptime with a low probability of failure. If the distribution function $F(x)$ of the time to failure is known, then the required time x is found as a solution to the equation $F(x) = p$. Obtaining reasonable conclusions about the value of x for small values of p is quite a difficult task. The following method is often used. According to the failure data, the function $F(x) = F(x, \theta)$ is selected, where $\theta = (\theta_1, \ldots, \theta_n)$ is a vector of unknown parameters. Next, they try to find estimates of the unknown parameter θ. After that, the desired quantile estimate is obtained from the equation $F\left(x, \widehat{\theta}\right) = p$.

The existing methodology offers an approach to the solution of the problem of constructing an unshifted estimate of the time of system uptime, during which the probability of uptime will not exceed a sufficiently small value p using the theory of β-expected limits [17].

A significant number of statistical problems that arise in the assessment of reliability based on the results of equipment samples tests are formed in such a scheme. Let the vector of time-to-failure observations be obtained as a result of testing n samples of equipment $X_1, \ldots, X_n$. The uptime distribution function $F(x) = F(x, \theta)$ is known with accuracy up to the distribution parameter θ. The results of observations are considered independent. It is necessary to determine the operating time of products in operation, during which the probability of failure will be equal to the given value p, based on the vector of observations $X_1, \ldots, X_n$. To solve such a problem, it is proposed to use the theory of parameter-independent tolerant p-limits. By definition, the equation holds for the lower tolerance p-limit.

$$P(Y < \tau(x)) = p. \tag{1}$$

If we denote by Y the random operating time of the product to failure, then $\tau(x)$ will represent the operating time during which the probability of failure will be equal to p. If the number of m products in operation is known, then Eq. (1) can be written in the form $P(Y(k) < \tau(x)) = p$, where $Y(k)$ is the operating life of the k-th product out of m in operation. The novelty of the proposed approach is as follows. From the dependence $P(Y < \tau(x))$ it is possible to obtain a non-shifted estimate for the operating life, during which the probability of failure will be equal to p. The proposed approach allows to take into account the number of products in operation. The possibility of using the theory of p-limits for estimation with a small sample arises due to the fact that the tolerance limits are interval estimates. The width of the interval will be random and depends on the number of observations. The expression $P(Y < \tau(x))$ should be understood so that the probability that the random variable Y will take a value less than some statistic $\tau(x)$ is equal to p for any value of the parameter of the distribution function. Here and further, the lower tolerance p-limit is meant.

Consider the statistical interpretation of tolerant p-limits. Let the tolerance p-limit $\tau(x)$ for the time of trouble-free operation be constructed based on the results of tests of equipment samples. Suppose that for N products in operation, a vector of observations of random work-up to failure $Y = Y_1, \ldots, Y_m$ is obtained. Let's introduce a random variable Z such that

$$Z = \begin{cases} 1, \mathit{if}\ Y_i < \tau(k), \\ 0, \mathit{if}\ Y_i \geq \tau(k), \end{cases} \text{where } i = 1, \ldots, n. \tag{2}$$

Then the fulfillment of Eq. (1) guarantees for any $n \to \infty : M[Z] = p$, where M is the symbol of mathematical expectation. That is, with a large N, the ratio of the number of cases when $Y_i < \tau(x)$ to N will tend to p. For one product, if the time of failure-free operation is specified, it is equal to the implementation of the tolerance p-limit. The p-limit is found based on the results of tests of n samples. In this case, the probability of the product failure during this time will exceed the value of p. This holds true for any value of the parameter of the law uptime distribution.

Thus, the formulas (1) and (2) for the tolerance p-limit can be used to calculate the time of failure-free operation of equipment.

Let's evaluate the time of failure-free operation of technical products with the Weibull's distribution of operating time to failure. One of the most widely used laws for the distribution of work between failures is the Weibull distribution law:

$$F(x) = 1 - exp\left(-\left(\frac{x}{\theta}\right)^{\beta}\right), \tag{3}$$

where β is a shape parameter; θ is the resource characteristic.

Using the transformation $Y = \ln(X)$, we pass from the random variable X, which has a Weibull distribution, to the random variable Y with the distribution function

$$F_{Y(x)}(y) = 1 - exp\left(-exp\frac{y - \theta_0}{\theta_1}\right), \tag{4}$$

where $\ln(\theta) = \theta_0$ is the shift parameter; $\frac{1}{\beta} = \theta_1$ is the scale parameter.

The distribution function (4) refers to distributions with a shift and scale parameter. Such a distribution is often called double exponential. Such a transition makes it possible to construct tolerant p-limits for random variables with the Weibull distribution.

Suppose that during the tests of n samples of the equipment, a vector of observations of operating conditions before failure $X_1, \ldots, X_n$ was obtained. It is known that there are m similar products in operation. It is necessary to determine the time of failure-free operation, during which k failures will appear in operation with a probability p. Let the random value Y represent the working life of products in operation. Then we can offer statistics that look like

$$G = \frac{Y(km) - \widehat{\theta_0}}{\widehat{\theta_1}}, \tag{5}$$

where $Y(km)$ is the operating time for failure before the appearance of the k-th failure; $\widehat{\theta_0}, \widehat{\theta_1}$ are estimates of the shift and scale parameter obtained during the tests.

From here we find the tolerance limit $Y(km)$, during which the probability of k failures out of m in operation will not exceed p

$$\tau(x) = \widehat{\theta_0} + \widehat{\theta_1} q(p), \tag{6}$$

where $q(p)$ is the quantile of level p of the distribution function G.

Thus, $q(p)$ is the quantile of level p of the distribution function (5), $\tau(x)$ represents the value of the working hours of m products in operation until the appearance of the k-th failure.

If the number of products in operation is unknown, then in this case the central random variable has the form

$$Q = \frac{Y - \widehat{\theta_0}}{\widehat{\theta_1}}, \tag{7}$$

where Y is the operating life of products in operation.

According to the observations $X_1, \ldots, X_n$, we determine the value of the operating time, during which the probability of failure of the product will be no more than p: $\tau(x) = \widehat{\theta_0} + \widehat{\theta_1} g(p)$, where $g(p)$ is the quantile of the level p of the distribution function τ.

For the Weibull distribution, the tolerance limits have the form

$$\tau(x) = exp\left\{\widehat{\theta_0} + \widehat{\theta_1} g(p)\right\}. \tag{8}$$

Let's consider the solution of the problem of determining the interval in which the assessment of the resource characteristic of m products in operation will be located. Let's construct a central random variable

$$S = \frac{\widehat{\theta_{0y}} - \widehat{\theta_{0x}}}{\widehat{\theta_1}}, \tag{9}$$

where $\widehat{\theta_{0y}}$ is the future estimate of the shift parameter that needs to be determined in the future for m products in operation, $\widehat{\theta_{0x}}$ is the estimation of the shift parameter, which needs to be estimated based on the results of tests of n products, $\widehat{\theta_1}$ is the scale parameter.

Hence, the limits in which the estimate $\widehat{\theta_{0y}}$ from Eq. (9) will be are determined from the equation

$$\widehat{\theta_{0x}} + \widehat{\theta_1} \delta(p) < \widehat{\theta_{0y}} < \widehat{\theta_{0x}} + \widehat{\theta_1} \delta(1 - p), \tag{10}$$

where $\delta(p)$, $\delta(1 - p)$ are the quantiles of the distribution function of the random variable S of level p and $1 - p$, respectively.

In order to estimate the time of failure-free operation of products with the Weibull's distribution of work-to-failure, it is necessary:

1. To test the hypothesis that the available observations X_n belong to the Weibull distribution.

A special correspondence criterion has been developed to determine the correspondence of Weibull's law statistics. The application of such a criterion is carried out as follows. In the obtained observations $X_1, \ldots, X_n$, we apply the transformation $Y_i = \ln(X_i)$ for $i = 1, \ldots, r$.

The statistics based on the criterion have the form

$$S = \frac{\sum_{i=\left[\frac{r}{2}\right]+1}^{r-1}\left(\frac{T_{i+1}-T_i}{M_i}\right)}{\sum_{i=1}^{r-1}\left(\frac{T_{i+1}-T_i}{M_i}\right)}, \tag{11}$$

where $\left[\frac{r}{2}\right]$ means the largest integer of the fraction $\frac{r}{2}$. The value of M_i, as well as the critical values of the statistic S are given in [18]. The hypothesis is rejected if the available value of the criterion statistic is greater than the critical value of the corresponding value level. Hypothesis testing is carried out for samples obtained before and after modifications.

2. It is necessary to estimate the parameters θ_0 and θ_1.

Let $X_1, \ldots, X_n$ order statistics of time-to-failure having a Weibull distribution. By transforming $T_i = \ln(X_i)$ for $i = 1, \ldots, n$, we pass to the distribution function $F_{Y(x)}(y) = 1 - exp\left(-exp\frac{y-\theta_0}{\theta_1}\right)$. Then the estimated parameters θ_0 and θ_1 are determined from the expressions

$$\theta_0 = \sum_{i=1}^{r} A_i \cdot T_i, \tag{12}$$

$$\theta_1 = \sum_{i=1}^{r} C_i \cdot T_i, \tag{13}$$

where A_i and C_i are linear multipliers for the corresponding values of r and n.

Such estimators are called best linear invariant estimates (BLIO).

3. Estimation of the time of the product's failure-free operation, the probability of failure during which it is given and is equal to p.

   Estimation is carried out according to the formula

$$\tau(x) = exp\left\{\widehat{\theta_0} + \widehat{\theta_1} g(p)\right\}, \tag{14}$$

where $\widehat{\theta_0}$ and $\widehat{\theta_1}$ are estimates of the shift and scale parameters obtained according to formulas (12) and (13); $g(p)$ is the quantile of the level p of the random variable distribution function (4).

4 Numerical Demonstration

Below are examples of reliability calculations for equipment samples for five and two failures.

Example 1. For example, failure data with a Weibull distribution was generated when: $\theta = 1200$ hours, $\beta = 2$.

An example of calculating the reliability indicators of aircraft radio electronic equipment (REO) samples.

Table 1 shows the performance data of 5 samples of radio equipment before failure in the testing process. Weibull distribution law determines the time-to-failure distribution function.

The weighting factors for calculating the parameters are given in Table 1.

Table 1. The operating hours and weighting factors.

Equipment number	1	2	3	4	5	Σ
X_i, hours	567	716	1020	1316	1522	
$Y_i = \ln(X_i)$	6,34	6.57	6.92	7.18	7.33	
A_i	0.0531	0.1035	0.1638	0.2460	0.4335	
B_i	–0.1581	–0.1557	–0.1118	–0.0056	0.4512	
$\Delta_i = Y_i A_i$	0.34	0.68	1.13	1.77	3.18	7.10
$\delta_i = Y_i B_i$	–1.0	–1.02	–0.77	–0.04	3.3	0.47
$E(LB)$						0.143
$E(CP)$						–0.03

Estimation of the shift parameter:

$$\theta = e^{\Delta_\Sigma} = e^{7.1} = 1212\,\text{h}.$$

Estimation of the scale parameter:

$$\beta = \frac{1}{\delta_\Sigma} = \frac{1}{0.47} = 2.12.$$

The obtained estimates are specified according to the methodology [2].

The coefficients for adjustment are given in Table 1.

$$\beta^* = \frac{\beta}{1 - E(LB)} = \frac{2.12}{0.97} = 2.18.$$

This is refined estimation of the scale parameter.

Then, a refined estimate:

$$\delta^* = \frac{1}{\beta^*} = \frac{1}{2.18} = 0.46,$$

$$\theta^* = \theta + \beta^*(E(CP)) = 1212 + 2.18 \cdot (-0.03) = 1211.9,$$

where θ^* is refined estimate of the shift parameter.

For example, let's determine the probability of failure-free operation $P(t)$ of the REO sample during 750 h of operation, we have:

$$P(t) = e\left(-\left(\frac{t}{\theta*}\right)^{\beta*}\right) = e\left(-\left(\frac{750}{1211.9}\right)^{2.18}\right) = 0.70.$$

The tolerant p-limit can be defined as a quantile of a given p level in two ways:

1. Analytically.
2. Monte Carlo method.

Analytically, with known estimates of the shift and scale parameters, the quantile is determined by the formula [18]:

$$L(p) = \Delta + \delta^* \left(ln \left(ln \left(\frac{1}{1-p} \right) \right) \right).$$

$$X(p) = e^{L(p)}.$$

For comparison, we define quantiles of several levels, we have:

$$L(0.1) = 7.1 + 0.46 \cdot \left(ln \left(ln \left(\frac{1}{0.9} \right) \right) \right) = 6.064$$
$$X(0.1) = e^{6.064} = 430 \text{ h}.$$

$$L(0.05) = 7.1 + 0.46 \cdot \left(ln \left(ln \left(\frac{1}{0.95} \right) \right) \right) = 5.73$$
$$X(0.05) = e^{5.73} = 309 \text{ h}.$$

$$L(0.01) = 7.1 + 0.46 \cdot \left(ln \left(ln \left(\frac{1}{0.99} \right) \right) \right) = 4.98$$
$$X(0.01) = e^{4.98} = 146 \text{ h}.$$

That is, for the conditions of example 1, with a probability of 0.9 the lower tolerance limit will not be worse than 430 h, with a probability of 0.95 the lower tolerance limit will not be worse than 309 h, with a probability of 0.99 the lower tolerance limit will not be worse than 146 h.

Example 2. An example of calculating the reliability indicators of two samples of radio equipment. Table 2 shows the operating data of two samples of radio equipment before failure during testing for 5 years (43,800 h).

Weibull distribution law determines the time-to-failure distribution function.
The weighting factors for calculating the parameters are given in Table 2.
Estimation of the shift parameter:

$$\theta = e^{\Delta} = e^{10.67} = 43045 \text{ h}$$

Estimation of the scale parameter:

$$\beta = \frac{1}{\delta} = \frac{1}{0.14} = 7.14.$$

The obtained estimates can be refined according to the method given in [2].

Table 2. The operating hours and weighting factors.

Equipment number	1	2	Σ
X_i, hours	32000	45000	
$Y_i = \ln(X_i)$	10.37	10.71	
A_i	0.1107	0.8892	
B_i	–0.4213	0.4213	
$\Delta_i = Y_i A_i$	1.15	9.52	10.67
$\delta_i = Y_i B_i$	–4.37	4.51	0.14
$E(LB)$			0.4158
$E(CP)$			0.0375

The coefficients for adjustment are given in Table 2.

$$\beta^* = \frac{\beta}{1 - E(LB)} = \frac{7.14}{0.5842} = 12.22,$$

where β^* is refined estimation of the scale parameter.

Then the refined estimate of δ is equal to:

$$\delta^* = \frac{1}{\beta^*} = \frac{1}{12.22} = 0.082,$$

$$\theta^* = \theta + \beta^*(E(CP)) = 43045 + 12.22 \cdot (0.0375) = 43045.5,$$

where θ^* is refined estimate of the shift parameter.

The refined estimate $\Delta = ln\,\theta^* = ln\,43045.5 = 10.67$ has practically not changed.

For example, let's determine the probability of failure-free operation $P(t)$ of the radio equipment sample during 40,000 h of operation. We have:

$$P(t) = e\left(-\left(\frac{t}{\theta^*}\right)^{\beta^*}\right).$$

$$P(t = 40000) = e\left(-\left(\frac{40000}{43045.5}\right)^{12.22}\right) = 0.67.$$

Analytically, with known estimates of the shift and scale parameters, the quantile is determined by the formula [18]:

$$L(p) = \Delta + \delta^* \left(ln\left(ln\left(\frac{1}{1-p} \right) \right) \right).$$

$$X(p) = e^{L(p)}.$$

For comparison, we define quantiles of several levels.

$$L(0.1) = 10,67 + 0.082 \cdot \left(ln\left(ln\left(\frac{1}{0.9} \right) \right) \right) = 10.485.$$

$$X(0.1) = e^{10.485} = 35774\,\text{h}$$

$$L(0.05) = 10,67 + 0.082 \cdot \left(ln\left(ln\left(\frac{1}{0.95} \right) \right) \right) = 10.426$$

$$X(0.05) = e^{10,426} = 33725\,\text{h}.$$

$$L(0.01) = 10,67 + 0.082 \cdot \left(ln\left(ln\left(\frac{1}{0.99} \right) \right) \right) = 10.29$$

$$X(0.01) = e^{10.29} = 29436\,\text{h}.$$

For the conditions of example 2, the following results were obtained: with a probability of 0.9 the lower tolerance limit will not be worse than 35724 h, with a probability of 0.95 the lower tolerance limit will not be worse than 33725 h, with a probability of 0.99 the lower tolerance limit will not be worse than 29436 h.

5 Conclusions

Thus, the paper proposes a simple and effective, compared to existing methods, method of calculating reliability indicators with a small number of failures for data with the Weibull distribution.

The novelty of the proposed approach lies in the possibility of evaluation with a small sample using the theory of tolerance limits. These tolerance limits are interval estimates. It is important to note that the width of these intervals is random and depends on the number of observations. At the same time, regardless of the number of observations, compliance with the specified dependencies is always guaranteed, which is essential in practical use.

The given methods of calculating reliability indicators make it possible to obtain unbiased estimates of reliability indicators with a small amount of statistical data on failures for highly reliable equipment.

References

1. Mencik, J.: Concise Reliability for Engineers. IntechOpen, Rijeka (2016). https://doi.org/10.5772/62372
2. Napiorkowskia, J., Gonera, J.: Analysis of failures and reliability model of farm tractors. Agric. Eng. **24**(2), 89–101 (2020). https://doi.org/10.1515/agriceng-2020-0020
3. Cui, L., Li, Y., Shena, J., Cong, L.: Reliability evaluation based on a dependent two-stage failure process with competing failures. Appl. Math. Model. **64**, 699–712 (2018). https://doi.org/10.1016/j.apm.2018.07.39
4. Praks, P., Kopustinskas, V., Masera, M.: Monte-Carlo-based reliability and vulnerability assessment of a natural gas transmission system due to random network component failures. Sustain. Resilient Infrastruct. **2**, 97–107 (2017). https://doi.org/10.1080/23789689.2017.1294881
5. Yang, J., Chen, J., Huo, Y., Liu, Y.: A novel cluster-based wireless sensor network reliability model using the expectation maximization algorithm. J. Sens. **2021**, 1–13 (2021). https://doi.org/10.1155/2021/8869544
6. Torres, E.S., Sriramula, S., Celeita, D., Ramos, G.: Reliability model and sensitivity analysis for electrical/electronic/programmable electronic safety-related systems. IEEE Trans. Ind. Appl. **56**(4), 3422–3430 (2020). https://doi.org/10.1109/TIA.2020.2990583
7. Zubarev, V.V., Kovtunenko, A.P., Vasilenko, A.V., Chepkov, I.B., Shishanov, M.A.: Foundations of the theory of complex substantiation of requirements for technical indicators of complex systems (in Russian). NAU, Kyiv, Ukraine (2010)
8. Kovtunenko, A.P., Zubarev, V.V., Lanetskiy, B.N., Zverev, A.A.: Mathematical modeling in problems of research of reliability of a technical systems (in Russian). NAU, Kyiv, Ukraine (2006)
9. Muscolino, G., Santoro, R., Sofi, A.: Reliability analysis of structures with interval uncertainties under stationary stochastic excitations. Comput. Methods Appl. Mech. Eng. **300**, 47–69 (2016). https://doi.org/10.1016/j.cma.2015.10.023
10. Tobias, P.A., Trindade, D.C.: Applied Reliability, 3rd edn. CRC Press, Boca Raton (2012)
11. Lanetskii, B.N., Lukyanchuk, V.V., Artemenko, A.A.: Comprehensive assessment of indicators of reliability and residual durability of complex technical systems operated according to their technical condition. Inf. Process. Syst. **2**(139), 40–43 (2016). fundamentals (in Russian)
12. Open'ko, P.V., Drannyk, P.A., Kobzev, V.V., Brovko, M.B., Zalevsky, G.S.: Substantiation of reliability requirements for mobility means of surface-to-air missile systems. Adv. Mil. Technol. **12**(1), 91–99 (2017). https://doi.org/10.3849/aimt.01122
13. Open'ko, P.V., Tyurin, V.V., Mironyuk, M.Y., Drannyk, P.A., Kobzev, V.V.: Reliability and maintainability requirements for mobility means of surface-to-air missile systems. Adv. Mil. Technol. **15**(2), 425–434 (2020). https://doi.org/10.3849/aimt.01402
14. Fernández, A.J.: Tolerance limits and sample-size determination using Weibull trimmed data. Axioms **12**(4), 351 (2023). https://doi.org/10.3390/axioms12040351
15. Chen, J., Li, K.-H., Lam, Y.: Bayesian single and double variable sampling plans for the Weibull distribution with censoring. Eur. J. Oper. Res. **177**(2), 1062–1073 (2017). https://doi.org/10.1016/j.ejor.2005.11.023
16. Dyptan, V.P., Tiurin, V.V., Yablonsky, P.M., Butenko, M.P., Klymchuk, V.P.: Reliability assessment of means of ground support for flights using the theory of tolerance limits (in Ukrainian). Air Power Ukraine **1**(2), 56–60 (2022). https://doi.org/10.33099/2786-7714-2022-1-1(2)-56-60
17. Zacks, S.: Examples and Problems in Mathematical Statistics. Wiley, New York (2014)
18. Kapur, K.C., Lamberson, L.R.: Reliability in Engineering Design. Wiley, New York (1991)

Management and Automation

Formation of a System for Optimizing Business Processes of Aviation Enterprises Based on Their Automation

Zarina Poberezhna(✉)

National Aviation University, Liubomyra Huzara Avenue, 1, Kyiv 03058, Ukraine
zarina.poberezhna@npp.nau.edu.ua

Abstract. The paper is devoted to optimization of business processes on the basis of their automation. The issue of managing data arrays and combining them into single databases is considered. When implementing the appropriate automated Customer Relationship Management (CRM) system, it is possible to successfully manage reducing the time for the flow of the main business processes. The study was carried out directly on the example of the business process "Cooperation with suppliers" of aviation enterprises, during which the basic requirements for an information system for interaction with suppliers were substantiated, including: general, functional and special business requirements. The general information model of the subsystem for managing interaction with suppliers of an aviation enterprise is presented and the standard business process of such interaction is characterized, indicating its main bottlenecks that delay the time of cooperation with suppliers. In order to form an idea about the business process "Cooperation with suppliers" of aviation enterprises, the author formalized a block diagram of this typical business process showing which of the processes are not basic and only burden and delay the implementation of the aviation enterprise's interaction with suppliers, which need optimization. Introduction of CRM systems in the management of business processes of aviation enterprises can provide an economic effect by speeding up interaction with suppliers, freeing up extra employees by reducing the time for intermediate business processes of the "Cooperation with Suppliers" process, and reducing the use of paper carriers, which will lead to both economic and social benefits.

Keywords: Aviation Enterprise · A System of Optimizing · Database · CRM systems · Business Process · Interaction with Suppliers

1 Introduction

1.1 Introduction to the Problem

The current business environment, characterized by trends in which supply significantly exceeds demand, plays a key role in the relations of enterprises with their main stakeholders: first of all, customers who are the end users of goods, works or services of the enterprise and suppliers who provide business processes with everything necessary

I. Ostroumov and M. Zaliskyi (Eds.): ACASD 2024, LNNS 992, pp. 325–338, 2024.
https://doi.org/10.1007/978-3-031-60196-5_24

in time to increase their efficiency. That is why the success of enterprises in the market environment is largely determined by the nature of relations with stakeholders, the effectiveness of cooperation with which is determined by the convenience of formalizing relations between the aviation enterprise and the stakeholder, the quality of management and use of its own customer base, the availability of services for remote users, etc. Traditional business process automation, robotic process automation, and hyperautomation enable software to perform the labor of human beings in a digital environment [1].

In order to automate stakeholder relations in the period of digitalization and digitalization of society and business, aviation enterprises should implement Customer Relationship Management (CRM) systems, which are a single tool for managing large databases of suppliers and customers. By using these systems, aviation enterprises can streamline the work with documents, which in turn will ensure transparency of business processes and reduce the time for information exchange between process participants; control the process of document approval and prevent document loss, increase the efficiency of access to management and accounting information [2, 3]. Several organizations have increasingly invested in business process automation to improve their business processes [4–6].

Although processes are widely automated, developing automation for a business process is a challenge. A business process is far more complex than a few manual repetitive tasks performed in software by a single person. It involves several elements such as data (inputs and outputs), workflows (tasks, steps, flows, rules, and exceptions), stakeholders (professionals, teams, and departments), and technologies [7].

Therefore, the automation of business processes has a positive effect on the indicators of the aviation enterprise's activity and increases the need for their assessment in terms of different areas of activity (purchasing, finance, personnel, sales, marketing). For this purpose, the need to form a system becomes highly relevant management of business processes taking into account the assessment of the existing potential of the aviation enterprise.

1.2 Motivation

In order to reduce costs, transfer routine tasks into services and applications, as well as make them more transparent and controllable, airlines are encouraged to implement business process automation, which will allow optimizing the management of resources and activities in general.

1.3 Contribution

The key contribution of this study is the developed methodology for the introduction of automation of business processes with the aim of their optimization, which provides the aviation enterprise with competitive advantages in the market, increases its profitability and significantly simplifies the work of employees of all levels, which is steadily reflected in the growth of their work productivity.

1.4 Organization of the Paper

The paper contains six sections. The first section describes the introduction. The second section highlights the key findings of the literature review. In the third section, the main features and methods for automating the business processes of the aviation enterprises are considered. The fourth section presents an example of calculation and modeling. The fifth section is the conclusion. Finally, the sixth section outlines the future scope.

2 Literature Review and Problem Statement

The issues of automation and digitalization of business processes and the use of modern information systems have been addressed by such scholars as: identified the digital economy as a key trend in the development of post-industrial society [8]. Studied the use of advanced information technologies for passenger traffic at airports [9].

In the research [10] considerable attention is paid to the digitalization of business processes in civil aviation. Identified the general directions of development of information systems and accounting technologies for small and medium-sized businesses in Ukraine [11]. The issues of using information systems tools to automate business processes of an enterprise are detailed [12].

Business process automation (BPA) refers to the technology to automate and optimize business processes, aiming to improve efficiency, reduce costs, and increase the performance of organizations [13, 14].

In general, despite the significant achievement of domestic and foreign research in the field of business process management of aviation enterprises on the basis of their automation, special attention should be paid to the issues of business process optimization, their formalization and detailing to improve the efficiency of interaction with key stakeholders.

The aim of this paper is to analyze the possibilities of optimizing business processes of aviation enterprises through the use of modern information systems.

3 Materials and Methods

Automation of aviation enterprises relations with suppliers can be achieved through the implementation of modern CRM systems, which are a tool for managing a single database. These systems help to streamline document management, ensure transparency of key business processes, thereby reducing the time required to exchange information between business process participants, manage document approval processes, prevent document loss, and increase the speed of access to information.

The need to automate the interaction of aviation enterprises with suppliers is that most airlines have centralized databases, but the work with the information contained in these databases is not automated at the appropriate level, because in most cases employees use only Excel, i.e., there is no unified data accounting and management system.

Let us formulate the main requirements for an information system for interaction with suppliers of aviation enterprises in Table 1.

Table 1. List of requirements for the information system of interaction between aviation enterprises with suppliers.

Requirements	Description of requirements
General business requirements	The ratio between the price of the system and the quality should be appropriate The system should be easily and quickly integrated into the overall enterprise management system and unite all its departments Storage of a large amount of data Automatic detection and elimination of duplicate information Ensuring the security of enterprise data from external influences Use of cloud technologies for work and data storage Combining data on accounting, supplier and customer records, operations, etc. into a single system
Functional business requirements	The system must be adapted to expand the array of entered and stored data The system must have strong technical support Functionality of searching by specified parameters User-friendly interface to ensure easy adaptation of personnel to the use of the system Generation of reports according to the specified parameters Management and control of access to documents Ability to remotely access and manage the system Protection of information in the system from destruction due to equipment failures (backup, etc.) and protection of information from unauthorized access for the purpose of misuse or destructive influence
Special requirements	Possibility to customize the system to meet the requirements of the aviation enterprise Possibility of training on methodical materials Ability to adapt the system to the requirements of the aviation enterprise without involving the system developers

Most of the business process for working with suppliers is called "Supplier Relations". Suppliers usually send contracts to the enterprise by mail or courier. Upon receipt, they are entered into the enterprise's general database as a rule, most modern aviation enterprises use the Excel program to streamline and organize their activities. The contract specialist registers the package of documents, marks it "Received", makes copies and sends them to the departments and divisions that are related to the document.

If the essential terms of the contract are not satisfactory to at least one of the heads of the relevant divisions or departments, it is returned by courier to the contract specialist. The specialist draws up a "Letter of Approval" and sends it back to all divisions and departments of the aviation enterprise. This procedure continues until all the essential terms of the contract are agreed upon.

The risks that arise at this stage of working with suppliers are the possible loss of a contract with one of the heads of a division or department. In this case, a new contract is requested, which once again burdens the contract specialist, who must find the contract, make copies and send them to the relevant managers.

There are situations when the contract is not signed, but a "Protocol of Disagreements" is drawn up and sent to the supplier for consideration with the amendments made. If the supplier accepts the differences, the contract is signed and sent to the aviation enterprise for review and signature. After all the discrepancies have been resolved, the contract is signed and archived, with copies kept by the contract specialist and the lawyer. That is why, in this study, we have detailed the issue that is part of the business process "Interaction with Suppliers" – the task of automating the process of approving contractual documentation in order to optimize it.

Having considered the described business process from the receipt of the contract and its approval to the moment of signing, we have identified a number of the following problems:

- documents can get lost;
- loss of working time for searching for the required document and forming a thematic collection of documents;
- creation of a large number of copies of the same document, which is primarily costly in terms of money, working time and reduces the social responsibility of the enterprise in terms of resource and energy saving;
- accumulation of a large number of documents;
- time spent on preparing and approving documents.

The information model of the business process of managing the "Interaction with Suppliers" of an airline company is detailed in Fig. 1.

The main purpose of automating the processes of interaction with suppliers is to generate data about them and prices for their goods and services, as well as to create a reporting system for receiving invoices for services received, automate the process of concluding contracts and optimize the overall circulation of contractual documents, and create a single electronic data repository.

Taking into account the above, let us outline the range of tasks that should be solved by automating the subsystem of contractual documentation for aviation enterprises:

- systematization of data on suppliers in various areas (equipment, types of work and services) into a single database of suppliers and contracts;
- creating a single standardized reporting for each supplier, including entering their details and contract numbers;
- automatic preparation of analytical payment tables (matrices) with costs for each supplier, volume of goods and services for the current and planned periods in order to identify deviations and eliminate them;
- ensuring more efficient management through automatic control over the execution of contracts;
- enabling parallel work with documents of different departments through simultaneous access to them;

- ensuring faster document search (by various parameters);
- improving the security of documents and the convenience of their storage, as they are stored electronically on the aviation enterprise's server or in the cloud storage;
- reducing the use of paper documents and reducing the cost of their storage due to the availability of a functional electronic archive.

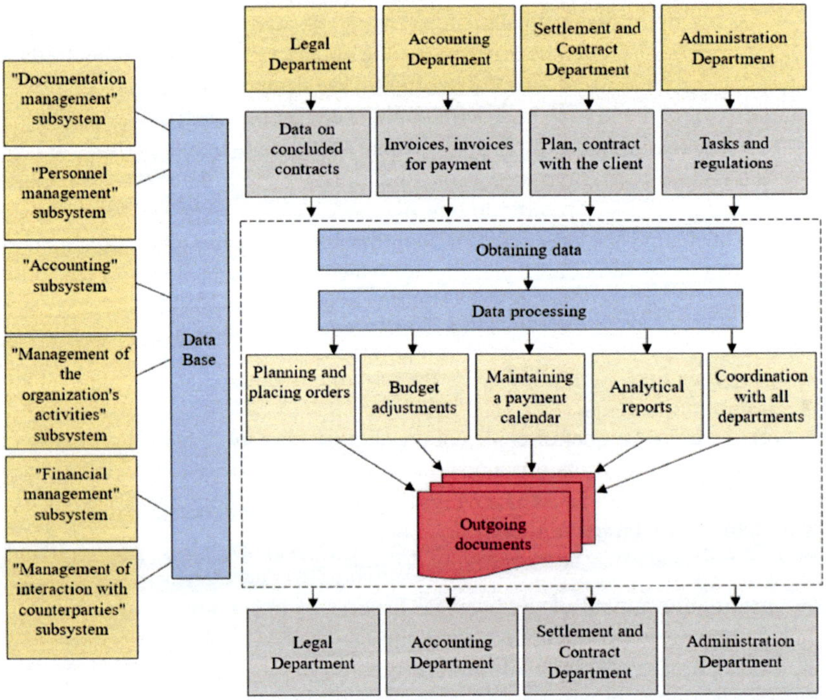

Fig. 1. Information model of the business process of managing "Interaction with suppliers" of an aviation enterprise.

Automation of work with suppliers requires the regulation of input and output information. The input information should include the following data: name, details, contract number, additional agreements, approval letter, protocol of discrepancies, orders, expenses, etc. The output information includes: data on each supplier by contract number, an agreed supplementary agreement, a completed letter of approval or protocol of disagreements, agreed orders, a list of expenses by priority.

Schematically, the input, intermediate and output databases of the aviation enterprise's information database on work with suppliers are shown in Fig. 2 and Fig. 3.

The databases can formalize the business process of cooperation between an aviation enterprise and a supplier of goods, works and services. It should be noted that the replacement of printed documents with electronic ones and the use of CRM information systems will lead to significant optimization and acceleration of work in the module "Interaction with Suppliers".

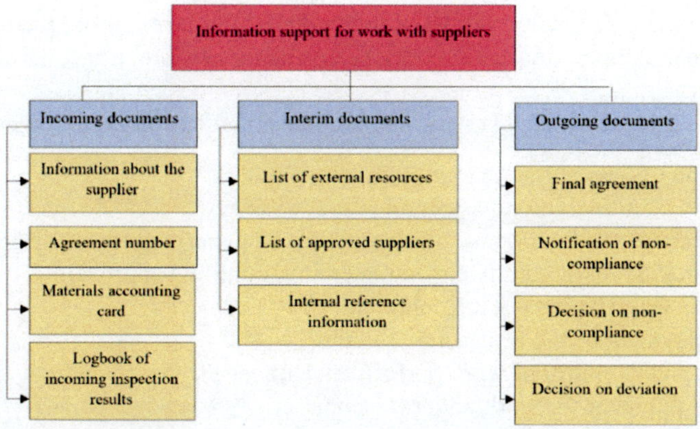

Fig. 2. Information support of work with suppliers when automating the module "Interaction with suppliers" of an aviation enterprise.

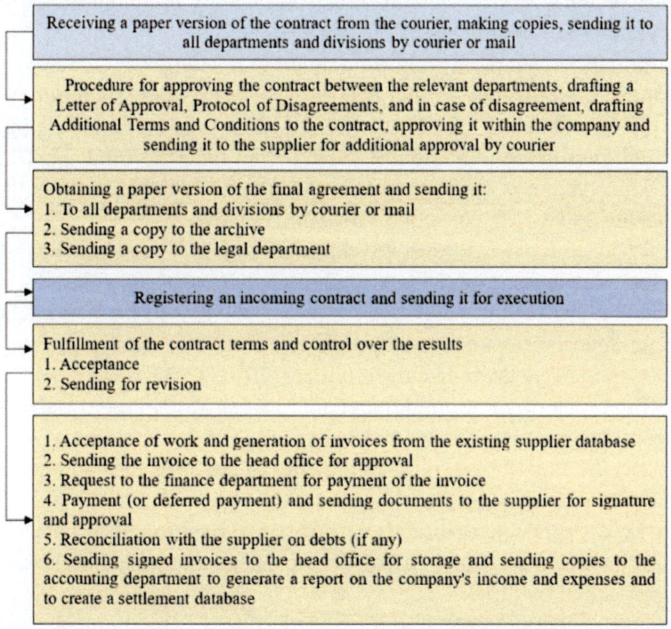

Fig. 3. Simplified flowchart of a typical business process "Interaction with suppliers" of an aviation enterprise.

As for the conceptual approaches to solving issues related to the automation of information databases of interaction between aviation enterprises and suppliers, we believe that the implementation of a CRM system that would eliminate all the identified shortcomings of the existing standard business process would be effective. By applying automation, the typical business process of concluding an agreement with an aviation enterprise's supplier will be simplified at many levels, from the conclusion to the execution of cooperation agreements, performance of work, its approval, acceptance and payment.

In general, the automated system will include a number of the following processes:

- preparation of the contract;
- entering the contract into the system;
- simultaneous consideration of the contract by all departments and divisions;
- parallel processing of comments on the contract until its approval;
- accounting (registration) of the contract;
- advance payment of the contract;
- accounting and control of work performance under the contract;
- control of payment for work under the contract;
- finalizing the contract and storing the contract;
- accounting reconciliation.

Today, there are a large number of CRM systems that contain subsystems for contractual documentation. The most well-known and widely used are Microsoft Dynamics CRM, DocsVision, and DIRECTUM.

Comparing these systems, it should be noted that Microsoft Dynamics CRM is the cheapest of these systems, but not the best in terms of its use. Its implementation will take from 2 to 4 months, and it is not flexible to other information systems used by the company, which requires duplication of data, i.e., in fact, the time spent on data entry will double. This system is characterized by an interface that is not flexible and rather difficult to master, and it does not support the built-in viewing of attached files. Therefore, although such a system exists, we would not recommend it to aviation enterprises.

DocsVision is a more expensive software, but it takes two months to learn and implement the system. The only drawback of the system is the lack of duplicate document checking when entering documents and registering them in the system.

The most expensive system is currently the DIRECTUM system, but it can take up to 5 months for an enterprise to implement and master it, which is a rather long period. However, even such a system has a number of drawbacks, including the following: it does not support the functions of built-in viewing of attached files and does not create a document based on an existing one.

However, before proceeding directly to the process of automating business processes, it is necessary to accurately establish how they proceed and what characteristic features they are accompanied by [15].

After the implementation of the information system, all data on suppliers and contracts concluded with them are automatically entered into a single database, which allows employees to track the stages of work on each individual order. As already mentioned, the system also contains information about potential suppliers that may be required by the company in the future, in case of non-fulfillment or improper fulfillment of contractual relations by existing suppliers. Employees of the aviation enterprise will be able to create a list of tasks in the information system in real time and control the time spent on a particular task. This is a very effective mechanism of self-management in the management system. The implementation of a CRM system will significantly reduce the time for completing tasks and calculations, because the distribution of tasks will be carried out automatically in the program: plans, tasks, standards, reporting, information for warehouses, which in turn will significantly reduce paperwork in the aviation enterprise and the time for transferring paper documentation from department to department.

It should be noted that when optimizing this business process, none of its links is eliminated, but a large number of operations will be performed by an automated system rather than a person, for example: launching the approval process in the system, approving the contract between departments, registering the contract, accounting for contracts, paying in accordance with the contract, submitting the results to the aviation enterprise's accounting department, closing the contract, entering the contract into the database, which will significantly speed up this business process of the aviation enterprise.

4 Results and Discussions

Consider an example of the implementation of software for automating business processes of an aviation enterprise, components, subsystems and systems. One of the approaches is to estimate the information productivity ratio (quick economic feasibility ratio):

$$REJ = \frac{\Delta EVA}{I_{CRM}} \rightarrow \max, \tag{1}$$

where ΔEVA is the incremental value added, I_{CRM} is the maintenance costs.

The payback process of investments in CRM systems can be described by the conditional effect function:

$$NPV(t) = \sum\nolimits_{t=1}^{n} \frac{EVA_t - I_t(CRM)}{(1+r)^t} - I_0(CRM), \tag{2}$$

where EVA_t is the added value at time t, $I_t(CRM)$ is the cost of maintaining CRM, $I_0(CRM)$ is the initial investment, r is a fixed interest rate of discount.

We propose to model the costs of CRM maintenance $I_t(CRM)$ on the basis of a statistical analysis of the structure of direct costs for the acquisition and operation of information systems of aviation enterprises (Table 1).

The payback period can be determined as follows:

$$PP = \frac{I_0}{\Delta Bp}, \tag{3}$$

where ΔBp is the reduction in transaction costs. Let PP_n be the standard payback period for investments in business process automation systems. Let ΔT_t be the reduction in time costs, and ΔBP_t be the reduction in support and transaction costs corresponding to the effective use of automation systems in time period t, the total cost of use in this period is denoted by I_t. If an investment is made at time t, the value of the conditional effect changes abruptly and becomes $NPV_t = NPV_{t-1} - I_t$, where NPV_{t-1} is the conditional effect of previous costs (investments).

Table 2. Structure of direct costs for the acquisition and operation of information systems.

Type of expenditure	Average value of share of total expenditure	Lower boundary of 95% confidence interval	Upper boundary of 95% confidence interval
Hardware and software	28%	26%	30%
Administration	24%	22%	26%
Technical support	21%	19%	23%
Project development, application development	13%	11%	15%
Payment for communication and communication means	10%	8%	12%
Other expenses	4%	2%	6%

In addition, by the time $t+PP_n$, the conditional effect, defined by ΔBP_t, must exceed the conditional effect ΔBP_{t-1}. Otherwise, the cost or investment of the amount of funds I_t is considered inefficient.

Therefore, the change in the conditional effect of business process automation costs can be described by the following recurrent relations:

$$NPV_{t+1} = \Delta Bp_t \cdot t + NPV_t = \Delta Bp_t + NPV_{t-1} - I_t. \tag{4}$$

The minimum reduction in related and transaction costs that corresponds to the efficient use of I_t funds can be defined as follows:

$$NPV_t(PP_n) > NPV_{t-1}(PP_n), \tag{5}$$

$$\Delta Bp_t > \Delta Bp_{t-1} + \frac{I_t}{PP_n}. \tag{6}$$

From the latter inequality, we can obtain an estimate of the amount of spending in time t that will be effective:

$$I_t < PP_n \cdot (\Delta Bp_t - \Delta Bp_{t-1}). \tag{7}$$

So, the mathematical model for optimizing investments and costs in business process automation systems can be presented as follows:

$$PV(t) = \sum\nolimits_{t=1}^{n} \frac{EVA_t - I_t}{(1+r)^t} - I_0 \to \max, \tag{8}$$

where $PV(t)$ is the model's target function, which is investigated for a conditional extremum (maximum) provided that the following system of constraints is met:

$$\begin{cases} I_t \leq PP_n(\Delta Bp_t - \Delta Bp_{t-1}), \\ I_t = \sum_{s=1}^{6} I_{ts}, I_{ts} = k_s \cdot I_t, k_s \in \{0.28; 0.24; 0.21; 0.13; 0.1; 0.04\}, \\ \Delta Bp_t \geq \Delta Bp_0, \forall t \in \{1, 2, \ldots, \mathrm{n}\}, \\ \Delta EVA_t \geq REJ \cdot I_t, \forall t \in \{1, 2, \ldots, \mathrm{n}\}, \end{cases} \tag{9}$$

where the model parameters are: ΔBp_0 is the minimum allowable value of reduction of related and transaction costs; I_0 is the initial investment (costs required for the purchase and initial implementation of the automation system), k_s is weighting factors determined by experts, according to Table 2; $REJ \geq 1$ is the information productivity ratio; PP_n is the standard payback period.

Based on the analysis of the cost structure of five aviation enterprises according to the data of 2023, we obtain the following results (Table 3).

Table 3. Results of the economic effect of the implementation of CRM systems with a planning during for 3 years (USD).

Index	Enterprise 1	Enterprise 2	Enterprise 3	Enterprise 4	Enterprise 5
I_0	7000	7000	5000	5000	5000
PP_n	2	2	2	2	2
REJ	1.25	1.25	1.2	1.2	1.2
ΔBp_0	3500	3000	2000	2500	2000
EVA_1	12300	10000	8000	5000	6000
EVA_2	14900	12500	10200	6500	7500
EVA_3	16800	14000	12500	7800	8500
ΔBp_1	4000	4000	3000	3000	3000
ΔBp_2	5000	5500	4000	3500	3500
ΔBp_3	7000	6500	5000	4500	4000
I_1	1000	2000	2000	1000	2000
I_2	2000	2000	1760	1000	1000
I_3	1520	1200	1840	1040	800
NPV_1	–4727.27	–6090.91	–5000.00	–3636.36	–5000.00
NPV_2	3785.12	520.66	157.02	–330.58	–867.77
NPV_3	13837.72	9010.52	6438.02	3771.60	4166.04

We can see that $NPV_1 < 0$ for all aviation enterprises, which is natural, since the standard payback period $PP_n = 2$ years. The value of $NPV_2 > 0$ for three aviation enterprises out of five, which shows the effect of conditional optimization performed by the model. That is, the payback period for three aviation enterprises was reduced to 1 year.

In addition, the use of the model makes it possible to estimate the effective cost volumes It, which allows for an additional economic effect. For five aviation enterprises, the initial parameters of the model differed: the amount of initial investment, the information productivity coefficient, and the minimum acceptable value of reducing related and transaction costs. Nevertheless, all aviation enterprises showed a significant increase in NPV and a reduction in the payback period (Fig. 4).

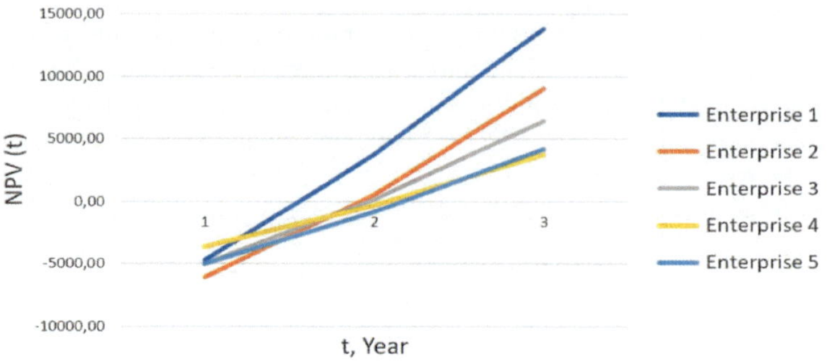

Fig. 4. Dynamics of NPV for five aviation enterprises (for three years).

Automating the business processes of the aviation enterprises leads to the improvement of existing business processes or the introduction of new business processes and systems based on innovative technologies.

The use of innovative technologies and processes, development and implementation of innovative types of products allows the enterprise to occupy a leading position on the market. This will provide products with a high degree of scientific content and novelty, thereby making them competitive on the world market [16].

It is worth noting that the latest IT technologies are the basis of innovative development in the management of aviation enterprises, which will help to improve the quality of service and correctly systematize, analyze the necessary important information, and monitor the work of personnel. Automation of business processes helps to study the target audience, forecast the demand for services, implement an effective marketing policy.

5 Conclusions

The feasibility of optimizing the business process of managing interaction with suppliers of aviation enterprises through informatization based on the implementation of CRM systems should be generally confirmed by relevant indicators. Namely, by freeing up employees' time or the total number of employees by reducing the time spent working with documents, reducing paper costs and increasing the social compliance of the enterprise, improving the quality of control over the fulfillment of contractual relations, etc. All of this suggests that the implementation of CRM systems or their individual modules at aviation enterprises will bring economic benefits to the aviation enterprise, so the automation project can be considered quite profitable. The implementation of software will be the basis for the formation of business opportunities for control, flexibility and operational optimization of all business processes of aviation enterprise operations in conditions of an unstable external environment.

Therefore, due to the automation of business processes, the quality and speed of customer service improves, which is a significant advantage in a highly competitive market. Within the scope of the study, it was established that that such automated systems exist business process management: project and task management systems, accounting, customers, employees, ESP-systems, CMS-systems and systems for building the organizational structure. The final choice regarding the automation system of business processes or their combination depends on the goals that the management sets for automation and on the financial capabilities of the aviation enterprise.

6 Future Scope

Future directions of research are related to:

- complex automation of all business processes of the aviation enterprise with the aim of increasing its efficiency and achieving economic effect;
- research on the effectiveness of planning and redistribution of operating costs;
- improvement of automated systems at the aviation enterprise based on the application of innovative technologies.

References

1. Menezes, T.: A review to find elicitation methods for business process automation software. Software **2023**(2), 177–196 (2023). https://doi.org/10.3390/software2020008
2. Gartner Says Worldwide Spending on Robotic Process Automation Software to Reach $680 Million in 2018. https://www.gartner.com/en/newsroom/press-releases/2018-11-13-gartner-says-worldwide-spending-onrobotic-process-automation-software-to-reach-680-million-in-2018. Accessed 03 Dec 2023
3. Lewicki, P., Tochowicz, J., van Genuchten, J.: Are robots taking our jobs? A RoboPlatform at a bank. IEEE Softw. **36**, 101–104 (2019)
4. Axmann, B., Harmoko, H.: Robotic process automation: An overview and comparison to other technology in Industry 4.0. In: 10th International Conference on Advanced Computer Information Technologies (ACIT), pp. 559–562. IEEE, Deggendorf, Germany (2020). https://doi.org/10.1109/ACIT49673.2020.9208907
5. Hofmann, P., Samp, C., Urbach, N.: Robotic process automation. Electron. Markets **30**(1), 99-106 (2020). https://doi.org/10.1007/s12525-019-00365-8
6. Gartner Forecasts Worldwide Hyperautomation-Enabling Software Market to Reach Nearly $600 Billion by 2022. https://www.gartner.com/en/newsroom/press-releases/2021-04-28-gartner-forecasts-worldwidehyperautomation-enabling-software-market-to-reach-nearly-600-billion-by-2022. Accessed 03 Dec 2023
7. IEEE Std 2755–2017. IEEE Guide for Terms and Concepts in Intelligent Process Automation. IEEE: Piscataway. USA, pp. 1–16 (2017)
8. Pantielieieva, N.M.: Digital economy as a key trend in the development of postindustrial society. DVNZ University of Bank Services, Kyiv, Ukraine (2019). (in Ukrainian)

9. Yanovskyi, P.O., Valko, A.M.: Use of advanced information technologies for passenger service at the airport. In: Conference on Computer technologies in urban and regional economy, pp. 38–39. Kharkiv, Ukraine (2015)
10. Oleshko, T.I., Popyk, N.V., Babych, M.O.: Digitization of business processes in civil aviation (in Ukrainian). Econ. Authority **4**, 43–46 (2021). https://doi.org/10.32702/2306-6806.2021.4.43
11. Bakharyeva, Y.V.: Directions of development of information systems and technologies of accounting of small and medium business enterprises in Ukraine (in Ukrainian). Effective Econ. **11**, 1–8 (2018). https://doi.org/10.32702/2307-2105-2018.11.208
12. Dolhova, L., Yamnenko, H.: Use of information system tools to automate business processes of the enterprise (in Ukrainian). Econ. Anal. Collect. Sci. Works **31**(2), 90–97 (2021). https://doi.org/10.35774/econa2021.02.090
13. Gartner Glossary. https://www.gartner.com/en/glossary. Accessed 03 Dec 2023
14. Aalst, W.M., Bichler, M., Heinzl, A.: Robotic process automation. Bus. Inf. Syst. Eng. **60**, 269–272 (2018). https://doi.org/10.1007/s12599-018-0542-4
15. Ma, Y., Lin, D., Chen, S., Chu, H., Chen, J.: System design and development for robotic process automation. In: International Conference on Smart Cloud (SmartCloud), pp. 187–189. IEEE, Tokyo, Japan (2019). https://doi.org/10.1109/SmartCloud.2019.00038
16. Smerichevskyi, S., Mykhalchenko, O., Poberezhna, Z., Kryvovyazyuk, I.: Devising a systematic approach to the implementation of innovative technologies to provide the stability of transportation enterprises. East. Eur. J. Enterp. Technol. **3**(13(123), 6–18 (2023). https://doi.org/10.15587/1729-4061.2023.279100

Optimizing Supply Chain Operations with Unmanned Aerial Vehicles

Haoyang Li and Volodymyr Kharchenko(✉)

National Aviation University, Liubomyra Huzara Avenue, 1, Kyiv 03058, Ukraine
vlkharch@gmail.com

Abstract. This paper delves into optimizing supply chain operations, with a special focus on the benefits and efficiencies of drone delivery systems. It highlights how drones can bypass congested road traffic and consume less energy compared to traditional transportation methods, thereby increasing delivery speeds and reducing operational costs. The paper provides comprehensive insights into fuel efficiency metrics and energy conservation strategies in transportation. Additionally, it discusses detailed mathematical models and optimization tasks that are designed to enhance the efficiency of these systems. The integration of GPS technology is also explored, emphasizing its role in improving the accuracy of drone deliveries. Furthermore, the paper touches on the use of experimental data analysis for evaluating system performance, with a particular focus on unmanned aerial vehicles (UAVs). This analysis is crucial for understanding and refining the capabilities of these systems. The paper concludes by highlighting the significant potential for cost savings and increased profitability for organizations that adopt innovative UAV technology in their supply chain management strategies, marking a shift towards more technologically advanced, efficient, and sustainable operations.

Keywords: Supply Chain Optimization · Drone Delivery · Logistic Efficiency

1 Introduction

Companies are looking to optimize their supply chains and reduce costs. Drone delivery has a number of advantages over traditional delivery methods, including shorter delivery times, greater efficiency, and lower overhead costs [1].

In terms of speed, drones have the potential to significantly reduce delivery times. Drones can fly directly from point A to point B without the need to navigate traffic or stop at different locations. This not only reduces delivery times, but also increases efficiency as it eliminates the need for multiple drivers. In addition, drones require less energy than traditional vehicles, resulting in lower fuel costs and reduced emissions [2, 3].

Optimizing supply chain operations and reducing costs depends on fuel efficiency [4, 5]. The fuel efficiency of an aircraft is determined by the amount of fuel it uses to cover a certain distance under certain conditions. This indicator is a fuel efficiency metric (FEM).

I. Ostroumov and M. Zaliskyi (Eds.): ACASD 2024, LNNS 992, pp. 339–350, 2024.
https://doi.org/10.1007/978-3-031-60196-5_25

Fuel efficiency is the distance traveled per unit of fuel used; for example, kilometers per liter (km/l) or miles per gallon (MPG), where 1 MPG (imperical) $\approx$ 0.354006 km/l. $J = kg \cdot m^2 \cdot s^{-1}$. The higher the value, the more fuel-efficient the vehicle is (the greater the distance it can travel on a given amount of fuel). Energy efficiency is similar to fuel efficiency, but consumption is usually measured in units of energy such as mega joules (MJ), kilowatt hours (kWh), kilocalories (kcal), or British thermal units (BTU). The inverse of "energy efficiency" is "energy intensity," or the amount of input energy required to produce a unit of output, such as MJ/passenger-kilometer (for passenger transport), BTU/ton-mile or kJ/t-km (for freight transport), GJ/t (for steel and other materials), Btu/(kWh) (for electricity), or liters/100 km. Liters per 100 km is also a measure of "energy intensity," where the input is measured by the amount of fuel and the output by the distance traveled.

The results show that to achieve the best fuel efficiency, a cargo drone based on the very light class of general aviation aircraft (VLA) with a traditional aircraft layout should have a composite structure and a wing aspect ratio of about 10–12. Assessment of fuel efficiency at distances from 500 to 2500 km shows that such UAV will be competitive with the current generation of large manned commercial cargo aircraft, achieving energy efficiency of about 7–8 MJ/(t.km). This is much better than the 14–24 MJ/(tkm) of utility turboprops. To a certain extent and in some cases, such a UAV will be competitive with road transport, whose energy intensity is estimated at 4–5 MJ/(t.km). This study should be considered as a preliminary step towards a comprehensive assessment and optimization of the transport systems of the UAV, including the aircraft itself, ground infrastructure, operation, etc.

The use of drones also provides cost savings [6–9]. By optimizing the supply chain, companies can reduce inventory costs as well as labor costs. Drones can also transport goods directly from the supplier to the customer, eliminating the need for intermediaries. This results in lower costs for the customer as well as increased profits for the business. In addition to cost savings, drone delivery also provides greater accuracy and improved customer service. By using GPS tracking technology, companies can ensure that the package arrives on time and in the right place. This helps reduce customer complaints and increase customer satisfaction. In addition, drones can be used to monitor inventory levels, allowing companies to better manage inventory and reduce wastage.

2 Problem Statement

When addressing the optimization of drone delivery, multiple methodologies can be employed. The authors propose a specific approach that stands as a viable option. This involves the process of either minimizing or maximizing a particular function within a set of additional parameters. Such a process is emblematic of the mathematical models commonly used in decision-making for the design and operation of technological systems. This includes areas like computer-aided design of equipment, vehicle control systems, updating models based on experimental data analysis, and in the broader spectrum, assessing the efficiency of complex systems, especially those involving unmanned vehicles and networks.

This approach extends beyond mere computational modeling; it involves a deep dive into data analysis, predictive modeling, and scenario planning. For drone delivery optimization, this encompasses variables such as adapting to changing weather conditions, optimizing flight trajectories, managing energy efficiency, especially in terms of battery usage, and adhering to payload constraints. The approach is further refined through continuous iterations, which might include both simulated scenarios and empirical testing, to ensure the system's adaptability and resilience in real-world conditions. Such a comprehensive and dynamic approach is crucial in the evolving field of drone delivery, ensuring that these systems are not only efficient but also scalable and adaptable to changing environmental and operational demands.

3 Proposed Method

The simplest type of such a problem assumes that the choice of a variant of the unmanned object being designed (i.e., the optimization problem will be interpreted as the choice of a solution in design) is characterized by the choice of the value of the vector of design parameters $y = (y_1, \ldots, y_N)$. At the same time, the conditions for the creation and operation of the vehicle impose some restrictions on the permissible values of the vector y. These are formally described as the requirement that the vector y belongs to some valid region Q in the N-dimensional parameter space R^N. The effectiveness of an object variant that corresponds to a given value of the vector y is described by the indicator $\varphi(y)$, which can be calculated based on the analysis of the mathematical model of this object. Such calculations will be called experiments. If we assume that a decrease in $\varphi(y)$ corresponds to an improvement of the project, then the choice of the value of y^*, which determines the best option, is reduced to an approximate solution of the minimization problem:

$$y^* = arg\ min\{\varphi(y) : y \in Q\}. \tag{1}$$

Based on the results $Z^i = \varphi\big(y^i\big), 1 \leq i \leq k$ of a finite number of experiments at points $y^i \in Q$.

The selection of experimental points can be carried out sequentially, i.e., when selecting the next point y^{i+1}, the already known results $Z^1, \ldots, Z^i$ of the experiments at the previous points $y^1, \ldots, y^i$ can be used. It is assumed that some selected points may not belong to the region Q. The mathematical model of the object should contain a test to detect such cases (with a negative test result taken as the result of the experiment). Thus, the procedure for selecting points from region Q can be determined indirectly – through the test for belonging to this region. It is also possible that the admissible region is empty, which corresponds to the incompatibility of the requirements for the object. Establishing this fact is also considered a solution to problem (1).

The accuracy of the approximate solution of problem (1), which has coordinate y_* and value $\varphi_* = \varphi(y_*)$, can be characterized by some accepted (for meaningful reasons) notion of proximity to the exact solution, for example, by the estimate $||y^* - y_*||$ or the estimate $|\varphi^* - \varphi_*|$, where $\varphi^* = \varphi(y^*)$.

Limited Differences in the Estimation of the Optimum

Any possibility of a reliable estimate of the global optimum in a multi-extreme problem is fundamentally based on the availability of certain a priori information that allows to relate possible values of the minimization function to known values at points where measurements have already been made.

Let's assume that the admissible region Q in problem (1) coincides with the search region D (i.e., there are no restrictions).

We will use the fact that in many applied problems, a change in the parameter vector y leads to a change in the characteristics of the object, limited to a certain degree of change in y. This fact can be interpreted as a reflection of the capacity constraint that generates changes. Of course, there are situations when opposite phenomena occur (shock effects, resonance phenomena, etc.), which should be modeled as discontinuities in characteristics. A formal model that describes the above property of boundedness of changes for the considered problem with one characteristic $\varphi(y)$ can be, for example, the following system of inequalities:

$$|\varphi(y') - \varphi(y'')| \leq K\rho(y', y''), y', y'' \in D, \tag{2}$$

where $\rho(y', y'')$ is some distance function in the parameter space, and K is a given constant. In the case when in (1)

$$\rho(y', y'') = (||y' - y''||)^{1/N}, \tag{3}$$

where the function $||\cdot||$ corresponds to the Euclidean metric, we say that the function φ satisfies the uniform condition with exponent $1/N$ and coefficient K.

When $N = 1$, this condition is called the Lipschitz condition with constant K. We will consider this case to explain the main ideas, and for the sake of clarity, we assume that the problem is one-dimensional, and, therefore, the search domain D is some segment $[a, b]$ of the real y-axis.

So, let the function $\varphi(y)$, $y \in [a, b]$ satisfy the condition with a given constant K:

$$|\varphi(y') - \varphi(y'')| \leq K|y' - y''|, y', y'' \in [a, b], \tag{4}$$

and, let the experiments be carried out at points $y^1, \ldots, y^k$ in the interval $[a, b]$. Renumber the points y^1 using the lower indices in ascending order of coordinate values:

$$a \leq y_1 < \ldots < y_i < \ldots < y_k \leq b.$$

Let denote the function values calculated in them as $z_i = \varphi(y_i)$, $1 \leq i \leq k$. Figure 1 illustrates the above. Let's introduce the functions:

$$\varphi_i(y) = z_i - K|y - y_i|, y \in D, 1 \leq i \leq k, \tag{5}$$

for which, in accordance with (4), the following applies:

$$\varphi(y) \geq \psi_i(y), y \in D, 1 \leq i \leq k,$$

or

$$\varphi(y) \geq \psi(y) = max\{\psi_i(y) : 1 \leq i \leq k\}, y \in D. \tag{6}$$

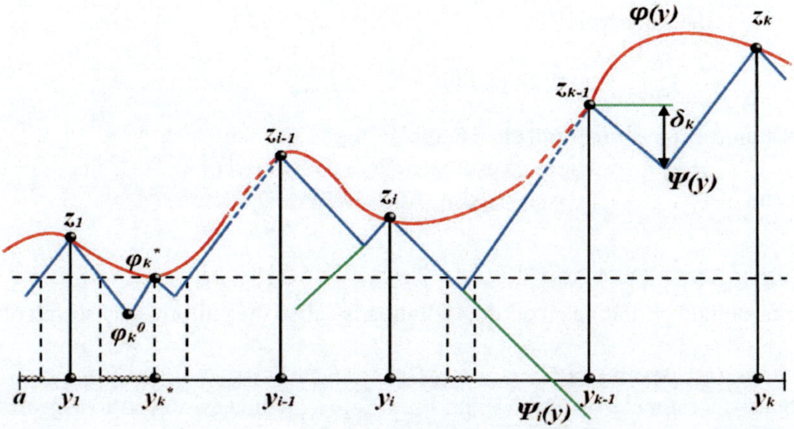

Fig. 1. Search for the global optimum.

That is, the experimental results and condition (4) allow us to construct the minorant $\psi(y)$ from (6) of the minimized function $\varphi(y)$ in the search domain. At the same time, the desired minimum value of $\varphi(y^*)$ is estimated:

$$\varphi_k^\circ \le \varphi\left(y^*\right) \le \varphi_k^*, \tag{7}$$

where:

$$\begin{aligned} \varphi_k^\circ &= min\{\psi(y) : y \in D\}, \\ \varphi_k^* &= min\{z_i : 1 \le i \le k\}, \end{aligned} \tag{8}$$

and the upper estimate of φ_k^* corresponds to the point y_k^* at which it is reached, i.e., $\varphi_k^* = \varphi(y_k^*)$. The lower bound φ_k° can be calculated quite simply. If we denote:

$$\varphi_{ki}^\circ = min\left\{\psi(y) : y \in \left[y_{i-1}, y_i\right]\right\}, 1 \le i \le k+1, \tag{9}$$

where $y_0 = a$ and $y_{k+1} = b$, then according to (8),

$$\varphi_k^\circ = \varphi_{kt}^\circ = min\left\{\varphi_{ki}^\circ : 1 \le i \le k+1\right\}. \tag{10}$$

In accordance with (4), (5), (9):

$$\varphi_{ki}^\circ = (z_i + z_{i-1} - K\delta_i)/2, 1 < i \le k, \tag{11}$$

$$\varphi_{k1}^\circ = z_1 - K\delta_1, \varphi_{k,k+1}^\circ = z_k - K\delta_{k+1}, \tag{12}$$

where $\delta_i = y_i - y_{i-1}, 1 \le i \le k+1$.

The search method. It follows from (9), (11), (12) that:

$$\delta_i = min(z_{i-1}, z_i) - \varphi_{ki}^\circ \le K\delta_i/2, 1 < i \le k, \tag{13}$$

$$\delta_i = z_1 - \varphi_{k1}^\circ = K\delta_1, \delta_{k+1} = z_k - \varphi_{k,k+1}^\circ = K\delta_{k+1}. \tag{14}$$

Since, at the same time:

$$\varphi_k^* - \varphi_k^\circ \min\{\delta_i : 1 \le i \le k+1\},$$

then the uniform location of the test points y_i, $1 \le i \le k$, at which $\Delta_i = 2\delta/K$, $1 \le i \le k$, $\delta_1 = \delta_{k+1} = \delta/K$, provides an estimate of (7) with an accuracy of δ, since, according to (3) and (4):

$$\varphi_k^* - \varphi_k^\circ \le \delta. \quad (15)$$

This method of finding an optimal solution is called the uniform grid search method.

Search on an Uneven Grid

To obtain an estimate of (15) by the brute force method, we need to perform $k = K(b-a)/2\delta$ experiments. This estimate can be achieved with a much smaller number of experiments, given that the desired solution y^* can only belong to a set:

$$D_k = \{y \in D : \psi(y)\} \le \varphi_k^*. \quad (16)$$

In Fig. 1, the corresponding set D_k is indicated by shading. Thus, it is possible to perform the search on a coarse grid, which corresponds to an accuracy of $\delta' > \delta$ and requires a small number of experiments k'. At the next stage of the search, the search is performed only in a subset of D_k for a higher accuracy $\delta'' < \delta'$, and so on (until the required accuracy δ is reached). By increasing the number of stages and decreasing the number of experiments at each stage, it is possible to build a fully sequential method in which exactly one experiment is performed at each stage.

One such method, which consistently generates a non-uniform mesh, is that each subsequent $k+1$ iteration (next experiment) is performed at a point:

$$y^{k+1} = (y_t + y_{t-1})/2 - (z_t + z_{t-1})/2K,$$

where the minimum of the minorant $\psi(y)$ is achieved, i.e., the lower estimate of φ_k° from (8), (10) is achieved. The expediency of conducting the next experiment at the point of minimum of the minorant ψ can be (informally) motivated by the fact that obtaining a small value of the minimized function φ at this point will lead to an improvement in the upper estimate $\left(\varphi_{k+1}^* < \varphi_k^*\right)$, and obtaining a large value will lead to an improvement in the lower estimate $\left(\varphi_{k+1}^\circ > \varphi_k^\circ\right)$, if the latter was achieved at a single point. In this case, each experiment improves the estimate (7). The starting points for the experiments can be, for example, $y^1 = a$, $y^2 = b$.

Estimation of the Constant and Stochastic Model

The previous discussion was based on the assumption that the constant K in conditions (2) or (4) is given. A rough upper estimate that can be obtained in applied problems leads, as can be seen from (15), to the need to use significantly denser networks (uniform or non-uniform), the nodes of which are used for experiments. With an underestimation of the value of the constant, inequalities (4) and the entire scheme for estimating the optimum based on them lose their validity.

A possible solution is as follows. Instead of condition (4), which implies reliable boundedness of the differences of the function, we assume that the differences $\varphi(y') - \varphi(y'')$ are interpreted as random variables with an absolute value of the mathematical expected value equal to $L/y' - y''/$, if the minimum point y^* does not lie between the points y', y''. In this case, it is possible to build current estimates of the unknown constant L based on the results of the experiments, i.e., based on the observed values of the relative differences of the function.

This approach leads to the fact that the estimates of the desired optimum based on the results of experiments are also stochastic in nature. At the same time, both estimates using probability distributions and precise estimates, such as maximum likelihood estimates for the global minimum point, can be entered, which allows you to build a sequential search method, according to which each subsequent experiment is carried out at the current point of maximum likelihood.

Maximum Likelihood Algorithm
The first experiment is performed at an arbitrary point $y^1 \in (a, b)$. The point of any subsequent $k + 1$ experiment is determined by the expression:

$$y^{k+1} = \frac{y_t + y_{t-1}}{2} - \begin{cases} (z_t - z_{t-1})/2r\mu, \ 1 < t \leq k, \\ 0, \ t = 1, t = k+1, \end{cases} \tag{17}$$

where the number t is determined from the condition:

$$R(t) = \max\{R(i) : 1 \leq i \leq k+1\}, \tag{18}$$

where for $1 < i \leq k$

$$R(i) = \delta_i + (z_i - z_{i-1})^2/\mu^2\delta_i - 2(z_i + z_{i-1})/r\mu, \tag{19}$$

$$R(1) = 2\delta_1 - 4z_1/r\mu, R(k+1) = 2\delta_{k+1} - 4z_k/r\mu, \tag{20}$$

$$\mu = max\{|z_i - z_{i-1}|/\delta_i : 1 < i \leq k\}. \tag{21}$$

If expression (21) is equal to zero, as well as when $k = 1$, it is assumed that $\mu = 1$. The number r from (17), (19), (20) is a parameter of the method, and the inequality $r > 1$ must be satisfied.

Convergence Conditions
Methods (17)–(21), as already mentioned, sequentially perform iterations at the points that are most likely to contain the global minimum, estimating their location by means of an appropriate interpretation of the experimental results within a certain stochastic model. However, the rules (17)–(21) for selecting iterations can be investigated beyond the initial assumptions. If this algorithm is used to minimize (with a constant k) the function φ, then for any boundary point y', the sequence $\{y^k\}$ generated by it is valid:

- the point y' is at least a locally optimal point of the function φ if this function has a finite number of local minima;

- if there is another boundary point y'' of the sequence $\{y^k\}$, then $\varphi(y'') = \varphi(y')$, which means that simultaneous convergence to different values of the function is impossible, and, therefore, the method generates an uneven mesh when minimizing functions other than a constant;
- $z^k = \varphi(y^k) \geq \varphi(y'), k = 1, 2, \ldots,$ i.e., the algorithm cannot generate convergence to points where the function value exceeds the result of any experiment;
- if at some stage the condition

$$r\mu > 2K. \tag{22}$$

Then y' is the point of global minimum of the function φ, and, moreover, the set of all boundary points of the sequence $\{y^k\}$ coincides with the set of points of global minimum of the function φ.

It is important to note that this algorithm, which does not require an a priori setting of the Lipschitz constant, has proven to be effective in solving many specific applied problems.

The Influence of Dimensionality

The considered scheme for estimating the global optimum based on the test results and the a priori assumption (2), which characterizes the limitation of the differences in the values of the minimized function to a certain extent of the differences in the corresponding argument values, remains true in multidimensional problems, i.e., when $N > 1$. However, the computational complexity of the evaluation increases significantly.

Exponential Growth of the Search Complexity

Let us assume that the minimized function φ satisfies the condition (with constant K) in the admissible domain $Q = D$ of (5). Similarly, to (5), we introduce a family of minorants ψ_i, i.e.:

$$\varphi(y) \geq \psi_i(y) z^i - K\left|\left|y - y^i\right|\right|, y \in D, \ 1 \leq i \leq k, \tag{23}$$

where y^i and $z^i = \varphi(y^i)$ denote the points and experimental results, respectively. It follows from (23) that in the closed ε-neighborhood $U(y^i)$ of the point y^i, there is an estimate:

$$z^i - min\{\varphi(y) : y \in D \cap U\left(y^i\right)\} \leq \delta,$$

if $\varepsilon = \delta/K$. By selecting the test points so that they form an ε-network in the search area D, we obtain an estimate of the desired solution:

$$\varphi_k^* - \varphi(y^*) \leq \delta, \tag{24}$$

where

$$\varphi_k^* = min\{z^i : 1 \leq i \leq k\}. \tag{25}$$

Since, for the solution point y^* and the coverage of the domain D by the neighborhoods of $U\left(y^i\right)$, it is true that:

$$\varphi\left(y^*\right) = \min_{1 \le i \le k} (min\{\varphi(y) : y \in D \cap U\left(y^i\right)\}). \tag{26}$$

The way in which the admissible region Q is set determines the search area for the global extremum $D = \left\{y \in R^N : a_i \le y_i \le b_i, 1 \le i \le N\right\}$, which describes the range of possible parameter changes, and additional constraints such as inequalities describing the conditions for the normal functioning of the object, which can be reduced to a single form $g_i(y) \le 0, 1 \le i \le m$.

In this case:

$$Q = \{y \in D : g_i(y) \le 0, 1 \le i \le m\}. \tag{27}$$

A uniform ε-network in a multidimensional domain D from Eq. (27) can be effectively constructed in various ways. In this case, the total number of nodes in such a uniform network exceeds $(b_1 - a_1 \cdot \dots \cdot (b_N - a_N)/\varepsilon^N$, i.e., grows exponentially with increasing dimension N.

Difficulty in Consistently Constructing Non-uniform Grids

Similarly to the one-dimensional case, it is possible to reduce the number of measurements required to obtain an estimate of (24) using the brute force method by first performing a low-fidelity search on a coarse mesh containing k nodes and, based on the measurement results, constructing the set D_k from (16), where $\psi(y)$ is the upper bending (6) minor of ψ_i from (23) and φ_k^* from (5). The set D_k should contain the desired solution y^*, and further search (with greater accuracy) should be performed in this set, and so on.

When $N = 1$, the set (16) is a set of segments on the real number axis (see shading in Fig. 2). When $N > 1$, the constructive task of this set becomes much more complicated. To illustrate the difficulties, let us consider a specific example for the case of $N = 2$ and $k = 4$ shown in Fig. 2, where in the rectangular region D the points of the four measurements are marked (by small black circles) and the values of the minimized function at these points are indicated, as well as the lines of a constant level of the minorant $\psi(y)$. The outline of the non-contour region D_k is highlighted (shading).

The difficulty of determining the region D_k, which should include a uniform grid corresponding to a higher accuracy, is removed if we switch to a fully sequential search scheme, when at each $k+1$ step in the corresponding region D_k exactly one measurement is performed, as discussed for the one-dimensional case. In this case, the point y^{k+1} of the next measurement that improves the estimate of (7) can, for example, be used as the point at which the minorant ψ(y) reaches the absolute minimum of (8). This point, of course, belongs to the domain D_k, but it can be determined by minimizing ψ in a simpler domain D.

In the one-dimensional case, according to (9)–(12), minimizing ψ is quite simple and reduces to choosing the minimum of $k+1$ values (10), each of which is the minimum value of (9) of the minor of ψ in the corresponding subinterval $[y_{i-1,}y_i]$,$1 \le i \le k+1$. This simplicity follows from the fact that for $N = 1$, each measurement point y_i falls into at most two intervals $[y_{i-1,}y_i]$, $[y_i, y_{i+1}]$, and the minimum value of ψ in each such

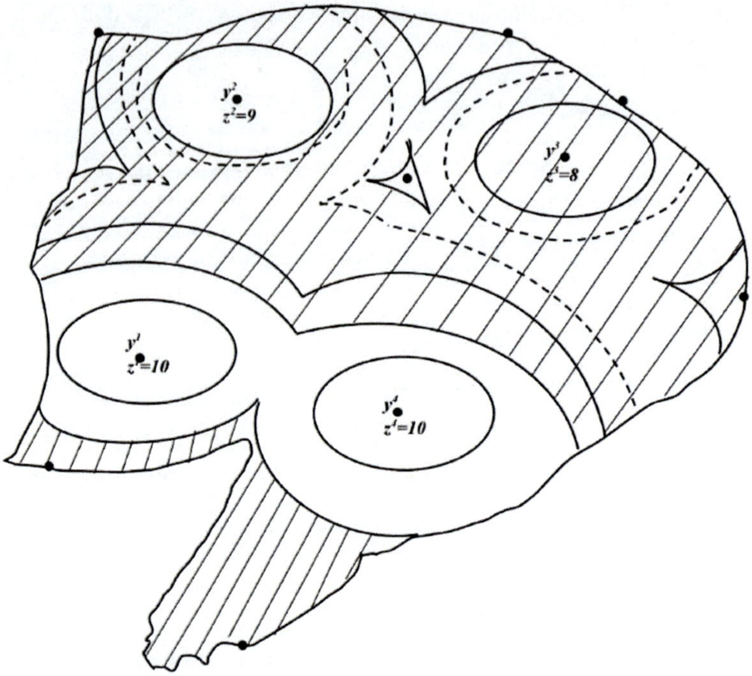

Fig. 2. Unlinked search area for parameters.

interval does not depend on the results of measurements in other intervals (and can be found analytically as a function of the coordinates of the endpoints of the interval and the values of the minimized function at these endpoints). When $N > 1$, the relationship between the measurement results and the minimum point is minor:

$$y^{k+1} = \arg\min\{\psi\left(y; y^1, \ldots, y^k; z^1, \ldots, z^k\right) : y \in D\}. \tag{28}$$

Indeed, the rule (28) for selecting the next measurement point itself includes a minimization problem similar to the original problem (2). The values of the function ψ from (6), (23) are usually much simpler to calculate than the values of the object characteristic φ from (2), which must be minimized, obtained by analyzing a complex mathematical model of this optimization object. However, problem (26) has to be solved at each iteration of problem (2). Note that this difficulty remains unchanged for approaches according to which the next iteration is carried out not at the point of minimum of the minority ψ, but at the point of maximum likelihood or at the point of minimum of the expected value of the function φ (both of these rules are based on probabilistic analogues of model (2), similar to the model for the case $N = 1$ discussed above), or by using another rule that ensures consistent improvement of the estimates (7) at each (or almost each) step by performing iterations at the point (26) that provides an extremum to some current estimate φ of the desired solution.

4 Conclusions

Let's summarize some conclusions. Solving multidimensional problems using a simple search method on a uniform grid requires a significant number of iterations and is actually possible only for small values of N and low solution accuracy. The use of parsimonious sequential methods that create a non-uniform grid requires solving an additional multi-extreme problem of type (26) at each iteration, which also needs to be solved by search methods (e.g., the brute force method), which dramatically increases the computational complexity of the iteration.

On the other hand, when $N = 1$, simple sequential search methods can be constructed for a multi-extreme problem based either on a given value of the constant or on estimates calculated during the solution process, which create an economical non-uniform grid whose nodes are used for measurements. In this regard, it is worth considering the possibility of reducing multidimensional, multi-extreme problems to some equivalent one-dimensional problems.

It should be noted that a targeted search for solutions to multi-extreme problems based on a priori assumptions of type (2) about the boundedness of differences generally requires more measurements than a local search for solutions to unimodal problems, since in the first case measurements are made at the nodes of a (non-uniform) grid in the multidimensional search area, and in the second case - at the nodes located on some one-dimensional descent path. This ratio can change under certain conditions, complementing model (2).

In general, drone delivery is an effective option for companies looking to optimize their supply chains and reduce costs. By reducing delivery times, increasing efficiency, and saving costs, drone delivery can help businesses stay competitive in the marketplace [10–12].

References

1. Rejeb, A., Rejeb, K., Simske, S.J., Treiblmaier, H.: Drones for supply chain management and logistics: a review and research agenda. Int. J. Log. Res. Appl. **26**(6), 708–731 (2023). https://doi.org/10.1080/13675567.2021.1981273
2. Alvarado, U.Y., Kotzab, H.: Supply chain management: the integration of logistics in marketing. Ind. Mark. Manag. **30**(2), 183–198 (2001). https://doi.org/10.1016/S0019-8501(00)00142-5
3. Kharchenko, V., Pawęska, M., Bugayko, D., Prusov, D.: The efficiency and effectiveness of remotely piloted aircraft systems used in logistics problems solving due to territorial infrastructure. Logist. Transp. **22**, 13–20 (2014)
4. Amouzegar, M., Drew, J., Tripp, R.S.: A simulation model for the analysis of end-to-end support of Unmanned Aerial Vehicles. Int. J. Appl. Decis. Sci. **3**(3), 239–258 (2010). https://doi.org/10.1504/IJADS.2010.036101
5. Carlsson, J.G., Song, S.: Coordinated logistics with a truck and a drone. Manag. Sci. **64**(9), 4052–4069 (2018). https://doi.org/10.1287/mnsc.2017.2824
6. Colicchia, C., Strozzi, F.: Supply chain risk management: a new methodology for a systematic literature review. Supply Chain Manag. Int. J. **17**(4), 403–418 (2012). https://doi.org/10.1108/13598541211246558

7. Frachtenberg, E.: Practical drone delivery. Computer **52**(12), 53–57 (2019). https://doi.org/10.1109/MC.2019.2942290
8. Kajba, M., Jereb, B., Obrecht, M.: Considering IT trends for modelling investments in supply chains by prioritising digital twins. Processes **11**(1), 262 (2023). https://doi.org/10.3390/pr11010262
9. Kharchenko, V.: The use of unmanned aircraft systems for fast delivery goods. Logist. Transp. **47**(3–4), 89–101 (2020)
10. Ma, Z., Chen, J.: Adaptive path planning method for UAVs in complex environments. Int. J. Appl. Earth Obs. Geoinf. **115**, 103–133 (2022). https://doi.org/10.1016/j.jag.2022.103133
11. Bangani, S., Jokonya, O.: Factors affecting the adoption of contactless technologies in supply chain and logistics. J. Transp. Supply Chain Manag. **16** (2022). https://doi.org/10.4102/jtscm.v16i0.751
12. Maharana, S.: Commercial drones. Int. J. Manag. Appl. Sci. **5**(3), 96–101 (2017)

Human Factor in Civil Aviation

Developing Future Aviation Specialists' Readiness for Professional Activity Through System-Synergetic Diagnostic Technologies

Elvira Luzik(✉), Nataliia Ladohubets, and Lesia Konoplianyk

National Aviation University, Liubomyra Huzara Avenue, 1, Kyiv 03058, Ukraine
elvira.luzik@npp.nau.edu.ua

Abstract. The internal and spiritual development of the country is significantly influenced by how well the educational system within higher technical education institutions can meet the demands of both society and individuals for high-quality educational services. However, the insufficient conceptual creation of the methodological foundations and mechanisms for aviation specialists' professional training results in increasing contradictions concerning the objective needs of the labour market for qualified specialists capable of solving intricate professional tasks, thus contributing to their sustained self-development and professional growth. The current state of the labour market, both domestically and internationally, is characterized by the growing competition among aviation specialists, whose professional competence and commitment to continuous self-development and self-improvement are generally acknowledged as the benchmark for measuring competitiveness. The main contribution of the research is the development of technologies for system diagnostics of future aviation specialists' readiness for sustained professional growth as well as social and professional mobility to become highly professional leaders with effective communication, critical thinking, teamwork skills, and creativity, – the qualities inherent to a modern specialist. Furthermore, the designed technology facilitates the quantitative and qualitative analysis of the effectiveness of aviation specialists' training in higher technical education institutions.

Keywords: System-Synergetic Diagnostic Technology · Didactic Modeling · Aviation Specialist · Didactic Invariant · Competence · Competency

1 Introduction

1.1 Introduction to the Problem

One of the most urgent tasks considered by state policy in higher professional education is creating conditions for the effective training of specialists able to solve multifunctional professional tasks. These individuals are anticipated to be socially active and competent specialists [1, 2], adapted to rapidly changing conditions of the labour market and prepared for professional development as well as social and professional mobility.

I. Ostroumov and M. Zaliskyi (Eds.): ACASD 2024, LNNS 992, pp. 353–363, 2024.
https://doi.org/10.1007/978-3-031-60196-5_26

Theoretical studies and empirical experience [3–5] affirm the significance of optimizing the structure and content of all training cycles in higher technical education institutions to unveil the comprehensive potential of young individuals. This optimization ensures their autonomy and responsibility in making objective managerial decisions [6–8].

The increase in information flows within a changing information educational environment, social, economic and political transformations at the contemporary stage of country development, and the implementation of a system-synergetic approach in higher technical education institutions "have contributed to the emergence of fundamentally new tasks for the educational system of higher technical education institutions as a whole, the product of the formation of the external world into the internal world of the individual as a professional subject, and directly for the system of higher professional education, as it plays a key role in shaping the scientific and intellectual potential of Ukraine" [9].

The paradigm shift in the system of higher professional education in Ukraine, as outlined in the European Commission's project "Tuning Educational Structures in Europe", signifies a transition from a process-oriented approach to a results-oriented approach to organizing professional training [10]. This shift contributes to creating a novel perspective of educational outcomes, wherein the essence of a results-oriented approach lies in students' proficiency to execute actions acquired during their training. Consequently, this facilitates the establishment of a strategic goal aligned with the prioritized social and economic development vectors of higher professional education, as outlined in key regulatory state documents [11–13]. The strategic goal entails training specialists capable of adapting to the changing conditions of the labour market, demonstrating autonomous and responsible execution of multifaceted tasks amidst dynamic uncertainty, and being ready for sustained professional growth, as well as social and professional mobility. In this aspect, emphasis is placed not only on the quantity of productive knowledge acquired from a variable and information-conflicting environment, but also on the ability and skills to actively employ this knowledge, engage in proactive decision-making, and enable future specialists to navigate effectively in an open and dynamically changing space.

1.2 Motivation

The paradigm shift in the training system for aviation specialists in Ukraine, driven by the growth of the multi-vector nature of tasks, sets goals for education and its outcomes as system-forming factors. According to these factors, the social and psychological components of educational activities are mutually organized and interrelated, thereby shaping the most critical tasks of modern higher professional education in Ukraine. These tasks include the identification of the educational outcomes (skills) by qualification levels (competencies), the modification of the structure and content of education according to ICAO standards, and the development of innovative technologies along with the search for valid and reliable means of assessing training outcomes.

1.3 Contribution

The principal contribution of this study lies in defining the optimality conditions of the elements of an open dynamic system of aviation specialists' professional training in higher education institutions and the criteria (as system-forming foundations) for the quality of this training. Optimal conditions are quantitative and qualitative relationships between didactic components of educational activities, enabling the subjects of these activities to transform acquired productive knowledge, integrated skills, and abilities into psychological formations of personality – manifesting as skills, experience, character traits, and abilities. This occurs under the condition that productive knowledge becomes the orientation and conviction of the individual, while integrated skills and intellectual abilities are refined to attain mastery and professional proficiency [14, 15].

2 Literature Review and Problem Statement

The traditional system of aviation specialists' training is primarily focused on subject-oriented learning, where operational-role components (knowledge and skills) play a decisive role, reflecting a single component of specialists' professional competency. Specifically, the renewal (expansion and deepening) of the structure and content of the future aviation specialists' competency takes place, while the success of subsequent professional development depends on both the level of the individual's personality development and their professional self-organization, forming the basis of soft skills. Therefore, according to Ukrainian and foreign researchers [2, 5, 16, 17], these personal (subjective) characteristics act as system-forming conditions (criteria) for the transition of a future specialist from one educational level to another. This is owing to their responsibility for the successful actualization of professional knowledge and skills, thereby ensuring professional harmonious development in the implementation of a creative learning trajectory [18]. However, it is possible only within flexible, adaptive educational systems anticipating aviation specialists' ability for self-education and self-development, rapid professional reorientation, and professional development [19]. "The development of such personal qualities creates integrated opportunities for implementing forward-looking learning through the cultivation of self-awareness, independence, and self-control, realized through effective lifelong self-learning" [9].

Based on key findings [10, 14], it can be asserted that:

- the methodological basis of the contemporary aviation specialists' training in a higher technical education institution, characterized by a multi-vector approach, is mainly determined by two vectors: 1) the profession-oriented content of the knowledge vector $\overrightarrow{C}$, which involves the development of a specialist within the specified parameters and standards (knowledge, abilities, skills); 2) personality-oriented vector of objectives $\overrightarrow{OJ}$, aimed at creating conditions for the individual's harmonious development during professional growth;
- professional competence becomes a system phenomenon that arises when there is an alignment between the individual-activity system and the professional-normative knowledge system;

- professional competency, representing "an open dynamic multi-level system of the future specialist's development – professional development of an individual, which involves solving a series of multi-level tasks, consists of a set of dynamic subsystems that encompass the specific interaction of a specialist as the subject of activity and a particular professional task, requiring self-organization for its solution" [15].

3 Materials and Methods

We have studied the issues related to determining the conditions for optimizing the social and psychological elements of educational activities and the criteria for the effectiveness of the professional training of future aviation specialists in higher technical education institutions through didactic modeling.

Taking into account that the objects of didactic modeling are both natural and artificial systems and the subject is the mental activity of future specialists, didactic modeling is considered as "a system of actions aimed at ensuring adequate assimilation of modeling properties, connections and relations of the cognitive and transforming object (both natural and artificial)" [9].

The implementation of new requirements for the professional training of future aviation specialists in higher technical education institutions is possible under various conditions of a system-personal approach [20]. However, the most important ones among them are changes in the structure and content of professional education within the professional external information space and the introduction of a synergetic information system of pedagogical technologies aimed at implementing the tasks of the educational process for a specialist who can effectively work in a dynamically changing open environment. Owing to the alignment of the educational strategy within Ukrainian higher professional education with an emphasis on personal development and a transition towards educational outcomes (e.g., effective communication skills, teamwork, development of critical thinking and creativity) constituting the "DNA of the 21st-century education", the assessment of learning outcomes plays a crucial role in the development of the professional subjectivity of future specialists. Consequently, the following criteria for optimizing the professional training of future aviation specialists in higher technical education institutions have been outlined:

- the goal (objectives) of training, which is interdependent and interconnected with the structure and content of professional training [11];
- the learning outcomes obtained during training, i.e., acquisition of professional subjectivity skills of a specialist [21];
- "efficiency factor" of learning, which relates the learning outcomes to the objectives of specialists' professional training set at the beginning of training process.

Given that any activity, including the educational one, commences with a clear understanding of the goal (objectives), acknowledging the multi-vector and multifaceted nature of the prospective professional activity of aviation specialists, the goal is considered as the main criterion for developing professional subjectivity of future aviation specialists, i.e., as a space-time vector, whose components are subject-specific, generic-cognitive, and educational objectives.

On the other hand, the goal of the professional training of a specialist in a higher technical education institution, based on the main conceptual principles of the competence (results-based) approach, can be presented as a vector-function comprising interdependent and interrelated competencies (skills) denoted as CP_1, CP_2, ..., CP_n. As prescribed by regulatory documents governing the professional training of aviation specialists, proficiency in these competencies is imperative for the specialist:

$$\overrightarrow{OJ} = \overrightarrow{OJ}(CP_1, CP_2, \ldots, CP_n). \quad (1)$$

For the effective implementation of the formulated objectives of aviation specialists' professional training, it is essential to choose the mechanisms for correlating objectives (competences) and actual learning outcomes (competencies). These outcomes encompass a set of acquired productive knowledge, integrated skills and developed professionally significant qualities.

The problem can be solved through system-didactic modeling, which is a "system of goal-directed actions that is robust and valid, ensuring the appropriate assimilation of modeled properties with a focus on both natural and social and cultural aspects, with the student's cognitive activity as a focal point" [15].

4 Results and Discussion

Based on the concept of competence-concentric integrity of the professional subjectivity of future aviation specialists, and taking into account regulatory documents (e.g., State Standards, National Qualifications Framework, and Educational and Professional Programs) that outline a set of their professional competencies, the algorithm for the system modeling of the educational process for training aviation specialists and the concentricity of their development within innovative management paradigms include three fundamental components. These three components – profession-oriented, profession-formative and practical professional-reflective-evaluative ones – constitute the core of the human-centered social, psychological and pedagogical essence of professional training. In addition, each component has several stages [15].

The practical (reflective-evaluative) stage, aimed at actualizing the application of productive knowledge, integrated skills, and developing mental formations (skills) in the process of quasi-practical professional activity of future specialists, includes the components as follows:

- diagnosis and assessment of the future specialists' practical readiness for professional activity;
- implementation of integral, information-operational and personal components of professional subjectivity of a specialist according to the international educational standards (modular organization of learning with interdisciplinary coordination of integrative training courses [10]);
- mastery of integrated competencies of the specialist's personality, including self-development and career advancement, competitiveness in the market of educational services, a high level of professional competence, opportunities for collaboration with colleagues, and strategic and critical thinking.

The first criterion for the development of professional subjectivity is the readiness to form individual-reflexive market technologies and psychological and pedagogical technologies of self-creation in an integrated dynamic-synergistic environment, and strategies for accelerating and enriching knowledge. This supports the idea that in a spiral learning process, the future specialist will follow an optimal individual didactic trajectory while moving to a new educational level [15].

The second criterion for developing the professional subjectivity of the future specialist is the learning outcomes as the accumulation of experience gained by the student during the educational process of optimizing the professional training of future specialists. This process has been created by using a system modeling algorithm and it comprises the sum of competencies (skills) that should be designed by specialists, i.e.

$$O = \sum\nolimits_{i=1}^{n} CP_i. \tag{2}$$

The spatiotemporal dynamic model of educational standards is used to introduce the third criterion of the development of professional subjectivity, specifically the "efficiency factor" in optimizing the professional training of future aviation specialists. This criterion establishes a connection between the learning outcomes and training objectives as outlined in relevant regulatory documents (e.g., State Standards, National Qualifications Framework, and Educational and Professional Programs). According to this model, the learning outcome is related to its objectives and content (skills) by:

$$O = \left(\overrightarrow{OJ} \cdot \overrightarrow{C}\right), \tag{3}$$

where $\overrightarrow{C}$ is a vector function of the educational content, which includes the didactic components of the educational environment (knowledge, abilities, skills) of future specialists' professional training as a multi-vector component of educational process.

In the organizational-operational structure of educational activity, the outcome is a multi-vector system of acquired integrated abilities and skills, productive knowledge, and professionally significant qualities [1]. However, a relevant question for employers is whether it is feasible to predict and provide didactic justification for the educational outcome (acquisition of pertinent skills) based on predetermined objectives (competences) [2]. Let us use didactic invariants to get an answer to the question. The didactic invariant of the educational level is an element of educational activity that does not change due to changes in research methods of educational activity. Based on this definition, the goal of the professional training of an aviation specialist has been chosen as the first didactic invariant, serving as a vector function of interdependent and interrelated components of competencies that correspond to the professional subjectivity of the future specialist:

$$\overrightarrow{D_1}(I) = \overrightarrow{OJ}\left(\overrightarrow{CP_1}, \overrightarrow{CP_2}, \ldots, \overrightarrow{CP_n}\right). \tag{4}$$

The learning outcome has been chosen as the second didactic invariant:

$$D_2(I) = O\left(\overrightarrow{CP_1}, \overrightarrow{CP_2}, \ldots, \overrightarrow{CP_n}\right), \tag{5}$$

which is related to the goal of professional training of future specialists and the corresponding content of this training:

$$\overrightarrow{C} = \overrightarrow{C}\left(\overrightarrow{C_{gc}}, \overrightarrow{C_s}, \overrightarrow{C_{ed}}\right), \tag{6}$$

where $\overrightarrow{C_{gc}}, \overrightarrow{C_s}, \overrightarrow{C_{ed}}$ are, respectively, interdisciplinary vectors of general-cognitive, subject-specific and educational knowledge and skills related by the ratio:

$$D_2(I) = O\left(\overrightarrow{CP_1}, \overrightarrow{CP_2}, \ldots, \overrightarrow{CP_n}\right) = \left(\overrightarrow{OJ} \cdot \overrightarrow{C}\right) = |OJ| \cdot |C| \cdot \cos(\alpha). \quad (7)$$

where α is an angle between the competence vector and objective vector.

Let us prove the ratio (7). So, if the relationship between formulas (1) and (4) is apparent, we proceed to establish a connection between formulas (5) and (7). From the perspective of determining the result of professional activity – work as a physical quantity, formula (7) is the result of the activity performed by the subjects of the educational process (both students and the teacher) to transform the acquired information into knowledge, abilities, skills, i.e., into mental formations of the student's personality (competencies), following the educational objectives (competences) formulated and outlined in regulatory documents.

Since the optimality of the learning process has been proven in studies [9, 15], according to which the deviation from zero of the second didactic invariant $D_2(I)$ is a necessary condition of the spiral learning process (identity to the dynamic principles of continuity and sequence). The student's transition to a higher educational level will be carried out through a didactic spiral along a personally optimal trajectory. At the same time, it should be noted that if the vector of the educational content is arbitrarily located in the information and educational environment of training and does not coincide with the direction of the vector of objectives (insufficient correspondence of the content of academic disciplines to the objectives and tasks of professional training of an aviation specialist), the educational process will be carried out along the didactic spiral under the condition of achieving the objectives determined by the given educational level of the National Qualifications Framework and the Educational and Professional Program, which are correlated with the level of professional subjectivity of the student.

The efficiency factor of the learning process, which characterizes this connection and has been chosen as the third criterion for the development of the professional subjectivity of future specialists, is determined by the equation:

$$d = \frac{C}{OJ} \cdot \cos(\alpha). \quad (8)$$

It confirms the existence of discrepancies (gap) between the state order (competences) that the future specialist must master and the competencies (knowledge, abilities, skills, mental formations) of the future specialist in the process of personality-oriented educational activities. To confirm the optimality of the training process for future aviation specialists in a higher technical education institution, the task of achieving a fuzzy goal has been considered, the formulation of which is determined by the need for the formalization of numerous variables, such as the development of key and professional competencies of a future aviation specialist. Let X be the universal set of alternatives (strategies) for achieving the set goal, i.e., the universal set of all possible choices of the decision-making subject. A fuzzy goal in X is a fuzzy subset of X, denoted by G, which is defined by the membership function:

$$\mu_G : X \to [0, 1]. \quad (9)$$

Suppose some alternative X ensures the achievement of the goal with a degree of $\mu_G(x)$ and satisfies the constraint with a degree of $\mu_C(x)$, then the membership degree of this alternative task solution is equal to the minimum of these numbers (i.e., the alternative that is permissible with a degree of 0.5, for instance, with the same degree belongs to the fuzzy solution as well, even though it ensures the achievement of the goal with a degree equal to 0.8). In our case, the membership function has the following form:

$$\mu_D(x) = \min\{\beta_1\mu_{G_1}(x), \ldots, \beta_n\mu_{G_n}(x), \sigma_1\mu_{C_1}(x), \ldots, \sigma_m\mu_{C_m}(x)\}, \tag{10}$$

where β_i is the coefficient of the relative importance of goals, σ_i is the coefficient of constraints ($i = 1, \ldots, n$). Let $X = \{1, 2, \ldots, 10\}$ and a fuzzy goal G and two constraints C_1, C_2 are given in Table 1 [22].

Table 1. Parameters of a fuzzy goal G, constraints C_1 and C_2, and fuzzy instruction (strategy).

X	1	2	3	4	5	6	7	8	9	10
$\mu_G(x)$	0	0.1	0.4	0.8	0.2	0.7	0.4	0.6	1.0	0
$\mu_{C_1}(x)$	0.3	0.7	0.4	0.9	0.8	0.6	1.0	0.7	0.2	0
$\mu_{C_2}(x)$	0.2	0.4	0.7	0.3	0.6	0.5	0.8	0.7	1.0	0.4
$\mu_D(x)$	0	0.1	0.4	0.3	0.2	0.5	0.4	0.6	0.2	0.4

The last raw of Table 1 contains the solution D to the given task. This solution can be considered as a fuzzy instruction for choosing strategies to overcome difficult situations while learning, the implementation of which will ensure the achievement of the set goal.

5 Conclusions

Three possible options for the successful professional training of future specialists in the aviation higher technical education institution have been determined:

1. If the vector of educational content ($\overrightarrow{C}$) is in the plane perpendicular to the vector of objectives of the educational level ($\overrightarrow{OJ}$), i.e., the content of the syllabi and course training programs of the academic disciplines does not meet the training objectives outlined in the regulatory standards, then, based on the ratio (7), the second didactic invariant will be zero. This means that the Educational and professional programs, and therefore students, do not perform educational activities successfully to transform the information acquired in the learning process into productive knowledge and integrated skills according to the objectives of the disciplines. Consequently, the necessary professionally relevant qualities are not developed. As a result, the required and sufficient conditions for developing professional competences (skills) will not be created. Even though the level of students' general knowledge is sufficient to broaden their horizons (abilities), their academic achievements after completing this level of training are not enough to allow their transferring to the higher educational and professional level.

2. If the vector of educational content ($\overrightarrow{C}$) is in the same plane as the vector of objectives ($\overrightarrow{OJ}$) and their directions coincide (the content of the syllabi and course training programs of the relevant disciplines meets the training objectives), the student, having a sufficient level of knowledge, skills and high positive internal motivation, can effectively perform educational activities to internalize productive knowledge into integrated skills and intellectual abilities, thereby developing the relevant components of the professional subjectivity of the future professional. As a result, the mental formations (skills) acquired by the student enable them to move to the next educational level by a didactic spiral along the optimal curve, and the educational process of intellectual and professional development of the individual is harmoniously creative.
3. If the vector of educational content ($\overrightarrow{C}$) is in the same plane but arbitrarily relative to the vector of objectives ($\overrightarrow{OJ}$), the achievements in the learning process along a didactic spiral will depend on the conditions when the transition to each subsequent level is carried out only after the student has achieved the professional educational and pedagogical objectives defined by the given educational level and has developed relevant professional qualities of character. The coefficient characterizing this dependence corresponds to the ratio (8). Thus, deviation from zero of the second didactic invariant $D_2(I) \neq 0$ is a required condition for the spiral (harmonious) process of developing the professional subjectivity of the future aviation specialist.

The student's activity in internalizing productive knowledge and intellectual skills into mental formations is most effective when there is a convergence between the "knowledge field" and the "field of objectives" during professional training. In this case, the learning process resembles a spiral harmonious curve and the student's intellectual and professional development becomes creative.

6 Future Scope

Our study has proved that the system modeling of the enhancement of future aviation specialists' professional training within a higher technical education institution can be achieved by incorporating didactic invariant criteria, denoted as $\overrightarrow{D_1}(I) = \overrightarrow{OJ}$ and $D_2(I) = O$, into the educational process. The criteria determine and interconnect education objectives $\overrightarrow{D_1}(I)$ (competences) and learning outcomes $D_2(I)$ (competencies).

In addition, didactic invariants can be applied not only in training bachelor's and master's students in higher education institutions but also for PhD candidates. Moreover, they might be valuable for retraining, recruitment, and team building (professional growth) in a professional environment. However, in this context, skills will replace competences and knowledge, and goals will take on a multi-vector nature.

The empirical investigation aimed at examining the behavioral and emotional reactions of aviation specialists, identifying and analyzing the influence of psychological stressors on the professional activity and emotional state of aviation specialists, and unveiling the distinctive features of their emotional reactions have confirmed the significance of the obtained findings of the theoretical study of future aviation specialists' readiness for professional activity. This serves as an additional affirmation of the effectiveness of employing didactic modeling technologies, due to which aviation specialists adapt to the specifics of professional activity.

Thus, the obtained findings allow considering the development of the personality not only as an ideal result of developing critical thinking and the educational and self-development process resulting from activity and communication but also as developing the personality's readiness (motivational, theoretical, practical, normative and adaptive pre-professional activities), as a state of mobilization of the psychological and psychophysiological systems of an individual, determined by a combination of various factors characterizing multiple levels of readiness aspects.

References

1. Khomiuk, I.V.: The teacher's work system in forming the basic level of future engineers' professional mobility. Pedagogical Sci. Theory Hist. Innov. Technol. **6**(24), 337–345 (2012). (in Ukrainian)
2. Petrenko, O.V.: The inner world of a professional team's professional and psychological space as resources for analyzing flight crew reliability. Actual Probl. Psychol. **18**, 388–401 (2011). (in Ukrainian)
3. Plachinda, T.S.: Normative requirements for professional training of aviation specialists. Pedagogy Dev. Creative Pers. High. Secondary Schools **32**(85), 373–380 (2016). (in Ukrainian)
4. Konoplianyk, L.M.: Experimental model of aviation engineers' training for foreign professional activity. Humanitarian Educ. Tech. Univ. **17**, 186–193 (2009). (in Ukrainian)
5. Lavrukhina, T.V.: Formation of future air traffic controllers' key competencies as a professional activity readiness condition. Actual Probl. Sociol. Psychol. Pedagogy **16**, 168–174 (2012). (in Ukrainian)
6. Lizzio, A., Wilson, K., Simons, R.: University students' perceptions of the learning environment and academic outcomes: implications for theory and practice. Stud. High. Educ. **27**(1), 27–52 (2002). https://doi.org/10.1080/03075070120099359
7. Mazurenko, Y.: Component structure of aircraft maintenance engineers' readiness for professional communication. Innov. Pedagogy **30**, 88–92 (2020). https://doi.org/10.32843/2663-6085/2020/30-1.18. (in Ukrainian)
8. Tarnavska, T., Glushanytsia, N., Akilli, E., Nahorna, O., Kostenko, D.: Professionally significant personality traits and soft power skills development in ESP teaching. Arab World Engl. J. (AWEJ) **9**, 202–218 (2023). https://doi.org/10.24093/awej/call9.14
9. Luzik, E.V.: Didactic modeling as a basis of the competence-based approach to optimizing the learning process in higher educational institutions. In: VIII International Conference on Quality Strategy in Industry and Education, Kyiv, Ukraine, pp. 159–164 (2012). (in Ukrainian)
10. Baranovska, L.V.: Aviation university: the educational environment of the university as a determinant of the quality of professional training of students. Sci. Bull. Flight Acad. Section: Pedagogical Sci. **1**, 18–24 (2017). (in Ukrainian)
11. Borets, I.V.: Organizational and pedagogical conditions of formation of key competencies of future specialists in aviation transport systems. Pedagogical Discourse **12**, 53–56 (2012). (in Ukrainian)
12. Kovtun, O.V.: Theoretical and Methodological Principles of Developing Future Aviation Specialists' Professional Speech, Odesa, Ukraine (2013). (in Ukrainian)

13. Kovtun, O., Kokarieva, A., Khaidari, N.: Modeling a repertoire of pilots' professional communication skills for meeting flight safety and aviation security challenges. In: Guda, A. (ed.) Networked Control Systems for Connected and Automated Vehicles. LNNS, vol. 510, pp. 1069–1078. Springer, Cham (2022). https://doi.org/10.1007/978-3-031-11051-1_109
14. Luzik, E.V.: Theory and methodology of general science training in engineering higher education. Doctoral thesis, Institute of Higher Education of the Academy of Pedagogical Sciences of Ukraine (1996). (in Ukrainian)
15. Luzik, E.V., Khomenko-Semenova, L.O., Kokareva, A.M., Hurska, O.: University education innovative models in Ukraine: foreign experience. Bull. Natl. Aviat. Univ. Ser.: Pedagogy Psychol. **2**(13), 19–25 (2018). https://doi.org/10.18372/2411-264X.13.13399
16. Lavrysh, Y., Leshchenko, M., Tymchuk, L.: Development of metacognitive skills through digital narratives in higher education. Revista Electrónica de Investigación Educativa **25**, 1–15 (2023). https://doi.org/10.24320/redie.2023.25.e07.5028
17. Fowler, R., Mathews, E., Lynch, J., Roberts, J.: Aviation English assessment and training. Collegiate Aviat. Rev. Int. **39**(2), 26–42 (2021)
18. Shawcross, P.: Proficiency requirements underscore the importance of teaching and testing. ICAO J. **59**(1), 18–19 (2004)
19. Pylypchuk, M., Kovtun, O., Rudina, M., Poliakova, O., Khaidari, N.: Teaching aviation English: enhancing translation skills through the content and language integrated learning method application. Conhecimento Diversidade **15**(40), 278–303 (2023). https://doi.org/10.18316/rcd.v15i40.11304
20. Ladohubets, N.V.: Professional training of aviation professionals in conditions of fundamentalization. In: VI International Conference on Strategy of Quality in Industry and Education, Varna, Bulgaria, pp. 345–347 (2013). (in Ukrainian)
21. Leftor, V., Ushakov, V.: Features of classroom training highly specialized aviation technical disciplines during distance education. Sci. Bull. Flight Acad. Section: Pedagogical Sci. **11**, 125–131 (2022). https://doi.org/10.33251/2522-1477-2022-11-125-131
22. Voloshyn, O.F., Mashchenko, S.O.: Models and methods of decision-making. Kyiv University, Kyiv, Ukraine (2010). (in Ukrainian)

Multi-criterion Evaluation of the Criteria of the Information Model Conformity of the Air Traffic Controller Simulator to the Real System

Volodymyr Kolotusha(✉), Tetiana Shmelova, and Dmytriy Bondarev

National Aviation University, Liubomyra Huzara Avenue, 1, Kyiv 03058, Ukraine
Kolotusha.vp@gmail.com

Abstract. The aviation industry plays an important role in the world economy. Shortly, thousands of highly experienced specialists will be required at the global level. Simulation training of aviation operators, including air traffic controllers, is an extremely important stage of professional training, which requires demonstrating the integration of several areas of knowledge and skills. The paper presents an approach in which a certain simulator is purposefully used for simulation a working environment, considering specific practical training tasks. For this purpose, it is proposed to use expert evaluation methods and fuzzy-logic methods for obtaining reliable results in the lack of accurate initial data and considering the uncertainty of information. For the evaluation of practical activities, it is proposed to use the multiplicative aggregation of the results of simulator training, which allows to conclude that the operator has captured the necessary knowledge and skills and acquired the necessary competencies under the training tasks. It is proposed to apply the outlined approaches to establishing the criteria for determining and using air traffic control simulators in the development of national technical standards for the evaluation of simulators for training air traffic controllers.

Keywords: Air Traffic Control Controller · Professional Environment · Information Model · Simulator · Skills · Practical Training · Taxonomy · Expert System · Method of Expert Evaluations · Fuzzy Logic · Adequacy

1 Introduction

1.1 Introduction to the Problem

Global civil aviation, which employs tens of millions of workers, plays a significant positive role in the global economy. Since 1977, the volume of world air transportation has doubled every 15 years and continues to grow [1]. Taking into account the forecast values, the aviation transport system in the coming years will need many highly educated specialists: flight crew, technical personnel, air traffic control operators (ATCOs). Table 1 presents the results of research by the International Civil Aviation Organization (ICAO), which compares the average number of specialists around the world who will need to be trained annually with the training capabilities of existing institutions [2]. This indicates an existing shortage of specialists, namely 160 000 pilots, 360 000 technical personnel and 40 000 ATCOs.

I. Ostroumov and M. Zaliskyi (Eds.): ACASD 2024, LNNS 992, pp. 364–384, 2024.
https://doi.org/10.1007/978-3-031-60196-5_27

Table 1. ICAO comparative analysis of the number of specialists who need to be trained annually with the training capabilities of existing institutions.

Type of personnel	Available number of specialists (for 2010)	Need for staff (by 2030)	The need for annual training	Possibilities of the training system	Annual shortage of specialists
Flight crew	463 386	980 799	52 506	44 360	8 146
Technical staff	580 926	1 164 969	70 331	52 260	18 071
ATCOs	67 024	139 796	8 718	6 740	1 978

A similar situation with personnel can be observed at the level of the European region [3]. According to forecast data, by 2036 the aviation sector needs [4]:

- 620 000 new pilots (67 pilots per day for aircraft with more than 100 seats);
- 125 000 ATCOs (13 new ATCOs every day).

Simulator training of aviation operators, in particular pilots and ATCOs, is an extremely important component of their acquisition of a certain range of professional competencies [5]. At the stage of simulator training, which is present at almost every stage of professional training of ATCOs [6], the theoretical knowledge obtained by the trainee is transformed into the necessary skills and abilities. Due to the effective use of exercise equipment, there is an opportunity to: 1) ensure the gradation "from simple to complex" formation of professional competencies in a safe, especially when practicing unpredictable and emergency flight conditions; 2) optimize the training resources (involve instructors who are not active in the training); 3) repeat training exercises several times; 4) develop skills and improve decision-making skills in conditions of uncertainty and lack of time (development of professional confidence); 5) teach effective interaction and coordination of activities when working in a team (with related air traffic management (ATM)/flight support services) and exchange of information flows; 6) conduct a wide range of experimental research: modeling of elements of the airspace structure, capacity determination, development of new procedures for organization and air traffic services (ATS), and much more. Significant changes in the ATM system "pass" through their simulation on simulators.

Thanks to the correct strategy of air traffic control (ATC) training simulators application (a significant role is given to the instructor of practical training), it is possible to design a systematic approach to the development of professional competencies or their combinations required by the customer of training.

1.2 Motivation

Practical training on simulators of real equipment is an integral and essential part of aviation specialist training. At the stage of simulator training, theoretical knowledge is transformed into practical skills. At present, there are no specific requirements for the use of simulators in training ATCOs and the extent to which they reflect the real ATS system. Such requirements exist only for technical flight simulation devices, i.e., flight simulators.

This means that the establishment of criteria for the efficiency of simulator use at the stage of practical training of ATCOs is entrusted entirely to the authorized state authority for issuing certificates. The decision-makers will not rely on specific criteria, but on qualitative values of simulator applicability, e.g., "the simulator should provide an opportunity to apply previously acquired knowledge when practicing practical exercises". These facts urgently require scientific research to develop criteria for the applicability of air traffic control (ATC) simulators at different stages of practical training of ATCOs.

1.3 Contribution

This study contributes to the development of reasonable quantitative criteria for the applicability of simulators at different stages of practical training of air traffic controllers based on the integration of expert judgment methods and fuzzy logic methods. The integrated use of these methods will allow more flexible and adaptive processing of various data concerning the degree of simulation of ATS system elements, which will allow considering the uncertainty and fuzzy information in real simulators.

1.4 The Organization of the Paper

The paper consists of four main sections. The first section is an introduction. This section concentrates on the analysis of related works and problem statements. The second part of this paper is devoted to the continuation of the analysis of related works and the definition and evaluation of the criteria for the compliance of the information model of the air traffic controller training simulator with the real ATS system. This section describes the illustrative example of the building of an expert system for evaluating the criteria for compliance of the Information Model of the ATC simulator with the real system based on the task of the design of the professional competence "Vectoring". The third part is results and discussions. The fourth section is the conclusion.

1.5 Formulation of the Problem

The simulator is a synthetic means of training that provides a representation of important aspects of the real operational environment and reproduces operational conditions in which the person (student) undergoing training has the opportunity to directly practice tasks that are performed in real-time [7]. According to the requirements of ICAO, each training simulator that is intended to be used for training must be submitted to the relevant authorized body for the issuance of a certificate to determine the expediency of using the simulator for training purposes before starting its use. Regarding the duties of the body appointed by the state for personnel licensing, it is noted that this body is responsible for the permission for the usage of ATC simulators and their use to gain practical experience [8, 9]. That is, it applies exclusively to means of practical training of pilots.

The authority for issuing the certificate is responsible for establishing criteria for determining the expediency of using such simulators and their ability to ensure the desired result of training air traffic controllers. Currently, qualification assessment criteria are needed for simulators, both flight and air traffic control [9]. For this, the authority responsible for licensing must consider the tasks of practical training that are supposed to be performed using this simulator, the prescribed form of reporting, and the capabilities of the simulation tool to achieve the desired training result [8–10].

The purpose of the work is to solve the following problems:

- Analysis of the information model of the ATC simulator of the real ATS system.
- Development of criteria for compliance of the information model of the ATC simulator with the real ATS system.
- Development of an expert system for quantitative assessment of the criteria of compliance of the information model of the ATC simulator with the real system.

2 Determination and Evaluation of the Criteria for Compliance of the Information Model of the ATC Simulator with the Real System

2.1 Information Model of a Real Air Traffic Service System

Relevance of the simulator to the working environment in which the ATCOs work, to perform practical training tasks, is a fundamental concept in the development of the required professional competences. Figure 1 shows a block-scheme of Information Model of the ATC Simulator (IMATCS) that adequately addresses the working environment.

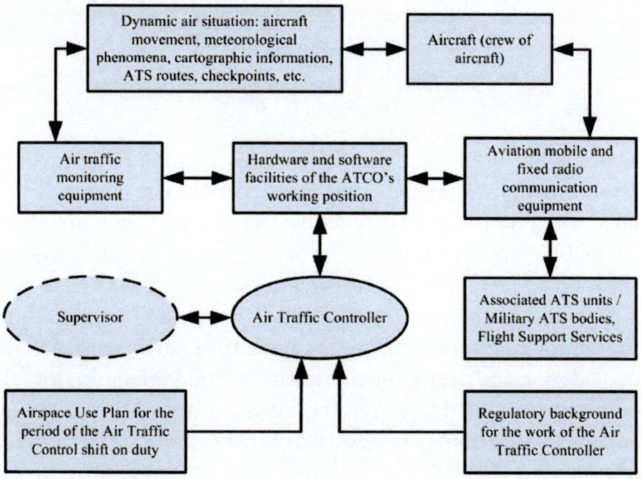

Fig. 1. Information model of the working position of the ATCO (the simplest outline of the ATS system).

The compliance consists of the following main criteria (requirements):

– adequacy of time-space modeling of the dynamic air situation;

- adequacy of cartographic and other static information;
- adequacy of procedures for coordination and interaction with “aircraft crews”, related units/ sectors and other flight support services.

The information model of the reproduction of the real ATS system is a more important component of the ATC simulator. The air traffic controller, based on available data, forms and makes decisions that ensure the safety, orderliness and efficiency of flights. The simplest contour of the ATS system can be called its element, which can perform the same functions and tasks as the entire system as a whole to provide air traffic service. If the information model, which is implemented at the ATC simulator, does not reflect the necessary realism needed to acquire relevant knowledge, skills, and abilities, according to the training goals, then it does not provide an opportunity to perceive the necessary information and form adequate management actions. Information flows in the IMATCS (Fig. 2) are:

- $\overrightarrow{X} = (x_1, \ldots, x_n)$ – a multidimensional vector that characterizes the real air traffic services system (dynamic air environment, operation of means of communication, navigation, surveillance, ATM processes, coordination, interaction, etc.);
- $\overrightarrow{Y} = (y_1, \ldots, y_n)$ – a multidimensional vector that characterizes simulator in terms of displaying the operation of a real ATS system with a certain level of realism;
- $\overrightarrow{U} = (u_1, \ldots, u_n)$ – a multidimensional vector that characterizes “obstacles”, i.e. those influencing factors that do not allow fully matching the processes that take place in the operational ATS system to be reproduced on the flighting simulator;
- $\overrightarrow{S}$ and $\overrightarrow{Q}$ – input (real system) and output (training system) for the expert system.

The influence of the “interference” vector on the reproduction of real ATM processes at the ATC simulator:

$$\overrightarrow{Y} = \overrightarrow{X} \pm \overrightarrow{U}. \tag{1}$$

“Disturbance” refers to the error of the simulator in the reproduction of air traffic service processes: tolerances for the conditional performance of technological operations by the air traffic controller, coordination, and interaction with adjacent air traffic control authority (flight support services), reproduction of the meteorological situation, dynamics of aircraft movement, etc. Information models, which are ATC simulators (with varying degrees of similarity to real ATS systems) are a more important element of the practical training of future ATCOs. The modern state of the hardware and software component, considering the development of information technologies, allows to bring ATC simulators to such a level that training on them becomes more effective than on real equipment, due to their ability to ensure a higher intensity of training, adding to this also economic attractiveness. Adequacy of perception is a generalized criterion in which the ATCO, subjectively for himself, perceives the training tool as his own professional environment.

In the process of simulating practical activities on the means for working out partial tasks the future air traffic controller makes decisions and develops the skills and abilities that will ensure him to perform his functional duties in compliance with the required level of flight safety. If the information model of the real ATM system and equipment does not adequately reflect the professional environment of the ATCO or outside the limits of permissible errors, then it cannot be used to practice the practical component of training. Quantitative evaluation of the criteria for the compliance of the information model of the ATC simulator with the real system was carried out with the help of an expert system (Fig. 3).

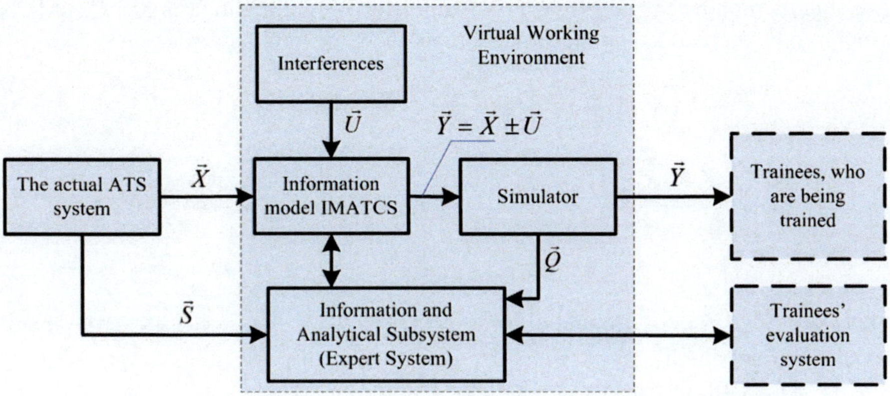

Fig. 2. The structural scheme of the simulation training system of ATCO.

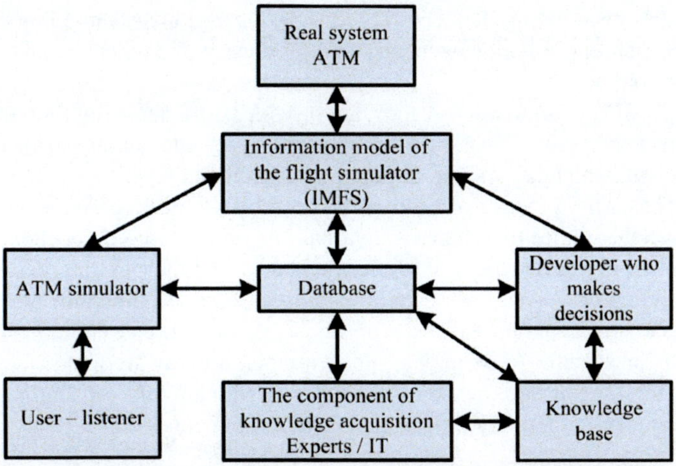

Fig. 3. Expert system of quantitative assessment of information model compliance criteria flighting simulator to a real ATM system.

2.2 A Simulator for Simulating a Certain Work Environment in Order to Work Out Certain Tasks of Practical Training

To obtain an adequate training tool for simulating a certain working environment in order to work out certain tasks of practical training, the problem of substantiating the accuracy of the simulation is important. This is due to the fact that certain processes that take place during ATS have different (in terms of impact on flight safety) significance. In addition, at the initial stages of the formation of basic practical skills and abilities (e.g., "vectoring" [11] of one aircraft without changes in the vertical flight profile), it is economically impractical to use ATC simulators with a high degree of reproduction of ATS processes (one hour of operation of such a simulator costs several thousand UAH). The authors propose the development of simulator and information model (IMATCS), which can be reproduced on the simulator and meet the training goals:

$$\vec{F} = F\left(\vec{X}, \vec{Y}, \vec{U}\right) \rightarrow min,\ \vec{X}^* \leq \vec{X} = (x_1, \ldots, x_n) \leq \vec{X}^{**}, \tag{2}$$

$$\vec{Y}^* \leq \vec{Y} = (y_1, \ldots, y_n) \leq \vec{Y}^{**},\ \vec{U}^* \leq \vec{U} = (u_1, \ldots, u_n) \leq \vec{U}^{**}, \tag{3}$$

$$x_i \geq 0, y_i \geq 0, u_i \geq 0, \tag{4}$$

where $\overrightarrow{X}, \overrightarrow{Y}, \overrightarrow{U}$ are information flows in IMATCS; $\overrightarrow{X}^*, \overrightarrow{Y}^*, \overrightarrow{U}^*$ are given quality indicators (lower limit); $\overrightarrow{X}^{**}, \overrightarrow{Y}^{**}, \overrightarrow{U}^{**}$ are given quality indicators (upper limit).

Based on [10], the means of simulator training of ATCOs can be characterized as follows:

1. A simulator with high-fidelity reproduction of the working environment (HI FI SIM – Hight-fidelity Simulator). It is a copy of the ATCOs workplace, including the hardware and corresponding software necessary to reproduce all the functions and tasks solved by this workplace.
2. Simulator (SIM - Simulator). A tool that provides a trainee with a representation of important features of a real situation and reproduces the conditions of the working environment, which allows you to practice tasks in real time.
3. Trainer for partial practice of training tasks (PTT - Part-Task Trainer). A simulator that allows the trainee to practice individual functional duties independently of other tasks, unlike in a real professional environment where they are interconnected.
4. Other training device (OTD - Other Training Device). A simulator that allows you to practice some operational tasks on copies of operational equipment and is not modified for specific ATM targets. For example, a hardware and software tool for working out the phraseology of radio exchange with aircraft, determining the space-time parameters of moving objects in a 3D environment, etc.).

There are five levels of taxonomy to complete training tasks and obtain the necessary knowledge and skills in the field of training ATCOs [6, 12, 13].

Taxonomy levels 1 and 2 require the ability to remember and reproduce data and the ability to demonstrate actions to specific situations and circumstances, respectively. For example, you need to practice aviation phraseology and the rules of radio exchange. It makes no sense to use the estimated ATC simulator itself for this level of taxonomy with a high degree of similarity to the real ATM system, to involve pseudo-pilots and ATCO instructors. It is enough to use a simulator and a computer base with a radiotelephone headset, which will allow you to practice: a) radio communication phraseology helps to remember and repeat certain sentences; b) communication skills: master radio communication skills, including clarity, accuracy, and evenness (lack of accents) of speech, eliminating "fear of the microphone on"; c) understanding the rules and procedures of coordination and interaction with the aircraft (ACFT), and related parties.

Tasks at taxonomy level 3 for the discipline "Air Traffic Management" of the basic and rating training stages of ATCO are practical tasks and require the integration of several areas of knowledge and skills. For example, such an element of simulator training as aircraft "vectoring" requires knowledge and skills in the field of aviation radiotelephone communication and phraseology, flight technical characteristics of aircraft, air navigation, and radar theory. The goals of taxonomy level 3 should be achieved with the help of a simulator for partial practice of training tasks or SIM - Simulator [6]. "Vectoring" – providing navigational guidance of aircraft in the form of providing a determined course based on the use of ATS surveillance systems [14].

Taxonomy level 4 requires the ability to determine the correct sequence of actions to the problematic situation that has arisen and to choose an adequate method of solving it. This ability is characterized by the ability to act correctly in a regular situation. The educational goals of taxonomy level 4 are achieved mostly using SIM - Simulator. Part-Task Trainer can be used in certain operational situations.

Taxonomy level 5 requires from the ATCO to demonstrate the ability to analyze a new situation (the situation does not have previously known algorithms for its solution) to develop and apply an appropriate management strategy to solve a complex problem. The defining feature of the fifth level of taxonomy is that the situation is significantly different from those previously encountered, there is no solution algorithm. The air traffic controller is required to make a very accurate assessment of events to be sure that the decision made by him in this situation is possible and optimal. The goals of taxonomy level 5 are achieved only using the SIM - Simulator. Level 5 of the taxonomy is related to the cognitive processes of a human operator: analysis, evaluation, development of an innovative solution, and its implementation with full control of the event in real time. The means of practical training for achieving taxonomy level 5 is aimed at the implementation of complex scenarios, non-standard situations, development of new approaches to organization and ATM. To practice the objectives of taxonomy level 5, it is necessary to use a simulator with HI-FI SIM. This will involve significant human, financial, and material resources.

The concept of using synthetic training tools is proposed for implementation: "a certain training tool for simulating a certain working environment to work out certain tasks of practical training". The difference between reality and simulation is that reality is a whole, and simulation is only a part of the reflection of the real world. Imitation of reality using modeling is approximate [15]. To what extent the simulation of reality is approximate is determined by modeling a certain work environment and evaluating the performance of certain tasks of practical training.

2.3 Simulation of a Certain Working Environment in Order to Work Out Certain Tasks of Practical Training

For the practical purposes of the educational institution of the initial training of ATCOs, the teaching effectiveness (as a whole, without breaking down into separate subsystems and simulated elements) of the information model of the simulator training tool (IM_{TR}) is proposed to evaluate the realism of the reproduction of the information model of the operating ATC system when practicing training tasks. The assessment of the adequacy of the synthetic training device for the training task is carried out by determining the value of the permissible discrepancy $Q_{app.prob}$ between IM_{TR} and (IM_{ATC}). Comparison of $Q_{app.dist}$ with a given limit value (norm) Q_{norm} will allow to decision making about the possibility of using this tool at a certain stage of practical training.

Each IM simulator consists of a significant number of subsystems (elements). Denoting the set of subsystems simulated by IM_{TR} by M_{TR}, and the set of subsystems of the information model of the current ATC system by M_{ATC}, the conditions of similarity of IM_{TR} can be written as follows:

$$M_{TR} - M_{ATC} \leq Q_{app.prob}. \tag{5}$$

In the ideal case, we have:

$$M_{TR} - M_{ATC} = 0. \tag{6}$$

The value of the permissible discrepancy is within the limits of:

$$0 \leq Q_{app.prob} \leq 1. \tag{7}$$

For consideration, based on [9], the following gradation regarding the accuracy of reproduction of the professional environment of the air traffic controller is presented:

- $0.01 \leq Q_{app.prob} \leq 0.1$ – S – Specific (high accuracy of reproduction of the working environment of the ATCO);
- $0.1 \leq Q_{app.prob} \leq 0.2$ – R – Representative (typical accuracy of reproduction of the working environment of the ATCO);
- $0.2 \leq Q_{app.prob} \leq 0.3$ – G – Generic (basic accuracy of reproduction of the working environment of the ATCO);
- $Q_{app.prob} \geq 0.3$ – N – Not Applicable (requirements for the accuracy of reproduction of the working environment of the ATCO are not put forward).

To determine the degree of adequacy (similarity) of the ATC simulator to the real ATS system ("high accuracy", "typical accuracy", "basic accuracy", "insignificant accuracy"), the following methods are proposed for consideration:

- functional conformity criterion: the ability of the ATC simulator to accurately and fully reproduce the functions and capabilities of the real ATS system is evaluated;
- detailed analysis of the simulator and the real ATS system to determine the degree of similarity and reproduction of the main operating characteristics required for a certain stage of practical training;
- expert assessment: a group opinion of experts familiar with the real ATS system is determined to assess the simulator's compliance with the selected indicators;
- comparative testing: "run-through" of test practical exercises is carried out on the simulator and their comparison with similar situations in the real ATS system;

- the criterion of informational similarity of the simulator: it is assessed how much the informational content of the situation "played out" on the simulator reflects typical, emergency, and non-standard scenarios that have (may have) a place in a real ATS system.

Each method has its advantages and disadvantages, and the choice of a particular method depends on the purpose of the assessment. At the same time, it should be considered that a combination of methods can give the most reliable result. The following methods are proposed for evaluating the criteria for the adequacy of the ATC simulator and generalizing the criteria: Expert Judgment Method (EJM), methods of fuzzy logic, and multiplicative aggregation of indicators [16]. The assessment of experts is determined regarding the correspondence of the perception of processes and competencies that must be acquired by the trainee, the quality of training, and the ability of the training device to contribute to the formation of the professional competencies required for this stage of training (this training exercise).

Quantitative assessment of the accuracy levels of the ATC simulator - the adequacy of the ATC simulator to the real ATS system using the EJM. This is one of the methods for solving multi-criteria decisions or determining the quantitative values of quality indicators ("high accuracy"; "typical accuracy"; "basic accuracy"; "low accuracy").

Algorithm for determining the required adequacy of reproduction of the professional environment of the ATCO according to the "accuracy" criterion using EJM consists of five steps:

1. Questioners for experts. Experienced 31 experts (work experience of 5 years or more) and ATCO-instructors were involved. Matrix of individual preferences is determined for opinion of the experts and their systems of individual preferences, R_i.
2. Matrix of group preferences R_{ij} is used to determine the experts' group opinion R_{grj}: $R_{grj} = \frac{\sum_{i=1}^{m} R_i}{m}$, where m is number of experts.
3. The coordination of experts' opinions. It is necessary to calculate variance for each quality indicator – levels of accuracy: $D_i = \frac{\sum_{i=1}^{m} (R_{grj}-R_i)^2}{m-1}$; standard deviation: $\sigma_j = \sqrt{D_j}$; coefficient of the variation for each quality indicators: $v_j = \frac{\sigma_j}{R_{grj}}$. If the coefficient of variation is $v_j < 0.33$, the opinion of the experts is coordinated. Otherwise, the opinions of the experts are not coordinated. In this case, it is necessary to calculate Kendal's coefficient of concordance (for number of indicators n, common variance S, and numbers of same ranks T_i) to provide an interrogation of the experts again:

$$W = \frac{12S}{m^2(n^3 - n) - m\sum_{j=1}^{m} T_j}. \tag{8}$$

4. The significance of the coefficient W using χ^2 criterion for known table value $\chi_t{}^2$:

$$\chi^2 = \frac{S}{\frac{1}{2}m(n+1) - \frac{1}{12(n-1)}\sum_{j=1}^{m} T_j} > \chi_t{}^2. \tag{9}$$

5. Accuracy of the ATC simulator according to adequacy criteria (weight coefficients w_j) for known medium mark of experts' indicators R_{ij}:

$$w_j = \frac{C_j}{\sum_{j=1}^{n} C_j}; \tag{10}$$

$$C_j = 1 - \frac{Rij - 1}{n}. \tag{11}$$

The results of the assessment are presented in Table 2.

Table 2. Results of assessment of the accuracy of the ATC simulator using EJM.

Accuracy	High accuracy	Typical accuracy	Basic accuracy	Low accuracy
Experts	S – Specific	R – Representative	G – Generic	N – Not Applicable
Expert 1	1	2	3	49
…	…	…	…	…
Expert 31	1	3	3.5	3.5
R_{grj}	1.21	2.16	2.89	3.79
D_j	0.15	0.21	0.16	0.15
$\sigma_j = \sqrt{D_j}$	0.38	0.45	0.40	0.38
v_j, %	31.61	21.02	13.93	10.09
C	0.95	0.71	0.53	0.30
w_j	0.38	0.29	0.21	0.12

During training on the simulator, professional competencies should be formed in stages: "from simple to complex". As can be seen, the results of the accuracy assessment of the ATC simulator next: high accuracy (0.38); typical accuracy (0.29); basic accuracy (0.21); low accuracy (0.12). The highest quantitative assessment, and consequently greater similarity to the real professional environment of air traffic controllers, corresponds to simulator S – Specific.

The first stage is preparation for training on the simulator (briefing-preparation). Stage taxonomy level is 1, 2, training device (OTD – Other Training Device). Trainees learn:

- the working environment in which they will work: the control area, the terminal control area, the control area, ATS routes, control points, air traffic take-over points, the flight plan for the training period: departing, arriving, transiting air traffic, aircraft types, the presence of prohibitions and restrictions on the use of airspace, etc.;
- ergonomics of the workplace: interaction procedures and means of communication with related ATS sectors, flight support services;
- other information relevant to the process of performing the training exercise.

At the second stage the trainee performs certain actions with controls and simple technological operations. The taxonomy level of the stage is 2, 3. The purpose of the stage is the formation of the trainee's initial skills in relation to air traffic control, for example, "vectoring" of one departing, arriving, or transiting aircraft in simple meteorological conditions and the absence of potentially conflicting situations (at least at the beginning of the stage). Specific professional competencies are practiced, and a simulator is used for partial practice of training tasks and SIM - Simulator, for example, competence: "coordination" is described as "coordinating the interaction between operational personnel and other interested parties" [17, 21]. An element (indicator) of the ATCO's behavior that he (she) must demonstrate:

- Identifies the need for coordination and coordinates in a timely manner with personnel at other workplaces and other stakeholders;
- Selects the method of coordination taking into account the circumstances, including the urgent nature of coordination, the state of facilities, and established procedures;
- Coordinates: movement, control of transfer of control, and modification of previously coordinated flight data using established coordination procedures; changes in the state of operational means such as equipment, systems, and functions; changes in the status of airspace and airfield resources;
- Uses: clear and concise terminology for verbal coordination; standard formats and protocols of ATM messages for non-verbal coordination;
- Conducts effective briefing during shift transfer.

The third stage of practical training. The taxonomy level of stage is 3, 4. It is characterized by the complex use of almost all professional competencies and all available technical means of the workplace. The key competencies of the air traffic controller are presented in Table 3 [17].

The work of the air traffic control team is practically simulated, with all necessary communications and cooperation. The realism of flight and technical characteristics of air traffic and meteorological conditions is high. It is recommended to use HI-FI SIM and SIM - Simulator. It should be noted that the high accuracy of reproduction of the working environment is not a criterion for higher quality practical training. It is necessary to compare the degree of compliance of the ATC simulator with the training goals. Individual components of the information model of the ATCO's working environment can significantly increase the cost of training without increasing its effectiveness, from the point of view of acquiring the relevant professional competencies.

Assessment of suitability/compliance of the ATC simulator for training purposes using fuzzy-logic methods.

Algorithm for determining the suitability/compliance of the ATC simulator for training purposes using "fuzzy logic" methods ("low suitability", "medium suitability", "most suitable") consists of six steps [16]:

1. Preparation for surveying experts. The maximum possible value of suitability is taken as equal to 100 units. The interval [0; 100] is divided into 10 segments, for each of which the number of experts who used a certain value of a linguistic variable to express their opinion about the compliance of the ATC simulator with the training goals was counted.

Table 3. Key competencies of the air traffic controller.

Competencies	Description of competence	Code
Awareness of the situation	Understand the current operational situation and predict the course of events	a_1
ATM and throughput	Ensure safe, orderly, and efficient air traffic and provide important information regarding the surrounding environment and potentially dangerous situations	a_2
Separation and resolution of conflict situations	Resolution of potential conflict situations in air traffic and support (approx. established standards) echeloning	a_3
Communication	Ensure effective communication in all operational situations	a_4
Coordination	Coordinate interactions between operational staff and other stakeholders	a_5
Management of freelance situations	Identify and respond to emergency situations in air traffic and ensure the operation of the air traffic control system in conditions of degradation (adv. deterioration of its operational capabilities)	a_6
Solving problems and decision-making	Finding and implementing solutions for identified threats and associated undesirable conditions	a_7
Self-management and continuous development	Demonstrate personal qualities that improve production activity and support active participation in self-education and self-development	a_8
Workload management	Use available resources, set priorities, and complete tasks efficiently and in a timely manner	a_9
Teamworking	Act as a team member, collaborating with others to achieve a common goal	a_{10}

2. Formation of the matrix with data obtained after conducting the first round of questioning of experts. Experienced ATCO and ATCO-instructors were involved as experts.
3. Formation of the first matrix for processing the received data and determining k_{max} for known results b_{ij} of the survey of experts (for example, in Table 5) and number m of results in column/row

$$k_j = \sum\nolimits_{i=1}^{m} b_{ij}. \quad (12)$$

4. Formation of second matrix and definition c_{imax}:

$$c_{ij} = \frac{b_{ij} k_{max}}{k_j}. \tag{13}$$

5. Construction of membership functions (MF) μ according to the formula:

$$\mu_{ij} = \frac{c_{ij}}{c_{imax}}. \tag{14}$$

6. The results obtained after the first round of examination are entered into the second matrix.

After conducting the second round of the experiment, the degree of agreement between the experts is compared according to the Kendall coefficient. Since quantitative values of variables are needed to formalize the choice of a training room, the qualitative indicators of the level of efficiency in the case of using a training tool and its characteristics regarding the "accuracy of the reproduction of the professional environment of the air traffic controller" are as follows:

- low suitability, application is not mandatory, 10 units;
- average suitability, low efficiency, 30 units;
- the most suitable, the efficiency is significant, 55 units.

Figure 4 illustrates that the highest qualitative indicators of the simulator's effectiveness in terms of "accuracy of reproducing the professional environment of an air traffic controller" correspond to the quantitative indicator of 55 units. The lowest qualitative indicators of the simulator correspond to the quantitative indicator of 10 units.

The application of membership functions in the conditions of unclear information allows to formalization of the qualitative characteristics of the level of suitability for a specific situation. Based on [10], Table 4 shows gradation of the use of training tools according to the characteristic that corresponds to the accuracy of the reproduction of the professional environment of the ATCO, depending on the type of practical activity.

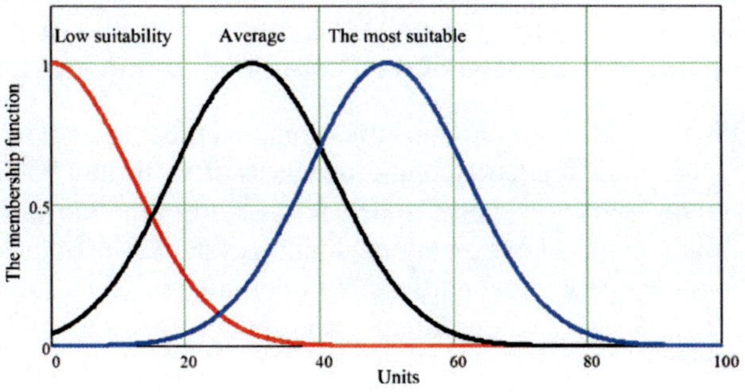

Fig. 4. The membership functions μ for determining the accuracy of the reproduction of the professional environment of the ATCO.

Table 4. Use of training equipment depending on the type of practical activity.

The type of practical activity	Means of training and its characteristics in relation to the accuracy of the reproduction of the professional environment of the air traffic controller/Taxonomy level			
Acquisition	OTD – Not Applicable Acquisition: knowledge/skills	PTT - Generic Acquisition: skills	SIM - Representative Acquisition: skills	HI-FI SIM - Specific Acquisition: skills
Taxonomy level	1, 2	2, 3	3, 4	4, 5
Acquisition of knowledge	The most suitable	Application is optional	Application is optional	Application is optional
Working out partial tasks	Efficiency insignificant (30 un.)	The most suitable (55 un.)	The most suitable (55 un.)	Application is optional (10 un.)
Individual training	Efficiency insignificant (30 un.)	Efficiency Insignificant, (30 un.)	The most suitable (55 un.)	The most suitable (55 un.)
Training as part of a group	Efficiency Insignificant	Efficiency Insignificant	The most suitable	The most suitable
Training as part of the team	Efficiency insignificant (30 un.)	Efficiency Insignificant (30 un.)	The most suitable (55 un.)	The most suitable (55 un.)

To obtain a quantitative value of the level of suitability/application of the simulator training tool, the apparatus of the theory of fuzzy logic was used, and information about the possibility of using the simulators was obtained from experts and interpreted quantitatively.

2.4 The Expert System for Evaluating the Criteria for Compliance of the Information Model of the ATC Simulator with the Real System Based on the Example of the Design of the Professional Competence "Vectoring"

"Vectoring" (guidance) is an important professional element of activity from the point of view of ensuring safety, regularity, and orderliness of air traffic. Each aircraft has its own "flight program", which is approved before the start of the movement, in the form of a submitted "flight plan". However, air navigation system is a complex "mechanism" with rapidly changing operating conditions: non-compliance (for many reasons) by air traffic participants with the declared flight plan, the development, and impact of dangerous meteorological phenomena, the introduction of bans and restrictions on the use of air space, the occurrence of emergency and unforeseen circumstances both in the air and on "earth". "Vectoring" allows the ATCO to navigate the aircraft along a certain time-space trajectory in the airspace, to adjust their route of movement outside of dangerous zones and obstacles, thereby preventing the emergence or further development of potentially dangerous situations, preventing their transformation into a catastrophic one. To effectively perform "Vectoring", the ATCO must demonstrate a wide range of complex applications of knowledge and skills, the main of which are:

- f_1 is knowledge of the structure of the airspace and flight rules: understanding of the elements of the airspace (control areas, control zones, prohibited, flight restriction zones, the structure of air traffic control routes by geographical distribution and height, the plan for the use of controlled air space and the air space of adjacent air traffic control units);
- f_2 is current and prospective aeronautical information, for the period of its activity, regarding the management of the existing airspace structure;
- f_3 is knowledge of the rules and the ability to use radio communication with all participants in air traffic;
- f_4 is planning, coordination, and interaction skills: the ability to quickly plan (replan) air traffic both along the air traffic control routes and beyond them, taking into account priorities in the use of airspace, meteorological conditions, and other restrictions, for example, from the military air traffic control units;
- f_5 is air traffic service: the skills of "guidance" of aircraft taking into account their flight technical characteristics and the capabilities of the crews, restrictions on the part of the air navigation infrastructure in the area of responsibility;
- f_6 is decision-making: analysis of the situation, assessment, decision-making in conditions of risk and uncertainty, implementation of management actions, monitoring of the development of events, and correction if necessary.

Theoretical knowledge of air traffic controllers is tested by demonstrating it at taxonomy levels one and two.

According to the training tasks, at the briefing stage, the ATCO-instructor checks the readiness of the trainee for practical activities. The pre-exercise evaluation system was considered using artificial neural networks [18, 19]. The authors propose to apply as one of the elements of the knowledge test at the briefing preparation, providing the trainee with a definition of the meaning of "Vectoring" at the first level of the taxonomy (level 1 of the taxonomy requires the availability of general knowledge and the ability to remember and restore information and data). According to [12] for the first level of taxonomy, the following actions are used: define, draw, list, name, cite, recognize, formulate. Table 5 shows the results of evaluating the significance of actions for taxonomy level 1 for the task "Vectoring", obtained using fuzzy logic methods (fuzzy logic):

1. Determine. Define the concept and describe the limits of its application.
2. Draw. Complete a picture, draw.
3. Count. Talk one after the other.
4. Call. Give the name of objects or procedures.
5. Quote. Repeat what was said or written.
6. Recognize Perceive, and distinguish something because one has seen it before, show understanding.
7. Formulate. Teach in a formal or prescribed manner.

The appropriateness functions for determining the significance of actions in the case of the description of the "Vectoring" task (for taxonomy level 1) is shown in Fig. 5.

Table 5. Determining the significance of the actions for the description of the task "Vectoring" (taxonomy level 1).

N	Determination of the most appropriate meaning of the verb of taxonomy level 1 for the description of "Vectoring"									
	Interval, units									
	0–0.1	0.1–0.2	0.2–0.3	0.3–0.4	0.4–0.5	0.5–0.6	0.6–0.7	0.7–0.8	0.8–0.9	0.9–1
1	1.00	0.67	0.20	0.05	0.00	0.00	0.00	0.00	0.00	0.00
2	0.00	0.42	1.00	0.72	0.07	0.00	0.00	0.00	0.00	0.00
3	0.00	0.00	0.00	0.37	0.94	1.00	0.67	0.00	0.00	0.00
4	0.00	0.00	0.00	0.00	0.00	0.00	0.87	1.00	0.53	0.00
5	0.00	0.00	0.00	0.00	0.00	0.00	0.54	0.86	1.00	0.00
6	0.00	0.00	0.00	0.00	0.00	0.00	0.00	0.72	1.00	0.26
7	0.00	0.00	0.00	0.00	0.00	0.00	0.00	0.00	0.14	1.00
Sum	1.00	0.67	0.20	0.05	0.00	0.00	0.00	0.00	0.00	0.00

Figure 5 clearly shows the results of assessing the importance of action verbs for experts at taxonomy level 1 for the task "Vectoring", obtained using fuzzy logic methods. A similar approach can be applied to other levels of taxonomy, depending on what competencies are required from an air traffic controller. Evaluation of practical activity on the ATC simulator is a separate, rather complex and responsible task [20].

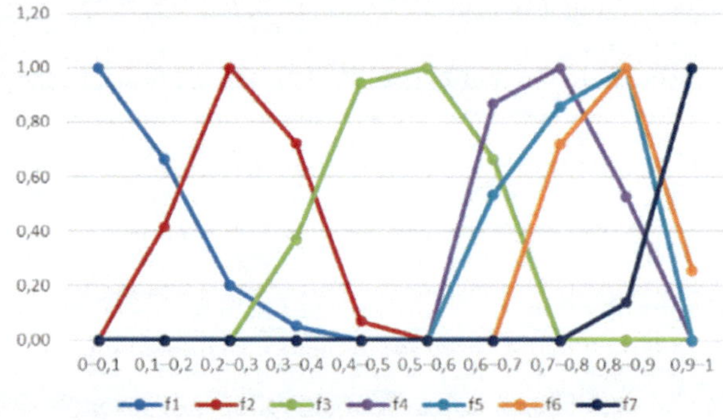

Fig. 5. The membership functions μ for determining the significance of actions in the case of the description of the "Vectoring" task (taxonomy level 1).

Multiplicative aggregation of activity results is used for comprehensive evaluation of the results of practical training at the ATC simulator. The results of the assessment of the professional competence "Vectoring", the comparison of complex assessments in the case of the application of additive and multiplicative aggregation are shown in Table 6. Two students take part in the experiment: student *A* who did not miss classes and student *B* who missed classes.

Table 6. An example of obtaining a comprehensive assessment of the professional competence "Vectoring": additive and multiplicative aggregation of component assessments.

Constituents task "vectoring"	Estimation of the students						Additive aggregation		Multiplicative aggregation	
	R	R_{av}	C	w	A	B	$W(A)$	$W(B)$	$W(A)$	$W(B)$
f_1	5.6	5.5	0.25	0.07	5	5	0.36	0.36	1.12	1.12
f_2	4	4	0.5	0.14	4	0	0.57	0.00	1.22	0.00
f_3	5.6	5.5	0.25	0.07	4	4	0.29	0.29	1.10	1.10
f_4	2.3	2.5	0.75	0.21	5	5	1.07	1.07	1.41	1.41
f_5	2.3	2.5	0.75	0.21	3	3	0.64	0.64	1.27	1.27
f_6	1	1	1	0.29	4	4	1.14	1.14	1.49	1.49
Sum/Aggregation	–	–	3.5	1	25	21	4.07	3.50	4.01	0.00

The results of the comparison of comprehensive assessments received by student B (additive and multiplicative aggregation of the comprehensive assessment), mark for option f_2 "Draw. Complete a picture" – 0 ("absent"). Comprehensive assessment of the B student's performance in the "Vectoring" practical task (additive aggregation):

$$W_{ad} = \sum\nolimits_{i=0}^{n} w_i f_i = 0.07 \cdot 5 + 0.14 \cdot 0 + 0.07 \cdot 4 + 0.21 \cdot 5 + 0.21 \cdot 3 + 0.29 \cdot 4 = 3.5.$$

Comprehensive evaluation of the B student's performance in the task of "Vectoring" (multiplicative aggregation):

$$W_m = \prod\nolimits_{i=1}^{n} f_i^{w_i} = 5^{0.07} \cdot 0^{0.14} \cdot 4^{0.07} \cdot 5^{0.21} \cdot 3^{0.21} \cdot 4^{0.29} = 0.$$

When training aviation operators, especially at the stage of simulator training, multiplicative aggregation of practical results should be used, because it does not assume mutual compensation due to positive (higher) results on other evaluation elements, as is the case with additive evaluation.

3 Results and Discussions

To obtain quantitative indicators of the complexity of ATC simulators, various methods of expert evaluation are used: the Expert Judgment method and the Fuzzy-Logic method for different tasks. For general quantitative assessment, the Expert Judgment method was used to solve the following problems:

- accuracy levels indicators in ATC simulator ("high accuracy"; "typical accuracy"; "basic accuracy"; "low accuracy");
- suitability of the ATC simulator for training ("low suitability", "medium suitability", "most suitable");
- the efficiency of actions in the case of the description of the task "Vectoring" (determine, draw, count, call, quote, recognize, formulate).

Table 7 shows the results of comparing two methods for simulators, with different degrees of similarity, in the real professional environment of air traffic controllers.

Table 7. An Expert Method and method of Fuzzy Logic.

Estimation	Accuracy				Suitability		
ATC simulator	Accuracy levels in ATC simulator (EJM)				Suitability of the ATC simulator (Fuzzy-Logic)		
	high	typical	basic	low	low	medium	most
	S	R	G	N	application is not mandatory	low efficiency	the efficiency is significant
	0.38	0.29	0.21	0.12	$\leq$10 units	$\leq$30 units	$\geq$55 units
ATCs actions	Efficiency of ATCs actions (Fuzzi-Logic)						
	determine	draw	count	call	quote	recognize	formulate
	0.11	0.31	0.55	0.75	0.91	0.93	0.98

An illustrative example shows how you can choose an ATC simulator according to the level of competence of the trainee (student) and get a comprehensive assessment of the trainee's readiness (practical air traffic control operations).

4 Conclusions

The key competencies of air traffic controllers, which must be practiced at the stages of simulator training and their gradation depending on the required level of taxonomy, are given. The proposed concept for implementation at the stage of application of synthetic training devices: "a certain training tool for simulating a certain working environment in order to work out specific tasks of practical training".

The gradation of ATC simulators, characteristics in terms of accuracy of reproduction of air traffic controller's professional environment, depending on the type of practical actions training, from acquisition of knowledge (skills) to full size training as part of a team (air traffic control shift) is developed.

In order to evaluate the criteria for the adequacy of the ATC simulator to the real ATS system and to generalize these criteria, expert evaluation methods and fuzzy-logic methods are proposed for use, which allow obtaining reliable results in the absence of accurate initial data and taking into account the uncertainty of information.

To evaluate the results of the practical activity of aviation operators, it is proposed to use multiplicative aggregation of the results of simulation training, which allows us to conclude with a guarantee that the operator has mastered the necessary knowledge and skills and acquired the necessary competencies, according to the training tasks.

The application of approaches based on the specified accuracy of the reproduction of the professional environment, depending on the type of practice of the practical activity of the ATCO, allows for the formation of a more scientifically and economically justified task for the development of appropriate means of support, taking into account the specified conditions.

It is suggested that the described approaches to establishing criteria for the definition and use of ATC simulators be applied in the development of national technical standards for the evaluation of simulators for training ATCO.

It is proposed to apply the applied approaches to determining taxonomy levels for a range of other competencies that are chosen for the content of the required level of practical training both for the training of ATCOs themselves and for other aviation specialists when forming educational and methodological packages.

In the future, the development of an adaptive automated expert system for managing the training process is planned [19].

References

1. ICAO Doc 9750-AN/963. Global Air Navigation Plan, 2016–2030, Fifth Edition: ICAO, Montreal, Canada (2018)
2. ICAO Study Reveals Strong Demand for Qualified Aviation Personnel up to 2030. https://www.icao.int/Newsroom/Pages/icao-study-reveals-strong-demand-for-qualified-aviation-personnel-up-to-2030.aspx. Accessed 17 Jan 2023
3. Europe needs 6,000 new pilots a year for next two decades, warns Boeing. https://www.telegraph.co.uk/business/2022/07/27/europe-needs-6000-new-pilots-year-next-two-decades-warnsboeng/#:~:text=Airlines%20in%20Europe%20must%20hire,both%20North%20America%20and%20China. Accessed 17 Jan 2023
4. About NGAP. https://www.icao.int/safety/ngap/pages/ngapinitiatives2.aspx: Author, ICAO, Montreal, Canada. Accessed 17 Jan 2023
5. Kolotusha, V., Bondarev, D.: Adequacy of the professional environment at the stage of simulation training of air traffic controllers. In: All-Ukrainian Scientific and Technical Conference on Sustainable development of Communication, Navigation, Surveillance and Air Traffic Management systems, pp. 126–130. NAU, Kyiv, Ukraine (2023). https://er.nau.edu.ua/handle/NAU/61947. Accessed 17 Jan 2024. (in Ukrainian)
6. EUROCONTROL Specification for the ATCO Common Core Content Initial Training. Edition 2.0, Edition date: 02/04/2015, Author, EUROCONTROL, Brussels, Belgium (2015)
7. Aviation Regulations of Ukraine: Technical Requirements and Administrative Procedures for Issuing Certificates and Certificates to Air Traffic Controllers. Order of the State Aviation Administration of Ukraine of 31.05.2019, No. 391. (in Ukrainian)
8. ICAO Doc 9841-AN/456. Manual on the Approval of Training Organizations, Third Edition: Author, ICAO, Montreal, Canada (2018)
9. ICAO Doc 9625. Manual of Criteria for the Qualification of ATC Simulation Training Devices, Volume I – Aeroplanes, Fourth Edition: ICAO, Montreal, Canada (2015)
10. Simulation Facilities for Air Traffic Control Training/Eurocontrol. Edition 1.0, Edition Date: 15/03/2000, Author, EUROCONTROL, Brussels, Belgium (2000)
11. ICAO Doc 4444. Procedures for Air Navigation Services-Air Traffic Management, Sixteenth Edition: ICAO, Montreal, Canada (2016)
12. Acceptable Means of Compliance (AMC) and Guidance Material (GM) explaining the provision of the Aviation Regulations of Ukraine. Technical Requirements and Administrative Procedures for Issuing Certificates and Certificates of Air Traffic Controllers. Order of the State Aviation Administration of Ukraine of 23.03.2018, No. 391. (in Ukrainian)

13. Acceptable Means of Compliance (AMC) and Guidance Material (GM) to Part ATCO Requirements for the licensing of air traffic controllers. Annex 1 to ED Decision2015/010/R, EASA. Issue 1, 13 March 2015, Author, EASA, Cologne, Germany (2015)
14. Aviation Regulations of Ukraine. Air Traffic Management. Order of the State Aviation Administration of Ukraine of 09.12.2022, No. 1920. (in Ukrainian)
15. Green, P.: The Importance of Simulator-based Training and its Role in ATC. https://www.skyradar.com/blog/importance-of-simulator-training-and-its-implications-in-atc. Accessed 17 Jan 2023
16. Research anthology on decision support systems and decision management in healthcare, Business, and engineering research anthology on decision support systems and decision management in healthcare, business, and engineering. Information Resources Management Association. IGI-Global, Pennsylvania, USA (2021). https://doi.org/10.4018/978-1-7998-9023-2
17. ICAO Doc 9868. Training, Third Edition: Author, ICAO, Montreal, Canada (2020)
18. Shmelova, T., Sikirda, Yu., Jafarzade, T.O.: Artificial neural network for pre-simulation training of air traffic controller. In: Cases on Modern Computer Systems in Aviation, pp. 20–58. IGI-Global, Pennsylvania, USA (2019)
19. Kaddoura, S.: Handbook of research on AI methods and applications in computer engineering. IGI-Global, Pennsylvania, USA (2023). https://doi.org/10.4018/978-1-6684-6937-8
20. Assessment of ATCO Students/Trainees. https://skybrary.aero/articles/assessment-atco-studentstrainees. Accessed 17 Jan 2023
21. Broumi, S.: Handbook of research on advances and applications of fuzzy sets and logic. IGI-Global, Pennsylvania, USA (2022). https://doi.org/10.4018/978-1-7998-7979-4

Using Time Series for Biomedical Signal Processing Under Uncertainties

Mikle Burichenko, Olga Ivanets, Maryna Arkhyrei(✉), Rimvidas Khrashchevskyi, and Oleg Melnykov

National Aviation University, Liubomyra Huzara Avenue, 1, Kyiv 03058, Ukraine
marina_arkhyrei@ukr.net

Abstract. The paper presents analysis of time series usage for the evaluation of biological systems' functioning with a specific emphasis on applications in aviation and human factors. Analysis of the functional state of aviation operators is an important component of aviation safety. The paper concentrates on the visualization of the functional activity and dynamics of the organism as a biological system based on time data. Through the employment of recurrent plots, the identification of properties and mechanisms inherent to the biological system generating these series is facilitated. Changes in the state of the biological system are discernible by analyzing the visual structures of these plots, both topological and structural, thereby diminishing the informational uncertainty regarding the state of the biological object and supplementing it with an informative component. The nature of processes occurring within the biological system is indicated by the appearance of the recurrent plots, revealing the presence and impact of noise, states' repetition and stagnation (laminarity), and the occurrence of abrupt state changes (extreme events). The stability of entire biological system functioning is assessable through the variability of heart rate, employing retrospective data for forecasting purposes. The necessity for intelligent analysis of biomedical data, particularly heart rate variability, in aviation contexts, is underscored by this analysis. The results of this study, employing multi-axis recurrence methods for the analysis of signals (time series) from the human cardiovascular system, are elucidated, providing crucial insights into human factors in aviation safety and performance.

Keywords: Time Series · Recurrence Plots · Recurrence Quantitative Analysis · Data Processing · Electrocardiography · RR-intervals · Time Series Analysis · Dynamic Systems

1 Introduction

1.1 Introduction to the Problem

Nonlinear processes in biological systems are inherently influenced by continual external destabilizing factors, coupled with the biological system's intrinsic feedback mechanisms to maintain stability. Biomedical signals are formed as a result of the complex

I. Ostroumov and M. Zaliskyi (Eds.): ACASD 2024, LNNS 992, pp. 385–394, 2024.
https://doi.org/10.1007/978-3-031-60196-5_28

mutual influence of all subsystems of the organism on each other. This leads to the formation of signals, the analysis of which must be carried out taking into account their nonlinearity. At the same time, the reaction of the biological system to the constant influence of external factors can contain a valuable informational component for assessing the adaptive reserves of the organism at a given time and forecasting the functional state under external loads.

The role of the cardiovascular system in assessing the functional state of the whole organism has been well-documented in numerous studies [1, 2]. Accordingly, this research employs biomedical signals from electrocardiograms and the time interval between the highest points of the electrocardiogram (RR intervals), which characterizes the constancy of heart contractions and makes it possible to calculate their frequency and determine heart rate variability. Heart rate variability is a frequent subject of study for analyzing the organism's functional stability of the organism [1] and can be used in the formation of phase trajectories in the m-dimensional phase space, where each state corresponds to a certain point in the phase space.

In this study, heart rate variability is represented as a time series, projected into a multidimensional space using recurrent plots. This methodology enables the visualization of the dynamic changes in heart rate variability. Furthermore, the application of Recurrence Quantification Analysis (RQA) allows for the derivation of quantitative metrics that substantiate conclusions drawn from visual assessments. Such an approach is particularly relevant in aviation contexts, where understanding human factors, especially the capacity to withstand and adapt to varying environmental and operational stressors, is crucial for ensuring safety and optimal performance.

1.2 Motivation

Physiological (biological) systems are non-linear dynamic systems that constantly change under the influence of external noises (disturbances) and changes in the physiological state. The resultant output signals of these systems emerge from intricate dynamic interactions among their components. Notably, many signals from these components elude direct experimental assessment. The state variables of the system in time form a trajectory in the m-dimensional phase space, and each state corresponds to a certain point in the phase space. The methodology of recurrent diagrams allows to project the results of measuring one signal (time series) into a multidimensional space using delay and embedding procedures. Recurrence plots are the basic tools for visualizing the recurrence of dynamic systems, and RQA is a method of quantitative assessment of features of recurrence plots.

Examples of the use of quantitative analysis of recurrence for biomedical signals has been discussed in various studies [2, 3]. Its application necessitates further detailed investigation and validation, particularly in the context of aviation. In aviation, understanding and analyzing these physiological signals are critical, especially given the unique environmental and operational stressors faced by aviation professionals. Such analysis contributes significantly to the field of human factors, enhancing the safety and efficiency of aviation operations by providing deeper insights into the physiological responses and adaptability of individuals in high-stress and complex environments.

1.3 Contribution

The objective of this research is to ascertain the utility of employing recurrent plots for augmenting the objectivity of diagnostic conclusions regarding the human cardiovascular system's state. Additionally, this study aims to identify the optimal parameters necessary for the construction of these plots, specifically focusing on the delay time, recurrence threshold, and embedding dimension. A quantitative analysis of the resultant recurrent plots of RR intervals has been conducted. This analysis is instrumental in providing a quantitative assessment of the functional state of the biological system. The observed regularities in the transformation of these measures are hypothesized to possess a universal character, suggesting their applicability in the analysis of other biomedical signals [1]. Furthermore, the findings of this research may serve as an auxiliary tool for evaluating the cardiovascular system's condition, and by extension, the overall state of the biological system. This is particularly pertinent for the identification of anomalies within the system, achieved through the examination of chaotic processes.

1.4 The Organization of the Paper

This paper is structured as follows. Section 1 elucidates the imperative of employing nonlinear dynamics methodologies in the analysis of biomedical signals. Section 2 delineates an overview of extant literature focusing on the utilization of recurrence plot construction for various applications. In Sect. 3, the methodology for constructing a recurrence plot pertaining to heart rate variability coupled with RQA is explicated. Section 4 introduces and elaborates on an innovative approach for the visual and quantitative evaluation of biomedical RR intervals. Finally, Sects. 5 and 6 are devoted to conclusions and potential directions for future research.

2 Literature Review

Recently, to solve problems of various types, recurrent plot has been used to visualize their dynamic transformations [3] including for processing biomedical signals [4].

The stability of the functional state of the body can be assessed using heart rate variability analysis, which, according to the authors [4, 5], demonstrates the characteristics of a dynamic system. It is complex, nonlinear and extremely sensitive to initial conditions. Changes in the dynamics of heart rate variability indicate a lesser ability to maintain homeostasis, which may predispose to a violation of the adaptive potential of the body [1, 4]. In [2, 6, 7], recurrent plots were used to analyze biomedical signals based on recurrent plots and it was concluded that this approach can represent a very promising method for monitoring the functional state of the body through the color and pattern of the generated images. Therefore, we can see vegetative signs of recovery during the observation period transplant patients.

In [8–10], the graphical analytical tool of recurrent plots was used, which represents cyclic processes in the form of diagonal lines. RQA tool analyzes the duration (mean diagonal line length, MDL), complexity (entropy) and determinism (proportion of diagonal elements) of cyclic processes to characterize the dynamic characteristics in elementary time series, which makes it possible to determine the impact of a decrease in cycle stability (determinism), duration (average diagonal length) and complexity (entropy) in the profiles of the effects of various substances on the stability of biological objects.

3 Materials and Methods

In this research, a quantitative analysis of the recurrence of RR intervals was used to evaluate the characteristics of the human cardiovascular system. The methodology used in this paper has the following stages.

1. The first stage is obtaining empirical electrocardiogram data and extracting from it RR intervals that characterize heart rate variability and presenting these intervals in the form of time series.
2. The second stage is calculation of the recurrent matrix containing information about the recurrent behavior of the biological system.
3. The third stage is the construction of a recurrent plot, which allows to analyze the nature of the processes taking place in this system, the presence and influence of noise, states of repetition and stagnation (laminarity), the occurrence of sudden changes in state (extreme events).
4. The last stage is the calculation of quantitative measures of recurrent plot for the determining a number of measures based on the computing the density of recurrent points and the construction of the frequency distribution of the lengths of diagonal lines [11]. The study of these measures allows us to discover bifurcation points, transitions from one state of the system to another [1, 7, 8].

Data for research are obtained from the biomedical database [12]. This database contains files of RR intervals obtained for 54 long-term electrocardiograms recordings of subjects with normal sinus rhythm (30 men and 24 women). The original electrocardiograms recordings of the base signals are in the MIT-BIH normal sinus rhythm database [13]. With the mutual dependence of data of other dimensions, it is possible to restore the phase trajectory of a dynamic system by delaying the data of existing time series and constructing multidimensional data from delayed copies (embedding) [14]. The phase trajectory reconstructed by this method preserves the topological properties of the original phase space.

To display the m-dimensional phase trajectory of the states of the system in which the black point corresponds to the repetition of the state at the same time and at some other time j, the distance matrix is calculated based on the state coordinates of the point system. Using the coordinates of the state of the system of points x in the m-dimensional phase space of the dynamic system, a distance matrix is calculated, the elements of which are:

$$R_{ij} = \|\boldsymbol{x}_i - \boldsymbol{x}_j\|; i, j = 1, 2, ...m. \tag{1}$$

Using recurrence, it is possible to obtain a recurrence plot based on data from a one-dimensional time series X, which is a two-dimensional portrait of the dynamics of the time series, expressed through its recurrent characteristics [2]. Then the elements of the recurrent matrix are given as

$$R_{ij} = \begin{cases} 1 : \|\boldsymbol{x}_i - \boldsymbol{x}_j\| \leq \varepsilon, \\ 0 : \|\boldsymbol{x}_i - \boldsymbol{x}_j\| > \varepsilon; i, j = 1, 2, ...m, \end{cases} \tag{2}$$

where $\| \ldots \|$ is the distance norm in the m-dimensional phase space; ε is radius that we use to determine whether points are close enough in phase space to be considered recurrent. The recurrence plot displays the values of the elements of the recurrence matrix on the graph with black and white dots: black dots (ones) mean the presence of recurrence, white dots (zeroes) – its absence. The time is postponed according to the schedule.

This paper investigates the relationship between quantitative measures of recurrent plots and the state of the cardiovascular system. Several such measures were calculated: recurrence rate (%REC), determinism (DET), maximum length of diagonal lines (L_{max}), entropy (ENT), laminarity (LAM), trapping time (TT), etc. Formulas for calculating quantitative measures are given, for example, in [15].

Data processing (time series of RR intervals) was performed using the functions of the software package Cross Recurrence Plot Toolbox for Matlab [16], the functions of the package programs, as well as the MATLAB software code developed by the authors [17].

4 Results

Important parameters of the quantitative analysis of recurrence are the time delay τ, the embedding dimension n and the threshold value (radius) ε. The time delay is usually determined by the method of average mutual information [18]. However, for discontinuous signals, such as RR intervals extracted from continuous signals of electrocardiograms, it is better to set the delay to 1 (points in the time series are not skipped) [19].

The embedding dimension is chosen according to the method of false nearest neighbors [20]. This is a method of estimating the optimal dimensionality of time series embedding, based on the idea that in a low-dimensional space some points may seem close to each other, but in fact they are far from each other in a higher-dimensional space. These points are called false nearest neighbors and can be detected by comparing the distance between two points before and after applying the higher dimensionality of the system dynamics. Graphs of the dependence of the relative number of false nearest neighbors on the embedding dimension for the studied signals (Fig. 1) show that the minimum of this dependence is approximately the same for all signals.

So, as can be seen from the graphs, the time delay $\tau = 1$ (count) and the embedding dimension $m = 6$ are selected for signal analysis. The threshold value (radius) is chosen according to the recommendations of [21]:

- the value of the radius ε should be in the range of the linear region of the double logarithmic graph of $log(\%REC)$ versus $log(\varepsilon)$;
- for the selected ε, the %REC recurrence value should be low (e.g., 0.1 to 2.0%).

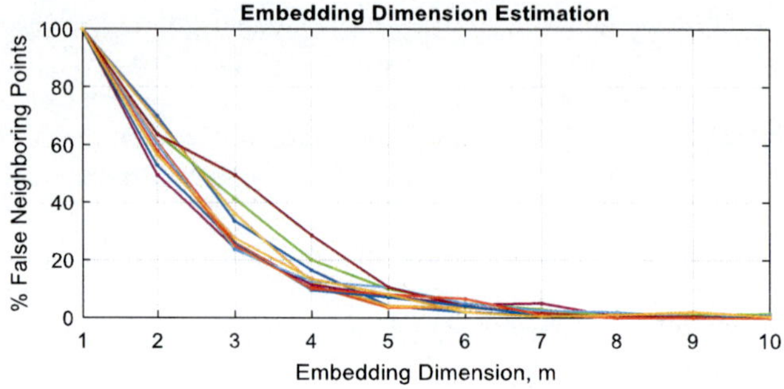

Fig. 1. Dependence of the relative number of false nearest neighbors from the size of the attachment.

To select the threshold according to recommendations corresponding graphs of $log(\%REC)$ versus $log(\varepsilon)$ dependences were constructed for the investigated signals. Examples of investigated RR signals of nsr001 and nsr002 intervals are shown in Fig. 2. Plots of $log(\%REC)$ versus $log(\varepsilon)$ (double logarithmic scale, Fig. 3) were plotted for these signals.

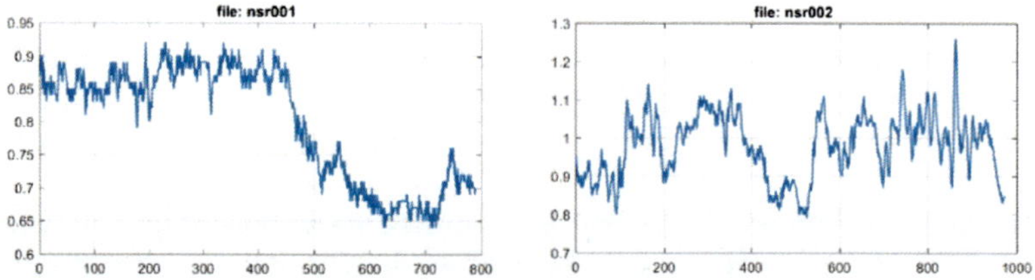

Fig. 2. Signal nsr001 and nsr002.

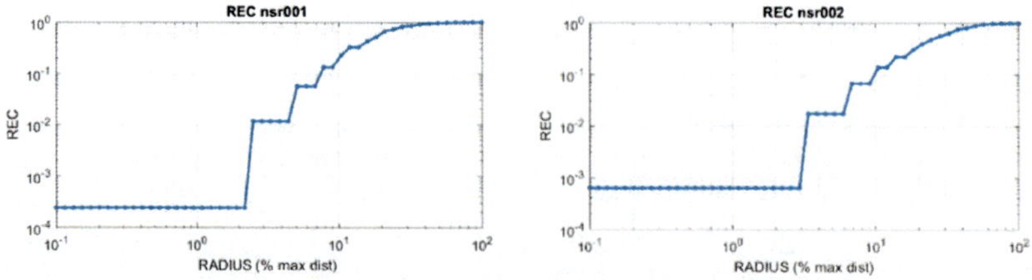

Fig. 3. Dependence of recurrence on the radius.

For signals of RR intervals, the obtained graphs of the dependence of the recurrence on the radius have several horizontal sections, on which the change in the radius ε does not affect the recurrence value. For many signals, the value $\%REC = 0.1 \ldots 5\%$ corresponds to the value $\varepsilon = 5 \ldots 10\%$, in view of the recommendation [2], a fixed value of the radius $\varepsilon = 5\%$ is chosen [22].

In the process of calculations, it was found that these values of the threshold (radius) correspond to the minimum values of the measures of clustering coefficients (CLUST)

[23] and transitivity (TRANS) (Fig. 4). To find out such behavior of the system, further research is needed, perhaps the dependence of such measures on the radius ε can be used.

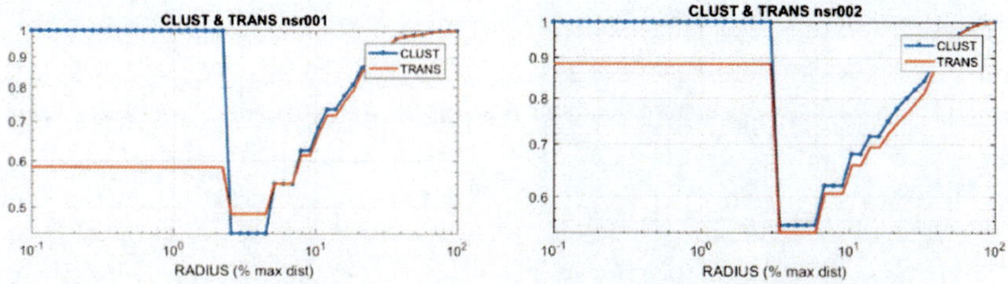

Fig. 4. Dependence of clustering coefficients and transitivity on the radius.

Examples of the received distance diagrams and recurrence plots, which are built according to the signals (Fig. 2) are shown in Fig. 5 and 6.

Quantitative measures were calculated for the received recurrent plots that are shown in Table 1.

Table 1. Quantitative measures of recurrent plot of electrocardiographic data.

Signal	*REC*	*DET*	L_{mean}	L_{max}	*ENTR*	*LAM*	*TT*	V_{max}	RT_{max}
nsr001	1.311	88.57	4.21	19	0.279	0.533	4.258	33	510
nsr002	1.818	89.33	4.32	47	0.285	0.692	3.359	21	883

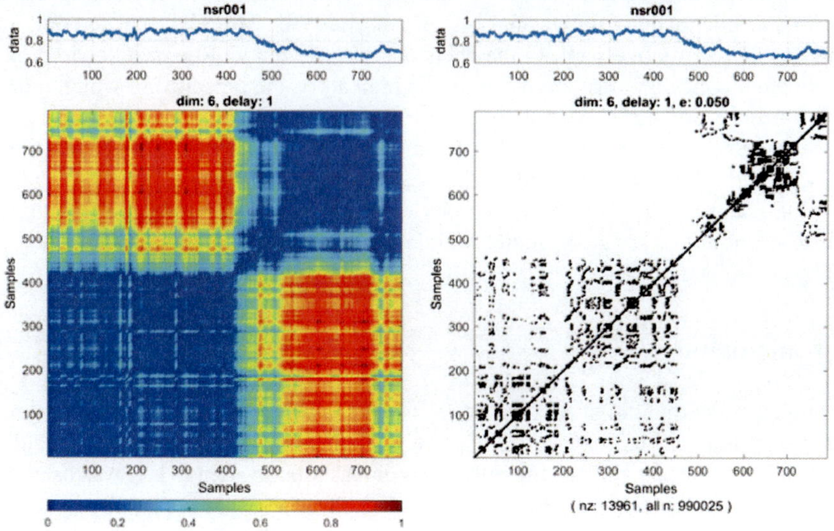

Fig. 5. Signal distance chart and recurrence plots of signal nsr001.

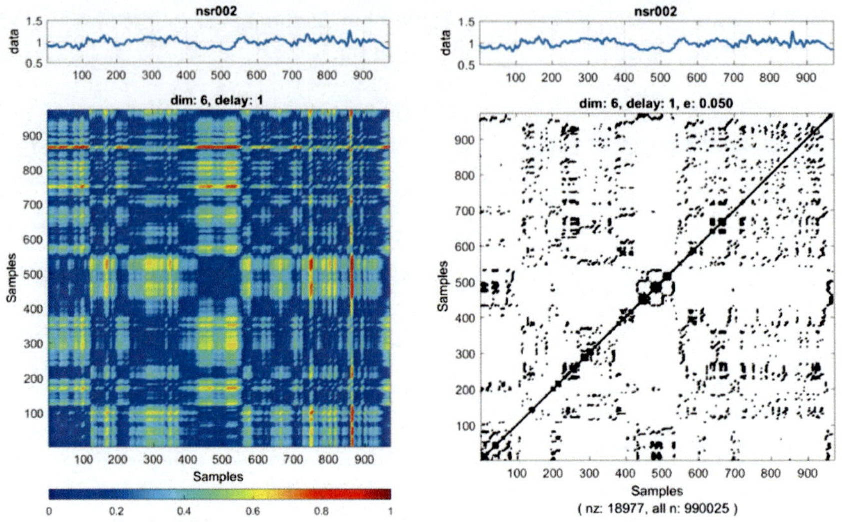

Fig. 6. Signal distance chart and recurrence plots of signal nsr002.

In this paper, to assess the stability of the functioning of the operator's body as a biological system, it is carried out by assessing the constancy of the functioning of the cardiovascular system as the most sensitive indicator of changes in the body [24].

Based on the analysis of data on heart rate variability in files nsr002, we can conclude that the functioning of the cardiovascular system is constant, at the same time, on the recurrent plot built on the basis of data in files nsr001, the absence of repetition of zones in the plots is clearly visible. In addition to the visual construction of recurrent plot, in the paper, a quantitative analysis of recurrence measures was carried out, which allows us to quantitatively check the decisions that are made. Recurrent charts made it possible to identify the formation of specific indicators, the presence of which indicates the occurrence of a certain event. At the same time, measures show different levels of sensitivity. So, the measures of TT, L_{max} and RT_{max} for the first and second files are TT: 4.24 and 3.36; L_{max}: 19 and 47; RT_{max}: 510 and 883, respectively. At the same time, DET and LAM measures change by only 15–20%. Therefore, the first group of indicators should be used to predict a violation of the functional state of a biological object due in the first place.

5 Conclusions

In the paper, a comparative analysis of recurrent plots of RR intervals of electrocardiograms signals is carried out from the point of increasing the objectivity of diagnostic conclusions about the state of the human cardiovascular system [1]. It was established that the following input parameters should be used to analyze the recurrence of RR intervals: embedding dimension $m = 6$ and delay time $\tau = 1$ (count). When choosing the threshold ε of recurrence, it was established that the graph of $log(\%REC)$ dependence on $log(\varepsilon)$ has horizontal sections, where the recurrence does not depend on the threshold.

For different signals, it is possible to detect areas corresponding to a fixed recurrence threshold $\varepsilon = 4 \ldots 10\%$ of the maximum value of the distance matrix. At the same time, the recurrence rate in most cases does not exceed 5%.

The sequence of RR intervals is a non-stationary process, so the recurrence plots and their quantitative measures for each time series depend on the studied epoch of the signal. The values of quantitative measures of recurrence significantly depend on the given threshold (radius) during calculation, as well as on the dimension of the embedding, so additional research is needed. Interpretation of changes in recurrence rates also require additional research. Quantitative analysis of recurrent plots of RR intervals can be used as an additional tool for assessing the state of the cardiovascular system, which allows to detect abnormalities.

6 Future Scope

Future scope of this research are as follows:

- using this approach on the empirical data of aviation operators to assess and forecast their functional state;
- intellectual analysis of biomedical data to obtain an additional informational component;
- the use of data mining for tasks of evaluating biomedical signals of various complexity;
- analysis of signals using recurrent plots;
- carrying out a comparative characterization of the results of recurrent plots with quantitative values.

References

1. Ivanets, O.B., Khrashchevskyi, R.V., Burichenko, M., Kulik, M.S.: Intelligence analysis of empirical data based on time series. Radio Electron. Comput. Sci. Control **61**(2), 38–49 (2023). https://doi.org/10.15588/1607-3274-2023
2. Burichenko, M., Onikienko, Yu., Arkhyrei, M.: Quantitative analysis of recurrence of kephalographic data. In: International Scientific and Technical Conference on Modern Technologies of Biomedical Engineering, Odesa, Ukraine, pp. 23–27 (2023)
3. Angus, D., Watson, B., Smith, A.E., Gallois, C., Wiles, J.: Visualizing conversation structure across time: insights into effective doctor-patient consultations. PLoS ONE **7**(6), e38014 (2012). https://doi.org/10.1371/journal.pone.0038014
4. Takakura, I.T., et al.: Recurrence plots: a new tool for quantification of cardiac autonomic nervous system recovery after transplant. Braz. J. Cardiovasc. Surg. **32**(4), 245–252 (2017). https://doi.org/10.21470/1678-9741-2016-0035
5. Austin, C., et al.: Dynamical properties of elemental metabolism distinguish attention deficit hyperactivity disorder from autism spectrum disorder. Transl. Psychiatry **9**, 238 (2019). https://doi.org/10.1038/s41398-019-0567-6
6. Marwan, N., Wessel, N., Meyerfeldt, U., Schirdewan, A., Kurths, J.: Recurrence-plot-based measures of complexity and their application to heart-rate-variability data. Phys. Rev. E - Statist. Phys. Plasmas Fluids Related Interdisciplinary Top. **66**, 1–16 (2002). https://doi.org/10.1103/PhysRevE.66.026702

7. de Godoy, M.F., Takakura, I.T., Correa, P.R., Machado, M.N., Miranda, R.C., Brandi, A.C.: Preoperative nonlinear behavior in heart rate variability predicts morbidity and mortality after coronary artery bypass graft surgery. Med. Sci. Monit. **15**(3), 117–122 (2009)
8. Billeci, L., Marin, D., Insana, L., Vatti, G., Varanini, M.: Patient-specific seizure prediction based on heart rate variability and recurrence quantification analysis. PLoS ONE **13**, e0204339 (2018). https://doi.org/10.1371/journal.pone.0204339
9. Schlenker, J., et al.: Recurrence quantification analysis: a promising method for data evaluation in medicine. EuroMISE **10**(1), 1–6 (2014). https://doi.org/10.24105/ejbi.2014.10.1.7
10. Lyby, M.S., et al.: Intervention and reductions in distress symptoms in women with breast cancer. Front. Appl. Math. Stat. **5** (2019). https://doi.org/10.3389/fams.2019.00037
11. Riley, M., Balasubramaniam, R., Turvey, M.: Recurrence quantification analysis of postural fluctuations. Gait Posture **9**(1), 65–78 (1999). https://doi.org/10.1016/s0966-6362(98)00044-7
12. Stein, P.: Normal Sinus Rhythm RR Interval Database, 3 March 2003. Version: 1.0.0. https://physionet.org/content/nsr2db/1.0.0/. Accessed 10 Jan 2024
13. https://archive.physionet.org/physiobank/database/nsrdb/. Accessed 10 Jan 2024
14. Eckmann, J.P., Kamphorst, S.O., Ruelle, D.: Recurrence plots of dynamical systems. Europhys. Lett. **4**(9), 973–977 (1987)
15. http://www.recurrence-plot.tk/. Accessed 10 Jan 2024
16. Cross Recurrence Plot Toolbox for Matlab. Reference Manual. Version 5.17, Release 28.17. http://tocsy.pik-potsdam.de. Accessed 10 Jan 2024
17. Simple MATLAB functions for calculating recurrence plots and recurrence quantification. https://github.com/pucicu/rp/. Accessed 10 Jan 2024
18. Wallot, S.: Calculation of Average Mutual Information and False-Nearest Neighbors for the Estimation of Embedding Parameters of Multidimensional Time Series in Matlab. https://www.ncbi.nlm.nih.gov/pmc/articles/PMC6139437/. Accessed 10 Jan 2024
19. Mohebbi, M., Ghassemian, H., Mohammadzadeh, B.: Structures of the recurrence plot of heart rate variability signal as a tool for predicting the onset of paroxysmal atrial fibrillation. J. Med. Signals Sens. **1**(2), 113–121 (2011)
20. Mohammadi, S.: FNN: MATLAB function to calculate corrected false nearest neighbors. Statistical Software Components T7415010, Boston College Department of Economics. https://ideas.repec.org/c/boc/bocode/t741510.html. Accessed 10 Jan 2024
21. Webber, C.L., Zbilut, J.P.: Recurrence quantification analysis of nonlinear dynamical systems. https://www.researchgate.net/publication/228680523_Recurrence_quantification_analysis_of_nonlinear_dynamical_systems. Accessed 10 Jan 2024
22. Golubnichy, O., Zalisky, M., Shcherbina, O., Ivanets, O.: Submission of the AR model through the Jimman-McClure function when predicting processes in telecommunications. Radioelectron. Commun. Syst. **65**(8), 420–432 (2022). https://doi.org/10.3103/S0735272722090023
23. Kuzmin, V.M., Khrashchevskyi, R.V., Kulik, M.S., Ivanets, O.B., Zaliskyi, M.Y., Petrova, Y.V.: Mathematical model for decision making system based on three-segmented linear regression. Radio Electron. Comput. Sci. Control **38**(3), 38–49 (2022). https://doi.org/10.15588/1607-3274-2022-3-4
24. Ivanets, O., Morozova, I.: Features of evaluation of complex objects with stochastic parameters. In: International Conference on Advanced Computer Information Technologiesm, Deggendorf, Germany, pp. 159–162. IEEE (2021). https://doi.org/10.1109/ACIT52158.2021.9548579

Sensors in Civil Aviation

Integration of Lightweight Sensors for Altitude Measurement in Compact Aviation Systems

Olena Kozhokhina[1], Yaroslav Yakovlev[2], Liudmyla Blahaia[1](✉), Svitlana Pavlova[3], and Aria Nazarparvar[4]

[1] National Aviation University, Liubomyra Huzara Avenue, 1, Kyiv 03058, Ukraine
b_ludmila@ukr.net
[2] Antonov Company, Akademika Tupoleva Street, 1, Kyiv 03062, Ukraine
[3] Shanxi Agricultural University, Longcheng Street, 81, Xiaodian, Taiyuan 030031, Shanxi, China
[4] Georgian Aviation University, Ketevan Depoli Avenue, 16, 0103 Tbilisi, Georgia

Abstract. A small aircraft system requires innovations in the altitude control that take into account its physical size and weight constraints. To provide accurate altitude measurements, this study presents an integrated altitude monitoring system designed specifically for small aircraft. This system combines miniaturized sensors with the advanced data fusion and filtering algorithms such as Kalman filters. Based on the maximum likelihood estimation and extended Kalman filtering, the proposed system improves measurement accuracy over traditional methods taking into account the unique challenges and limitations posed by small aircraft systems. As a result of its compact size, the researched system can be used for smaller, lighter aircraft systems that often cannot accommodate larger and heavier systems. Experimental results have been conducted to validate the system's effectiveness. The results of this study could have a significant impact on the design and operation of small aircraft by providing strategies for maintaining responsive and stable flight conditions.

Keywords: Compact Aviation Systems · Altitude Control System · Sensor Data Fusion · Kalman Filter · Maximum Likelihood Estimation · Mean Squared Error · Autopilot System · Data Reliability · Atmospheric Pressure Sensors · Altitude Measurement

1 Introduction

1.1 Introduction to the Problem

Developing effective and efficient altitude control mechanisms is becoming increasingly critical in the rapidly developing segment of small and light aircraft systems. However, it is critical to note that such systems, while essential for navigation and stability, must also accommodate the limitations of size and weight inherent in small aircraft system designs. While traditional altitude control systems are reliable, they are too heavy and bulky for small aircraft. An additional gram or centimeter can negatively impact compact aircraft performance and fuel efficiency.

I. Ostroumov and M. Zaliskyi (Eds.): ACASD 2024, LNNS 992, pp. 397–409, 2024.
https://doi.org/10.1007/978-3-031-60196-5_29

As a result, it is necessary to develop altitude measurement systems that have high accuracy and reliability of measurements without significantly increasing their physical dimensions, since small aviation systems require compact and accurate components. Small aircraft have limited space and payload capacities, thus the need for this equipment arises. There is a significant impact on aerodynamic performance, operating costs and payload when large or heavy equipment is used in this case.

Lightweight altitude control systems are particularly critical in small aircraft applications. These include recreational and business aviation, as well as remotely piloted aircraft systems and scientific research platforms. Such a system must provide accurate altitude information under environmental conditions and flight scenarios. As a result, design and implementation may be further complicated.

This research aims to design a lightweight integrated altitude control system customized for small aircraft to address the above problems. To provide accurate altitude measurements, the system combines advanced sensor technology with sophisticated data processing algorithms, including maximum likelihood estimation (MLE) and extended Kalman filtering (EKF). In this approach, precision and reliability are balanced with small aircraft systems' size and weight constraints to provide the most suitable possible solution.

1.2 Motivation

Small and light aircraft systems focus on improving performance without increasing size and weight. In this study, the problem of precisely controlling altitude is addressed in the context of lightweight compact aircraft systems. Weight and size constraints are critical. It should be noted that for such systems, accurate real-time operational altitude data is critical to navigation accuracy, as well as maintaining structural integrity and optimizing fuel consumption. Small aircraft systems require lightweight and compact control mechanisms. For this to be achieved, a system must be designed that satisfies the above requirements without compromising the system's functionality and security.

1.3 Contribution

In this paper, a compact altitude control system for small aircraft is designed and validated. Microsensor data is integrated into this system using advanced processing algorithms, such as Kalman filters. A combination of MLE and EKF is used in this system to enhance altitude measurement accuracy. A lightweight and compact framework is presented in this study suitable for small aircraft systems. In this way, the resulting system provides accurate navigational data, which is critical for small aircraft safety and efficiency.

1.4 The Organization of the Paper

The paper has six sections. The first section of this study explores motivation for altitude control systems for small aircraft and provides an introduction to the problem. An analysis of the literature is conducted in the second section in order to formulate a problem statement. Third section, materials, components and methods used to design a control altitude system are described. In the fourth section, simulations and data analysis was conducted on the researched system. Sections five and six present the conclusions and future scope.

2 Literature Review and Problem Statement

Due to the development of hybrid aircraft, aircraft improvements in the modern world tend to reduce weight. It should be mentioned that an electric hybrid aircraft is planned to be released by Faradair by 2025 for regional transport in the United Kingdom. In accordance with [1], also by the year 2050.

In order to reduce the weight of an aircraft, it is necessary to change not only the design [2, 3], but also the power system [4, 5], as well as all the components, including the avionics [6]. Additionally, in accordance with modern trends, the equipment on board that provides information regarding the primary flight parameters should be reduced in size and weight. To control aircraft flight dynamics, a control algorithm is being developed [7, 8].

The proportional-integral-differentiating controller (PID controller) was developed in the first half of the twentieth century to provide accurate and efficient control of automatic systems [9]. In addition, with the advent of digital technologies, many control schemes based on PID controllers have been developed. Their relative simplicity and high functionality explain this [10, 11]. The paper [12] presents a comparative evaluation of traditional PID feedback tuning methods and modern control techniques. Additionally, classical tuning methods and modern control approaches are compared in various systems discussed in this paper.

In various areas of industrial economics, such controllers are widely used to resolve control problems [13] and offer a number of advantages over other control devices. It is discussed in [14] how classical PID controllers have been developed and how they have been improved by using fuzzy logic. It is conclusively shown that the traditional PID controller is the most popular controller due to its ease of operation and low cost [15]. As a result, this type of regulator was chosen. PID control can be used to solve a flight control problem, as illustrated in [16, 17]. As a result of the large amount of data input from the sensors, the controller gains must be adjusted iteratively [18, 19].

There is currently a gap in the literature regarding the integration of these systems within the framework of small aircraft despite the wealth of knowledge available. In particular, the references [20–22] provide methods for statistical data processing that might be useful in refining sensor data fusion methods necessary for small aircraft control.

As a result, the primary challenge addressed in this study is the integration of a robust altitude control system into a small aircraft. This is achieved by taking advantage of the convergence of sensor data fusion techniques and PID control mechanisms. Data fusion methods such as [12–14] are being used to enhance precision and reliability of altitude measurements. Simulation and optimization techniques are also being used to refine the performance of PID controllers.

In this study, the problem statement revolves around the need to develop a control system that incorporates the advantages of existing methods. In addition, it should take into account the unique challenges and limitations posed by small aircraft systems. This requires meticulous analysis of sensor data to minimize noise and inaccuracies. Additionally, it involves the application of advanced control strategies to maintain a responsive and stable flight.

3 Materials and Methods

The research focused on the development and validation of a highly integrated altitude control system designed specifically for small and light aircraft. It is critical to note that the Arduino Uno Rev 3 microcontroller is the central element of the system architecture. Microcontrollers serve as the link between signal processing and control logic. In this study, digital pressure and temperature sensors were used, BMP180 and BME280. By using barometric pressure readings, these sensors can provide altitude data compactly and accurately.

A significant aspect of the method is the calibration and synchronization of the BMP180 and BME280 sensors. As a result, accurate data collection was ensured. The data from these sensors was subjected to a rigorous preprocessing protocol to eliminate any bias or inconsistency from manufacturing variations or environmental conditions.

An algorithm based on classical control theory was developed as part of the study. In this algorithm, the input data from the above-mentioned sensors is translated into actuation commands. Furthermore, the SG90 servo drive was selected because of its light weight and responsiveness. As a result of this drive, the control system output is converted into mechanical movement.

Moreover, a PID controller was implemented in the study. Through continuous adjustments to its control gains, this component dynamically regulated system parameters: proportional (K_p), integral (K_i), and derivative (K_d) [9, 10]. Using the PID control algorithm (1), the system's response to altitude variations was refined. Within the control system, the transfer function (2) defines how input commands are translated into action within the system.

$$u(t) = K_p e(t) + K_i \int e(\tau)d\tau + K_d \frac{de(t)}{dt} \quad (1)$$

where K_p is a non-negative coefficient determining the current error value (the larger the number, the more force the controller applies); K_i is a non-negative coefficient for integral terms, which are used to calculate the past value of error (the smaller the number, the faster the controller will react to load changes, but the greater the oscillation risk); K_d is a non-negative coefficient that computes the possible future value of the error based on the current chance rate (the higher the number, the greater the damping effect); $u(t)$ are control variables; $e(t)$ is error value.

In the case of a PID controller, the transfer function $T(s)$ can be expressed as follows:

$$T(s) = K_p + K_i/s + K_d s = (K_d s^2 + K_p s + K_i)/s. \quad (2)$$

Figure 1 shows a PID controller and $r(t)$ is the desired process variable (PV) or set point (SP), $y(t)$ is the measured PV.

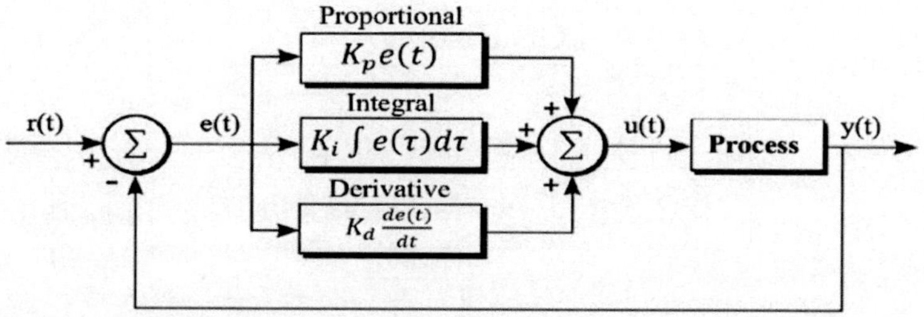

Fig. 1. A block diagram of a PID controller in a feedback loop.

For tuning the PID controller in a controlled simulation environment, a Simulink model was incorporated into the experimental setup. This allowed for a pre-deployment assessment of system performance. For operational efficiency, the SG90 servo motor was incorporated as an actuator within the control system. Its selection was based on a balance between its operational capabilities and the imperative of maintaining a minimal weight profile.

Additionally, an ultrasonic sensor HC-SR04, was employed to supplement altitude information, which proved useful during takeoff and landing. Sensor outputs had to be incorporated into the control algorithm, which necessitated a pre-processing regimen to improve the data fidelity.

Figure 2 shows an altitude control system for a compact aircraft, including a circuit layout and wiring diagram. The following schematic illustrates the connection between the Arduino Uno Rev 3 microcontroller, the LCD display, the I2C serial interface module, the BMP180 and BME280 sensors, the SG90 micro servo, the HC-SR04 ultrasonic sensor, and the power supply unit. As can be seen from the diagram, the system's functioning is underpinned by the systematic configuration of electronic components.

In this study, the methodology was characterized by the systematic integration of compact yet sophisticated sensors. Moreover, the system was preemptively tested and validated using refined PID control logic and simulation models. To ensure that the final altitude control system was accurate and congruent with small aircraft weight and size limitations, this comprehensive approach was taken.

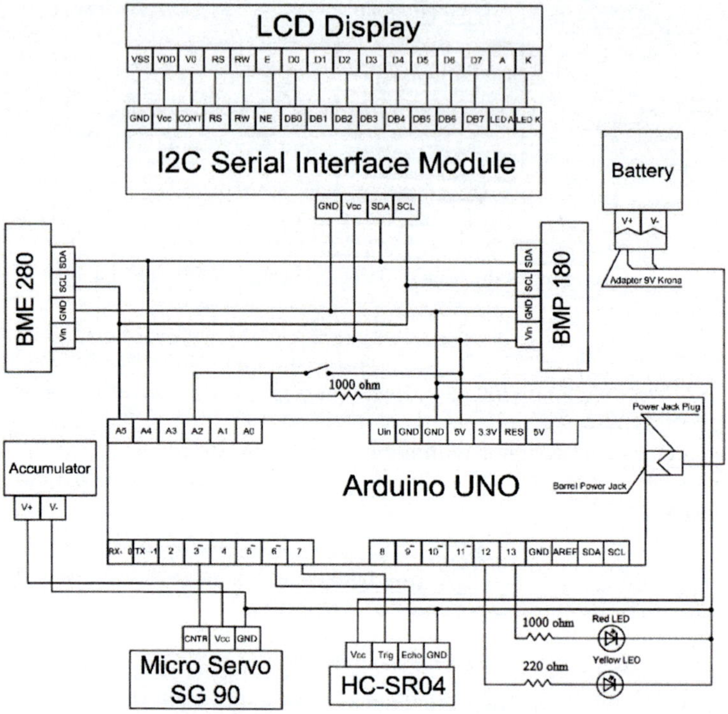

Fig. 2. Integrated circuit layout and wiring diagram for an altitude control system.

4 Results and Discussions

In this study, the primary objective was to evaluate the performance of an altitude control system for small aircraft systems. For accurate altitude data acquisition, MEMS pressure sensors were integrated. The system's response, governed by a sophisticated PID control algorithm, was subjected to various simulations to ascertain its efficiency and reliability under different operational conditions.

The schematic layout of the simulation model (Fig. 3(a)) provides a comprehensive visual representation of the interconnected components of the system. As shown in Fig. 3(b), the system's response is influenced by the transfer function, which is crucial to understanding system behavior under dynamic conditions. Graphical representations of the system's steady-state and transient responses demonstrate the PID controller's capability to minimize overshoots and maintain stability.

An extensive analysis of the impact of tuning PID coefficients on system performance was conducted through the use of an iterative simulation process. According to the transfer function graph (Fig. 4), initial simulations with only the proportional gain (P) revealed a tendency for the system to exhibit a steady-state error, as evidenced by the stabilization of output at a value less than the desired set point equal to (1).

In subsequent simulations, the integral coefficient (I) was included, reducing the steady-state error; however, oscillations persisted, indicating that the control parameters

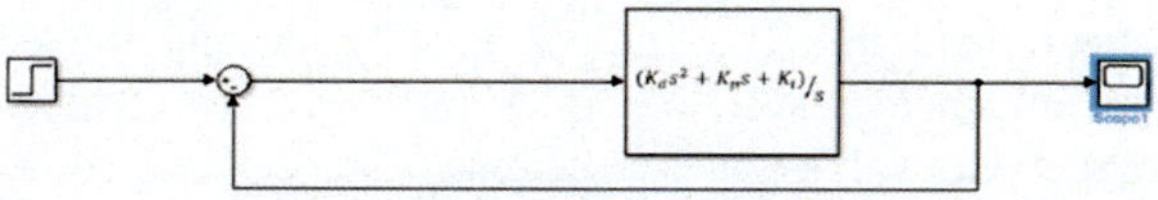

a)

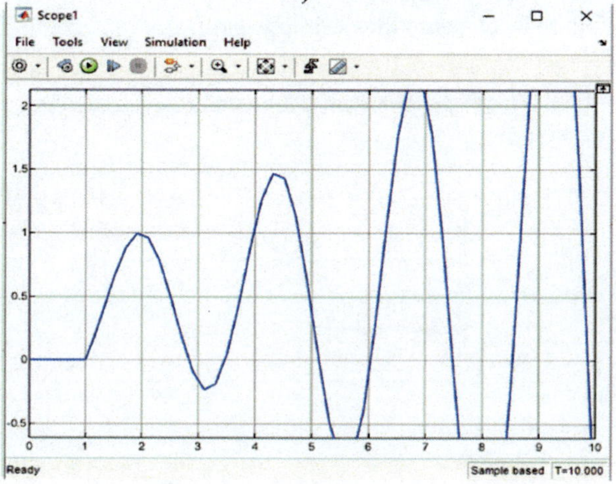

b)

Fig. 3. Diagrams of the simulation model (a) and the transfer function graph (b).

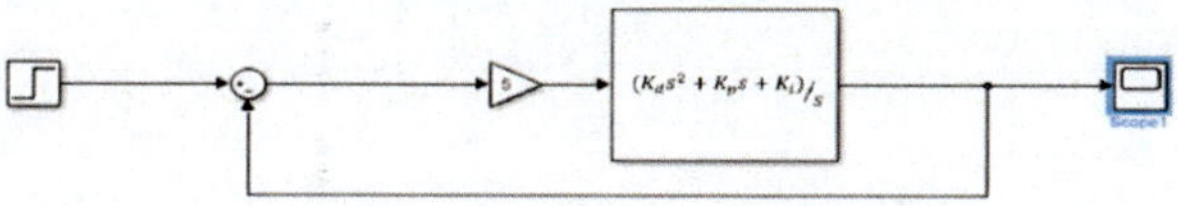

a)

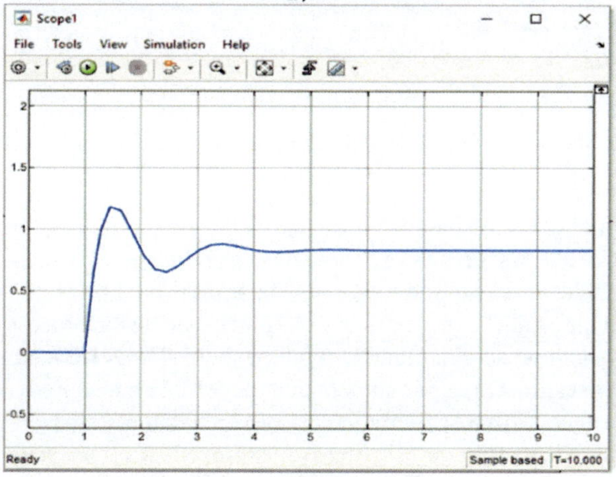

b)

Fig. 4. Diagram of the simulation model with P coefficient (a) and transfer function graph (b).

needed to be further refined (Fig. 5). A discussion of the introduction of the derivative (D) coefficient and its influence on system performance was then conducted. According to the simulation model with P, I, and D coefficients (Fig. 6(a)) and corresponding transfer function graphs (Fig. 6(b)), overshoot was reduced and settling time was improved, enabling the output to be closer to the desired value.

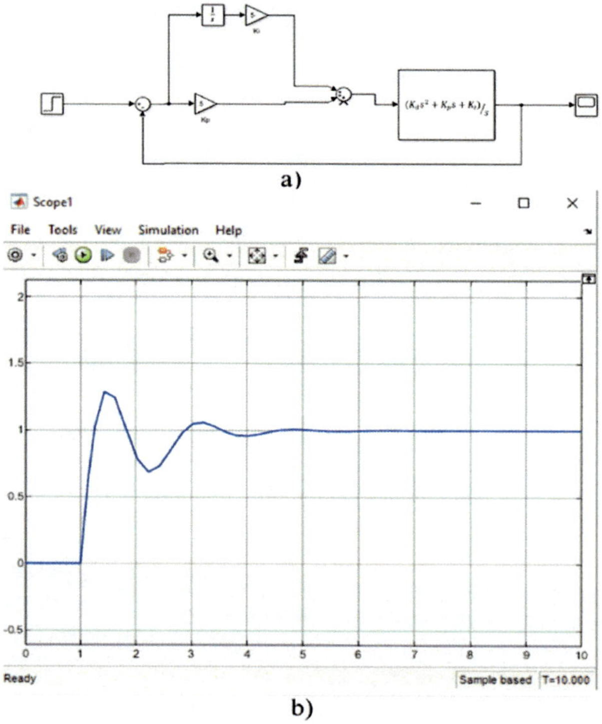

Fig. 5. Simulation model block diagram (a), and transfer function graph (b).

In order to integrate the data from the two sensors, the Indirect Feedforward EKF Method was selected. As a result of the method's relative independence from the state equation, this choice was made. As the state equation is not utilized in this context, this aligns well with the project's requirements. As part of the fusion process, MLE is applied to the data from both sensors in order to combine them. The MLE method is a statistical technique used to estimate the parameters of a model. MLE is employed in this context in order to calculate the most likely values of the parameters based on the observed data. In this way, the likelihood function is maximized. A robust statistical foundation is provided by MLE in this setting for the fusion of data from two independent sources. As a result, accurate altitude measurements can be obtained.

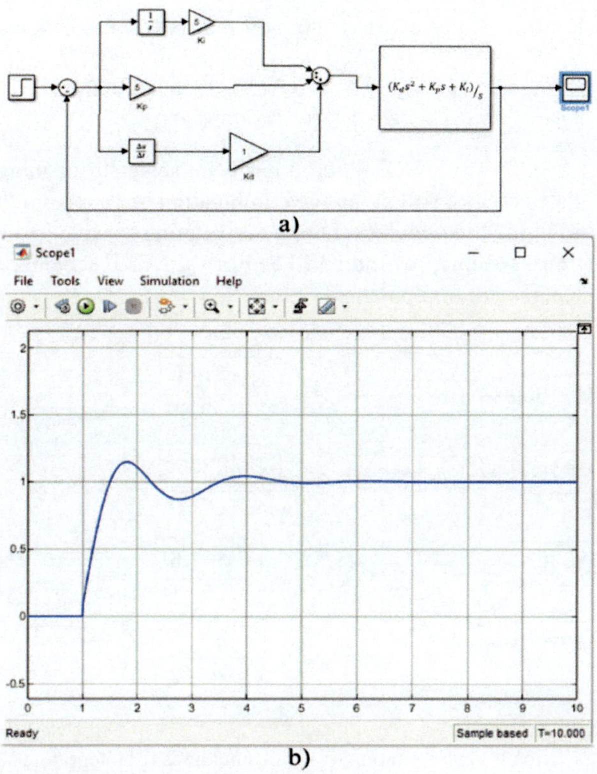

Fig. 6. Diagram of the simulation model with coefficients P, I, and D (a), and graph of the transfer function (b).

By using the maximum likelihood method, the parameter is estimated as follows:

$$\widehat{X} = \left(H^T R_z^{-1} H\right)^{-1} H^T R_z^{-1} Z \tag{3}$$

where $H = \begin{vmatrix} x \\ y \end{vmatrix}$ is the measurement vector ($x = 1,\ y = 2$ are data from sensors 1 and 2); $R_z = \begin{vmatrix} \sigma_x^2 & \sigma_{xy}^2 \\ \sigma_{xy}^2 & \sigma_y^2 \end{vmatrix}$ is a matrix that relates the state of the system to the measurements and $\sigma_{xy} = 0$ since there is no correlation between x and y; $Z = \begin{vmatrix} Z_1 \\ Z_2 \end{vmatrix}$ is matrix of our measurements, where Z_1 and Z_2 are the individual measurements from sensor 1 and sensor 2.

Therefore, $R_z = \begin{vmatrix} \sigma_1^2 & 0 \\ 0 & \sigma_2^2 \end{vmatrix}$ is the correlation matrix of measurement errors for two sensors, where σ_1^2 and σ_2^2 are the mean squared errors for sensor 1 and sensor 2, respectively.

The expression can also be expressed in non-matrix form as follows:

$$\widehat{X} = \frac{\sigma_2^2}{\sigma_1^2 + \sigma_2^2} Z_1 + \frac{\sigma_1^2}{\sigma_1^2 + \sigma_2^2} Z_2 \quad (4)$$

The Indirect Feedforward EKF Method has been selected for its relative independence from the state equation, allowing the combination of data from the BMP180 and BME280 sensors (Fig. 7), in which (a) Direct Pre-filtering scheme, (b) Direct Extended Kalman Filter (EKF) scheme, (c) Indirect Feedforward EKF scheme, and (d) Indirect Feedback EKF scheme are implemented.

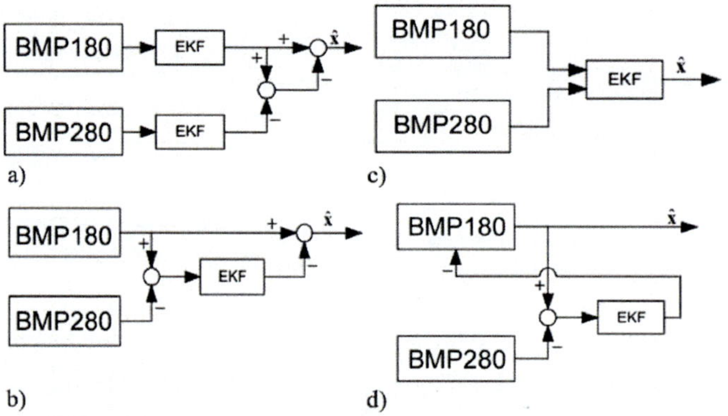

Fig. 7. A schematic representation of various methods for integrating sensor data.

Based on a comparative analysis of the mean squared error (MSE) between the BMP180 and BME280 sensors, the BME280 sensor was found to be more accurate than the BMP180 sensor. Based on the post-Kalman filter MSE, there was a significant improvement in data accuracy, which was crucial for operational effectiveness (Fig. 8). Using Kalman Filter refinement, this figure illustrates how the mean squared error (MSE) can be reduced for altitude measurements obtained from BMP180 and BME280 sensors, demonstrating the enhanced accuracy and effectiveness of sensor data in altitude control systems for small aircraft. Blue represents data from BMP180, orange represents data from BME280, and green represents refined estimates. This demonstrates the improvement in data precision critical to system functionality.

In order to refine data, the Kalman filter cannot incorporate acceleration parameters, which is a limitation of the current setup. In the context of rapid maneuvers, this is a factor that may affect the precision of altitude data.

As a result of this study, an integrated sensor system and a refined PID controller have been demonstrated to be useful for small aircraft systems. Simulating and adjusting the PID coefficients successfully demonstrated the adaptability and effectiveness of the control strategy for achieving precise altitude control. As a result of the discussion, there is a possibility that the system might be enhanced by integrating additional sensors and refining the Kalman filter in the future. This will include a wide variety of operational parameters that are relevant to real-world applications.

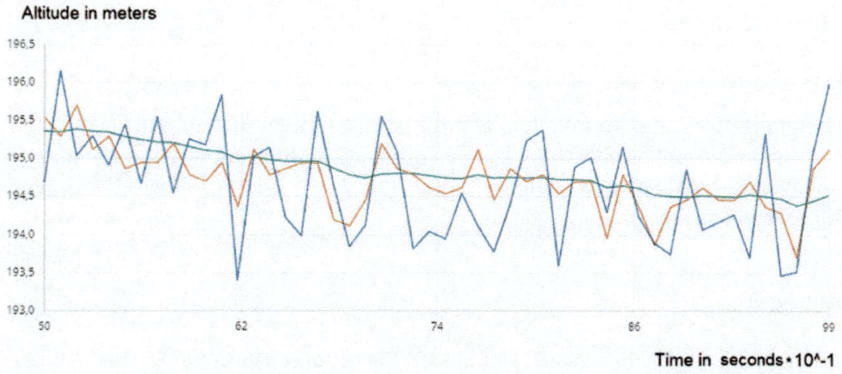

Fig. 8. Comparative analysis of altitude data precision.

5 Conclusions

In order to achieve lightweight and compact aircraft systems, innovative approaches to altitude control are required. This paper presents the design and testing of a compact altitude control system for small aircraft. In this system, miniature pressure and temperature sensors BMP180 and BME280 are combined with algorithms for data fusion and filtering, including Kalman filters. The proposed system is based on maximum likelihood estimation and extended Kalman filtering. With this method, measurement accuracy is enhanced over traditional methods while taking into account the unique challenges and limitations associated with small aircraft.

Implementing an attitude system controlled by a digital PID controller demonstrated the practical applicability of the theoretical concepts examined in this study. As a result of the successful integration of these components, advanced altitude control systems have demonstrated the ability to enhance small aircraft systems' safety and functionality. Experimental and simulation results demonstrate significant improvements that have been made in the accuracy of sensor data. This improves the altitude control system's ability to provide accurate, real-time navigation data. This is especially important when maneuvering at low altitudes or under difficult conditions.

As a result of this paper, the importance of continuous innovation in sensor technology and data filtering methods has been highlighted. In addition to sensors becoming more accurate and compact, as well as new filtering techniques being developed, altitude control systems continue to be highly prospective.

To conclude, it should be noted that the altitude control system presented in this paper achieves its intended objectives. Additionally, it opens up opportunities for further research and development into small aircraft systems. These will be critical in a variety of fields, ranging from commercial endeavors to critical military operations. In light of the continuing demand for sophisticated and reliable small aircraft systems, the findings of this thesis will undoubtedly play a crucial role in shaping the trajectory of this dynamic field.

6 Future Scope

Future research directions include:

- designing smaller, more efficient control hardware for small aircraft;
- testing the system's robustness and performance in different weather conditions;
- enhancing weather and altitude adaptability by adding other high-precision, lightweight sensors.

References

1. Darecki, M., et al.: Flightpath 2050 Europe's Vision for Aviation. Publications Office of the European Union, European Commission, 28 (2011)
2. Zhu, L., Li, N., Childs, P.: Light-weighting in aerospace component and system design. Propul. Power Res. **7**(2), 103–119 (2018). https://doi.org/10.1016/j.jppr.2018.04.001
3. Poll, I.: On the application of lightweight materials to improve aircraft fuel burn – reduce weight or improve aerodynamic efficiency. Aeronaut. J. **2014**, 903–934 (2014)
4. Sahoo, S., Zhao, X., Kyprianidis, K.: A review of concepts, benefits, and challenges for future electrical propulsion-based aircraft. Aerospace **7**(4), 44 (2020)
5. Wahler, N., et al.: An integrated framework for energy network modeling in hybrid-electric aircraft conceptual design. In: AIAA AVIATION 2022 Forum, Chicago, USA, pp. 1–20. American Institute of Aeronautics and Astronautics (2022). https://doi.org/10.2514/6.2022-3741
6. Sabatini, R., Kramer, K., Blasch, E., Roy, A., Fasano, G.: From the editors of the special issue on avionics systems. Future challenges. IEEE Aerosp. Electron. Syst. Mag. **36**, 5–6 (2021). https://doi.org/10.1109/MAES.2021.3064616
7. Nelson, R.: Flight Stability and Automatic Control. McGraw-Hill, New York (1998)
8. Hamid, T., Babar, M., Hassan, M., Hussain, M.: Pitch attitude control for an aircraft using linear quadratic integral control strategy. In: International Multitopic Conference (INMIC), Islamabad, Pakistan, pp. 1–6. IEEE (2019). https://doi.org/10.1109/INMIC48123.2019.9022787
9. Ang, K., Chong, G., Li, Y.: PID control system analysis, design, and technology. IEEE Trans. Control Syst. Technol. **13**(4), 559–576 (2005)
10. Li, Y., Ang, K., Chong, G.: PID control system analysis and design. IEEE Control. Syst. Mag. **26**(1), 32–41 (2006)
11. Deepa, S., Sudha, G.: Longitudinal control of aircraft dynamics based on optimization of PID parameters. Thermophys. Aeromech. **23**(2), 185–194 (2016)
12. Jaiswal, R., Prakash, O.: Classical and modern gain estimation approach of PID controller for the pitch control of the RCTA aircraft. INCAS Bull. **14**, 39–56 (2022)
13. Song, Y.: Control of Nonlinear Systems via PI, PD and PID: Stability and Performance. CRC Press, Boca Raton (2018)
14. Kumar, V., Nakra, B., Nakra, C, Mittal, A.: A review of classical and fuzzy PID controllers. Int. J. Intell. Control Syst. **16**, 170–181 (2011)
15. Shaji, J., Aswin, R.: Pitch control of flight system using dynamic inversion and PID controller. Int. J. Eng. Res. Technol. (IJERT) **4**(7), 604–608 (2015). https://doi.org/10.17577/IJERTV4IS070502
16. Solomentsev, O., Zaliskyi, M., Kozhokhina, O., Herasymenko, T.: Efficiency of data processing for UAV operation system. In: 4th International Conference Actual Problems of Unmanned Aerial Vehicles Developments (APUAVD), Kiev, Ukraine, pp. 27–31. IEEE (2017). https://doi.org/10.1109/APUAVD.2017.8308769

17. Lungu, R., Lungu, M., Grigorie, L.: Automatic control of aircraft in longitudinal plane during landing. IEEE Trans. Aerosp. Electron. Syst. **49**(2), 1338–1350 (2013). https://doi.org/10.1109/TAES.2013.6494418
18. Islam, M.T., Alam, M.S., Laskar, M.A.R., Garg, A.: Modeling and simulation of longitudinal autopilot for general aviation aircraft. In: International Conference on Informatics, Electronics and Vision (ICIEV), Dhaka, Bangladesh, pp. 490–495. IEEE (2016). https://doi.org/10.1109/ICIEV.2016.7760051
19. Vishal, Ohri, J.: GA tuned LQR and PID controller for aircraft pitch control. In: 6th India International Conference on Power Electronics (IICPE), Kurukshetra, India, pp. 1–6. IEEE (2014). https://doi.org/10.1109/IICPE.2014.7115839
20. Zaliskyi, M., Solomentsev, O., Kozhokhina, O., Herasymenko, T.: Statistical data processing for condition-based maintenance. In: Signal Processing Symposium (SPSympo), Krakow, Poland, pp. 232–235. IEEE (2019). https://doi.org/10.1109/SPS.2019.8882103
21. Kozhokhina, O., Yutskevych, S., Radchenko, O., Gribov, V., Chuzha, O.: Method of statistical data processing for two-stage fatigue tests. In: 10th International Conference on Advanced Computer Information Technologies (ACIT), Deggendorf, Germany, pp. 160–164. IEEE (2020). https://doi.org/10.1109/ACIT49673.2020.9208864
22. Kuzmin, V., Zaliskyi, M., Kozhokhina, O., Shcherbyna, O., Odarchenko, R.: Statistical analysis of wind turbine operational data. In: 15th International Conference on the Experience of Designing and Application of CAD Systems (CADSM), Polyana, Ukraine, pp. 1–4. IEEE (2019). https://doi.org/10.1109/CADSM.2019.8779354

Researching Influence of Vortex Generators on Aircraft Aerodynamic Characteristics

Oleksander Zhdanov, Valerii Orlianskyi, and Olha Sushchenko(✉)

National Aviation University, Liubomyra Huzara Avenue, 1, Kyiv 03058, Ukraine
sushoa@ukr.net

Abstract. This paper deals with researching the influence of vortex generators on the aircraft's aerodynamic characteristics. The principle of operation of vortex generators is explained. The full characteristics of blowing models of profiles are given. The process of designing blowing models is described including technological features. Samples of models of vortex generators printed on a 3-D printer are shown. The algorithm for creating the shape of the volumetric vortex generator ensures its formalization. The process of developing the shape of the vortex generators is explained. The profile models were studied in the wind tunnel. Different angles of attack during these experiments were used. The interconnection between the intensity of the vortex generator and the angle of attack is explained. The results of researching the influence of vortex generators on the aerodynamic characteristics of the aircraft are represented. The graphical dependences of the influence of the turbulator and vortex generator on the aerodynamic characteristics of aircraft are given. The obtained experimental results can be useful for the aircraft of the wide class.

Keywords: Vortex Generators · Blowing models · Profiles · Aerodynamic Characteristics · Angle of Attack · Experimental Results

1 Introduction

Nowadays, improving of safety and flight quality is a topic of the day. The aerodynamic characteristics of an aircraft influence greatly these factors. This research is aimed at expanding the range of flight angles of attack, eliminating the phenomenon of aerodynamic hysteresis, stalling and corkscrew, global separation of the airflow on the wing and control bodies of aircraft in subsonic flight modes with the help of volumetric vortex generators on the leading edge of the wing.

Volumetric vortex generators of a streamlined shape accelerate the flow around the leading edge of the wing, and compress the flow, which breaks off at the upper sharp edge, generating stable longitudinal vortex bundles along the chord of the wing [1–3]. Composite vortices in the form of a vortex slat of the type with several vortex generators create a vortex curtain, which, interacting with the pressure gradient on the wing and reaching the trailing edge, leads to a change in the integral aerodynamic characteristics of both lift force, drag force and longitudinal moment. The traditional model of flow

I. Ostroumov and M. Zaliskyi (Eds.): ACASD 2024, LNNS 992, pp. 410–422, 2024.
https://doi.org/10.1007/978-3-031-60196-5_30

around the wing at high angles of attack, which is based on the development of transverse vortices [4], is replaced by a longitudinal vortex model. The research of a new type of flow on the wing is necessary for the optimization of the parameters of the vortex generators and the creation of the optimal model of the vortex-active wing.

The designed three-dimensional vortex generators implement the idea of "inflow" on the leading edge of the wing in the form of mini inflows installed behind the wing span in the form of a vortex slat.

The vortex slat organizes the formation of an unsteady flow in the form of a scale-ordered model of a system of longitudinal vortices, which can reach the trailing edge of the wing and, thereby, leads to the destruction of the traditional vortex formation in the form of transverse large-scale vortices, which cause the phenomena of global separation of the airflow of an explosive nature and the phenomenon of aerodynamic hysteresis.

Volumetric vortex generators create a new type of unsteady longitudinal vortex flow around the wing, which is poorly researched. Studying the features of the longitudinal vortex flow around the wing will allow us to optimize vortex generators, and improve the stability and controllability of aerodynamic structures.

In this paper, it is proposed to use the energy of the oncoming flow at the leading edge of the wing with the help of volumetric vortex generators, which provide more effective acceleration of the flow due to the streamlined shape. Such an approach ensures the improvement of aerodynamic characteristics of an aircraft of the wide class.

The paper is organized as follows. In Sect. 2 we consider principles of creating models of aerodynamic profiles. In Sect. 3. Features of developing volumetric vortex generators of aerodynamic profiles are described. The graphical dependencies representing experimental results are represented in Sect. 4.

2 Blowing Models of Aerodynamic Profiles

The object of research is a set of blowing models of the compartment of the aerodynamic surface of various profiles. The chosen airfoils include the well-known and used in experimental studies symmetric airfoil NACA0012, asymmetric airfoils NACA4412 and AV216mod2 of increased thickness, which are used on stabilizers, asymmetric airfoil NACA23012, which is widely used on the blades of the main helicopter system, s-shaped airfoil BOING B-106, which is often used in aircraft of the flying wing type and the promising high-carrying profile WORTMANN FX63-137.

At the first stage, experiments have been carried out with the profile model NACA0012, which is frequently used in experiments. The blowing model has a rectangular shape in the plane, the span of the model is 350 mm, the chord is 175 mm, the area is 0.06125 m^2, the geometric elongation is $\lambda = 2$, the longitudinal axis of the model, relative to which its angular position changes, is located at a distance of 25% of the length of the chord or 0.04375 m from the toe. Vertical screens are installed on the sides of the profile model, which ensures a large effective elongation of the model approximately equal to 10.

The developed design of removable models of wing compartments provides weight studies in the wind tunnel at the angle of attack when turning the transverse support on which the model is installed [5, 6].

Research can be carried out for the tube and flight positions of the model in the range of angles of attack $0 \div 360°$ on separate sections with a length of 40°. Designed structural elements of models of wing compartments. Ribs for models of wing compartments are made on laser equipment [7]. In the same way, templates were made to determine the initial angle of installation of the model on the balance handle.

Structurally, the models have a power set, which consists of two tubular spars and a set of ribs in the amount of 8 pieces, which are strung onto the spars through the holes in the ribs. When assembling the power set, the gaps between the ribs are filled with foam, and all elements are fixed with glue [8]. In the elements of the power set, a through hole is provided for the passage of a transverse set of ribs in the amount of 8 pieces, which are strung on the spars through the holes in the ribs.

When assembling the power set, the gaps between the ribs are filled with foam and all elements are fixed with glue. In the elements of the power set, a through hole is provided for the passage of the transverse weight holder. The end ribs of the model are pasted with fan rings that protrude beyond the side plane of the rib and in which holes are made in the radial direction for fastening the model to the weight holder against rotation. Next, the profile models are covered with fiberglass impregnated with epoxy resin, which ensures acceptable mechanical properties of the model surface, as well as solvent resistance. Finally, the surface of the model is puttied, sanded, providing the shape of the profile, and painted. The set of blowing models for experimental researches is represented in Fig. 1.

Fig. 1. The set of six blowing models of the different profile for experimental researches.

3 Development of Volumetric Vortex Generators and Their Models

Models of aerodynamic profiles are composed of models of volumetric vortex generators, which are installed in their nose part. Based on literature information [9–11] on the use of various types of vortex generators, turbulators, aerodynamic inflows, and other aids of influence and control of vortex structures, the basic forms of three types of volumetric vortex generators shown in Fig. 2 in three projections were created.

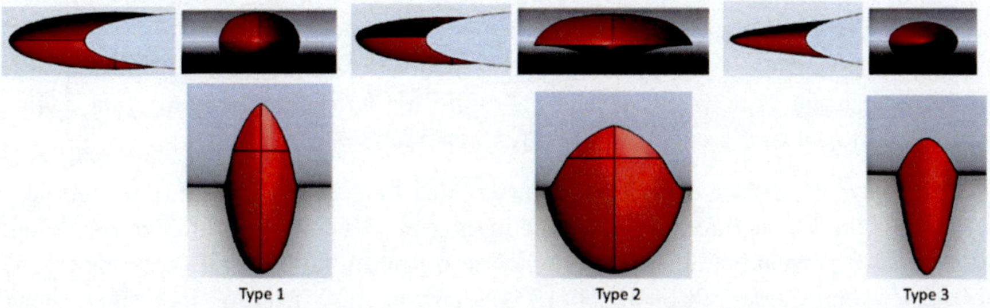

Fig. 2. Volumetric vortex generators of three types for the model of the profile NACA0012.

The first type of vortex generator has a spindle-like shape, and the source of vortex generation is the junction of the vortex generators and the profile; the second type of vortex generators has a vertically flattened shape with a sharp front and side edge of an elliptical shape in plan, which, at the same time as the joint, is also a source of vortex formation, undercuts are made on the lower surface to increase the load-bearing properties of the vortex generators; the third type of vortex generator also has a flattened shape, and when viewed from above, it has a cone outline with a rounded nose and a rounded side edge [6, 12, 13].

The opening angle of the cone is 28°, the source of the generation of longitudinal vortices is the junction of the vortex generators with the profile and the side edge of the vortex generators. The dimensions of volumetric vortex generators installed on the profile model can be changed by pushing the body of a volumetric vortex generator on the body of the profile or by scaling the dimensions of the volumetric vortex generator installed on the profile. In the first case, the extension of the vortex generator from the nose part of the profile by a certain amount is simulated, and in the second case, the installation of the vortex generator of a certain size is simulated [6, 7]. The algorithm for creating the shape of the volumetric vortex generator, which ensures its formalization, was as follows:

1. We create contours of the shape of the vortex generator in 3 projections. The profile of the vortex generator (side view) has a certain shape of the upper and lower surfaces, and has a chord and curvature when the middle line does not coincide with the chord. The contours of the profile can be specified by the coordinates of the upper and lower contours. The initial contour can be taken as the contour of the tip of the aerodynamic surface profile, on which the vortex generator is installed. Next, we modify the profile of the vortex generator by changing its dimensions, changing the curvature of the profile, changing the coordinates of the upper and lower contours, and describing them with a parabola, hyperbola, or other flat line.
2. We describe the shape of the vortex generator when viewed from the front with the help of ellipses (or segments of a circle), which can be different for the upper and

lower surfaces in the size of the vertical axis and the same in the size of the horizontal axis.
3. We form the contours of the vortex generator when viewed from above using flat lines of the second order, s-shaped lines, straight lines with a certain opening angle, or arc-shaped or elliptical.
4. We form rails at the junctions of vortex generators with the surface of the profile to ensure a certain smoothness of the transitions, we determine the radii of the rounding of the vortex-forming edge of the vortex generators.

Thus, its contours are described mainly by flat lines of the second order on the projections of vortex generators. The change in the external contour of the vortex generator in the section perpendicular to the chord line or the middle line of the vortex generator profile along the length of the chord is determined by its contours in the other two projections [6, 7]. For example, if the contour is defined by ellipses, then the change in the length of the axes of the ellipse occurs in accordance with the change in the external contours of the vortex generator viewed from above and from the side.

The dimensions of the vortex generators when installed on the aerodynamic profile can be changed by pushing the body of the vortex generators on the body of the profile or by scaling the dimensions of the vortex generators installed on the profile. In the first case, the extension of the vortex generators from the nose part of the profile by a certain amount is simulated, and in the second case, the installation of the vortex generators of a certain size is simulated [14]. The general geometric parameters of the layout of the vortex generators of various shapes and sizes will be the area of the vortex generators that protrudes beyond the front edge of the profile, the angle of the vortex generators installation (the angle between the vortex generators chord and the profile chord), the vortex generators installation step according to the span of the model. The creation of the shape of the vortex generators is aimed at ensuring the maximum intensity of the longitudinal vortex when it is installed on the surface of the profile model with minimal energy consumption of the external flow, ensuring the length of the vortex along the entire chord of the profile, eliminating the separation of the longitudinal vortex under the conditions of a change in the pressure gradient on the surface of the profile [15].

The source of vortex formation can be the point of contact between the aerodynamic surface and the volumetric vortex generator (the phenomenon of interference, which can be negative or positive depending on whether it causes an increase or decrease in frontal drag), or it can be the body of vortex generators itself, for example, the edge of the protruding part of vortex generators outside the profile and located at an angle to the airflow, or certain protrusions or depressions on the surface of the vortex generators. To reduce the negative effect of interference, and possibly to create a positive effect at the joints, it is necessary to provide for smooth transitions (glands) between the body of the vortex generators and the profile [16, 17].

The intensity of the vortex generator should depend on the angle of attack. In cruising modes, the influence of vortex generators should be minimal, and in the vicinity of critical angles of attack should have the greatest effect. This can be achieved by choosing the angular position of the vortex generator relative to the chord, and choosing the curvature of its chord, or the angular position should be variable depending on the airflow conditions. It is possible to use vortex generators only in case of going to critical angles of attack and for this constructively make them retractable. Another way of reducing or compensating for the increase in frontal drag from the installation of the airfoil is to increase the suction force both on the nose of the profile and on the surface of the airfoil due to the corresponding curvature of the shape of the airfoil surface and the acceleration of the airflow in the boundary layer on the surface of the nose of the airfoil and the profile.

It is in the indicated directions that the modification of the external shape of the vortex generator will take place in order to achieve the set goals in terms of improving aerodynamic characteristics of the aircraft at near-critical and post-critical angles of attack by ensuring the controlled separation of the boundary layer on the surface of the profile due to the generation of large longitudinal vortices by volumetric vortex generators, as well as in terms of ensuring acceptable flight characteristics in cruising flight modes [6]. Vortex generators for blowing models printed by 3-D printer are represented in Fig. 3. The view of the model with vortex generators in the working part of the wind tunnel in the presence of side screens during tests is shown in Fig. 4.

Fig. 3. Models of Vortex generators of 3 types, on blowing model, printed on the 3-D printer.

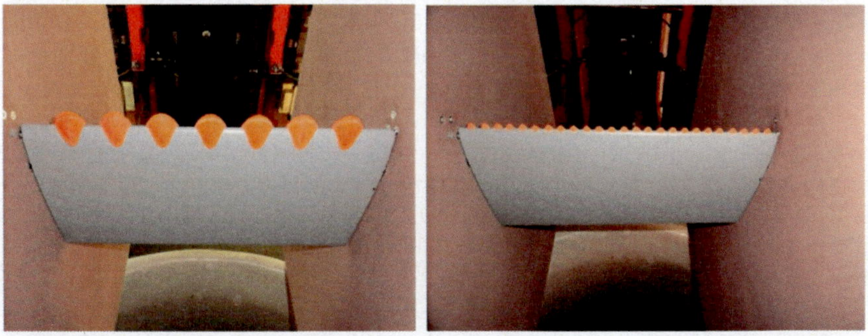

Fig. 4. View of the blowing model with VG in the presence of side screens during tests.

4 Experimental Results of Researching Influence of Vortex Generators on Aerodynamic Characteristics

According to the results of the tests, graphs of the dependencies of the aerodynamic characteristics of the profile model on the angle of attack were constructed in a form convenient for comparative analysis of the influence of the installation of vortex generators with different layout parameters. Based on the results of the tests, we will illustrate and analyze the influence of the studied factors on the aerodynamic characteristics of samples NACA 0012 profile model with volumetric vortex generators [1–3]. When analyzing the results to determine the optimal layouts, attention was primarily paid to the behavior of the dependence of the aerodynamic coefficient on an angle of the attack $C_{y\alpha} = f(\alpha)$ in the area of near-critical and post-critical angles of attack, on values and signs of C_{ymax}, α_{cr}, and the gradient of change $C_{y\alpha}$ in the super-critical area. The desired result of the application of vortex generators was to increase the magnitude of the lifting force, and eliminate its sharp decrease in the critical region due to the restructuring of the global separation at critical angles of attack by an ordered system of longitudinal vortices that generate vortex generators [6, 7].

The analysis of the effect of the installation of the volumetric vortex generator on the aerodynamic characteristics of the NACA0012 profile model was carried out by comparing aerodynamic coefficients as a function of the angle of attack for a clean profile and a profile with installed vortex generators, plotted on the same graph. Although the dependencies of the aerodynamic coefficients characterize the change in the total loads [18, 19], they are quite informative in order to clarify the differences in the behavior of the dependencies and put forward assumptions about the change in the structure of the flow around the upper surface of the model when reaching critical angles of attack and flow disruption. Graphs of aerodynamic characteristics for the layout of the vertex generator model (type 1) in the amount of 5 pieces, with size 20%, installed with a step of 70 mm are represented in Fig. 5 and 6. These Figs allow us to consider and analyze basic specific influences of mounting vortex generators on the aerodynamic characteristics of the profile models $C_{y\alpha} = f(\alpha)$. As we can see, the application of vortex generators significantly changes the behavior of dependencies $C_{y\alpha} = f(\alpha)$ in the near-critical and post-critical areas, depending on the parameters of the composition and shape of the vortex generators. Due to mounting vortex generators, it is possible to achieve such a behavior of the lifting force in the critical area, where there will be no critical angle of attack. In the area of larger than critical angles of attack, more than 24°, the increase in lifting force occurs mainly due to the increase in support on the lower surface of the profile, as well as in the case of the absence of vortex generators. The graphs in Figs. 5, and 6 make it possible to consider and analyze the main effects of the installation of vertex generators on the aerodynamic characteristics of the profile model. The analysis of the effect of the installation of the air conditioner on the aerodynamic characteristics of the NACA0012 profile model was carried out by comparing aerodynamic coefficients as a function of the angle of attack for a clean profile and a profile with installed vortex generators, plotted on the same graph.

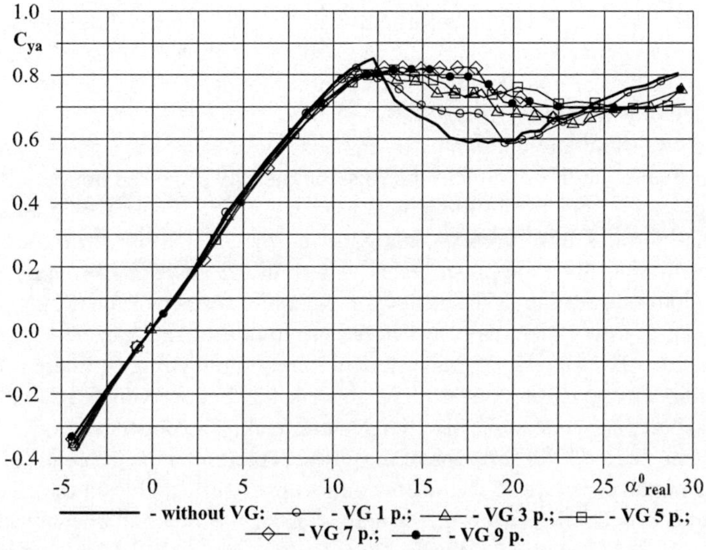

Fig. 5. Influence of the quantity and the step of location of the vortex generator of the type 1 with the relative dimension of the chord 7.5% and the angle of installation $\gamma = 0$ on the dependence $C_{y\alpha} = f(\alpha)$ on the profile model NACA0012 in conditions of the air flow $V = 18$ m/s (Reynold's number $Re \approx 200000$).

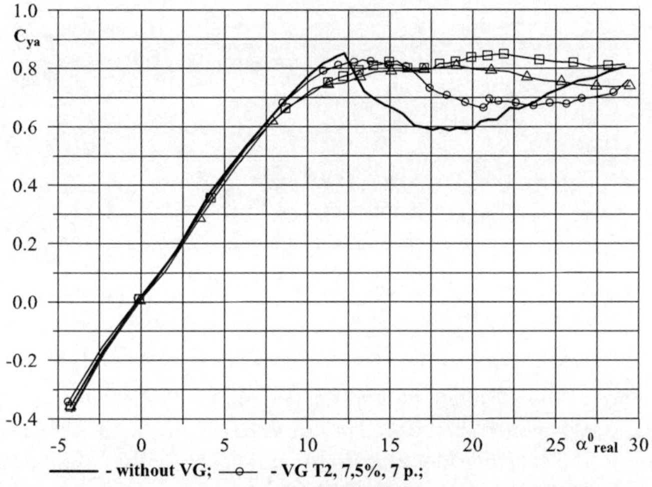

Fig. 6. The influence of parameters of layout of the vortex generator (type 2) with different dimensions and an angle of installation $\gamma = 0$ on the dependence $C_{y\alpha} = f(\alpha)$ on the profile model NACA0012 in conditions of the air flow $V = 18$ m/s (Reynold's number $Re \approx 200000$).

Nevertheless, they are quite informative in order to clarify the differences in the behavior of the dependencies and put forward assumptions about the change in the structure of the flow around the upper surface of the model when reaching critical angles of attack and flow disruption [6].

With vortex generators, nonlinearity in behavior of $C_{y\alpha} = f(\alpha)$ begins to manifest itself from smaller angles of attack. But this nonlinearity is caused not by the beginning of flow separation, but by its reorganization under the influence of interaction with longitudinal vortices from vortex generators and their departure from the surface. In the region of the critical angles of attack, where in the absence of a vortex generator there is a global separation of the airflow on the upper surface of the profile, in the case of the use of a vortex generator, the global separation of the transverse vortex of an explosive nature is suppressed by a system of longitudinal vortices. There is a structure of longitudinal vortices ordered in the transversal direction, which gradually departs from the airfoil surface at subcritical angles of attack. The dependence $C_{y\alpha} = f(\alpha)$ is smooth by the form. The description of the physical picture of the influence of the vortex generators on the structure of the airflow on its upper surface during the change of the angle of attack is confirmed by the visualization of the flow pattern with the help of mulberries. In Fig. 7, you can see the differences in the change in the structure of the flow around the upper surface of the profile model without vortex generators and with vortex generators. It should be noted that the change in the structure of the flow in the absence of vortex generators occurs suddenly and instantaneously, and for layout with vortex generators – smoothly and gradually.

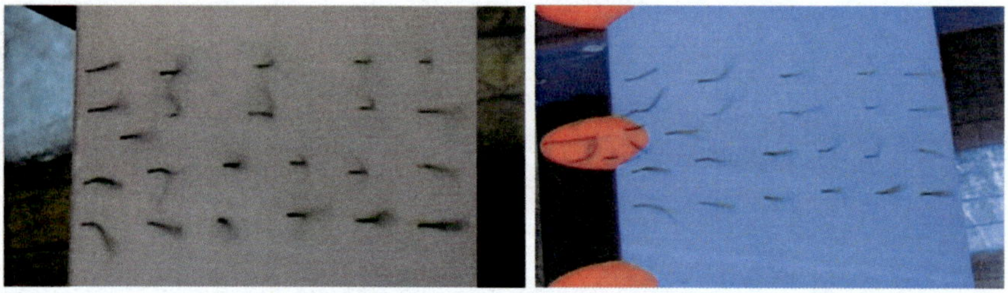

Fig. 7. Photo visualization of the flow around the upper surface of the profile model at post critical angles attacks without vortex generators (at left) and with vortex generators (at right).

The influence of the installation of vortex generators on the value of the frontal drag of the profile model is generally reduced to its increase in the area of the angles of attack corresponding to the cruising mode of flight and, accordingly, to the reduction of the aerodynamic quality, but the change of the frontal resistance in the area of the critical angle of attack occurs smoothly and its value becomes smaller in the future compared to the clean profile. Dependencies that describe the influence of vortex generators on the behavior of the drag coefficient $C_{x\alpha}$, aerodynamic quality K, and coefficient of the longitudinal moment m_{z1} on angle of the attack are shown in Fig. 8, 9, 10. The influence of the aerodynamic characteristics of vortex generators on the aerodynamic characteristics profile can be increased by increasing the area of the vortex generator.

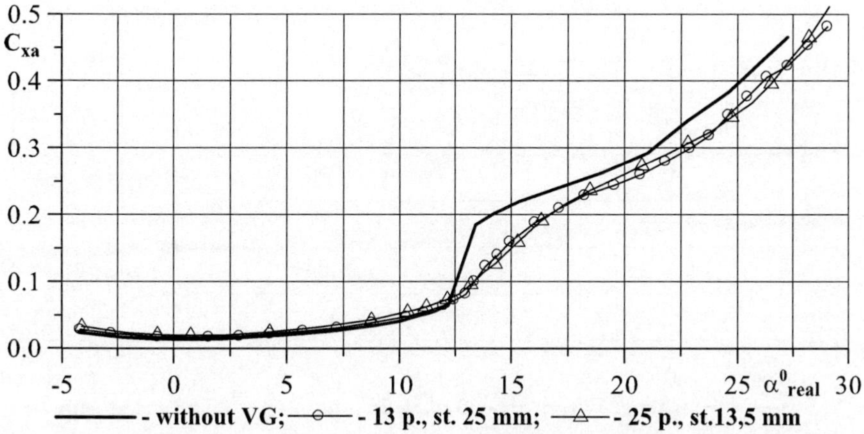

Fig. 8. The influence of the composition of small-sized 3.5% vortex generators (VG) of the third type on the drag coefficient $C_{x\alpha} = f(\alpha)$ of the profile model NACA0012 for $V = 18$ m/s.

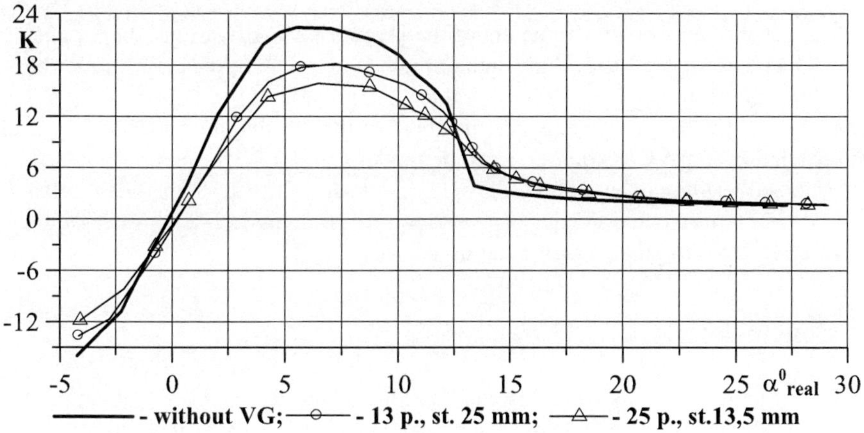

Fig. 9. The influence of the composition of small-sized 3.5% VG of the third type on the aerodynamic quality $K = f(\alpha)$ of the profile model NACA0012 for $V = 18$ m/s.

For example, insert a section of a certain thickness with a constant profile between two side surfaces of the vortex generator.

Small-sized vortex generators generate low-power longitudinal vortices, which at critical angles of attack do not prevent the complete separation of the airflow from the leading edge on the upper surface of the profile. The vortices do not reach the trailing edge of the profile. But still, small vortex generators increase the critical angle of attack and provide a smoother change in the critical area. Increasing the number of vortex generators from 13 to 25 worsened their effect on aerodynamic characteristics.

Probably due to the mutual influence of the vortex tracks (the tracks behind the vortex generators intersect with each other within the surface of the profile).

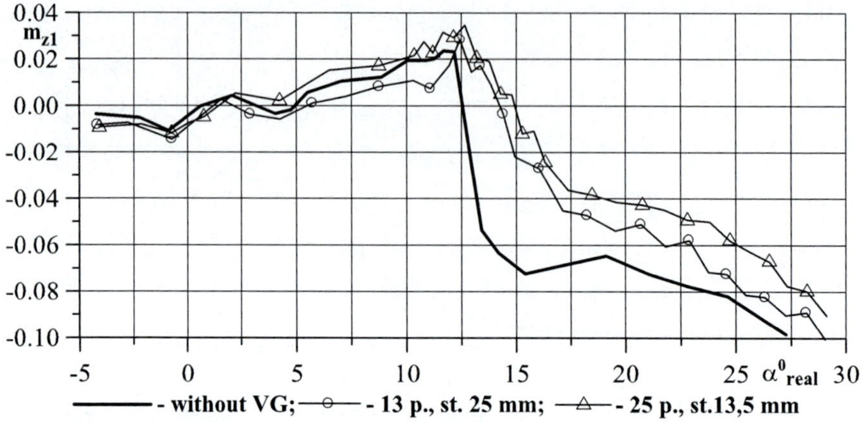

Fig. 10. The influence of the composition of small-sized 3.5% VG of the third type on the coefficient of the longitudinal moment $m_{z1} = f(\alpha)$ of the profile model NACA0012 for $V = 18$ m/s.

The general geometric parameters of the layout of VG of various shapes and sizes can be the chord of the VG and the area of the VG in the plan, which protrudes beyond the front edge of the profile in relative values, the volume of the VG body, the angle of installation of the VG (angle between the chord of the VG and the chord of the profile), the installation step VG on the scope of the model.

Obtained results can be useful for designing robust control systems for aircraft equipment [19–21], automated design of linear control systems [22–24], processing flight information [25, 26], and tests of aviation devices [27–29].

5 Conclusions

Different blowing models were designed and manufactured by means of 3-D printers. The influence of vortex generators on the aerodynamic characteristics of profile models has been researched.

Systematized multifactorial studies have laid the groundwork and determined the directions for further improvement of the parameters of layouts using vortex generators. First of all, it is necessary to improve the shape of the vortex generators in terms of increasing the power of vortex generation, creating conditions for the vortex to stick to the surface, and increasing the speed of the airflow that flows around the nose of the profile and the vortex generator itself. Providing such requirements in the nose part of the layout should increase the suction force and, accordingly, reduce the pressure resistance and increase the lifting force. Secondly, it is considered promising to use the own characteristics of the vortex generators to improve the aerodynamic characteristics of layouts of perspective aircraft.

In future work, it is proposed to describe features of the experiment in the wind tunnel and to improve the technique of processing measurement information. Also, other types of the profile models in addition to NACA0012 will be considered.

References

1. Udartsev, E.P., et al.: Vortex slat. Patent on useful model. No. 94179 (2014)
2. Udartsev, E. P., et al.: The generator of longitudinal vortices. Patent on useful model. No. 111701 (2016)
3. Udartsev, E.P., Zhdanov, O. I., Shcherbonos, O. I.: Vortex generator. Patent of Ukraine. No. 49403 (2010)
4. Lin, C.J.: Review of research on low-profile vortex generators to control boundary-layer separation. Prog. Aerosp. Sci. **38**, 389–420 (2002). https://doi.org/10.1016/S0376-0421(02)00010-6
5. Udartsev, E.P., Alekseenko, S.I., Zhdanov, O.O.: Unsteady aerodynamics of a vortex jet wing of an unmanned aerial vehicle at large and closed angles of attack. Electron. Control Syst. **4**(46), 40–45 (2015)
6. Shcherbonos, A.G.: Aerodynamic characteristics of a wing with vortex generators in the conditions of non-stationary flow (in Ukrainian). Dissertation of Ph.D. NAU, Kyiv, Ukraine (2012)
7. Oleksienko, S.I.: Control of flow separation using longitudinal vortices generators (in Ukrainian). Dissertation of Ph.D. NAU, Kyiv, Ukraine (2018)
8. Mode of Direct Access. https://airfoiltools.com/airfoil/details?airfoil=boe106-il. Accessed 11 Jan 2024
9. Johari, H., Henoch, C., Custodio, D., Levshin, A.: Effects of leading-edge protuberances on airfoil performance. AIAA J. **45**(11), 2634–2642 (2007). https://doi.org/10.2514/1.28497
10. Delinero, J.S., Di Leo, J.M., Camocardi, M.E: Vortex generators effect on low Reynolds number airfoils in turbulent flow. Int. J. Aerodyn. **2**(1) 1–14 (2020). https://doi.org/10.1504/IJAD.2012.046539
11. Johari, H., Henoch, C. H., Custodio, D.: Visualization of flow on hydrofoils with leading edge protuberances. In: ISFV, vol. 13 (2008)
12. Custodio, D., Henoch, C.W., Johari, H.: Aerodynamic characteristics of finite span wings with leading-edge protuberances. AIAA J. **53**(7), 1878–1893 (2015). https://doi.org/10.2514/1.J053568
13. Nguyen, T.V., Loan, P.T.K., Dung, D.V.: Numerical study of the effect of leading edge protuberances on wing hydrodynamic performance. In: National Conference on Fluid Mechanics. Vietnam (2011)
14. Malipeddi, A.K., Mahmoudnejad, N., Hoffmann, K.A.: Numerical analysis of effects of leading-edge protuberances on aircraft wing performance. J. Aircr. **49**(5), 1336–1344 (2012). https://doi.org/10.2514/1.C031670
15. Divekar, P.S., Ekbote, T.: Design of aerodynamic of an airplane wing. IJSDR **4**, 59–64 (2019)
16. Aftab, S.M.A., Razak, N.A., Mohd Rafie, A.S., Ahmad, K.A: Mimicking the humpback whale: an aerodynamic perspective. Progr. Aerosp. Sci. **84**, 48–69 (2016). https://doi.org/10.1016/j.paerosci.2016.03.002
17. Sushchenko, O.A., Bezkorovainyi, Y.N.: Improvement of UAV positioning by information of inertial sensors. In: IEEE 5th International Conference on Methods and Systems of Navigation and Motion Control (MSNMC), pp. 151–155. IEEE, Kyiv (2018). https://doi.org/10.1109/MSNMC.2018.8576307
18. Sushchenko, O.A., Tunik, A.A.: Robust stabilization of UAV observation equipment. In: 2nd International Conference on Actual Problems of Unmanned Air Vehicles Development (APUAVD), pp. 176–180. IEEE, Kyiv (2013). https://doi.org/10.1109/APUAVD.2013.6705318

19. Sushchenko, O.A., Bezkorovainyi, Y.N., Novytska, N.D.: Dynamic analysis of nonorthogonal redundant inertial measuring units based on MEMS-sensors. In: IEEE 38th International Conference on Electronics and Nanotechnology (ELNANO-2018), pp. 464–469. IEEE, Kyiv (2018). https://doi.org/10.1109/ELNANO.2018.8477553
20. Voliansky, R.S., Sadovoi, A.V.: The transformation of linear dynamical object's equation to Brunovsky canonical form. In: IEEE 4th International Conference Actual Problems of Unmanned Aerial Vehicles Developments (APUAVD), pp. 196–199. IEEE, Kyiv (2017). https://doi.org/10.1109/APUAVD.2017.8308808
21. Voliansky, R., et al.: The symmetry principle usage to design the previously disturbed linear control systems. In: IEEE 17th International Conference on Computer Sciences and Information Technologies, pp. 531–534. IEEE, Lviv (2022). https://doi.org/10.1109/CSIT56902.2022.10000725
22. Voliansky, R., et al.: Sliding mode control for DC generator with uncertain load. In: IEEE 15th International Conference on Advanced Trends in Radioelectronics, Telecommunications and Computer Engineering (TCSET), pp. 313–316. IEEE, Lviv-Slavske (2020). https://doi.org/10.1109/TCSET49122.2020.235446
23. Sushchenko, O.A., Bezkorovainyi Y.M., Golitsyn, V.O.: Fault-tolerant inertial measuring instrument with neural network. In: 40th International Conference on Electronics and Nanotechnology (ELNANO), pp. 797–801. IEEE, Kyiv (2020). https://doi.org/10.1109/ELNANO50318.2020.9088779
24. Sushchenko, O.A., Bezkorovainyi, Y.N., Novytska, N.D.: Theoretical and experimental assessments of accuracy of nonorthogonal MEMS sensor arrays. Eastern-Eur. J. Enterp. Technol. **39**(93), 40–49 (2018). https://doi.org/10.15587/1729-4061.2018.131.945
25. Zhiteckii, L.S., Azarskov, V.N., Solovchuk, K.Y., Sushchenko, O.A.: Discrete-time robust stead-state control of nonlinear multivariable system. IFAC Proc. Vol. **47**(3), 8140–8145 (2014). https://doi.org/10.3182/20140824-6-ZA-1003.01985
26. Solomentsev O., et al.: Data processing through the lifecycle of aviation radio equipment. In: IEEE 17th International Conference on Computer Sciences and Information Technologies, pp. 146–151. IEEE, Lviv (2022). https://doi.org/10.1109/CSIT56902.2022.10000844
27. Ostroumov, I., et al.: Relative navigation for vehicle formation movement. In: 3rd KhPI Week on Advanced Technology, pp. 1–4. IEEE, Kharkiv (2022). https://doi.org/10.1109/KhPIWeek57572.2022.9916414
28. Hryshchenko, Y., Romanenko, V., Zaliskyi, M.: Quality assessment of aircraft glide path entrance. In: CEUR Workshop Proceedings, vol. 2711, pp. 649–660 (2020)
29. Sushchenko, O., Averyanova, Y., Ostroumov, I., Kuzmenko, N., Zaliskyi, M., Solomentsev, O., et al.: Algorithms for design of robust stabilization systems. In: Gervasi, O., Murgante, B., Hendrix, E.M.T., Taniar, D., Apduhan, B.O. (eds.) ICCSA 2022. LNCS, vol. 13375, pp. 198–213. Springer, Cham (2022). https://doi.org/10.1007/978-3-031-10522-7_15

Modified Chua's Circuit in Different State Spaces

Roman Voliansky(✉) and Nina Volianska

National Technical University of Ukraine "Igor Sikorsky Kyiv Polytechnic Institute", Beresteiskyi Avenue, 37, Kyiv 03056, Ukraine
volianskyua.fm

Abstract. The paper deals with the development of theoretical backgrounds to design chaotic systems. These backgrounds are based on the use of coordinate transformations to represent system dynamics of known chaotic systems in various state spaces. The use of such an approach allows us to combine one state variable in different state spaces. In this case novel chaotic motions occur as result of changing inner system state variables. The main benefit of such an approach is a fact that if chaotic oscillations is generated in one state space then they occur in others too. This fact does not require to check Lyapunov exponents and other chaotic indicator for the transformed system. As a basis to perform coordinate transformation we consider defining series of Lie derivatives for the selected output state variable. Because of necessity to differentiate this variable one should consider chaotic system with differentiable nonlinearities. In case of the usage piecewise linear functions one should perform their smoothing by using various differentiable functions. The transformed chaotic system can be implemented in both continuous and discrete time domains by using various elements.

Keywords: Chua's Circuit · Chaotic Dynamic · Canonical Equations · Discrete-Time Model · Modal Control System

1 Introduction

Modern aviation cannot be imagined without communications [1, 2]. These communications are performed in both manual and automatic modes [3] to perform all flight stages and make them safe and reliable [4, 5]. It can be done only by transmition to the plane big amount of information [6, 7] including but not limiting landing [8–10] and satellite [11] navigation, air traffic control [12–14], traffic collision [15–17] and much more that cannot be imagined without information and its transmission. Due to the huge importance of flight safety problem various authors develop and implement various models [18] and algorithms [19, 20] to predict possible traffic conflicts and, thus, make flights safe.

All above-given methods and approaches allows us claim the importance of information and its authencity. Since the wrong generated information can cause catastrophic consequences, one should take into account security of information exchange channels.

I. Ostroumov and M. Zaliskyi (Eds.): ACASD 2024, LNNS 992, pp. 423–440, 2024.
https://doi.org/10.1007/978-3-031-60196-5_31

Analysis of modern ways to establish communication channels shows that some authors suggest using optical communications [21], including light [22] and laser [23], and ultra-short wave communications [24]. It is clear that from secure viewpoint such approaches are more secured then using conventional radio channel but it requires a direct visibility of both transmitter and receiver. So, it can be used only for some very short flights, for example the training on small aerial vehicles and so on.

Traditional radio communication remains the main channel to communicate between plane and various services. It is clearly understood that this channel can be intercepted and transmitted data can be changed by some third person with necessary technical possibility. Since it can disrupt the flight safety, we offer to encrypt transmitted information by using various dynamical systems with chaotic dynamic [25–27].

Study such type dynamical systems is one most rapidly developed area of modern control theory and its applications. Great interest to chaotic system is caused by their unique properties from one hand and great area to apply different control methods [28–30] and approaches [32, 33] to study and design systems which motions cannot be predicted.

Nevertheless, a huge attention to the chaotic systems the design of such systems is performed by using some empirical methods which do not allow to make any generalization about these systems structure or features. That is why we offer to use coordinate transformations to transform known system which usage can be compromised by known studies into various phase domain where system dynamic is not studied and designed in such a way system can be used as true random generator to establish secured data transmission.

Our paper is organized as follows: at first, we transform well-known dimensionless Chua's equations to design model in the discrete time domain. Then, we perform smoothing of Chua's diode characteristic by replacing its piecewise linear dependencies with polynomial ones near the fracture points. At third, we apply based on Lie groups differential geometry approach to rewrite Chua's circuit equations with smoothed nonlinearity into canonical space in both continuous and discrete-time domains. At last, we make a conclusion.

2 Discrete-Time Chua's Circuit Modeling

2.1 Chua's Circuit Model in the Normal State Space

Let us consider a dynamic of well-known Chua's circuit [31]

$$\dot{y}_1 = \alpha_1(-y_1 - g(y_1) + y_2); \quad \dot{y}_2 = y_1 - y_2 + y_3; \quad \dot{y}_3 = -\alpha_2 y_2, \tag{1}$$

here α and β are circuit factors, which depends on its parameters

$$\alpha_1 = \frac{C_2}{C_1}; \; \alpha_2 = \frac{R_1^2 C_2}{L_1}, \tag{2}$$

where C_1, C_2, R_1, and L_1 are circuit capacities, resistance and inductance, and $g(y_1)$ is a Chua's diode R_N characteristic (Fig. 1), which is given by following expression

$$g(y_1) = \begin{cases} a_0 y_1 + a_0 - a_1 & \text{if} \quad y_1 \le -1; \\ a_1 y_1 & \text{if} \quad -1 < y_1 < 1; \\ a_0 y_1 + a_1 - a_0 & \text{if} \quad y_1 \ge 1, \end{cases} \tag{3}$$

here a_0 and a_1 are Chua's diode characteristics factors.

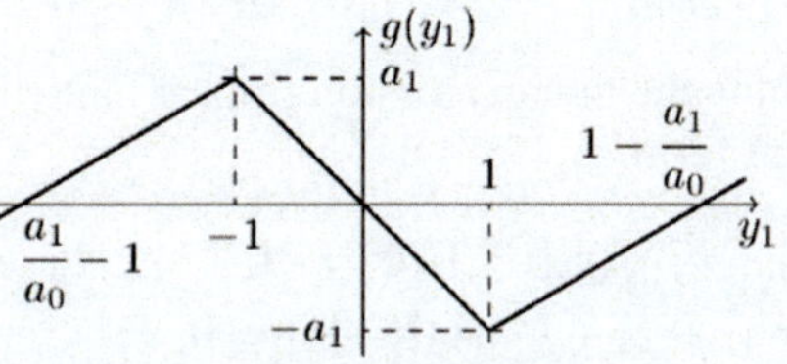

Fig. 1. Chua's diode characteristic.

The use of digital devices to implement the considered Chua's equations requires transformation of (1) into discrete-time domain. Since various transformation schemes are known, we use here the generalized Tustin transformation here, which allows to approximate continuous-time derivative in the discrete-time domain as follows

$$\frac{d}{dt} \approx \frac{2}{T}\frac{1 - z^{-1}}{1 + z^{-1}} \tag{4}$$

where z^{-1} is a shift operator and T is a discretization period.

If one applies (4) to (1), he can rewrite Chua's equation for discrete-time domain in such a way

$$\begin{gathered} (1 - z^{-1})y_1 = \tfrac{T}{2}(1 + z^{-1})a_1(-y_1 - g(y_1) + y_2); \\ (1 - z^{-1})y_2 = \tfrac{T}{2}(1 + z^{-1})(y_1 - y_2 + y_3); \quad (1 - z^{-1})y_3 = -\tfrac{T}{2}(1 + z^{-1})a_2 y_2 \end{gathered} \tag{5}$$

or by taking into account that shift operator means shifting j-th state variable on the one discretization period

$$y_j[i - 1] = z^{-1} y_j[i], \tag{6}$$

the following finite difference equations can be written down

$$\begin{gathered} y_1[i] = y_1[i - 1] + \tfrac{a_1 T}{2}(-y_1[i] - y_1[i - 1] - g(y_1[i]) \\ - g(y_1[i - 1]) + y_2[i] + y_2[i - 1]); \\ y_2[i] = y_2[i - 1] + \tfrac{T}{2}(y_1[i] + y_1[i - 1] - y_2[i] - y_2[i - 1] + y_3[i] + y_3[i - 1]); \\ y_3[i] = y_3[i - 1] - \tfrac{a_2 T}{2}(y_2[i] + y_2[i - 1]). \end{gathered} \tag{7}$$

Equations (7) define interrelations between current value of Chua's state variables $y_j[i]$ and their previous values $y_j[i - 1]$. . One can find appropriate formulas by solving (7) for state variables of the considered circuit.

It is clear that the first equation in (7) is a nonlinear and the others are linear ones. That is why we offer to split their solutions and, at first, solve the second and third equations to define $y_2[i]$, and $y_3[i]$ as follows

$$\begin{aligned} y_2[i] &= b_{21c}y_1[i] + b_{21p}y_1[i-1] + b_{22p}y_2[i-1] + b_{23p}y_3[i-1]; \\ y_3[i] &= b_{31c}y_1[i] + b_{31p}y_1[i-1] + b_{32p}y_2[i-1] + b_{33p}y_3[i-1], \end{aligned} \tag{8}$$

$$\begin{aligned} b_{21p} &= \frac{2T}{\beta T^2+2T+4}; \; b_{22p} = -\frac{\beta T^2+2T-4}{\beta T^2+2T+4}; \; b_{23p} = \frac{4T}{\beta T^2+2T+4}; \\ b_{31p} &= -\frac{\beta T^2}{\beta T^2+2T+4}; \; b_{32p} = -\frac{4\beta T}{\beta T^2+2T+4}; \; b_{33p} = -\frac{\beta T^2-2T-4}{\beta T^2+2T+4}. \end{aligned} \tag{9}$$

After that we substitute the first expression from (8) into the first equation of (7) and rewrite it into an implicit form

$$\begin{aligned} & b_{110}y_1[i] + b_{11}g(y_1[i]) + b_{111}y_1[i-1] + b_{11}g(y_1[i-1]) \\ & + b_{121}y_2[i-1] + b_{131}y_3[i-1] = 0, \end{aligned} \tag{10}$$

$$\begin{aligned} b_{11(0)} &= \tfrac{1}{2}a_1T(b_{21}-1) - 1; \quad b_{111} = \tfrac{1}{2}a_1T(b_{21}-1) + 1; \\ b_{121} &= \tfrac{1}{2}a_1T(b_{22}+1); \quad b_{131} = \tfrac{1}{2}a_1Tb_{23}; \quad b_{11} = \tfrac{1}{2}a_1T. \end{aligned} \tag{11}$$

Since in the most general form the Chua's diode characteristic can be more complex then given by piece-wise linear function (3), we offer to solve (10) in the general case by using Newton's method

$$\begin{aligned} y_1[i]_j = y_1[i]_{j-1} &- \frac{b_{11(0)}y_1[i]_{j-1} + b_{11}g(y_1[i]_{j-1} + b_{11}g(y_1[i-1]) + b_{111}y_1[i-1]}{b_{110} + b_{11}\left.\frac{\partial g(y_1)}{\partial y_1}\right|_{y_1=y_1[i]_{j-1}}} \\ &- \frac{b_{121}y_2[i-1] + b_{131}y_3[i-1]}{b_{110} + b_{11}\left.\frac{\partial g(y_1)}{\partial y_1}\right|_{y_1=y_1[i]_{j-1}}}, \end{aligned} \tag{12}$$

here for the considered case

$$\left.\frac{\partial g(y_1)}{\partial y_1}\right|_{y_1=y_1[i]_{j-1}} = \begin{cases} a_0 \text{ if } \left|y_1[i]_{j-1}\right| \geq 1; \\ a_1 \text{ if } \left|y_1[i]_{j-1}\right| < 1; \end{cases} \tag{13}$$

Thus, the simulation of Chua's circuit in the discrete-time normal state space can be considered as performing calculations according following algorithm:

1. Taking the initial conditions $y_{1(0)}, y_{2(0)}, y_{3(0)}$ and using them to assign values of state variables in (12)

$$y_1[1]_0 = y_{1(0)}; \; y_1[0] = y_{1(0)}; \; y_2[0] = y_{2(0)}; \; y_3[0] = y_{3(0)}. \tag{14}$$

2. Setting up counter variable i in one.
3. Defining $y_1[i]$ by using iterative expression (12). This expression has iterative solution which interrupts on j-th step, when following condition

$$\epsilon > |y_1[i]_j - y_1[i]_{j-1}|, \tag{15}$$

where ϵ is some predefined accuracy, becomes true.

4. Using value $y_1[i]$ to define inner circuit state variables by performing calculations according to (8).
5. Storing defined values of the state variables in an appropriate memory blocks, assigning initial value for the $y_1[i]_0$.
6. Incrementing counter by one and continue from step 3.

The above-given algorithm gives us possibility to define Chua's circuit dynamic in the normal state space which coordinate basis is defined by using state variables as coordinate axes. If this algorithm is used to simulate Chua's circuit it terminates when some condition like finish time is being reached. In case of using the considered algorithm to produce chaotic oscillations it can be interrupted when other conditions is reached. For example, it stops when it is no need to produce the chaotic signals.

We use the proposed algorithm to get simulation results similar to well-known ones (Fig. 2).

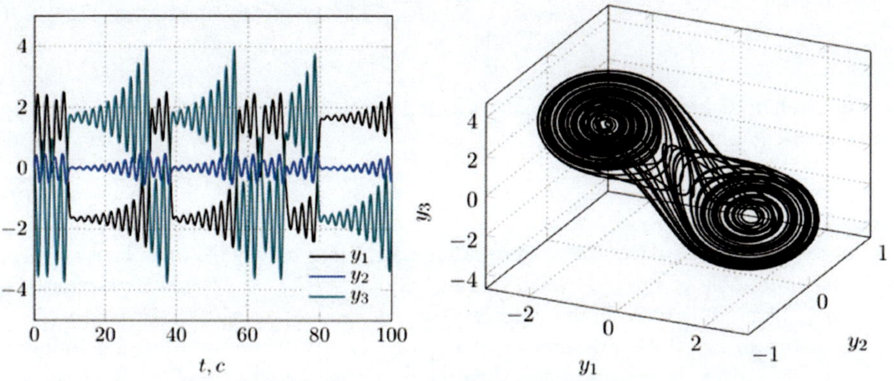

a) Chaotic oscillations   b) 3D chaotic attractor

Fig. 2. Chua's circuit simulation results.

Since these results are well-known, their usage can cause some data leakage while secured data transmission channels are being established. That is why we offer to avoid system discredit by considering it in another state space where system dynamic is not clearly studied.

2.2 Smoothing Chua's Diode Characteristic

It is clear that the number of coordinate axes where system dynamic is defined can be as large as you want and canonical state space is among them. The orths of this state space are defined as derivatives of the output variable. So, if one assumes that the system output is y_1 variable, he can define other state variables in as follows

$$x_1 = y_1;\ \dot{x}_1 = \dot{y}_1 = x_2;\ \dot{x}_2 = \ddot{y}_1 = x_3. \tag{16}$$

In case when system dynamic is defined by differentiable nonlinear functions, it is a very convenient to use the differential geometry methods and Lie group theory to rewrite (1) in such a way

$$\dot{x}_1 = x_2;\ \dot{x}_2 = x_3;\ \dot{x}_3 = h(x_1, x_2, x_3), \tag{17}$$

$$h(x_1, x_2, x_3) = L_{\boldsymbol{F}}^3 x_1 \tag{18}$$

where **F** means matrix-vector of right-hand expressions of (1)

$$\boldsymbol{F} = \begin{pmatrix} a_1(-y_1 - g(y_1) + y_2) \\ y_1 - y_2 + y_3 \\ -a_2 y_2 \end{pmatrix}, \tag{19}$$

$$L_{\boldsymbol{F}}^j x_1 = \begin{cases} L_{\boldsymbol{F}}^{j-1} x_1 \ \mathit{if}\ j \geq 2; \\ \nabla x_1 \boldsymbol{F} \ \ \mathit{if}\ j = 1; \\ \boldsymbol{F} \ \ \ \mathit{if}\ j = 0. \end{cases} \tag{20}$$

Interrelations between state variables y_i and derivatives x_i can be defined as solution of algebraic equation's system

$$x_i = L^{i-1} x_1,\ 1 \leq i \leq 3. \tag{21}$$

In the considered case, the system nonlinearity is a piecewise one. That is why it has discontinuous first derivative in fracture points $y_{f(1,2)} = \pm 1$ and its higher-order derivatives are not defined in these points.

We offer to avoid the problem of such function differentiating by using following mathematical trick: we assume that there is a neighborhood ε near the fracture point $y_{f(i)}$ where the considered piecewise linear function (3) becomes polynomial function. One can use the simplest second order polynomial here (Fig. 3).

Also, we assume that the neighborhood ε is so small that the difference between piecewise linear function y_{pwl} and piecewise polynomial one y_{poly} is negligible small and thus it does not affect on system dynamic.

Let us define the terms of the polynomials which smooth the considered piecewise Chua's diode characteristic. We define these polynomials by assuming that values of piecewise linear function and polynomial function in the intersection points y_{p1} and y_{p2} equal each other and also, they have the same derivatives in these points. It should be mentioned here, that if one defines parameters for polynomial which order is higher than two, he should assume that the second and higher derivatives for this high-order polynomial in these points equals to zero.

Due to the symmetry of the Chua's diode characteristic, we consider only left its part and assume that the similar results can be obtained for right parts of this characteristic as well.

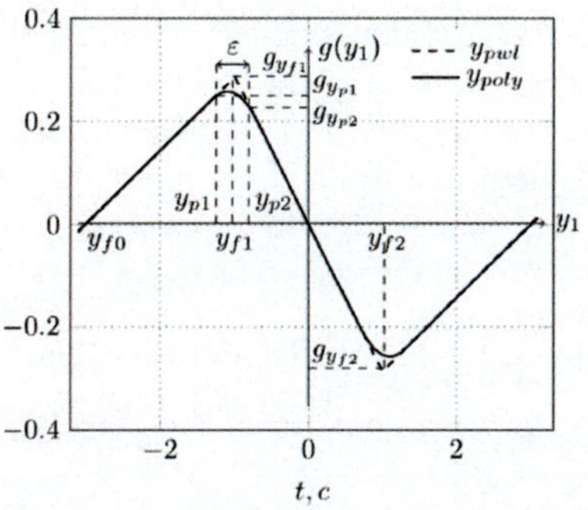

Fig. 3. Smoothed Chua's diode characteristic.

Let us write down equations for parts of piecewise linear characteristic in Fig. 3 as follows

$$f_1 = a_0 y_1 + a_0 - a_1; \quad f_2 = a_1 y_1. \tag{22}$$

The polynomial which smooths these characteristics is defined as follows

$$y_p = a y_1^2 + b y_1 + c; \ \dot{y}_p = 2a y_1 + b \tag{23}$$

and its first derivative can be written in such a way

$$\dot{y}_p = 2a y_1 + b. \tag{24}$$

We define unknown polynomial terms by substituting into (22)–(24) coordinates of points $\left(y_{f1}, g_{y_{f1}}\right)$ and $\left(y_{f2}, g_{y_{f2}}\right)$ and writing down following equations

$$\begin{gathered} a_0 - a_1 - a y_{p1}^2 + (a_0 - b) y_{p1} - c = 0; \\ -a y_{p2}^2 + (a_1 - b) y_{p2} - c = 0; \\ a_0 - 2a y_{p1} - b = 0; \quad a_1 - 2a y_{p2} - b = 0. \end{gathered} \tag{25}$$

If one solves (25) for unknown terms a, b, and c, he can write down following expressions for the right side of Chua's diode characteristic

$$\begin{aligned} a_{12} &= \frac{a_0 - a_1}{4(y_{p1}+1)}; \ b_{12} = \frac{(a_0+a_1)y_{p1}+2a_0}{2(y_{p1}+1)}; \\ c_{12} &= \frac{(a_0-a_1)y_{p1}^2+4(a_0-4a_1)y_{p1}+4(a_0-a_1)}{4(y_{p1}+1)}. \end{aligned} \tag{26}$$

Similar expressions can be written down for the right side of Chua's diode characteristic

$$\begin{aligned} a_{23} &= \frac{a_0 - a_1}{4(y_{p3}-1)}; \ b_{23} = \frac{(a_0+a_1)y_{p3}-2a_0}{2(y_{p3}-1)}; \\ c_{23} &= \frac{(a_0-a)y_{p3}^2+4(a_1-a_0)y_{p3}+4(a_0-a_1)}{4(y_{p3}-1)}. \end{aligned} \tag{27}$$

Analysis of terms (26) and (27) shows that smoothing polynomial can be defined by known piecewise linear function parameters and the points y_{pi} where this polynomial intersect with piecewise linear function. It is clear that these points defines the length of the above-mentioned neighbourhood ε

$$|\varepsilon| = y_{p2} - y_{p1}; \quad |\varepsilon| = y_{p4} - y_{p3} \tag{28}$$

and one can define them by using desired length of neighbourhood ε in such a way

$$y_{p1} = y_{f1} - 0.5|\varepsilon|;\ y_{p2} = y_{f1} + 0.5|\varepsilon|;\ y_{p3} = y_{f2} - 0.5|\varepsilon|;\ y_{p4} = y_{f2} + 0.5|\varepsilon|. \tag{29}$$

Resuming above-given formulas we offer following algorithm to define parameters of the smoothed polynomials:

1. Parameters of piecewise linear function are taken into account as initial information.
2. Length of neighbourhood ε is assumed to be known a priori as some small number.
3. Intersection points y_{pi} are defined by using (29).
4. These points are used to define the polynomials terms according to (27).

The use of function (23) with terms (26) and (27) gives us possibility to rewrite (3) in following way

$$g(y_1) = \begin{cases} a_0 y_1 + a_0 - a_1 & \textit{if} \quad y_1 \le y_{p1}; \\ a_{12} y_1^2 + b_{12} y_1 + c_{12} & \textit{if} \ y_{p1} < y_1 < y_{p2}; \\ a_1 y_1 & \textit{if} \ y_{p2} \le y_1 \le y_{p3}; \\ a_{23} y_1^2 + b_{23} y_1 + c_{23} & \textit{if} \ y_{p3} < y_1 < y_{p4}; \\ a_0 y_1 + a_1 - a_0 & \textit{if} \quad y_1 \ge y_{p4}. \end{cases} \tag{30}$$

We call (30) as polynomial-smoothed Chua's diode characteristic.

Simulation results of Chua's circuit with such characteristic are shown in Fig. 4. We study our system with assumption that ε neighbourhood length equals to 0.01.

As one can see from comparison simulation results in Fig. 4 and Fig. 2, our approach does not change system dynamic dramatically, but allows to form slightly different oscillations, which are caused by polynomial branches in Chua's dioide characteristic. This fact proves correctness of our approach.

Contrary to known polynomial Chua's diode characteristic the proposed approach allows us get the Chua's dioide characteristic which can be very close to well-known piecewise linear one if coordinates n pairs y_{p1}, y_{p2} and y_{p3}, y_{p4} are very close each other. But contrary to well-known piecewise linear characteristic of Chua's diode, the proposed characteristic is defined as differentiable function in all points.

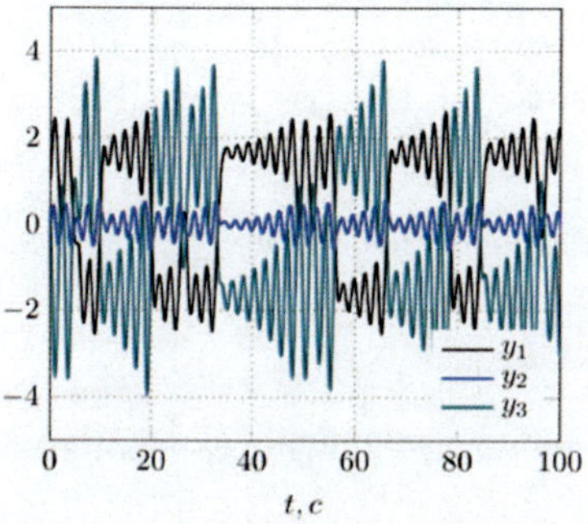

a) Chaotic oscillations

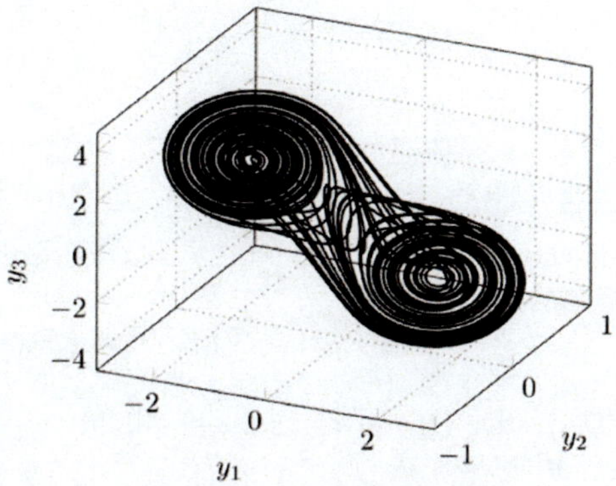

b) 3D chaotic attractor

Fig. 4. Chua's circuit simulation results with smoothed Chua's diode characteristic.

2.3 Chua's Circuit Model in the Canonical State Space

This fact gives us possibility to use expressions (18)–(21) to transform Chua's circuit equations into canonical form. We perform this transformation by assuming that output variable equals to y_1 state variable.

At first, we rewrite function (19) in the following way

$$\boldsymbol{F} = \begin{pmatrix} -\alpha y_1 + \alpha y_2 - \begin{cases} \alpha a_0 y_1 + \alpha a_0 - \alpha a_1 & \textit{if} \quad y_1 \leq y_{p1}; \\ \alpha a_{12} y_1^2 + \alpha b_{12} y_1 + \alpha c_{12} & \textit{if } y_{p1} < y_1 < y_{p2}; \\ \alpha a_1 y_1 & \textit{if } y_{p2} \leq y_1 \leq y_{p3}; \\ \alpha a_{23} y_1^2 + \alpha b_{23} y_1 + \alpha c_{23} & \textit{if } y_{p3} < y_1 < y_{p4}; \\ \alpha a_0 y_1 + \alpha a_1 - \alpha a_0 & \textit{if} \quad y_1 \geq y_{p4}. \end{cases} \\ y_1 - y_2 + y_3 \\ -\beta y_2 \end{pmatrix}. \tag{31}$$

Then, we use (31) to define the first Lie derivative

$$L_F x_1 = -\alpha y_1 + \alpha y_2 - \begin{cases} \alpha a_0 y_1 + \alpha a_0 - \alpha a_1 & \textit{if} \quad y_1 \leq y_{p1}; \\ \alpha a_{12} y_1^2 + \alpha b_{12} y_1 + \alpha c_{12} & \textit{if } y_{p1} < y_1 < y_{p2}; \\ \alpha a_1 y_1 & \textit{if } y_{p2} \leq y_1 \leq y_{p3}; \\ \alpha a_{23} y_1^2 + \alpha b_{23} y_1 + \alpha c_{23} & \textit{if } y_{p3} < y_1 < y_{p4}; \\ \alpha a_0 y_1 + \alpha a_1 - \alpha a_0 & \textit{if} \quad y_1 \geq y_{p4}. \end{cases} \tag{32}$$

According to (21), the first Lie derivative of system output x_1 is a x_2 "new" state variable

$$x_2 = \begin{cases} f_{111} y_1 + f_{121} y_2 + f_{10} & \textit{if} \quad y_1 \leq y_{p1}; \\ f_{212} y_1^2 + f_{211} y_1 + f_{221} y_2 + f_{20} & \textit{if } y_{p1} < y_1 < y_{p2}; \\ f_{311} y_1 + f_{321} y_2 & \textit{if } y_{p2} \leq y_1 \leq y_{p3}; \\ f_{412} y_1^2 + f_{411} y_1 + f_{421} y_2 + f_{40} & \textit{if } y_{p3} < y_1 < y_{p4}; \\ f_{511} y_1 + f_{521} y_2 + f_{50} & \textit{if} \quad y_1 \geq y_{p4}, \end{cases} \tag{33}$$

$$\begin{gathered} f_{111} = -\alpha(1 + a_0);\ f_{121} = \alpha;\ f_{10} = -\alpha a_0 + \alpha a_1;\ f_{212} = -\alpha a_{12}; \\ f_{211} = -\alpha(b_{12} + 1);\ f_{221} = \alpha;\ f_{20} = -\alpha c_{12};\ f_{311} = -\alpha(1 + a_1); \\ f_{321} = \alpha;\ f_{412} = -\alpha a_{23};\ f_{411} = -\alpha(b_{23} + 1);\ f_{421} = \alpha; \\ f_{40} = -\alpha c_{23};\ f_{511} = -\alpha(1 + a_0);\ f_{521} = \alpha;\ f_{50} = -\alpha a_1 + \alpha a_0. \end{gathered} \tag{34}$$

Now we take into account (20) and define the second Lie derivative

$$L_{\boldsymbol{F}}^2 x_1 = \nabla L_{\boldsymbol{F}} x_1 \boldsymbol{F} = \begin{pmatrix} \frac{dL_{\boldsymbol{F}} x_1}{dy_1} & \frac{dL_{\boldsymbol{F}} x_1}{dy_2} & \frac{dL_{\boldsymbol{F}} x_1}{dy_3} \end{pmatrix} \boldsymbol{F}. \tag{35}$$

Derivative (35) gives us the possibility to write down following expression for x_3 state variable

$$x_3 = \begin{cases} d_{111}y_1 + d_{121}y_2 + d_{131}y_3 + d_{10} & \textit{if} \quad y_1 \le y_{p1}; \\ d_{213}y_1^3 + d_{212}y_1^2 + d_{211}y_1 + d_{221}y_2 \\ \quad + d_{231}y_3 + d_{22}y_1y_2 + d_{20} & \textit{if} \ y_{p1} < y_1 < y_{p2}; \\ d_{311}y_1 + d_{321}y_2 + d_{331}y_3 & \textit{if} \ y_{p2} \le y_1 \le y_{p3}; \\ d_{413}y_1^3 + d_{412}y_1^2 + d_{411}y_1 + d_{421}y_2 \\ \quad + d_{431}y_3 + d_{44}y_1y_2 + d_{40} & \textit{if} \ y_{p3} < y_1 < y_{p4}; \\ d_{511}y_1 + d_{521}y_2 + d_{531}y_3 + d_{50} & \textit{if} \quad y_1 \ge y_{p4}, \end{cases} \tag{36}$$

$$\begin{gathered} d_{111} = \alpha^2(1 + a_0)^2 + \alpha;\ d_{121} = -\alpha^2(1 + a_0) - \alpha;\ d_{22} = -2\alpha^2 a_{12}; \\ d_{10} = \alpha^2(1 + a_0)(a_0 - a_1);\ d_{212} = \alpha^2(b_{12} + 1)a_{12};\ d_{413} = 2\alpha^2 a_{23}^2; \\ d_{221} = -\alpha^2(b_{12} + 1) - \alpha;\ d_{211} = \alpha^2(b_{12} + 1)^2 + 2\alpha^2 a_{12}c_{12} + \alpha; \\ d_{20} = \alpha^2(b_{12} + 1)c_{12};\ d_{311} = \alpha^2(1 + a_1)^2 + \alpha;\ d_{321} = -\alpha^2(1 + a_1) - \alpha; \\ d_{412} = 3\alpha^2(b_{23} + 1)a_{23};\ d_{421} = -\alpha^2(b_{23} + 1) - \alpha;\ d_{213} = 2\alpha^2 a_{12}^2; \\ d_{44} = -2\alpha^2 a_{23};\ d_{411} = \alpha^2(b_{23} + 1)^2 + 2\alpha^2 a_{23}c_{23} + \alpha; \\ d_{40} = \alpha^2(b_{23} + 1)c_{23};\ d_{511} = \alpha^2(1 + a_0)^2 + \alpha;\ d_{521} = -\alpha^2(1 + a_0) - \alpha; \\ d_{50} = \alpha^2(1 + m_0)(a_1 - a_0);\ d_{131} = d_{231} = d_{331} = d_{431} = d_{531} = \alpha. \end{gathered} \tag{37}$$

Thus, expressions (36) and (33) determine dependencies between state variables in canonical one x_i and normal state space y_i. We offer to transform these dependencies into clear interrelations which allows to define state variables in one state space by using state variables from another state space. We perform such a transformation by appending the system of equations which contains (33) and (36) with expression, which is obtained by using above-made assumption about output variable

$$x_1 = y_1. \tag{38}$$

Solution of Eqs. (38), (33), and (36) make it possible to write down following formulas

$$
\begin{gathered}
y_1 = x_1;\ y_2 = \begin{cases} h_{111}x_1 + h_{121}x_2 + h_{10} & \text{if} \quad x_1 \le y_{p1}; \\ h_{212}x_1^2 + h_{211}x_1 + h_{221}x_2 + h_{20} & \text{if } y_{p1} < x_1 < y_{p2}; \\ h_{311}x_1 + h_{321}x_2 + h_{30} & \text{if } y_{p2} \le y_1 \le y_{p3}; \\ h_{412}x_1^2 + h_{411}x_1 + h_{421}x_2 + h_{40} & \text{if } y_{p3} < x_1 < y_{p4}; \\ h_{511}x_1 + h_{521}x_2 + h_{50} & \text{if} \quad x_1 \ge y_{p4}; \end{cases} \\
y_3 = \begin{cases} e_{111}x_1 + e_{121}x_2 + e_{131}x_3 + e_{10} & \text{if} \quad x_1 \le y_{p1}; \\ e_{214}x_1^4 + e_{213}x_1^3 + e_{212}x_1^2 + e_{211}x_1 + e_{222}x_2^2 & \\ \quad + e_{221}x_2 + e_{22}x_1x_2 + e_{231}x_3 + e_{20} & \text{if } y_{p1} < x_1 < y_{p2}; \\ e_{311}x_1 + e_{321}x_2 + e_{331}x_3 & \text{if } y_{p2} \le x_1 \le y_{p3}; , \\ e_{414}x_1^4 + e_{413}x_1^3 + e_{412}x_1^2 + e_{411}x_1 + e_{422}x_2^2 & \\ \quad + e_{421}x_2 + e_{42}x_1x_2 + e_{431}x_3 + e_{40} & \text{if } y_{p3} < x_1 < y_{p4}; \\ e_{511}x_1 + e_{521}x_2 + e_{531}x_3 + e_{50} & \text{if} \quad x_1 \ge y_{p4}, \end{cases}
\end{gathered} \tag{39}
$$

$$
\begin{gathered}
h_{111} = -\tfrac{f_{111}}{f_{121}};\ h_{121} = -\tfrac{1}{f_{121}};\ h_{10} = -\tfrac{f_{10}}{f_{121}};\ h_{311} = -\tfrac{f_{311}}{f_{321}};\ h_{321} = -\tfrac{1}{f_{321}}; \\
h_{30} = 0;\ h_{212} = -\tfrac{f_{212}}{f_{221}};\ h_{211} = -\tfrac{f_{211}}{f_{221}};\ h_{221} = \tfrac{1}{f_{221}};\ h_{20} = -\tfrac{f_{20}}{f_{221}}; \\
h_{412} = -\tfrac{f_{412}}{f_{421}};\ h_{411} = -\tfrac{f_{411}}{f_{421}};\ h_{421} = \tfrac{1}{f_{421}};\ h_{40} = -\tfrac{f_{40}}{f_{421}};\ h_{511} = -\tfrac{f_{511}}{f_{521}}; \\
h_{521} = -\tfrac{1}{f_{521}};\ h_{50} = -\tfrac{f_{50}}{f_{521}};\ e_{413} = \tfrac{d_{42}f_{412}f_{421} - 2d_{412}f_{411}f_{412}}{f_{421}^2(d_{413}+d_{431})};
\end{gathered} \tag{40}
$$

$$
\begin{gathered}
e_{111} = \tfrac{d_{121}f_{111} - d_{111}f_{121}}{d_{131}f_{121}};\ e_{121} = -\tfrac{d_{121}}{d_{131}f_{121}};\ e_{131} = \tfrac{1}{d_{131}};\ e_{231} = \tfrac{1}{d_{213}+d_{231}}; \\
e_{10} = \tfrac{d_{121}f_{10} - d_{10}f_{121}}{d_{131}f_{121}};\ e_{214} = -\tfrac{d_{212}f_{212}^2}{f_{221}^2(d_{213}+d_{231})};\ e_{222} = -\tfrac{d_{212}}{f_{221}^2(d_{213}+d_{231})}; \\
e_{213} = \tfrac{d_{22}f_{212}f_{221} - 2d_{212}f_{211}f_{212}}{f_{221}^2(d_{213}+d_{231})};\ e_{22} = \tfrac{2d_{212}f_{211} - d_{22}f_{221}}{f_{221}^2(d_{213}+d_{231})};\ e_{321} = -\tfrac{d_{321}}{d_{331}f_{321}}; \\
e_{212} = \tfrac{2d_{212}f_{212}}{f_{221}^2(d_{213}+d_{231})} - \tfrac{2d_{212}f_{20}f_{212} + d_{212}f_{211}^2 - d_{22}f_{211}f_{221} - d_{221}f_{212}f_{221}}{f_{221}^2(d_{213}+d_{231})}; \\
e_{211} = \tfrac{d_{22}f_{20}f_{221} + d_{221}f_{211}f_{221} - d_{211}f_{221}^2 - 2d_{212}f_{20}f_{211}}{f_{221}^2(d_{213}+d_{231})};\ e_{531} = \tfrac{1}{d_{531}}; \\
e_{221} = \tfrac{2d_{212}f_{20} - d_{221}f_{221}}{f_{221}^2(d_{213}+d_{231})};\ e_{20} = \tfrac{d_{221}f_{20}f_{221} - d_{20}f_{221}^2 - d_{212}f_{20}^2}{f_{221}^2(d_{213}+d_{231})};\ e_{331} = \tfrac{1}{d_{331}}; \\
e_{311} = \tfrac{d_{321}f_{311} - d_{311}f_{321}}{d_{331}f_{321}};\quad e_{414} = -\tfrac{d_{412}f_{412}^2}{f_{421}^2(d_{413}+d_{431})};\ e_{431} = \tfrac{1}{d_{413}+d_{431}}; \\
e_{422} = -\tfrac{d_{412}}{f_{421}^2(d_{413}+d_{431})};\ e_{421} = \tfrac{2d_{412}f_{40} - d_{421}f_{421}}{f_{421}^2(d_{413}+d_{431})}; \\
e_{42} = \tfrac{2d_{412}f_{411} - d_{42}f_{421}}{f_{421}^2(d_{413}+d_{431})};\ e_{40} = \tfrac{d_{421}f_{40}f_{421} - d_{40}f_{421}^2 - d_{412}f_{40}^2}{f_{421}^2(d_{413}+d_{431})}; \\
e_{412} = \tfrac{2d_{412}f_{412}}{f_{421}^2(d_{413}+d_{431})} - \tfrac{2d_{412}f_{40}f_{412} + d_{412}f_{411}^2 - d_{42}f_{411}f_{421} - d_{421}f_{412}f_{421}}{f_{421}^2(d_{413}+d_{431})}; \\
e_{411} = \tfrac{d_{42}f_{40}f_{421} + d_{421}f_{411}f_{421} - d_{411}f_{421}^2 - 2d_{412}f_{40}f_{411}}{f_{421}^2(d_{413}+d_{431})}; \\
e_{511} = \tfrac{d_{521}f_{511} - d_{511}f_{521}}{d_{531}f_{521}};\ e_{50} = \tfrac{d_{521}f_{50} - d_{50}f_{521}}{d_{531}f_{521}};
\end{gathered}
$$

Analysis of (39) shows some general patterns in defining factors for linear and polynomial parts of the considered characteristic. These patterns allows us to define y_2 and y_3 state variable in i-th linear part of characteristic (30) in such a way

$$\begin{aligned} y_{2ilin} &= h_{i11}x_1 + h_{i21}x_2 + h_{i0}; \\ y_{3ilin} &= e_{i11}x_1 + e_{i21}x_2 + e_{i31}x_3 + e_{i0}, \end{aligned} \tag{41}$$

here i means i-th linear interval of the considered function.

Similar expression can be written down for circuit coordinates in nonlinear part of the considered characteristic

$$\begin{aligned} y_{2ipoly} = h_{i12}x_1^2 + h_{i11}x_1 + h_{i21}x_2 + h_{i0};\ y_{3ipoly} = e_{i14}x_1^4 + e_{i13}x_1^3 + e_{i12}x_1^2 \\ + e_{i11}x_1 + e_{i22}x_2^2 + e_{i21}x_2 + e_{i2}x_1x_2 + e_{i31}x_3 + e_{i0}. \end{aligned} \tag{42}$$

If one analyzes (40), he finds that it also have patterns in factors definition. So, for linear part of the considered characteristic these factors can be defined as follows

$$\begin{aligned} h_{i11} = -\tfrac{f_{i11}}{f_{i21}};\ h_{i21} = -\tfrac{1}{f_{i21}};\ h_{i0} = -\tfrac{f_{i0}}{f_{i21}};\ e_{i11} = \tfrac{d_{i21}f_{i11}-d_{i11}f_{i21}}{d_{i31}f_{i21}}; \\ e_{i21} = -\tfrac{d_{i21}}{d_{i31}f_{i21}};\ e_{i31} = \tfrac{1}{d_{i31}};\ e_{i0} = \tfrac{d_{i21}f_{i0}-d_{i0}f_{i21}}{d_{i31}f_{i21}}. \end{aligned} \tag{43}$$

In the nonlinear part of the smoothed Chua's diode characteristic these factors are defined in such a way

$$\begin{aligned} h_{i12} = -\tfrac{f_{i12}}{f_{i21}};\ h_{i11} = -\tfrac{f_{i11}}{f_{i21}};\ h_{i21} = \tfrac{1}{f_{i21}};\ h_{i0} = -\tfrac{f_{i0}}{f_{i21}};\ e_{i14} = -\tfrac{d_{i12}f_{i12}^2}{f_{i21}^2(d_{i13}+d_{i31})}; \\ e_{i13} = \tfrac{d_{i2}f_{i12}f_{i21}-2d_{i12}f_{i11}f_{i12}}{f_{i21}^2(d_{i13}+d_{i31})};\ e_{i21} = \tfrac{2d_{i12}f_{i0}-d_{i21}f_{i21}}{f_{i21}^2(d_{i13}+d_{i31})}; \\ e_{i2} = \tfrac{2d_{i12}f_{i11}-d_{i2}f_{i21}}{f_{i21}^2(d_{i13}+d_{i31})};\ e_{i22} = -\tfrac{d_{i12}}{f_{i21}^2(d_{i13}+d_{i31})};\ e_{i31} = \tfrac{1}{d_{i13}+d_{i31}}; \\ e_{i12} = \tfrac{2d_{i12}f_{i12}}{f_{i21}^2(d_{i13}+d_{i31})} - \tfrac{2d_{i12}f_{i0}f_{i12}+d_{i12}f_{i11}^2-d_{i2}f_{i11}f_{i21}-d_{i21}f_{i12}f_{i21}}{f_{i21}^2(d_{i13}+d_{i31})}; \\ e_{i11} = \tfrac{d_{i2}f_{i0}f_{i21}+d_{i21}f_{i11}f_{i21}-d_{i11}f_{i21}^2-2d_{i12}f_{i0}f_{i11}}{f_{i21}^2(d_{i13}+d_{i31})}; \\ e_{i0} = \tfrac{d_{i21}f_{i0}f_{i21}-d_{i0}f_{i21}^2-d_{i12}f_{i0}^2}{f_{i21}^2(d_{i13}+d_{i31})}. \end{aligned} \tag{44}$$

The use of patterns (41)–(44) allows to define interrelations between circuit coordinates in different state spaces without performing any algebraic transformations for both linear and nonlinear parts of Chua's diode characteristic. These interrelations make it possible to rewrite (45)

$$\begin{aligned} y_1 = x_1;\ y_2 &= \begin{cases} h_{i11}x_1 + h_{i21}x_2 + h_{i0} & \text{if } x_1 \in X_{lin}; \\ h_{i12}x_1^2 + h_{i11}x_1 + h_{i21}x_2 + h_{i0} & \text{if } x_1 \in X_{poly}; \end{cases} \\ y_3 &= \begin{cases} e_{i11}x_1 + e_{i21}x_2 + e_{i31}x_3 + e_{i0} & \text{if } x_1 \in X_{lin}; \\ e_{i14}x_1^4 + e_{i13}x_1^3 + e_{i12}x_1^2 + e_{i11}x_1 + e_{i22}x_2^2 & \\ \quad + e_{i21}x_2 + e_{i2}x_1x_2 + e_{i31}x_3 + e_{i0} & \text{if } x_1 \in X_{poly}, \end{cases} \end{aligned} \tag{45}$$

here X_{lin} and X_{poly} mean intervals where linear and polynomial parts of the considered function are defined

$$X_{lin} = (-\infty, y_{p1}] \cup [y_{p2}, y_{p3}] \cup [y_{p4}, \infty);\ X_{poly} = (y_{p1}, y_{p2}) \cup (y_{p3}, y_{p4}). \tag{46}$$

Intervals (46) as well as expressions (45) make backgrounds to build calculations procedures based on the arrays usage, which simplifies nonlinear functions definition and system dynamic calculations. That is why we offer to rewrite (3) in such a way

$$g(y_1) = \begin{cases} k_i y_1 + y_i & \textit{if } \ y_1 \in X_{lin}; \\ a_i y_1^2 + b_i y_1 + c_i & \textit{if } \ y_1 \in X_{poly}. \end{cases} \tag{47}$$

Now we define the third Lie derivative

$$\begin{aligned} L_F^3 x_1 &= \begin{pmatrix} \begin{cases} d_{i11} & \textit{if } \ y_1 \in X_{lin}; \\ 3d_{i13}y_1^2 + 2d_{i12}y_1 + d_{i2}y_2 + d_{i11} & \textit{if } \ y_1 \in X_{poly}, \end{cases} \\ \begin{cases} d_{i21} & \textit{if } \ y_1 \in X_{lin}; \\ 2d_{i12}y_2 + d_{i21} + d_{i2}y_1 & \textit{if } \ x_1 \in X_{poly}, \end{cases} \\ d_{i31} \end{pmatrix}^T \\ &\times \begin{pmatrix} -\begin{cases} (k_i + \alpha)y_1 - \alpha y_2 + y_{0i} & \textit{if } \ y_1 \in X_{lin}; \\ a_i y_1^2 + (b_i + \alpha)y_1 - \alpha y_2 + c_i & \textit{if } \ y_1 \in X_{poly}. \end{cases} \\ y_1 - y_2 + y_3 \\ -\beta y_2 \end{pmatrix} \\ &= \left(\begin{cases} q_{i1}y_1 + q_{i2}y_2 + q_{i3}y_3 & \textit{if } \ y_1 \in X_{lin}; \\ p_{i14}y_1^4 + p_{i13}y_1^3 + p_{i12}y_1^2 + p_{i0321}y_2y_3 + p_{i31}y_3 & \\ \quad + p_{i11}y_1 + p_{i0212}y_2y_1^2 + p_{i0211}y_1y_2 & \\ \quad + p_{i0311}y_1y_3 + p_{i22}y_2^2 + p_{i21}y_2 + p_{i0} & \textit{if } \ y_1 \in X_{poly}; \end{cases} \right) \end{aligned} \tag{48}$$

$$\begin{gathered} q_{i1} = d_{i21} - d_{i11}(k_i + \alpha); \ q_{i2} = d_{i11}\alpha - d_{i21} - d_{i31}\beta; \ q_{i3} = d_{i21}; \ q_{i0} = -d_{i11}; \\ p_{i13} = 2d_{i12}a_i + 3d_{i13}(b_i + \alpha); p_{i12} = d_{i1}a_i + 2d_{i12}(b_i + \alpha) + 3d_{i13}c_i + d_{i2}; \\ p_{i0212} = a_i d_{i2} - 3\alpha d_{i13}; p_{i11} = d_{i1}(b_i + \alpha) + 2d_{i12}c_i + d_{i21}; p_{i14} = 3d_{i13}a_i; \\ p_{i0211} = d_{i2}(b_i + \alpha) - 2d_{i12}\alpha + 2d_{i12} - d_{i2}; p_{i0311} = d_{i2}; p_{i22} = -\alpha d_{i2} - 2d_{i12}; \\ p_{i21} = -\alpha d_{i1} - \beta d_{i31} + c_i d_{i2} - d_{i21}; p_{i0321} = 2d_{i12}; p_{i31} = d_{i21}; \ p_{i0} = d_{i1}c_i; \end{gathered} \tag{49}$$

This derivative depends on circuit state variables in normal state space. At the same time, the use of high derivatives of output variable according to (16) and (17) makes backgrounds to consider the system in canonical state space. That is why we substitute (45) into (48) and rewrite it as follows

$$L_F^3 x_1 = \begin{cases} r_{i1}x_1 + r_{i2}x_2 + r_{i3}x_3 + r_{i0} & \textit{if } \ x \in X_{lin}; \\ s_{i14}x_1^4 + s_{i13}x_1^3 + s_{i12}x_1^2 + s_{i122}x_1^2 x_2 & \\ + s_{i11}x_1 + s_{i0} + s_{i112}x_1x_2 + s_{i113}x_1x_3 & \\ \quad + s_{i22}x_2^2 + s_{231}x_2x_3 + s_{i31}x_3 & \textit{if } \ x \in X_{poly}, \end{cases} \tag{50}$$

$$
\begin{gathered}
r_{i1} = e_{i11}q_{i3} + h_{i11}q_{i2} + q_{i1};\ r_{i2} = e_{i21}q_{i3} + h_{i21}q_{i2};\ r_{i3} = q_{i3}e_{i31};\\
r_{i0} = e_{i0}q_{i3} + h_{i0}q_{i2};\ s_{i14} = p_{i14};\ \ s_{i13} = p_{i0212}h_{i11} + p_{i13};\\
s_{i122} = p_{i0212}h_{i21};\ \ s_{i231} = p_{i0321}e_{i31}h_{i21};\ s_{i113} = p_{i0321}e_{i31}h_{i11} + p_{i0311}e_{i31};\\
s_{i12} = p_{i0321}e_{i11}h_{i11} + p_{i22}h_{i11}^2 + p_{i0211}h_{i11} + p_{i0212}h_{i0} + p_{i0311}e_{i11} + p_{i12};\\
s_{i11} = p_{i0321}e_{i0}h_{i11} + p_{i0321}e_{i11}h_{i0} + 2p_{i22}h_{i0}h_{i11} + p_{i0211}h_{i0} + p_{i0311}e_{i0}\\
+ p_{i21}h_{i11} + p_{i31}e_{i11} + p_{i11};\ \ s_{i22} = p_{i0321}e_{i21}h_{i21} + p_{i22}h_{i21}^2;\\
s_{i112} = p_{i0321}e_{i11}h_{i21} + p_{i0321}e_{i21}h_{i11} + 2p_{i22}h_{i11}h_{i21} + p_{i0211}h_{i21} + p_{i0311}e_{i21};\\
s_{i21} = p_{i0321}e_{i0}h_{i21} + p_{i0321}e_{i21}h_{i0} + 2p_{i22}h_{i0}h_{i21} + p_{i21}h_{i21} + p_{i31}e_{i21};\\
s_{i31} = p_{i0321}e_{i31}h_{i0} + p_{i31}e_{i31};\ s_{i0} = p_{i0321}e_{i0}h_{i0} + p_{i22}h_{i0}^2 + p_{i21}h_{i0}\\
+ p_{i31}e_{i0} + p_{i0}.
\end{gathered}
\tag{51}
$$

We consider (50) as the right-hand expression of the last equation of (17). This fact allows us to describe Chua's circuit dynamic in the canonical state space by system of differential equations

$$
\begin{gathered}
\dot{x}_1 = x_2;\ \dot{x}_2 = x_3;\\
\dot{x}_3 = \begin{cases} r_{i1}x_1 + r_{i2}x_2 + r_{i3}x_3 + r_{i0} & \textit{if}\ \ x \in X_{lin};\\ s_{i14}x_1^4 + s_{i13}x_1^3 + s_{i12}x_1^2 + s_{i122}x_1^2x_2 + s_{i11}x_1 + s_{i0} & \\ + s_{i112}x_1x_2 + s_{i113}x_1x_3 + s_{i22}x_2^2 + s_{231}x_2x_3 + s_{i31}x_3 & \textit{if}\ \ x \in X_{poly}, \end{cases}
\end{gathered}
\tag{52}
$$

or single high-order differential equation

$$
\dddot{x}_1 = \begin{cases} r_{i1}x_1 + r_{i2}\dot{x}_1 + r_{i3}\ddot{x}_1 + r_{i0} & \textit{if}\ \ x \in X_{lin};\\ s_{i14}x_1^4 + s_{i13}x_1^3 + s_{i12}x_1^2 + s_{i122}x_1^2\dot{x}_1 + s_{i11}x_1 & \\ + s_{i0} + s_{i112}x_1\dot{x}_1 + s_{i113}x_1\ddot{x}_1 + s_{i22}\dot{x}_1^2 & \\ + s_{231}\dot{x}_1\ddot{x}_1 + s_{i31}\ddot{x}_1 & \textit{if}\ \ x \in X_{poly}. \end{cases}
\tag{53}
$$

It is clearly understood that similar to system in normal state space, the designed systems in canonical one contains both piecewise linear and polynomial functions. At the same time, the transformation into canonical form increases order of smoothed polynomial. Thus, one can claim that transformation into canonical form makes nonlinear system part more complex and lives complexity of linear one unchanged.

2.4 Chua's Circuit Model in the Canonical Discrete-Time State Space

Similar to system in normal state space (1), the considered systems (52) and (53) are defined in continuous-time domain. We offer to apply known approximations for derivative operator to transform it into discrete-time. The use of such transformation allows us to simulate the designed system and implement it by using various digital devices.

Although systems (52) and (53) are identical in the continuous-time domain, they differ each other in the discrete-time one. This difference occurs because of using inner state variables x_2 and x_3 which calculations make additional time shift into system dynamic. That is why we use (53) to define Chua's circuit motions in the discrete time.

To perform its transformation into discrete-time domain we offer to use well-known feedback difference approximations for different order derivative operators

$$\begin{aligned} \dot{x}_1 &\approx \frac{x_1[i]-x[i-1]}{T};\ \ddot{x}_1 \approx \frac{x_1[i]-2x_1[i-1]+x_1[i-2]}{T^2}; \\ \dddot{x}_1 &\approx \frac{x_1[i]-3x_1[i-1]+3x_1[i-2]-x_1[i-3]}{T^3}. \end{aligned} \tag{54}$$

If one substitute (54) into (53) and solves the obtained equation for the current value of state variable $x_1[i]$, he can write down following expression

$$\begin{aligned} x_1[i] = &-\frac{(3-T^2r_{i2}-2Tr_{i3})x_1[i-1]}{T^3r_{i1}+T^2r_{i2}+Tr_{i3}-1} - \frac{(Tr_{i3}-3)x_1[i-2]}{T^3r_{i1}+T^2r_{i2}+Tr_{i3}-1} \\ &- \frac{x_1[i-3]}{T^3r_{i1}+T^2r_{i2}+Tr_{i3}-1} - \frac{T^3r_{i0}}{T^3r_{i1}+T^2r_{i2}+Tr_{i3}-1}. \end{aligned} \tag{55}$$

We call (55) as Chua's circuit finite difference equation or discrete-time equation in canonical form. The using of this form allows us to define only x_1 state variable, to define x_2 and x_3 we use (54)

$$x_2[i] = \frac{x_1[i] - x[i-1]}{T};\ x_3[i] = \frac{x_1[i] - 2x_1[i-1] + x_1[i-2]}{T^2}. \tag{56}$$

Simulation results for the designed Chua's circuit in the canonical state-space are shown in Fig. 5.

Analysis of simulation results shows that only variable x_1 is similar to the state variable $\dot{y}_1$ from the system in normal state space. The simulation error between these variables are neclectible small. Other state variables x_2 and x_3 differ from variables which are defined in a normal state space. This difference produce new chaotic oscillations and novel chaotic attractor which can be used for to reach various goals including secured communications and data encryption.

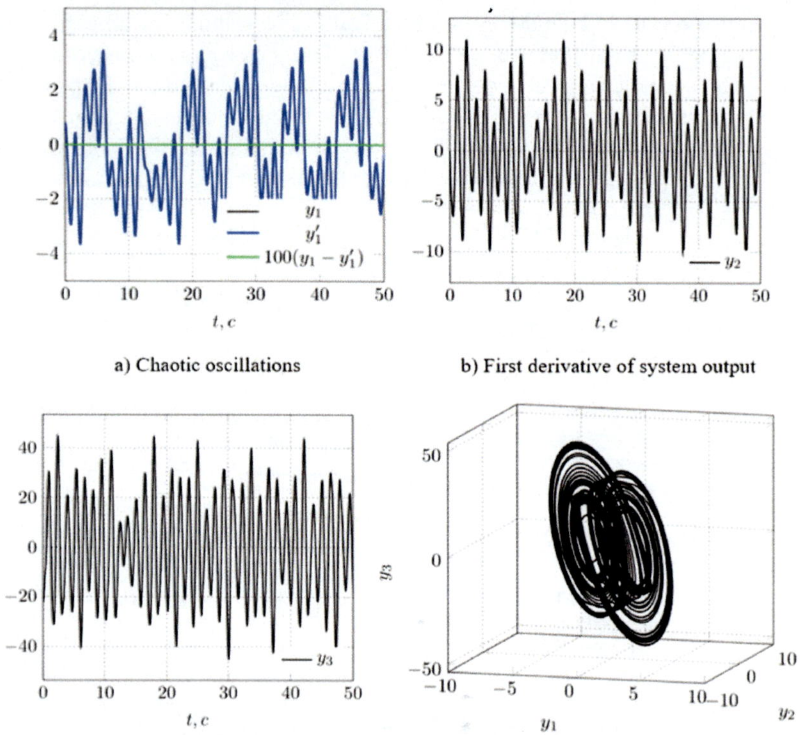

a) Chaotic oscillations

b) First derivative of system output

c) Second derivative of system output

d) Novel 3D chaotic attractor

Fig. 5. Chua's circuit simulation results in canonical state space.

3 Conclusions

Above-given formulas and simulation results allows us to claim that coordinate transformations give us possibility to design novel chaotic system which is based on the known one. Since in the most cases these transformations are differential ones, one should to take into account possible nonlinearities in the initial system and if they are described by piecewise linear functions some smoothing procedure should be applied.

Since we design model for transformed system in both continuous-time and discrete-time domains one can use different analog and digital elements to implement the proposed dynamical system. Future developing of our work can be found in designing controller for the above-given system to perform its synchronization.

References

1. Gräupl, T.: L-band digital aeronautical communications system (LDACS) flight trials in the national German project MICONAV. In: 2018 Integrated Communications, Navigation, Surveillance Conference (ICNS), Herndon, VA, USA, pp. 4A2-1–4A2-7 (2018)

2. Bellido-Manganell, M.A.: LDACS flight trials: demonstration and performance analysis of the future aeronautical communications system. IEEE Trans. Aerosp. Electron. Syst. **58**(1), 615–634 (2022)
3. Bae, J., Lee, H.: A study on communication technologies for urban air mobility. In: 13th International Conference on Information and Communication Technology Convergence (ICTC), Jeju Island, Korea, pp. 2235–2240 (2020)
4. Ding, Y., Sun, P., Cai, J.: A preliminary study on civil aircraft post-flight automatic maintenance test philosophy and design. In: 2018 IEEE CSAA Guidance, Navigation and Control Conference, pp. 1–5. IEEE, Xiamen (2018)
5. Sang, Y., Cai, Z., Wang, Y.: An exploration strategy based on Frontier and safe corridor for indoor flight vehicle. In: 2013 Chinese Automation Congress, Changsha, China, pp. 244–249 (2013)
6. Sushchenko, O.A., Golitsyn, V.O.: Data processing system for altitude navigation sensor. In: 2016 IEEE 4th International Conference on Methods and Systems of Navigation and Motion Control (MSNMC 2016), pp. 84–87. IEEE, Kyiv (2016)
7. Sushchenko, O., Chikovani, V., Tsiruk, H.: Redundant information processing techniques comparison for differential vibratory gyroscope. Eastern-Eur. J. Enterp. Technol. **4**(7–82), 45–52 (2016)
8. Grekhov, A., Vyshnyakova, Y.: Simulation of navigation and flight control systems traffic for UAS/RPAS. In: 6th International Conference on Methods and Systems of Navigation and Motion Control (MSNMC), pp. 43–46. IEEE, Kyiv (2020)
9. Ostroumov, I.V., Marais, K., Kuzmenko, N.S.: Aircraft positioning using multiple distance measurements and spline prediction. Aviation **26**(1), 1–10 (2022). https://doi.org/10.3846/aviation.2022.16589
10. Oh, J., Sung, C.K., Lee, J.S.: A new method to calculate relative distance of closest terrain point using interferometric radar altimeter output in real flight environment. In: 2018 IEEE/ION Position, Location and Navigation Symposium (PLANS), Monterey, CA, USA, pp. 141–148 (2018)
11. Xu, C., Huang, X., Li, M.: Autonomous navigation using natural landmark measurements for earth orbit satellites. In: 2019 Chinese Control Conference (CCC), Guangzhou, China, pp. 4078–4082 (2019)
12. Yang, X., Zhao, X., Yuan, X.: Distributed air traffic flow management at terminal control area. In: 2016 International Conference on Cyber-Enabled Distributed Computing and Knowledge Discovery (CyberC), Chengdu, China, pp. 119–122 (2016)
13. Ostroumov, I., Kuzmenko, N.: Risk assessment of mid-air collision based on positioning performance by navigational aids. In: 2020 IEEE 6th International Conference on Methods and Systems of Navigation and Motion Control (MSNMC), pp. 34–37. IEEE, Kyiv (2020). https://doi.org/10.1109/MSNMC50359.2020.9255506
14. Ostroumov, I.V., Marais, K., Kuzmenko, N.S.: Triple probability density distribution model in the task of aviation risk assessment. Aviation **24**(2), 57–65 (2020). https://doi.org/10.3846/aviation.2020.12544
15. Baptist, L. J., Raj, R. A., Kumar, S.: Air quality improvement by employing smart traffic management system controlled by internet of things for Botswana in the sub Saharan region of Africa. In: 2022 3rd International Conference on Communication, Computing and Industry 4.0 (C2I4), Bangalore, India, pp. 1–6 (2022)
16. Yu, L., Le, Y. Rui, Z.: Safety assessment of air traffic control support for night operation of plateau airport. In: 2022 IEEE 4th International Conference on Civil Aviation Safety and Information Technology (ICCASIT), pp. 358–362. IEEE, Dali (2022)
17. Ntombela, N., Umenne, P.: Access control with automated on duty notification tool in air traffic services. In: 2020 International Conference on Artificial Intelligence, Big Data, Computing and Data Communication Systems (icABCD), Durban, South Africa, pp. 1–5 (2020)

18. Mitici, M., Blom, H.A.P.: Mathematical models for air traffic conflict and collision probability estimation. IEEE Trans. Intell. Transp. Syst. **20**(3), 1052–1068 (2019)
19. Wen, H., Li, H., Wang, Z. Hou, X.: Application of DDPG-based collision avoidance algorithm in air traffic control. In: 2019 12th International Symposium on Computational Intelligence and Design (ISCID), Hangzhou, China, pp. 130–133 (2019)
20. Costea, M.L.: Automatic aircraft collisions algorithm development for civil aircraft. In: 2020 10th International Conference on Advanced Computer Information Technologies (ACIT), pp. 98–103. IEEE, Deggendorf (2020)
21. Lee, J.H., Park, K.H., Ko, Y.C.: A UAV-mounted free space optical communication: trajectory optimization for flight time. IEEE Trans. Wireless Commun. **19**(3), 1610–1621 (2020)
22. Singla, A., Sharma, D., Vashisth, S.: Data connectivity in flights using visible light communication. In: 2017 International Conference on Computing and Communication Technologies for Smart Nation (IC3TSN), Gurgaon, India, pp. 71–74 (2017)
23. Moll, F.: Demonstration of high-rate laser communications from a fast airborne platform. IEEE J. Sel. Areas Commun. **33**(9), 1985–1995 (2015)
24. Gao, C., Qi, X., Li, G.: Planning and optimization of ultra-short wave communication network resources in sea-air cooperative operations. In: 2021 IEEE International Conference on Electronic Technology, Communication and Information (ICETCI), Changchun, China, pp. 290–295 (2021)
25. Wang, S.F.: Dynamical analysis of memristive unified chaotic system and its application in secure communication. IEEE Access **6**, 66055–66061 (2018)
26. Jalilian, M, Ahmadi, A., Ahmadi, M.: Hardware implementation of a chaotic pseudo random number generator based on 3D chaotic system without equilibrium. In: 2018 25th IEEE International Conference on Electronics, Circuits and Systems (ICECS), pp. 741–744. IEEE, Bordeaux (2018)
27. Voliansky, R., Sadovoi, O.: Arduino-based Implementation of the dual-channel chaotic generator. In: 2019 3rd International Conference on Advanced Information and Communications Technologies (AICT), pp. 278–281. IEEE, Lviv (2019)
28. Dong, J., Zhang, G.J., Wang, J.: The chaotic characteristic and control of chaotic synchronization of nonlinear dynamic model of fractional-order. In: 2012 8th International Conference on Natural Computation, Chongqing, China, pp. 962–966 (2012)
29. Zhiteckii, L.S., Azarskov, V.N., Solovchuk, K.Y., Sushchenko, O.A.: Discrete-time robust steady-state control of nonlinear multivariable systems: a unified approach. IFAC Proc. Vol. **47**(3), 8140–8145 (2014)
30. Voliansky, R. Sadovoi, A.: Chua's circuits interval synchronization. In: 2017 4th International Scientific-Practical Conference Problems of Infocommunications. Science and Technology (PIC S&T), pp. 439–443. IEEE, Kharkov (2017)
31. Voliansky, R. Sadovoi, A.: Chua's circuits synchronization as inverse dynamic's problem solution. In: 2016 Third International Scientific-Practical Conference Problems of Infocommunications Science and Technology (PIC S&T), pp. 171–172. IEEE, Kharkiv (2016)
32. Yuan, J., Shi, B., Wang, Y.: Control and synchronization of fractional unified chaotic systems with a single sliding mode controller. In: Proceedings of the 33rd Chinese Control Conference, Nanjing, China, pp. 1800–1805 (2014)
33. Pan, Y., Ji, W., Liang, H.: Adaptive predefined-time control for Lü Chaotic Systems via backstepping approach. IEEE Trans. Circuits Syst. II Express Briefs **69**(12), 5064–5068 (2022)

Author Index

I. Ostroumov and M. Zaliskyi (Eds.): ACASD 2024, LNNS 992, pp. 441–442, 2024.
https://doi.org/10.1007/978-3-031-60196-5

Thomas W. Herminghaus
Feuerwehrhelme
Eine Entwicklungsgeschichte

Thomas W. Herminghaus

Feuerwehrhelme

Eine Entwicklungsgeschichte

EFB-Verlag · Hanau

1982

1. Auflage 1982

Herausgeber: TOTAL Feuerschutz GmbH, Ladenburg

Gesamtherstellung: Druckerei Haus GmbH, 6450 Hanau/M. 1
Printed in Germany

ISBN 3-88776-008-5

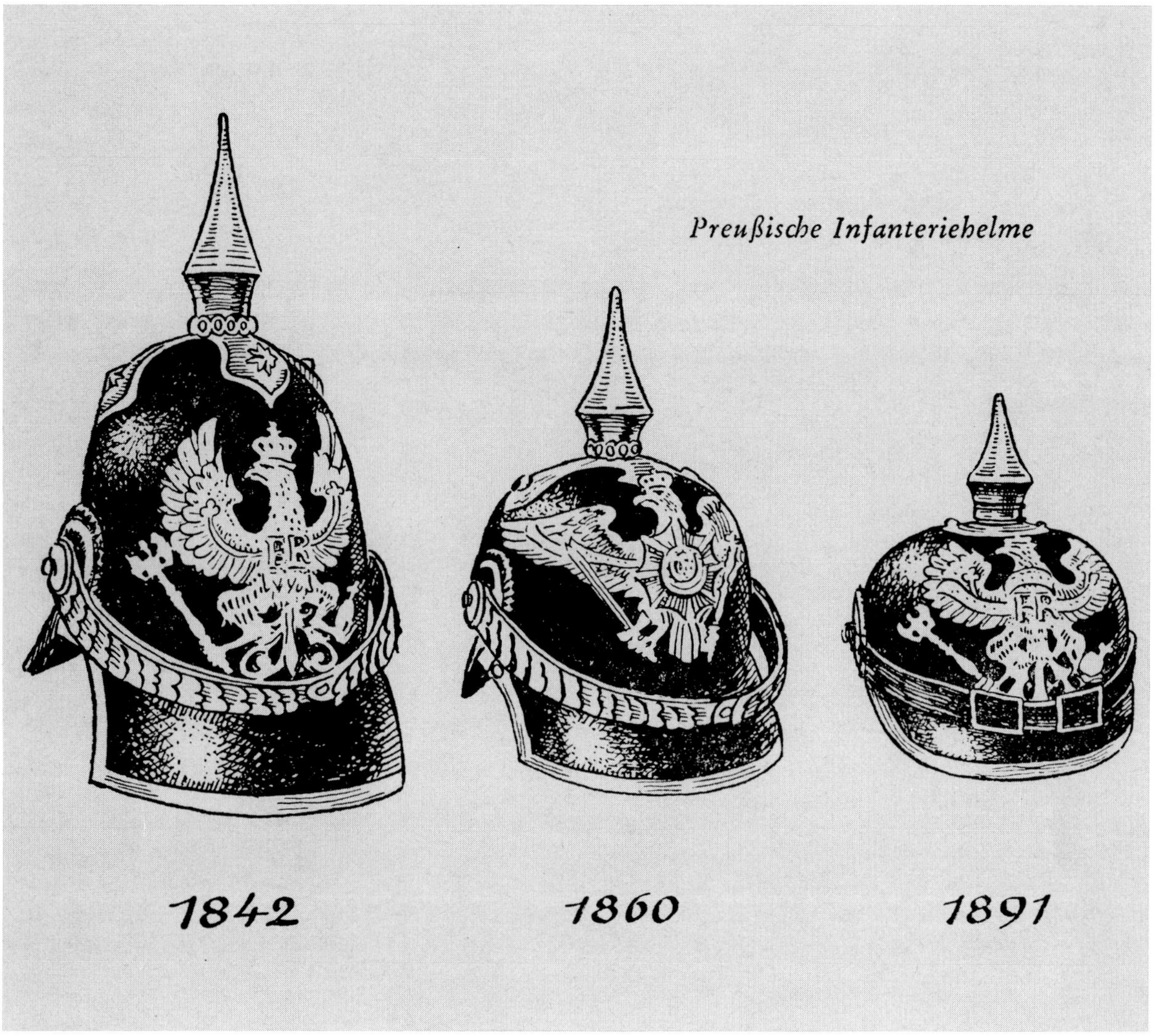

Historische Abhandlungen über staatliche und städtische Einrichtungen oder Militaria beginnen üblicherweise bei den „alten Griechen und Römern". Allerdings scheitert der Versuch, auch die Geschichte des Feuerwehrhelmes in die graue Vorzeit zurückzuverfolgen, denn organisierte spezielle Löschmannschaften sind im klassischen Altertum weitgehend unbekannt. Das soll nicht bedeuten, daß eine Brandbekämpfung, auch in größerem Umfang, unbekannt war. Eine der ältesten Darstellungen von etwa 850 v.Chr., ein Alabasterrelief aus einem Palast bei Niniveh, zeigt assyrische Soldaten bei Löschversuchen unter kriegerischen Bedingungen, und ein Dokument aus China, das sich auf das Jahr 564 v.Chr. datieren läßt, berichtet über Soldaten, die sich bei Großfeuern an der Brandbekämpfung beteiligen sollen. Die ältesten und bekannten organisierten „Brandschützer" waren also offensichtlich Armeeangehörige, die ihre Aufgabe als Feuerwehr sozusagen „nebenberuflich" erfüllten und bei diesen Einsätzen, die oft mit kriegerischen Ereignissen zusammenhingen, ihre soldatische Schutzkleidung einschließlich des Helmes trugen. Wir können also davon ausgehen, daß die ersten Feuerwehrhelme gleichzeitig Militärhelme waren, und daran hat sich, bis in das zwanzigste Jahrhundert, nicht allzu viel geändert.

Entwicklung des Legionärshelms im 1. Jh. n. Chr.

1.– 3.: *„Jockeymützen-Typ" (verschwindet um die Mitte des 1. Jahrhunderts n. Chr.).*
4.–6.: *Kaiserzeitlich-gallischer Typ*
7.–11.: *Italische Typen.*

Das Thema „Helmentwicklung" ist nicht eine Erfindung der Neuzeit, wie diese Tafel von P. Connolly aus seinem Werk „Die römische Armee" beweist. Es ging langsamer voran als heute: man brauchte immerhin ein paar Jahrhunderte, um so wichtige Attribute wie Nacken- und Wangenschutz durchzubilden und optimal zu gestalten. International war die Entwicklung schon damals, wie die Fundorte beweisen.

Auch von den römischen Vigiles, der wohl ersten organisierten Brandschutztruppe, ist ein spezieller Feuerwehrhelm nicht bekannt. Falls sie mit einer schützenden Kopfbedeckung ausgerüstet waren, handelte es sich vermutlich um Helme nach militärischem Vorbild.

Mit dem Untergang des Römischen Reiches geraten auch die damals recht fortschrittlichen Maßnahmen des abwehrenden Brandschutzes in Vergessenheit. Erst als im späten Mittelalter einzelne Großstädte, zur Pflege und Bedienung der primitiven Feuerspritzen, besoldete Feuerknechte fest anstellen, gibt es wieder eine Art Uniformierung dieses Personals. Man trug primitive Wachstuchkappen, oder, Mitte des 18. Jahrhunderts, hohe steife Filzhüte mit dem Stadtwappen (z.B. Wien, Kopenhagen) oder mit dem Zeichen der Versicherungsgesellschaft (London), die das Löschkorps unterhielt. Kopfbedeckungen dieses Typs haben sicherlich nicht viel Schutz geboten; sie waren wohl eher hinderlich bei anstrengender Arbeit, und hatten, wie die restlichen, bunten Uniformierungsstücke dieser ersten Berufsfeuerwehrmänner, hauptsächlich dekorativen Charakter, ähnlich wie die Felduniformen der damaligen Armeen. Da in der Regel aus gebührender Entfernung auf das Feuer gespritzt wurde, war diese Art der Schutzkleidung vielleicht auch vertretbar. Jan van der Heyden hat auf seinen dramatischen Abbildungen den Innenangriff zwar propagiert, mangels geeigneten Schlauchmaterials und Atemschutzgeräts ließ er sich aber damals in der heute üblichen Form nicht verwirklichen; seine um 1735 abgebildeten Löschmannschaften, in Zivil, tragen ,,Schlapphüte'' aus Filz oder Leder, nach der damaligen Mode.

Entwürfe aus dem Jahr 1915 für den deutschen Militärstahlhelm. Hieraus entstand später der jetzige Feuerwehrhelm nach DIN 14940.

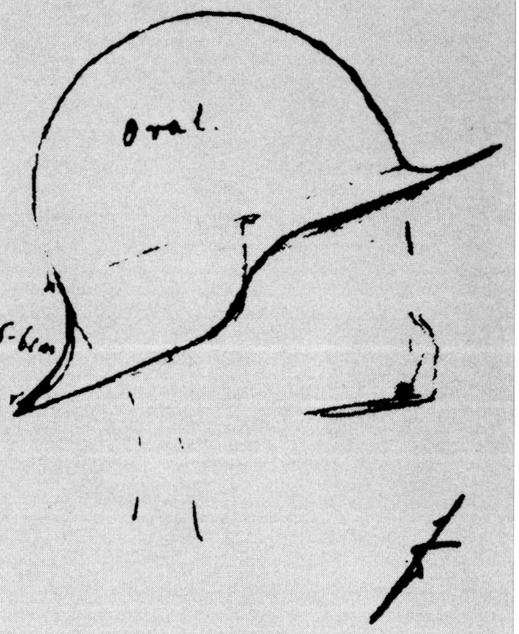

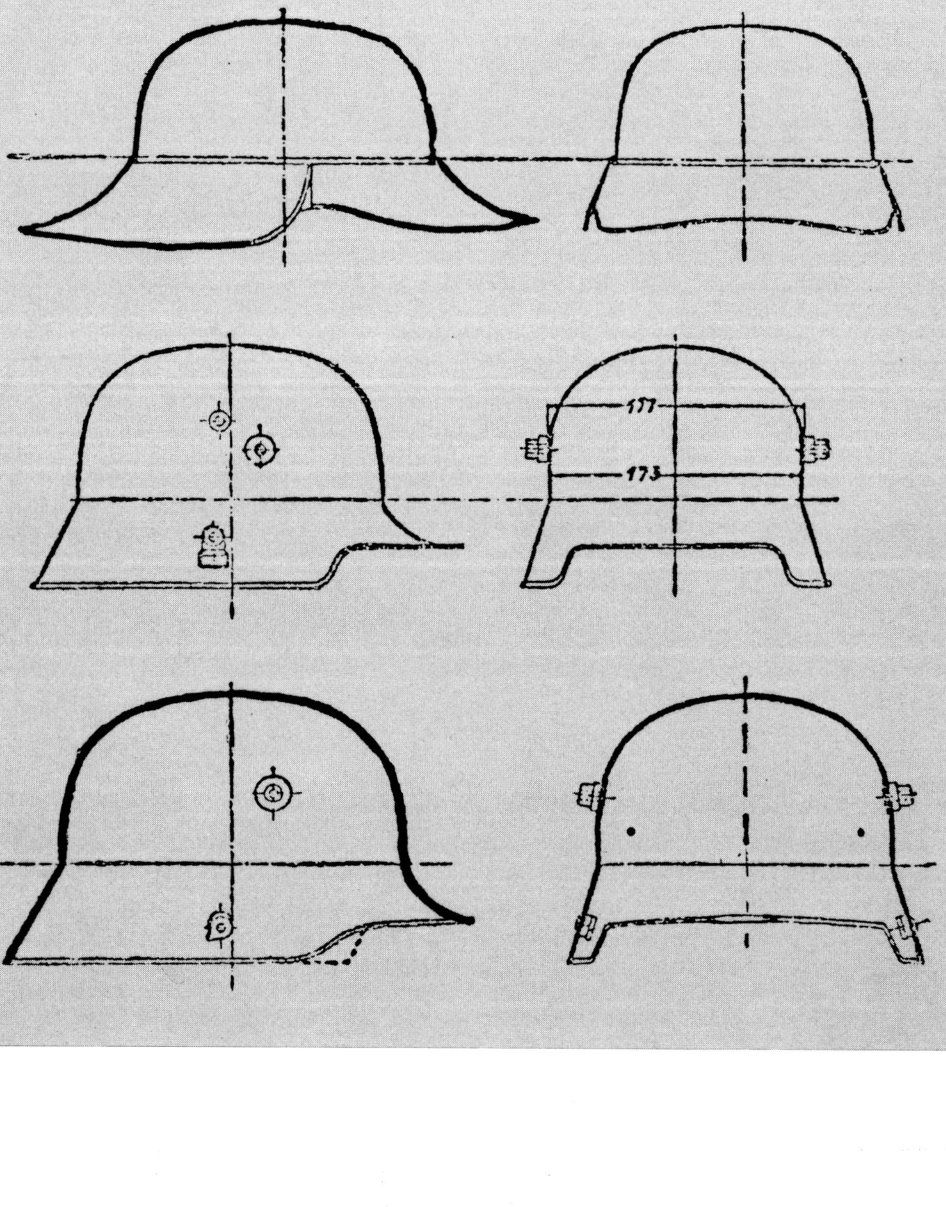

Der Schutzbekleidung der Brandbekämpfer wird erst wieder größere Aufmerksamkeit gewidmet, als das Feuerlöschwesen auf breiter Basis organisiert wird. Französische Sapeurs-Pompiers setzen schon um 1750 eine spezielle Brandschutzkappe aus Leder und Leinen auf, mit großem Messingschild, das den Anfangsbuchstaben der Gemeinde zeigt, und 1786 tragen die Gardes-Pompe einen Ganzmetallhelm mit Kamm. Von jetzt an orientieren sich die Feuerwehren am Militär, soweit es um Uniformen und Kopfschutz geht. Auch wenn wir uns die deutschen Armeehelme der vergangenen 150 Jahre ansehen, finden wir diese These bestätigt, denn diese Helme gibt es fast ohne Ausnahme auch in Feuerwehrausführung. Die Verwandtschaft läßt sich nicht leugnen.

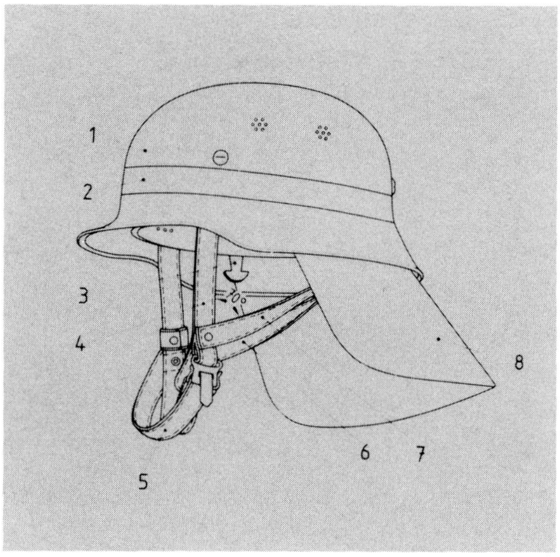

Vollständiger Feuerwehrhelm, DIN 14940

1 *Helmschale*
2 *Reflexstreifen*
3 *Aufsteckknoten*
4, 5, 6, 7 } *Teile des Kinn-Nacken-Riemens*
8 *Nackenschutz*

Für diese Entwicklung finden sich mehrere einleuchtende Gründe. Das Militär mit seinen straffen Organisationsformen war lange Zeit das Vorbild der Feuerwehren, die sich als neugebildete Formationen ihren eigenen Stil erst langsam suchen mußten. Und Militärhelme, besonders ausgemusterte, waren billig zu beschaffen: ein wichtiger Faktor für die unter chronischem Geldmangel leidenden Feuerwehren. Erst in jüngster Zeit haben Industrie und Regierung erhebliche Geldmittel in wissenschaftlich betriebene Helmentwicklungsprojekte für die Feuerwehr investiert.

Daß die Materie offensichtlich sehr komplex ist, zeigt sich in der Tatsache, daß „der ideale Feuerwehrhelm" bis heute nicht erfunden wurde. Falls es ihn eines Tages geben wird, sollte seine Einführung nicht an einem „zu teuer" scheitern, denn der Brandschutzhelm ist das wichtigste Ausrüstungsstück jedes Feuerwehrmannes: hier kann es um Leben oder Tod gehen, und der materielle Einsatz, die Anschaffungskosten des Helmes, dürfen bei dieser Kalkulation nur eine untergeordnete Rolle spielen.

Solide eleg. Ausführung. **Messinghelme** Muster gerne zu Diensten.

Sämtliche Helme besitzen kräftiges, schwarzes Lederfutter.

Kämme aufgeschraubt — ohne vorstehende Schrauben.

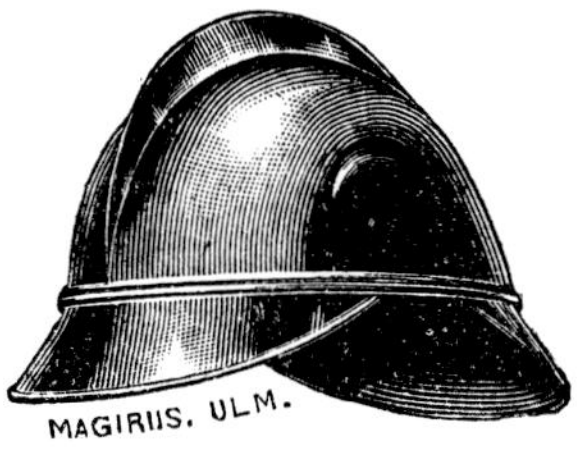

201

202

202a

203

204

205

205a

205b

206

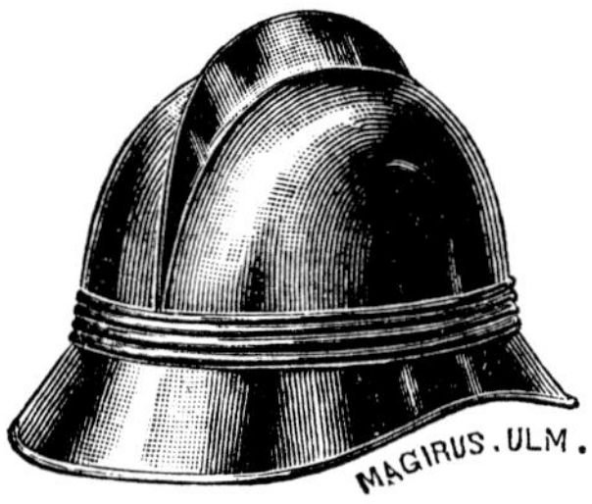

207

208

209

210

211

214

215

Die Garnierungsstücke können nach Wunsch des Bestellers in beliebiger Weise montiert bezw. vertauscht werden.

Ausser den oben verzeichneten Helmsorten wird auf Verlangen tunlichst jede beliebige Form nach einzusendenden Mustern oder Zeichnungen hergestellt.

Streitgespräch 1850: Messinghelm oder Lederhelm?

In den Gründungsjahren der freiwilligen Feuerwehren in Deutschland, nach 1850, gab es nicht viel Auswahl in Bezug auf Helmmaterialien. Leichte Filzhelme, ,,dauerhaft lackiert'', wurden zwar angeboten, schieden aber wegen ihrer geringen Schutzfähigkeit bald aus. Übrig blieben die vom Militär bekannten Lederhelme, und Messinghelme, wie sie bei der Pariser Feuerwehr eingeführt waren. In Anbetracht der politischen Einstellung der ersten deutschen ,,Pompiers-Corps'' und der bewußten Abgrenzung vom Militär war es daher nicht verwunderlich, daß besonders im ,,revolutionären'' Süddeutschland, der Wiege der deutschen freiwilligen Feuerwehr, Messinghelme zum Einsatz kamen. C.D. Magirus hat 1877 auch einleuchtende sachliche Gründe dafür angeführt: ,,Messing muß mehr geputzt werden als Leder und Filz, hat aber diesen gegenüber den Vorzug, daß der Helm nach dem Putzen stets wie neu ist, während der Lack der Leder- und Filzhelme sich nach und nach verliert, und der Helm hierdurch unansehnlich wird.''

Messinghelm Ulmer Form, Süddeutschland, ca. 1860

Messinghelm Badische Form, Salem, 1870

,,Als Uniformstück betrachtet hat der Messinghelm einen Vorzug, der unter erfahrenen Fachmännern stets anerkannt wird; er läßt den Feuerwehrmann auch auf große Entfernung und selbst bei Nacht auf den ersten Blick als solchen erkennen. Zudem gibt der Messinghelm dem Gesamtcorps ein stattliches Aussehen, was auch nicht als zu unterschätzender Vorzug bezeichnet werden darf.

Der Messinghelm muß somit als der praktischere bezeichnet werden.'' Natürlich lehnte man die französische Form des Messinghelmes, mit hohem Kamm, grundsätzlich ab, denn man war ja, trotz allem, sehr national gesinnt. Der Patriotismus ging sogar soweit, daß man jeder Landschaft ein eigenes Modell zugestand: Pfälzer, Badener, Württemberger, Elsäßer Helmformen waren ebenso anzutreffen wie der ,,Ulmer'' oder ,,Strassburger'' Helm. Da es aber keine Institutionen gab, die den Feuerwehren die Anschaffung bestimmter Helmmodelle verbindlich vorschreiben konnte, war jeder Kommandant in der Lage, die Helme zu beschaffen, die ihm am besten gefielen, oder die ihn von der Nachbarwehr unterschieden. Wir treffen sogar innerhalb größerer Feuerwehren auf die Tatsache, daß für jede Einsatzgruppe ein anderer Helmtyp gekauft wurde. Das Ergebnis: typisch ,,Ulmer'' Helme werden auch in Hessen angetroffen, und das Württembergische Modell kann durchaus in Bayern auftreten.

Messing-Helme, nickelplattirte Stahlblech- und Neusilber-Helme.

Die nickel**plattirten** Helme laufen nicht an, sind sehr leicht, elegant und verlieren den **Nickel** nicht durchs Putzen.

140
Messing M. 3.60

140 a
Messing M. 3.80
Nickelplattirt „ 5.40
Neusilber „ 7.—

141
Messing M. 4.—

142
Messing M. 4.—
Nickel „ 5.60
(ohne Schild M. 3.80/5.40)

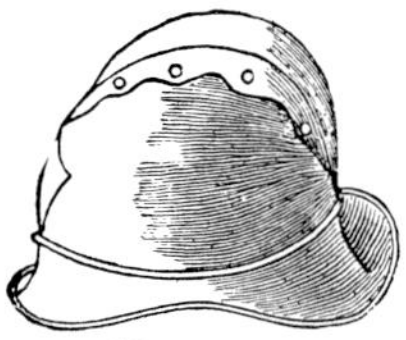

143 Elsässer
Messing M 5.50

146 a
Pfälzer M. 4.80

144
für Steigerlieutenant, roth
Busch mehr M. 3.—

147
Messing M. 4.80

146
Messing M. 4.80

146 b
Messing M. 4.80

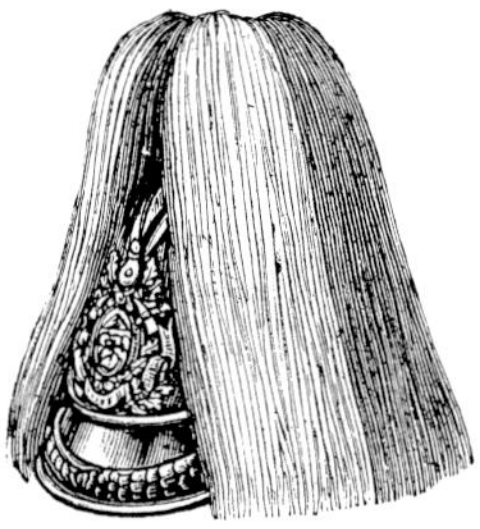

147 württ. Commandant mit
Wappen u. Schuppenbd. M. 12.—
Adjutant wssroth Busch „ 12.—

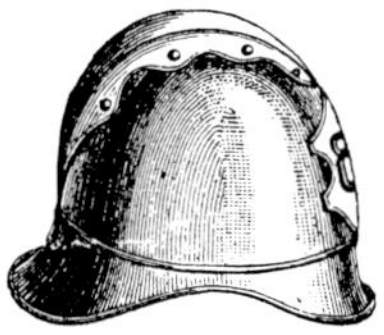

147
württbg. Form mit
Buchstabenschild M. 5.—

141 a
Messing M. 4.80

142 a
Messing M. 4.80

150
Elsass-Lothringer
Messing M. 7.50
Neusilb. m. gold. Beschl. „ 16.—

151
Messing, compl. M. 9.—
Nickelplatt. mit Gold „ 16.—
(sehr elegante Façon.)

152
Nickelplattirt mit vergoldeter ff. Garnitur in Lorbeer oder Renaissance
M. 25.—

153
Nickelplattirt mit vergoldetem Eichlaub, feinste Garnirung M. 30.—

148
Messing M. 7.—
(Wappen extra.)

149
Messing M. 6.50
ohne Busch, so lange
Partie von 125 St. reicht.

Helmschachteln, Wachsledertuch mit durchlaufenden Riemen

für Helme bis 22 cm und 30 cm Höhe
M. 2.— M. 2.20

Putzmaterial.

Putzseife per Stück . . . . M. —.25
Putzpomade in grösseren Büchsen . . . . . „ —.30

Messinghelm, Hessen-Darmstadt, 1865

Messinghelm, Bayern, ca. 1880

Messinghelm, Ulmer Form, ca. 1880

Messinghelm, Württemberg, ca. 1880

Messinghelm, aufgelötetes Emblem „B"
im Hufeisen, Süddeutschland, ca. 1870

Tombak-Helm für Offiziere
ähnlich Elsässer Form ca. 1880

Messinghelm, Württemberger Form, aufgelötetes Buchstaben-Emblem, ca. 1870

Messinghelm, Pfälzer Form, im Zweiten Weltkrieg schwarz lackiert, Verwendung bis 1945

Übte schon der glänzende Messinghelm eine besondere Faszination auf den Betrachter aus, so wurde dieser Eindruck noch gesteigert durch den langherabfallenden Roßhaarbusch, gemeinhin auch als ,,Schweif'' oder gar ,,Wedel'' bezeichnet, der besonders bei badischen Feuerwehren überall anzutreffen war. Dieser Roßhaarbusch, in einem Trichter befestigt, konnte oben auf den Helmkamm aufgeschraubt werden, und das (unverbindliche) Reglement sah vor, daß Kommandanten einen weißen, und Adjudanten einen weiß-roten Busch zu tragen haben.

Weithin eingeführt um 1870, ergab sich um die Jahrhundertwende dann eine heiße Diskussion um Vor- und Nachteile des Wedels, d.h. einerseits gute Sichtbarkeit des Kommandanten auch nachts auf der Brandstelle, andererseits lästige Dekoration und Lächerlichkeit in der Öffentlichkeit, und mit der schrittweisen Einführung von Achselstücken auf den Uniformen wurden die Roßhaarbüsche dann abmontiert, ähnlich wie die seitlich getragenen kleineren ,,Wedel'' der Hauptleute, oder die roten wollenen Raupen der Spritzenmeister.

Messinghelm, Badische Form,
Roßhaarbusch für Offiziere, Bodenseegebiet,
ca. 1870

Messinghelm mit Visier,
Ausgemusterter Militärhelm mit seitlichem Halter
für Federbusch, Frankreich, ca. 1830

Helme 1811 und 1830, Frankreich

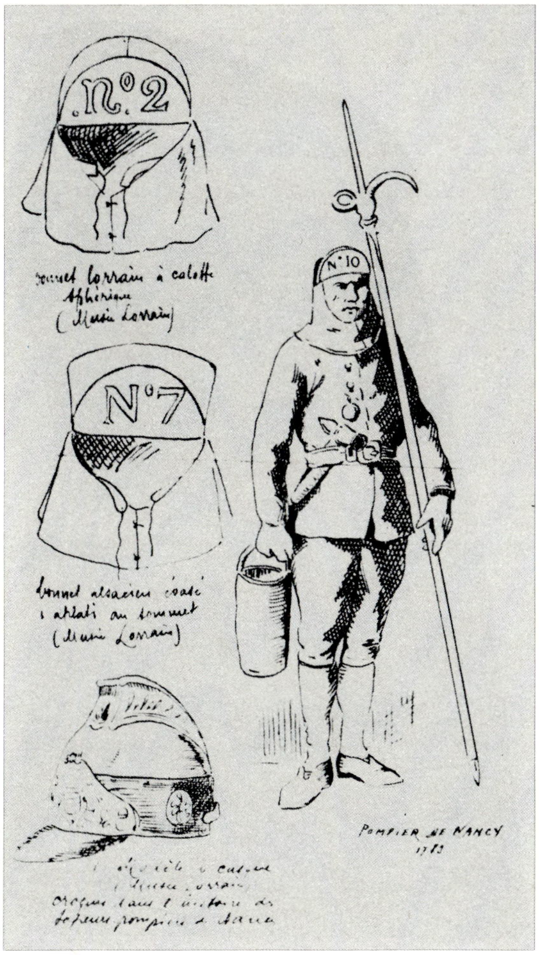

Lederkappe mit Metallschild
Nancy, Frankreich, 1783

Metallhelm, Frankreich, ca. 1810

Messinghelm,
Wappen: Sapeurs Pompiers de Lutterbach
Frankreich, ca. 1880

Messinghelm, Grundform ca. 1880.
Mit neuem Kamm und neuem Emblem
modernisiert nach 1900, Frankreich

Messinghelm,
Standardmodell von 1930
Frankreich

Frankreich

Frankreich hat sicher die älteste und ausgeprägteste Feuerwehrtradition der Neuzeit, und um 1850, als das organisierte Feuerwehrwesen in Deutschland erst langsam in Gang kam, hatten die Franzosen schon ihre 3. Feuerwehrhelmgeneration hinter sich; feststellbar um 1780, besonders im Elsaß und in Lothringen, sind feste gepolsterte Stoffhauben mit weitreichendem Nackenschutz, der vorne zusammengeknöpft werden konnte. Die Nummer der Spritzengruppe wurde, auf einem großen halbrunden Messingschild geprägt, vorne an der Feuerkappe getragen. Diese Kappe war weitverbreitet; dank ihrer soliden Konstruktion haben etliche Muster die vergangenen 200 Jahre ohne Schaden überlebt. Anfang des 19. Jahrhunderts wurde dann ein richtiger Metallhelm in Messing, oder auch mit Eisenblechkomponenten, eingeführt, mit hohem Kamm, und dem kaiserlichen Adler als Emblem, bzw. dem Stadtwappen. Dieser Helm war relativ schwer, und vom Design her nicht sehr ansprechend, und schon 1812 wurde ein neuer attraktiver Messinghelm vorgeschlagen, der sich eng an die ,,neueste Militärmode'' der französischen Armee anlehnte. Obwohl erheblich teurer als das alte Modell, wurde der neue Helm bald eingeführt. Erstmalig mit einem Klappvisier eingerichtet, wurde der neue ,,Pariser Feuerwehrhelm'' jahrzehntelang zum Neidobjekt; in Deutschland und der Schweiz zaghaft kopiert, wurde er in der 2. Hälfte des 19. Jahrhunderts in abgewandelter Form sogar zum Londoner Standardhelm. Weiterentwicklungen des Pariser Feuerwehrhelmes, in flacheren Ausführungen, wurden bis nach dem 2. Weltkrieg besonders bei freiwilligen Feuerwehren getragen.

Und als 1915 der Grabenkrieg auch in Frankreich die Forderung nach einem splittersicheren Stahlhelm erhob, wurde in Anbetracht der erforderlichen Eile der ,,Poilu Helm'' (Modell Adrian) nach dem Vorbild des französischen Feuerwehrhelmes gebaut, inklusive Helmkamm!
(siehe auch Seite 60 und Seite 70/71)

Paris le 29 février 1812

Messieurs,

D'après le décret de S. M. l'E. & R. qui veut que Chaque pompier, Soit Coiffé d'un Casque d'une forme dont le détail suit, Je prends la liberté de vous faire offre de cette fourniture, au prix le plus modéré, et avec toutes les facilités possibles, pour les payemens; Toute fois Je vous prierai, Si vous m'honorez de votre Confiance de me faire Savoir à quelle Epoque et de qu'elle manière Je Serai Soldé de cette fourniture.

Je Suis avec le plus profond Respect,

Messieurs,

Votre très humble et très obéissant Serviteur,
Daupley, fabricant de Casques
Hôtel de M. le Comte de Tessé,
rue Danjou S.t honoré, N.o 24

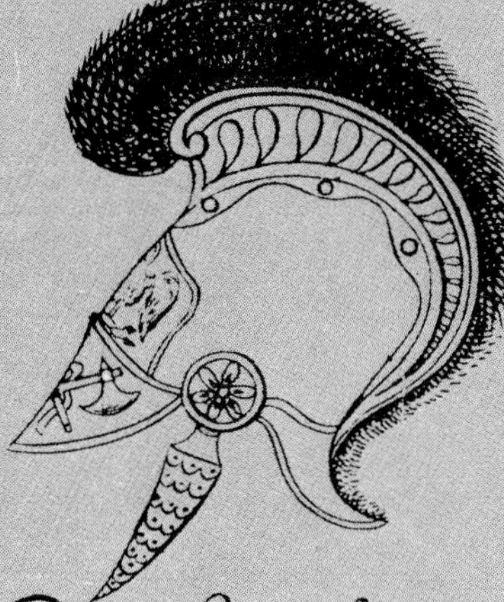

Nouvel Uniforme
de rigeur pour Paris.

Il se trouve devant une Visière qui met la figure à l'abri des dangers et se relève à volonté; il reste une petite visière pour garantir les yeux; derrière le Casque il se trouve un Cache-nuque; il se rapporte une Crinière qui se retire à volonté lorsque l'on va au feu.

Casque d'Officier doré . . 120. f.r
Casque de Sous officier . . . 40.
Casque de Simple pompier. . . . 27.

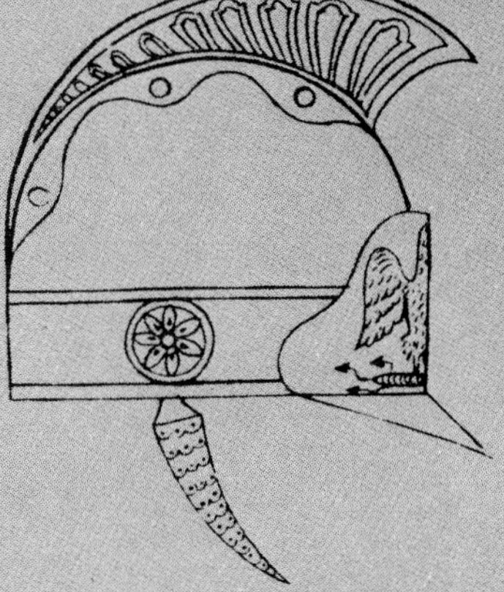

Ancien Uniforme.

Casque d'Officier, doré . . . 70 f.r
Casque de Sous Officier 20
Casque de Simple pompier . . 14

Angebot des französischen Helmherstellers Daupley vom Februar 1812 an die Pariser Feuerwehr über einen neuen Feuerwehrhelm (links). Das alte Modell wird rechts gezeigt.

HIGHLY FINISHED BRASS, PLATED, SILVER, AND LEATHER HELMETS, WITH FITTINGS.

(Fig. 183.)

LONDON BRIGADE.

Brass—No. 1. London make 25/
„ No. 2. Paris make... ... ... 20/
German Silver or White Metal, for Officers ... ... from 50/

(Fig. 177A.)

VOLUNTEER (No. 1).

Leather—With Cross Axe and Torch or other ornament, and Brass Mountings ... 15/
„ With German Silver ditto 21/

(Fig. 180.)

LONDON SALVAGE CORPS.

Leather—With Brass Mountings ... 18/

(Fig. 184.)

STAR CREST.

Leather—With Brass Mountings ... 30/
„ With White Metal Mountings 40/

(Fig. 176.)

PARISIAN.

Brass—Plain 17/6
„ Burnished 20/
German Silver or White Metal, for Officers ... ... from 42/

(Fig. 178.)

METROPOLITAN.

Leather—With Brass Mountings ... 25/
„ Officers, with White Metal Mountings 32/6

(Fig. 181.)

VOLUNTEER (No. 2)

Leather—With Brass Mountings ... 15/

(Fig. 186.)

OLD PATTERN.

Leather—With Brass Mountings ... 20/
„ With German Silver Mountings 30/

(Fig. 177.)

MILITARY.

Leather—With Brass Mountings... ... 15/
„ With German Silver Mountings 21/

(Fig. 179.)

LION CREST.

Leather—With Brass Mountings ... 25/
„ Officers, with White Metal Mountings 37/6

(Fig. 182.)

VOLUNTEER (No. 3).

Leather—With Cross Axe and Torch Ornament, or other design to order ... 15/

(Fig. 186A)

FELT.

With Ornament and Chin Strap ... 12/6

MERRYWEATHER & SONS, Fire Engine Makers and Hydraulic Engineers.

England

Messinghelm, Freiwillige Feuerwehr, England, ca. 1900

Messinghelm
MFB-Metropolitan Fire Brigade Melbourne, Australien, ca. 1900

Organisierte Feuerwehren finden wir in England nach 1700. Es waren die verschiedenen Brandversicherungsanstalten, die, jede für sich, eigene Löschmannschaften ausrüsteten, hübsch anzusehen mit bunten Uniformen und hohen Zylinderhüten. Das alles änderte sich nach 1833, als eine vereinigte Feuerwehr aufgebaut wurde, und nach 1866 treffen wir auch auf den markanten, nach Pariser Vorbild hergestellten Messinghelm, zuerst mit dem Schild ,,MFB'' (Metropolitan Fire Brigade), später dann mit den Buchstaben LCC/LFB (London City Council / London Fire Brigade). Bei der Entscheidung über den Messinghelm standen übrigens auch Lederhelme zur Diskussion (siehe Seite 20, Merryweather-Katalog), aber Branddirektor Shaw wollte ,,etwas Besonderes'' haben, um sich von den benachbarten freiwilligen Feuerwehren zu unterscheiden. Das Ergebnis: innerhalb weniger Jahre trugen auch die Nachbarwehren den neuen Messinghelm .... Er war übrigens in London bis nach 1930 in Gebrauch, und als Australien 1940 wegen der Kriegslage keine Helme mehr aus England bekam, erhielt die Firma Rider & Bell in Sydney den Auftrag, schnellstens nachzuproduzieren, auch für die Berufsfeuerwehren. Erst nach 1964, nach 100 Jahren ,,Modelltreue'', wurden diese großen Messinghelme dort offiziell ausgemustert.

Luxemburg Schweiz

Messinghelm, Modell Luxemburg, ca. 1880

Messinghelm, Schweiz
französisches Modell, mit Visier
Basel, 1880—1891

Messinghelm, französische Produktion
Schweiz, ca. 1920

Da Messing nicht nur attraktiv aussah, sondern als Material auch leicht zu verformen war, wurden Messinghelme in verschiedenen Abwandlungen auch für die kleineren Staaten in Europa hergestellt. Für Luxemburg war lange Zeit die ,,Bettpfostenkugel'' typisch, hier auf einem Helm süddeutscher Produktion. Die Schweiz orientierte sich beim französischen Nachbarn, wählte aber in jedem Fall eine niedrigere Kammform.
Die leichte Verformbarkeit vom Messinghelm, ein Produktionsvorteil, führte aber auch im Einsatz oft zu schweren Beschädigungen. Die Gefahr war um so größer, je komplizierter der Helm aufgebaut war; einfache Helme konnte man ziemlich stabil aus einer einzigen Messingplatte formen, aber dekorative Modelle, wie in Paris oder London getragen, mußten aus mehreren Einzelteilen (Kuppe, Kamm, Vorderschiene, Hinterschiene, Bund) zusammengelötet oder geschraubt werden, mit entsprechendem Verlust an integraler Stabilität. Dieses Problem, die hohe elektrische Leitfähigkeit und last not least die hohen Produktionskosten führten letzthin zum Verschwinden des Messinghelmes: Die Entscheidungskriterien der Feuerwehren wurden sachlicher.

Messinghelme nach englischem Muster, bei der Feuerwehr in Irland, ca. 1900.

Mannschafts-Lederhelme

aus bestem Kernleder, fein lackiert, mit Messingbeschlägen

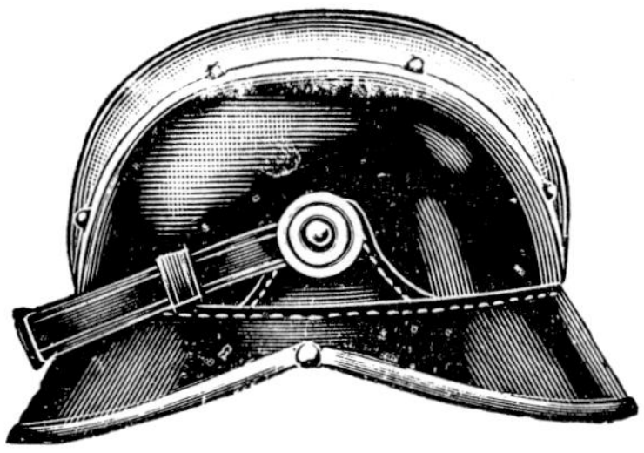

Nr. 40

Nr. 41

Nr. 42

Nr.	**40**	**Mannschaftslederhelm,** bayr. Normal, mit durchgehenden Schirmen, mit Schienen, durchgehendem, aufgesplintetem Kamm und Blankleder-Rosettensturmband	M. **14.50**
„	**40a**	**Derselbe Helm,** mit Kugelaufsatz, für Führer	„ **16.50**
„	**41**	**Mannschaftslederhelm,** Württbg. Vorschrift, mit kurzem, hoh. Messingkamm, mit Wulst, Vorder- u. Hinterschirmschienen, farbigem Lederbundstreifen und Blankleder-Rosettensturmband mit Wappen	„ **15.50**
„	**41a**	**Derselbe Helm,** mit flacher Messing-Schuppenkette für Zugführer, mit Wappen	„ **18.—**
„	**42**	**Lederhelm,** flache hohe Form, mit kurzem, hohem Messing-Spitzkamm, mit Vorderschirmschiene und Blankleder-Rosettensturmband, ohne Wappen	**13.50**

Lederhelme nach preußischer Vorschrift

Nr. 43b

Nr. 43

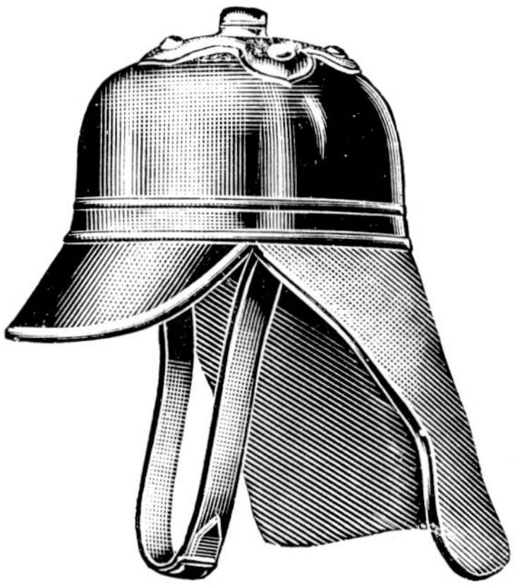

Nr. 43a

Nr.	**43**	**Berliner Normalkappe,** mit $7^{1}/_{2}$ cm breitem Vorderschirm, angenähtem Nackenleder, rotem oder beliebigem Lackleder-Bundriemen, mit Kopfpolster und Vorderkokarde . . .	M. **15.—**
„	**43a**	**Derselbe Helm,** mit Neusilber-Kreuzblatt, ohne Rückenschiene mit abschraubbarer Ventilkapsel	„ **23.—**
„	**43b**	**Berliner Feuerkappe,** neue Art, mit $8^{1}/_{2}$ cm breitem Vorder- und $6^{1}/_{2}$ cm breitem Hinterschirm, nach Vorschrift, mit beliebig farbigem Lackleder-Bundriemen, Stoffkopfpolster, abknöpfbarem Nackenleder, Neusilber-Kreuzblatt, mit abschraubbarer Ventilkapsel, Parierschiene und Vorderkokarde	„ **27.50**
		ohne Nackenleder weniger	„ **1.80**

Lederhelme

Berliner Normalkappe mit preußischem Feuerwehremblem, 1880

Feuerkappe für Mannschaften, Typ Danzig Baden, ca. 1910

Lederkappen als ,,Arbeitsschutzhelme'' sind uns in Deutschland seit dem Mittelalter bekannt; aus Abbildungen kennen wir lederne Jagdkappen, oder auch einfache ,,Kopfschützer'' aus den Anfängen des Untertagebergbaus. Ab Mitte des 19. Jahrhunderts finden Lederhelme auch zunehmend Verwendung bei den neugegründeten Feuerwehren, und zwar hauptsächlich im norddeutschen Raum, aber auch in Bayern. In der Regel waren es ausgemusterte und umgerüstete Militärhelme, die in großen Mengen kurzfristig verfügbar waren, bzw. mit bestehenden Werkzeugen kostengünstig herzustellen waren. Statt der Infanteriespitze oder der Artilleriekugel wurde ein Messing- oder Nickelkamm aufgeschraubt, und der Helmadler wurde durch das Ortswappen ersetzt. Falls durch die Umrüstung unschöne Löcher im Helmkörper sichtbar wurden, konnten sie als ,,Belüftungsöffnungen'' nachträglich gerechtfertigt werden. Für Offiziere wurde sogar die militärische Schuppenkette beibehalten, obwohl sie bei der Feuerwehr keine Schutzfunktion hatte. Ausgerechnet aus Preußen kam eine Alternative zu dem Helm militärischer Herkunft, nämlich eine ziemlich große ,,Feuerkappe'' mit eingebautem Kopfpolster und angenähtem großem Nackenleder, die sogenannte ,,Berliner Kappe'', die auch bei anderen großen Berufsfeuerwehren Eingang fand. Der Helmkopf wurde nicht durch einen Kamm, sondern einen Kreuzblattbeschlag nach Armeemodell verstärkt, auf den für Paradezwecke sogar eine Spitze aufgeschraubt werden konnte, wie bei der Infanterie. Magirus in Deutschland und Shaw in England bestätigten übrigens übereinstimmend und fast gleichzeitig, daß gut gebaute Lederhelme dem weitverbreiteten Messinghelm in Festigkeit nicht nachstünden, zumindest nach den damals üblichen primitiven Testkriterien.

Lederhelm, Preußische Form, Emblem: Hessischer Löwe, Mannschaftskokarde: Hessen, Hessen-Darmstadt, ca. 1920

Lederhelm, Preußische Form, Emblem: Nassauischer Feuerwehr-Verband, Mannschaftskokarde: Preußen, Hessen-Nassau, ca. 1920

Lederhelm, Preußische Form Emblem: Bad Soden, Taunus, ca. 1900

Lederhelm, Preußische Form, Geprägter Kamm Stadtwappen Mainz, ca. 1900

Lederhelm, Preußische Form, Emblem: Bad Soden, Taunus, Offiziersschuppenkette, Offizierskokarde Preußen, ca. 1900

Lederhelm, preußisches Modell
Reichskokarde schwarz/weiß/rot
Provinzwappen Westfalen, ca. 1920

Lederhelm, preußisches Modell
Kokarde Preußen
Provinzwappen Niedersachsen, ca. 1920

Lederhelm, preußisches Modell,
Rheinprovinz, ca. 1920

Lederhelm, bayerisches Modell
Wappen Rheinprovinz, ältere Form
Militär-Kinnriemen, ca. 1910

Der Feuerwehrhelmkamm, der mittlerweile international zum Kennzeichen, ja fast zum Symbol der Feuerwehrhelme geworden ist, war beim Lederhelm nicht nur Dekoration oder Ersatz militärischen Zierats, sondern eine absolute Notwendigkeit, den Helmen Festigkeit zu verleihen. Es hatte sich nämlich herausgestellt, daß Hitze am Brandort und extreme Feuchtigkeit, etwa durch den Löschwasserstrahl, den Lederhelm erweichen und seinen Schutzwert erheblich herabsetzen. Durch den Kamm und die Einfassung der Vorder-und Hinterschirme konnte dieses Manko reduziert werden. Um allen landsmannschaftlichen und örtlichen Wünschen gerecht zu werden, konnten die Standardhelmkörper mit einer Vielfalt von Kammgarnituren ausgerüstet werden; ein namhafter Offenbacher Hersteller bietet nicht weniger als 18 Versionen an.

Übrigens: fast alle in Deutschland und im benachbarten Ausland getragenen Lederhelme stammten aus der traditionsreichen Lederwarenproduktion in Offenbach am Main.

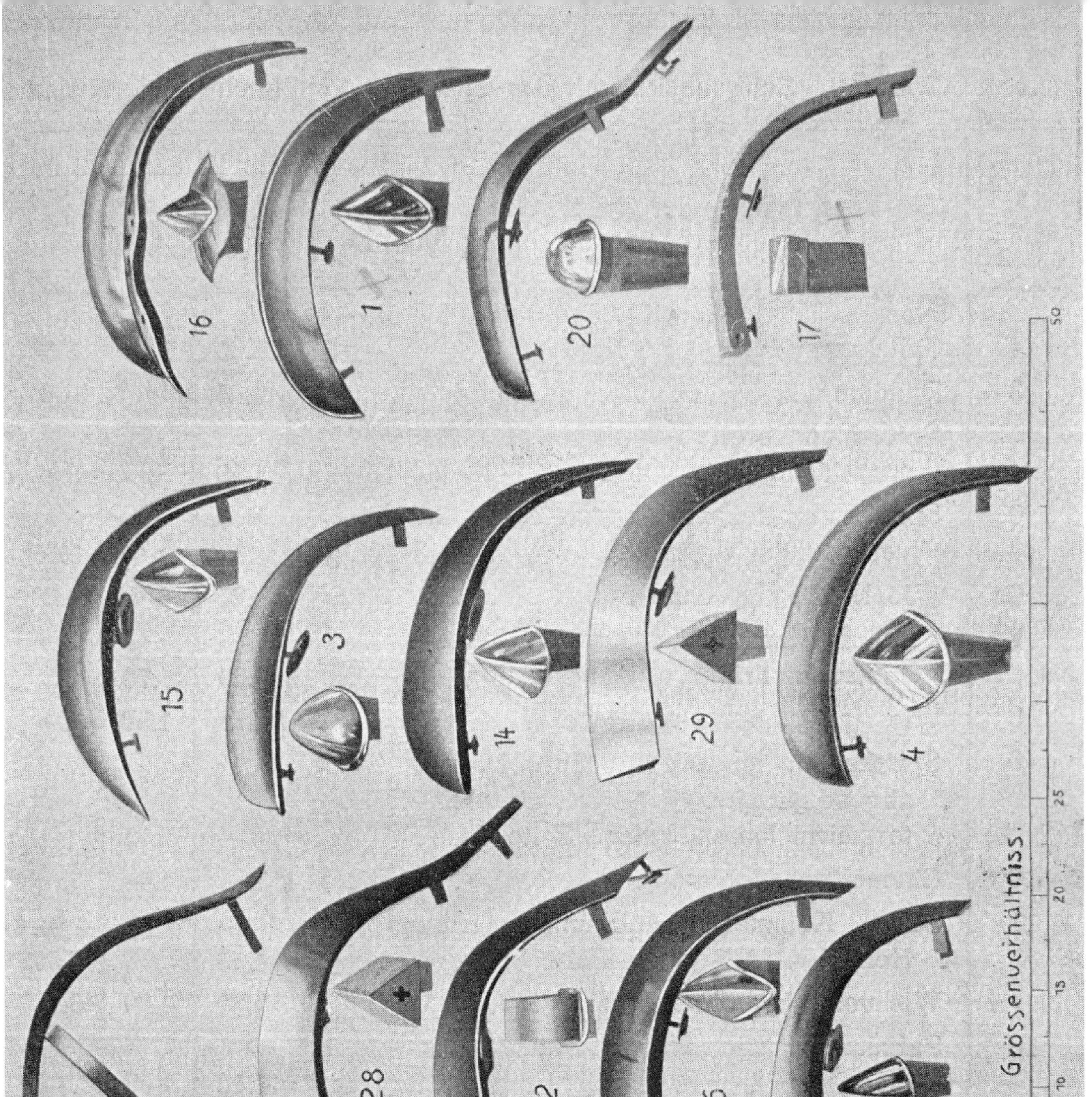

Kammgarnituren für Lederhelme. *Katalog der Firma Maury, Offenbach, 1911*

Helmkokarden:

Württemberg

Sachsen-Coburg-Gotha

Schwartzburg-Rudolstadt

Hamburg

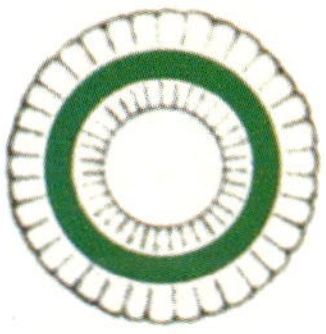
Sachsen

Oldenburg

Waldeck

Bremen

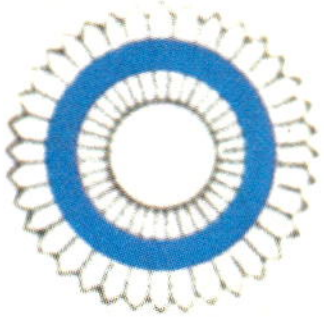
Bayern

Mecklenburg

Lippe

Lübeck

Preußen

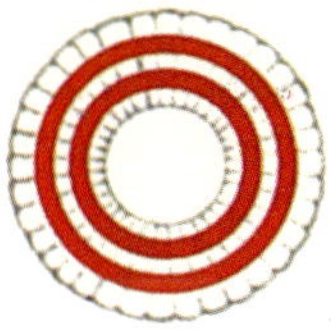
Hessen

Anhalt

Reuss

Deutsches Reich

Baden

Braunschweig

Schwartzburg-Sonderhausen

Lederhelm, Preußische Form
Buchstaben „M", ca. 1890

Lederhelm, Preußische Form
ausgemusterter Militärhelm,
Norddeutschland, ca. 1900

Lederhelm, Preußische Form
Buchstabenschild „H" (Hydrantengruppe)
Preußen, ca. 1890

Lederhelm, Preußische Form,
mit rotem „Lederbund"
Württemberg, ca. 1926

Lederhelm, Preußische Form
rot lackiert, Bayern, ca. 1890

Messingemblem OFF, Kokarde Oldenburg
ca. 1900

Messingemblem: Gardestern mit Inschrift
„Suum Cuique" („Jedem das Seine")
Gelöteter Messingkamm
Gebiet Bremen, ca. 1890

Messingemblem SF,
(Schröttinghauser Feuerwehr)
Gebiet Osnabrück, ca. 1900

Nickelemblem „Fabrikfeuerwehr"
Bremer Wollkämmerei, ca. 1900

Lederhelme, preußische Bauart

Zur Ausrüstung des Lederhelmes gehörten in der Regel auch die mehrfarbigen Kokarden, die auf runden gelochten Metallscheiben konzentrisch angeordnete Landesfarben zeigten. Sie wurden unter den Befestigungsschrauben oder -splinten des Kinnriemens getragen, rechts die schwarz-weiß-rote Reichskokarde, links die Landesfarben (in Blickrichtung des Helmträgers). Von dieser Regelung gab es aber in der Praxis häufig Abweichungen, offensichtlich nahm man es bei den damaligen freiwilligen Feuerwehren nicht allzu genau mit den Empfehlungen der Landesfeuerwehrverbände.

Helm-Wappen (1/2 natürl. Grösse.)

Nr. 1
fein vergold. od. versilb. ℳ 1.70

Nr. 2
fein vergoldet ℳ 1 70

Nr. 3
Messing ℳ 1.— f. bronc. ℳ 1 25
vergoldet ℳ 1.50

Nr. 4
Messing ℳ 1.— f. ℳ 1.25
vergoldet ℳ 1.50

Nr. 5
Messing broncirt ℳ 1—
fein vergold. (od vers. 1.25) „ 1.50

Nr. 6
auch mit „Aug und Hand“ „Reichsadler“ „Sanitätsz“ zu haben,
Messing ℳ —.70
vergoldet „ 1.—

Nr. 7.
Messing ℳ —.70
fein vergoldet „ 1.25

Nr. 8
Messing ℳ —.50
vergoldet „ 1.—
Stadtwappen extra.

Nr. 9 (mit Buchstaben)
Messing broncirt —.40 ₰
vergold. od. versilb. —.80 ₰.

Nr. 10
versilb. od. vergoldet ℳ 1.—
mit reichem Eichlaub-Kranz „ 2 50

Nr 11
Messing ℳ 1— } inclus
vergold. „ 1.80 } 2 Buchstaben

Nr. 12 gross und klein
ohne Cocarde ℳ —.50 u. 30 ₰.
mit Cocarde 50 ₰. mehr.

Nr. 13 vergoldet ℳ 2.— (Bei 25 Stck.)

Nr. 14
45 mm Messg. 35, vergld. 50

Monogramm 45 mm
Messg. 1.—, vergld. 1 50

Nr 15 bronc. ℳ 1.—, vergold. ℳ 1.30

Nr. 16 bronc. ℳ —.80 verg. ℳ 1.20.

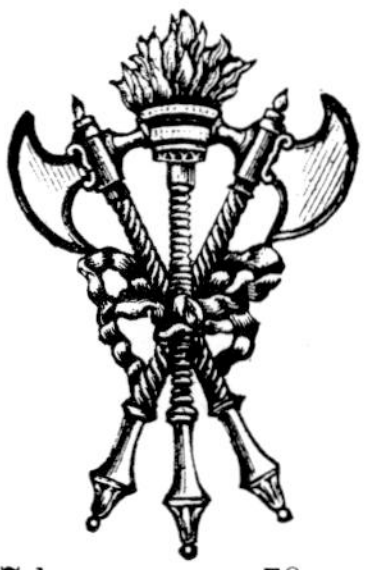

Nr. 17 bronc. ℳ —.70, verg. ℳ 1.—

Nr 18 { broncirt ℳ —.50
(7 cm) { vergoldet „ 1.—
(kleinere billiger)

Nr. 19 Messg. ℳ —.25

Katalog der Firma Lieb, Biberach, 1890

Offiziershelm, Infantriespitze mit
Schuppenkette
Preußischer Feuerwehrverband, ca. 1920

Auf alten Gruppenbildern läßt sich klar erkennen, daß oftmals, besonders in preußischen Landstrichen, vom Kommandanten und auch den anderen ,,Chargierten'' Militär-Lederhelme mit Spitze (Pickelhaube) oder gar Artilleriekugel getragen wurden. Das entsprach zeitweise den preußischen Uniformierungsvorschriften für Feuerwehrführer, konnte aber auch mit der Waffengattung zusammenhängen, in welcher der nunmehrige Feuerwehrhauptmann früher seinen Militärdienst abgeleistet hatte. Das dazugehörige Helmadleremblem wurde durch das Provinzwappen oder durch ein aufgesetztes preußisches Feuerwehremblem (Helm mit gekreuzten Beilen) ,,entmilitarisiert''.

Offiziershelm, Artillieriespitze
Preußischer Feuerwehrverband, ca. 1920

Offizier-Stellvertreter, Artilleriespitze
Oldenburg, ca. 1910

schwarzer Lederhelm mit Kreuzblatt
Emblem der Berufsfeuerwehr
Frankfurt, ca. 1890

Lederhelm, Modell Sachsen mit Lederkamm
ca. 1900

Lederhelm, preußische Form
breite Kammunterlage, für Offiziere
ca. 1880

Lederhelm, mit rotem Lederbund und
Militär-Kinnriemen, Sachsen, ca. 1890

Lederhelm, preußisches Modell
flache Militärausführung
für „Einjährige" (Offiziersanwärter)
Gardestern „Suum Cuique"
mit rheinisch-westfälischem
Flachkamm, ca. 1890

Lederhelm, ohne Metallbeschlagteile
für Feuerwehrelektriker
Dietzenbach, Hessen, ca. 1920

Obwohl es auch bei Lederhelmen, durch Form des Helmkörpers oder der Kamm-Type, ausgesprochene Regionalmodelle gab, fühlte sich die Feuerwehr bei ihren Beschaffungswünschen ziemlich ungebunden. Keine Seltenheit sind typisch preußische Lederhelme, umgerüstet auf den umlaufenden bayrischen Normalkamm und rot lackiert in Bayern getragen; und umgekehrt werden Helme vom Modell „Bayrisch Normal" auch an der „Waterkant" gefunden. Der leicht erkennbare „rheinisch-westfälische Flachkamm" ist in Oberhessen nachweisbar, und der Lederkamm nach sächsischer Vorschrift wurde auch anderswo geschätzt. Durch den „Feuerwehrelektriker-Helm", der keine Metallteile enthalten durfte, wurde er überall bekannt. Lederhelme wurden von deutschen Feuerwehren bis Anfang der dreißiger Jahre beschafft, und bei freiwilligen Feuerwehren auch noch während der Kriegszeit verwendet, besonders für Ersatzkräfte, Jugendfeuerwehr usw.. Sogar in den sechziger Jahren finden wir sie noch vereinzelt bei Löschgruppen auf dem Lande.

Lederhelm, ehem. bayrischer Militärhelm
Emblem: Hessischer Löwe, Oberhessen, ca. 1925

Lederhelm, schwere Ausführung
ehem. Modell „Wiesbaden", ca. 1910

Feuerwehrhelme, die um 1870 aus ausgemusterten bayrischen Militärhelmen des Modelljahrgangs ca. 1840 entstanden sind.

Man beachte die unterschiedlichen Kammformen und Kinnriemenbefestigungen.

Modell „Bayrisch-Normal"
wurde 1940 schwarz matt lackiert
Einführung ca. 1920

Modell „Bayrisch-Normal", rot lackiert
ca. 1920

Ehem. Militär-Lederhelm, bayrische Form,
mit Bayrischem Normalkamm, ca. 1920

Ehem. Militär-Lederhelm, bayrische Form
Kamm mit Unterlage, für Offizier-Stellvertreter
ca. 1900

Ehem. Militär-Lederhelm, Preußische Form
mit Bayrischem Normalkamm, ca. 1920

Ehem. Militär-Lederhelm, Bayrische Form
Ortsemblem „G", mit Inschrift:
„Gott zur Ehr, dem Nächsten zur Wehr".

THE AMERICAN FIREMAN

L. Maurer, delineator, Currier & Ives   New York, 1858

Während sich im Europa des 19. Jahrhunderts die Form der Feuerwehrhelme weitgehend an militärischen Vorbildern orientierte, ging man in Nordamerika ganz andere Wege. Zwar waren die ersten „fire hats", ähnlich den Hüten der Londoner Versicherungsbrigaden, noch in der traditionellen „Zylinderhut-Ausführung", aus Leder, und für Paradezwecke farbig bemalt mit dem Namen der Fire Company, aber mit der Loslösung vom Mutterland 1776 machte man sich auch bei Feuerwehrhelmen vom europäischen Vorbild frei. Nach und nach entwickelte sich im Raum New York ein Helmtyp, der im wesentlichen aus einem mit Rippen verstärkten Helmkopf bestand, und dessen angenähter Lederschirm, aus einem Stück geschnitten, im rechten Winkel an den Helmkopf aufgesetzt wurde, wie die Krempe eines Cowboy-Hutes. Besonders hinten war der breite Lederschirm weit hinuntergezogen, um den Nacken des Trägers zu schützen. Die Stabilität des Helmkörpers und der „Krempe" wurde durch eingenähte Drahtversteifungen erreicht. Statt des in Europa üblichen, relativ kleinen, Stadt- oder Landeswappen trugen diese amerikanischen Helme große Lederschilde mit der Nummer der Spritze und dem Namen der Fire Company (Löschverein). Gehalten wurde dieses „front piece" von einem dekorativen Messinghalter, der oben am Helm befestigt war. Neben einem stilisierten Amerikanischen Adler sind Messinghalter in Form eines Hahnenkopfes, einer Seeschlange, eines Fuchses usw. bekannt.

Weiterentwicklungen dieses Lederhelmtypes sind noch heute in der Produktion bei Cairns in New Jersey; die New Yorker Berufsfeuerwehr ist mit ihnen auch jetzt noch ausgerüstet.

Lederhelm von Cairns, New York, Lederschild „Posen" Fire Dept., Messing-Adlerkopf als Schildhalter, USA, ca. 1900

Roter Lederhelm, österr. Modell typisch türkischer Kamm Istanbul, 1950

Lederhelm, österr. Modell, Offiziersmodell, Griechenland, 1930

Vulkanfiber Helm, Tschechoslowakei, 1930

Lederhelm, österr. Modell Tschechoslowakei, 1950

Nicht zu verleugnen ist der österreichische Einfluß auf Feuerwehrhelme des Balkans, in Südosteuropa. Länder, die zur K.u.K. Donaumonarchie gehörten, wurden im allgemeinen auch in Bezug auf die Uniformierung von Wien aus gesteuert, sodaß meistens österreichische Standardmodelle zum Einsatz kamen, mit regionalen Unterschieden. Im weiteren Sinne traf das auch auf Griechenland und die Türkei zu, deren Feuerlöschwesen z.B. von Österreichern vom Grund auf neu organisiert wurde. Die hier gezeigten Helmtypen, durchweg „traditionsbeladene" Modelle, waren besonders bei freiwilligen Feuerwehren dort bis vor wenigen Jahren im Einsatz.

Lederhelm, preußisches Militärmodell
Schweden, ca. 1900

Lederhelm der Feuerwehr Riehen, 2. Gruppe
(Kanton Basel—Stadt) 1924—1946
(Hersteller: Maury, Offenbach)

Lederhelm, Schweizer Normaltyp
Rosetten mit Schweizer Kreuz
H 1 = 1. Hydrantengruppe, ca. 1910

Lederhelme aus Offenbach am Main wurden nicht nur in deutschen Landen getragen, sondern auch in Schweden, Dänemark, den Niederlanden und in der Schweiz. Da die Metallgarnituren beliebig austauschbar waren, konnte der typisch schwedische Metallstern mit eingelegtem Uniformknopf genauso angeboten werden wie die Kinnbandbefestigung mit Schweizer Kreuz statt den militärischen Löwenköpfen.

Lederhelme waren bei ihren Trägern sehr beliebt, denn in jahrelangem Feuerwehreinsatz, durch Hitze und Feuchtigkeit zeitweise erweicht, paßten sie sich der Kopfform des Besitzers an, und waren bequem ,,wie ein alter Schuh''. Übrigens: man findet heute nur wenige alte Lederhelme in der jetzt üblichen Standardkopfgröße 56/57. Der Grund, durch medizinische Untersuchungen der letzten 100 Jahre belegt: Körpergröße und Kopfumfang sind erheblich gestiegen, im Vergleich zur Gründungsperiode der Feuerwehren. Die Schlußfolgerung, ob wir wirklich auch klüger geworden sind, sei dahingestellt....

Offiziershelme aus Österreich

Freiw. Feuerwehr Klingfurth,
Lederhelmtyp Oberösterreich, ca. 1925

Lederhelmtyp der bayr. Armee
Messingschild: Gott zur Ehr, dem Nächsten zur Wehr, ca. 1900

Vulkanfiberhelm mit Blechschirm, Kamm Wiener Modell, Schild FF (Freiw. Feuerwehr), Oberösterreich, ca. 1890

Lederhelm, mit Blechschirm FF Brauneck, Modell Oberösterreich ca. 1925

Lederhelm, mit blankem Blechschirm, Offiziersstellvertreter Oberösterreich, ca. 1925

Lederhelm, Messingschild mit St. Florian Statue Oberösterreich, ca. 1925

Vulkanfiberhelm, mit rotem Lederbund Österreich, ca. 1900

Lederhelm, breiter Messingkamm Österreich, ca. 1900

Lederhelm des 2. Zuges der Rettungsabteilung, Freiw. Feuerwehr Wiesbaden, ca. 1892

Lederhelm der 1. Wiesbadener Freiw. Feuerwehr (Pompiers), ca. 1870

Blechhelm, Typ Bayrisch-Normal
Bayern, ca. 1920

Blechhelm, Form des bayr. Militärhelmes
mit „Normal" Kamm und Belüftungsöffnungen,
Bayern, ca. 1920

Blechhelm, Form des bayr. Militärhelmes
Orts-Messingschild, Bayern, ca. 1920

Der Schritt in das 20. Jahrhundert — neue Materialien — neue Formen —

Leder- und Messinghelme bewährten sich in der Anfangsphase der freiwilligen Feuerwehren recht gut, aber es gab Probleme, die dazu führten, nach anderen Materialien für die Helmherstellung zu suchen. Der Anschaffungspreis z.B. für die weitgehend handwerklich gefertigten Leder- und Messinghelme war recht hoch, sodaß selbst begüterte Feuerwehren zuerst oft nur die Steiger/Retterabteilungen mit Helmen ausrüsten konnten, während sich die „Pumper-Mannschaften" mit Stoff-oder Wachstuchmützen bescheiden mußten. Billige ausrangierte Militärhelme waren oft „ausverkauft", und so wurde mit für die damalige Zeit erstaunlicher Aktivität und Innovationsbereitschaft mit neuen Helmwerkstoffen experimentiert. Als Alternative zum Messinghelm gab es schon um 1900 Aluminiumhelme, besonders beliebt wegen ihrer glänzenden Oberfläche, aber sie waren noch teurer als die Messinghelme und wurden daher selten angeschafft. Billiger waren da schon die neuen Kunstlederhelme, denen im Lieb-Katalog von 1890 erstaunliche Eigenschaften zugeschrieben wurden: „Dieselben sind aus demselben Material und in derselben Weise hergestellt wie die ebenfalls in Amerika erzeugten Eisenbahnräder, Fässer ect. Sie sind so widerstandsfähig wie Metallhelme, leicht, und dem Kopf sich anschmiegend wie Filz- und Lederhelme, werden von der Sonne nicht erweicht, und eignen sich daher besonders auch für heiße Länder, gehen nicht ein und widerstehen auch der Nässe; — Kurz sie vereinigen mit ihrer Billigkeit alle Vortheile der sonstigen Helme, ohne deren theilweise Nachtheile." Trotz dieser hervorragenden Eigenschaften fanden diese Helme nicht allzubreite Verwendung. Ähnlich erging es den Filzhelmen, die in allen gängigen Helmformen, mit Kamm und Wappen versehen, angeboten wurde, lackiert wie die Lederhelme gleicher Bauart. Für den harten Feuerwehreinsatz, an vorderster Linie, waren sie nicht geeignet, und auch in Museen findet man sie heute nur vereinzelt, ein Beweis ihrer mangelhaften Dauerhaftigkeit.

Messinghelm, weiß lackiert, Modell 1870
Bayrisches Feuerwehrhelmschild,
Offiziershelm, nach 1945 getragen.

Blechhelm, rot lackiert, Modell 1900
Kamm bayrisch normal

Messinghelm, rot lackiert
Form bayrisches Militärmodell 1868
Unterfranken

Als echte Alternative zu den teuren Messinghelmen blieben nur Stahlblechhelme, schon vor der Jahrhundertwende angeboten, die eine weite Verbreitung allerdings erst durch die Materialknappheit nach 1915 erlebten, als sogar die Kaiserliche Armee versuchsweise zu Filz- und Stahlblechhelmen überging, da Leder Mangelware wurde. Um die Feuerwehren bei Ergänzungsbeschaffungen nicht mit gänzlich neuen Helmtypen zu konfrontieren, wurden fast alle Leder- und Messinghelmformen auch in Stahlblech hergestellt, wobei Qualität und Verarbeitung, Blechstärke usw. direkte Rückschlüsse auf die Verschlechterung der wirtschaftlichen Lage, nach 1914 und besonders in den 20iger Jahren, ziehen lassen. Um das Jahr 1930 gibt es dann noch einmal eine kurze „Renaissance'' sehr stabil gebauter Lederhelme des preußischen Militärmodells, die nunmehr nur noch für die Feuerwehr gebaut wurden, da die Reichswehr mittlerweile den Stahlhelm, und die Polizei statt der Pickelhaube den Jäger-Tschako trägt; aber nach 1933 beginnt dann endgültig „eine neue Zeit'', auch für die Helme der Feuerwehr.

Blechhelm, Kamm rheinisch-westfälisch, Kinnriemenrosetten mit „FF", Oberhessen, ca. 1910

Blechhelm, dreieckiger Kamm, aufgelöteter Messingbuchstabe „B", Federbuschhalter an Kinnriemenrosetten, Unterfranken, ca. 1900

Blechhelm, Messingkamm, Württemberger Form, ca. 1870, Messingschild „Sch"

Blechhelm, Kamm mit Unterlage, Aufgelötetes Schild „K", Gegend Würzburg, ca. 1900

Blechhelm, Messingkamm, Hessische Form, ca. 1900, Bereich Darmstadt

Blechhelm, flachrunder Kamm, Schild silberfarbig Hessen-Nassau, ca. 1900

Blechhelm, Wiener Modell, Gamischdorf/Österreich, ca. 1926

Die wirtschaftliche Situation, der industrielle Fortschritt und der akute Ledermangel führten nach dem 1. Weltkrieg auch in Österreich zum Übergang auf Kunstleder-, dann auf Stahlblechhelme. Vorerst wurden die gleichen Helmformen weiterverwendet, auch die im Vergleich zu Deutschland sehr dekorativen, hohen Kämme und andere Beschlagteile. Um den Schutz des Helmkörpers auch seitlich zu verbessern, entwickelte man einen neuartigen, spinnenförmigen Helmbeschlag, der seither zum Erkennungszeichen der österreichischen Feuerwehrhelme geworden ist.

Blechhelm, Nickelkamm Wappen von Kärnten, ca. 1910

Blechhelm, innen Filzbeschichtet, Pöchlarn/Donautal/Österreich, Vorgänger des modernen österr. Feuerwehrhelmes, Offizierslorbeerkranz

Helm der deutschen Feuerschutzpolizei Wels/Niederösterreich. Nach 1945 mit verchromter „Spinne" statt Kamm

Blechhelm, Schweizer Standardform, ca. 1920

Stahlhelm, Schweizer Armeehelm, Stadt Zürich, 1974

Stahlhelm, Alu-Kamm, Schweizer Armeehelm (Eidgen. Militärdepartement), Stadt Basel, 1938

Auch in der Schweiz fand der Stahlblechhelm weite Verbreitung, besonders bei neugegründeten Feuerwehren oder als Ersatzbeschaffung. Als in den dreißiger Jahren durch die drohende Luftkriegsgefahr das Arbeitsspektrum der Feuerwehren erweitert wurde, fand der Schweizer Armeehelm seinen Weg in die Feuerwehrmagazine. Als schwerer, splittersicherer Stahlhelm basierte er auf einer amerikanischen Entwicklung des 1. Weltkrieges, führte aber damals in den Schützengräben Flanderns zu folgenschweren Verwechslungen mit dem deutschen Stahlhelm, und wurde vom Fronteinsatz zurückgezogen. Erst seit den siebziger Jahren wurde er bei schweizer Feuerwehren durch Kunststoffhelme amerikanischer Herkunft ersetzt.

Lederhelm, aus einem Stück gefertigt,
La Cumbre, Argentinien, ca. 1940

Aluminiumhelm, Modell Cairns Senator,
Lederschild: OFD, USA, ca. 1930

Lederhelm, Messingkamm, Spanien, ca. 1930

Kunststoffhelm, Phenol-Harz
Schild: WFD (Westfield Fire Department)
New Jersey, USA, ca. 1930

Kunststoffhelm, Material Jupiterite,
Emblem CFB (Country Fire Brigade),
Victoria, Australien, ca. 1940

Lederhelm, Modell Cairns.QMC = Quarter
Master Corps (US-Armee Intendantur)
Californien, USA, ca. 1943

Lederhelm, Modell Merryweather Volunteer, Freiwillige Feuerwehr, England, ca. 1900

Die zwanziger und dreißiger Jahre sind auch im übrigen Ausland gekennzeichnet durch das Suchen nach neuen Helmformen und neuem Material für den Feuerwehrdienst. Neben traditionellen Formen, jetzt aber, einschließlich Kamm und „Krempe'', aus einem Lederstück gepreßt, oder aus den damaligen spröden Kunststoffen hergestellt, finden wir leichte Aluminiumhelme (z.B. USA) oder die Einführung des Korkhelmes in England, der in den dreißiger Jahren auch in London den traditionellen Messinghelm ersetzt. Fast als Rückschritt ist es zu betrachten, daß Ende der dreißiger Jahre in Europa dann vielerorts Helme des Militärtypes für die Feuerwehren Vorschrift wurden. Vom Material her sicherlich besser als die traditionellen Leder- und Messinghelme, waren sie in der Konzeption und vom Tragekomfort her für den Feuerwehreinsatz nur bedingt geeignet.

Korkhelm, WFB — Wantagh Fire Brigade England, ca. 1935

Lederhelm, Modell Hendry, Emblem: LSC (London Salvage Corps, Bergungscorps der Feuerversicherungen), London, ca. 1960

Stahlhelm, Militärmodell 1936, Emblem NFS, (National Fire Service), Helm der staatl. Feuerwehr, 1944

In Deutschland wurde nach 1933 sogleich auch das Thema „Feuerwehrhelm" politisiert. Obwohl offiziell gar nicht vorgeschrieben, machten sich besonders aktive Parteigenossen alsbald daran, Helmsymbole zu entfernen, die angeblich kirchliche- oder Freimaurer-Embleme enthielten. Sie wurden abgeschliffen oder unkenntlich gemacht. Und besonders stramme S.A.-Mitglieder bastelten sich den Parteiadler an den Feuerwehrhelm, um den anderen, in der Mehrheit politisch neutralen Feuerwehrmännern, die „Zeichen der Zeit" klarzumachen. Daß diese Absicht teilweise erfolgreich war, zeigt sich vielleicht in der Tatsache, daß auch nach 1936, entgegen ausdrücklicher offizieller Anweisung, bei einzelnen freiwilligen Feuerwehren sogar die Lederhelme alten Typs mit den Abzeichen der Feuerschutzpolizei (FSchPol) und der Partei versehen wurden. War das ein Zeichen besonderer Parteibegeisterung, oder wollte man damit demonstrieren, daß sich die ehemals sehr demokratisch organisierten freiwilligen Feuerwehren nicht so genau an die Weisungen des Reichsführers SS (als Chef der Deutschen Polizei und oberster Dienstherr der Feuerwehr) hielten? Wir wissen es heute nicht mehr.

Um 1930, als auch bei der Reichswehr Versuche liefen, den Stahlhelm des Jahres 1916 endlich durch ein zeitgemäßes Modell zu ersetzen, wurden bei den Berufsfeuerwehren Dortmund und Berlin erstmalig Feuerwehr-

Lederhelm,
Emblem teilweise unkenntlich gemacht
Hessen, 1935

Blechhelm, Emblem der Deutschen Polizei/
Feuerlöschpolizei, Ostmark, 1939

Versuchtyp, Kunststoffhelm, Phenol-Harz,
ähnlich Wehrmacht-Modell, 1935

Lederhelm, preussische Form,
Emblem der Feuerschutzpolizei
Hessen, 1936

helme in Stahlhelmform getestet. Aus dieser Zeit sind Versuchsmuster bekannt aus Stahl, Leichtmetall, Vulkanfiber, Leder, Kunstleder, Bakelit und harzgetränktem Gewebe. Bei praktischen Trageversuchen im Einsatzdienst bewährten sich die Kunststoffe nicht, und die Anordnung des Ministers des Innern in Preußen von 6.2.1934 schrieb daher für Neuanschaffungen vor:

Helme aus Leder oder Ganzmetall nach Form des Stahlhelmes, schwarz mit Kinnriemen, Schnalle, Kamm, Eichenlaubverzierung, abknöpfbarem Nackenleder und Stadtwappen ... vorhandene Kopfbedeckungen dürfen aufgetragen werden.

Überraschend, aus der heutigen Sicht, war die Zulassung von Lederhelmen in Stahlhelmform. Vielleicht wollte man der Lederindustrie nicht über Nacht ein erhebliches Absatzpotential nehmen. Prompt rüsteten sich auch einige Feuerwehren mit Lederhelmen in Stahlhelmform aus, aber diese Helme waren schwierig herzustellen und entsprechend teuer. Schon ein paar Monate später, im Mai 1934, wurden daher Lederhelme in der Bekleidungsordnung gestrichen, auch das vorne anzubringende Wappen mit Eichenlaubverzierung entfiel für den Ganzmetallhelm, der jetzt nur noch aus Stahl oder Leichtmetall hergestellt werden durfte. „Ein Musterhelm in der neuen Form liegt in der Hauptfeuerwache Berlin SW 19, Lindenstraße 40, aus.'' Die Helmform entsprach in ihren Konturen fast exakt dem Reichswehrver-

Versuchstyp, Kunststoffhelm, Bakelit, Vergitterte Luftlöcher, Modell 1930, ähnlich Reichswehr

Kunststoffhelm, Vulkan-Fiber, Wappen rechts: schwarz/weiß/rot Wappen links: schwarz/weiß, Preußen, 1933

Lederhelm, preussische Form Parteiadler auf Feuerwehr-Stern Oldenburg, 1935

Kunststoffhelm, Vulkan-Fiber, Wappen rechts: schwarz/weiß/rot, Wappen links: Schwarz/weiß Bauhaus, Hessen, 1933

suchshelm von 1933 (leichtes Modell), der für die Wehrmacht dann nicht in Produktion ging, und es ist im nachhinein müßig zu fragen, ob nicht wieder der Feuerwehr ein Militärhelm „anempfohlen" wurde, der für Feuerwehreinsätze weder konzipiert noch besonders geeignet war. Immerhin hatten sich Generationen von Feuerwehrmännern vordem schon mit ausgedienten kaiserlichen Pickelhauben begnügen müssen, deren be-

Feuerwehr-Stahlhelm, Modell ca. 1933, ohne Kamm, Belüftung durch Drahtgitter

Feuerwehr-Stahlhelm, Grundmodell 1934, Belüftung durch gelochte Einsätze Aluminium-Kamm

Feuerwehr-Stahlhelm, Modell ca. 1937 glänzender Tombak-Kamm

grenzter Einsatzwert im Feuerwehrdienst heute wohl unumstritten ist. Zumindest sahen die Feuerwehren auch jetzt wieder recht militärisch aus, und das war ja auch erklärter Sinn der neuen Ausrüstungsverordnung.

Unerklärlich war allerdings das „Hin und Her", das sich für die Anbringung der Helmabzeichen zwischen 1933 und 1936 ergab.

Während die Lederhelme in Stahlhelmform, nach preußischer Regelung vom 28.4.1933, noch rechts ein schwarz/weiß/rotes aufgesetzes Wappenschild trugen, links das schwarz/weiße Preußische Schild (zuerst geradestehend, kurze Zeit später schräg angebracht), wurden die neuen Ganzmetallhelme des öfteren umdekoriert:

ab 26.4.1934

rechts: großes weißes Hakenkreuz
links: Wappenschild schwarz/weiß/rot (schräg)

ab 10.7.1934
rechts: Wappenschild schwarz/weiß/rot (schräg)
links: großes weißes Hakenkreuz

ab 28.7.1934
rechts: Parteischild mit Hakenkreuz
links: Polizeiadler mit Rahmen, alufarbig

Daß diese häufigen Änderungen bei den Feuerwehren auf wenig Verständnis stießen, dürfte wohl klar sein. Noch in den siebziger Jahren konnte man bei freiwilligen Feuerwehren auf Helme stoßen, die die Originalbemalung vom April 1934 trugen, mit notdürftig entferntem großen Hakenkreuz. Das schwarz/weiß/rote Schild hatte man sicherheitshalber erstmal draufgelassen...

Die ursprünglich vorne vorgesehenen Stadtwappen waren schon 1934 fortgefallen. Es ist nicht bekannt, ob sie auf Stahlhelmen je getragen wurden, obwohl bei den Erstlieferungen die entsprechenden Lochungen vorhanden waren.

Eine weitere Kuriosität, die sich bei der Einführung des Feuerwehrganzmetallhelmes ergab, ist die Anordnung, daß der Helm „... mit einem Kamm in der Farbe der Uniformknöpfe versehen sein" soll. Diese Regelung bestand seit 3.5.1934, wurde im Sommer 1936 und im Sommer 1938 bestätigt, im November 1938 für die Berufsfeuerwehr (FSchPol) aufgehoben, für die Freiwilligen aber noch im Januar 1940 aufrecht erhalten.

Aluminium-Helm, mit geschwärztem Kamm, ca. 1942, linke Seite: Abzeichen der Deutschen Polizei

Stahlhelm, mit abgedeckten Kammlochungen ca. 1943, rechte Seite: Parteiemblem

Stahlhelm der Wehrmacht, Modell 1943 im Einsatz bei der FSchPol 1944

Luftschutzhelme, mit Emblem der FSchPol 1945

Aluminium-Helm, aus einem Stück gefertigt
Hersteller: Brissman, Fliegerhorst-Feuerwehr, Schweden, ca. 1945

Schwedischer Aluminium-Helm, nach deutschem Muster, leicht geänderte Schirmkonturen, ca. 1943

Feuerwehr-Stahlhelm, Modell 1944
Nach 1945 bei Falck & Zonen in Dänemark verwendet, Kreuzblatt-Beschlag

Allerdings wurden im weiteren Verlauf des Krieges dann nur noch Helme ohne Kamm geliefert, und vorhandene Kämme mußten in der Praxis den Altmetallsammelstellen zugeführt werden. Daß die Kämme der höheren Feuerwehrführer goldfarben sein mußten („in der Farbe der Uniformknöpfe"), sei nur am Rand bemerkt.

Die Einführung eines Helmkammes für den neuen Ganzmetallhelm wurde bei Feuerwehren und Stadtverwaltungen nicht widerspruchslos hingenommen. Man war sich einig, daß der Kamm zwar bei relativ weichen Leder- und Messinghelmen ein wichtiger Stabilitätsfaktor und somit vertretbar war, aber der äußerst widerstandsfähige Ganzmetallhelm benötigte eine solche Versteifung nicht. Im Gegenteil, durch die Bohrungen zur Befestigung des Kammes (Alu oder Tombak) ergab sich eine erhebliche Schwächung des Helmkörpers, die man zu vermeiden suchte. Außerdem bedeutete die Anschaffung der Kämme eine beträchtliche Geldausgabe für schwer beschaffbares Material, das für die Flugzeugproduktion viel dringender gebraucht wurde.

Entsprechende Vorstöße und Anträge der Feuerwehren und des Deutschen Gemeindetages in den Jahren 1935 und 1936 auf Abschaffung der Helmkämme waren jedoch nicht erfolgreich, denn es stellte sich heraus, daß die Einführung des Kammes „auf einer persönlichen, mündlichen Entscheidung des Führers und Reichskanzlers" vom 20.10.1934 beruhte, und somit, zumindest vorläufig, als endgültig zu betrachten war.

Sei es aufgrund seiner tatsächlichen Vorteile gegenüber Lederhelmen, sei es aufgrund seiner preisgünstigen Verfügbarkeit: Der neue deutsche Ganzmetallhelm fand auch bald in Nachbarländern Verwendung, entweder im Original, oder in leicht abgewandelter Form. Sogar der Kamm wurde auf breiter Basis akzeptiert, wohl mehr als traditionelles Symbol der Feuerwehr, als aus Stabilitätsgründen. In Dänemark ging (und geht) man sogar so weit, den Helmen der Falck-Feuerwehren den „Kreuzblatt-Beschlag" aufzuschrauben, nach Art der kaiserlichen Pickelhauben, und zur Abgrenzung zu den kommunalen Feuerwehren, die zum gleichen Helm den Kamm (in Messing!) tragen. Die Falck-Organisation versichert aber glaubhaft, daß die eigentlich dazugehörigen Helmspitzen nicht einmal zu Paradezwecken getragen werden!

Während des 2. Weltkrieges waren die Abgrenzungen zwischen Feuerschutzpolizei (FSchPol), Feuerwehr, Luftschutz, Luftschutzpolizei, Reichluftschutz-Bund (RLB), Sicherheits- und Hilfsdienst (SHD), Werksluftschutz usw. fließend, zumindest in Bezug auf die Ausrüstung. Besonders gegen Kriegsende wurden Ergänzungskräfte notdürftig mit dem Material ausgerüstet, das sich irgendwo ,,organisieren'' ließ, Beutestücke, längst ausgemusterte Uraltbestände, Monturen von anderen Organisationen. Besatzungen von Löschfahrzeugen boten mitunter ein bunt durcheinandergewürfeltes Bild, und man war oft zufrieden, überhaupt eine schutzspendende Kopfbedeckung zugeteilt bekommen zu haben.

Nachfolgemodell 1944, ohne Emblem

Am häufigsten anzutreffen war der ,,behördlich genormte Luftschutzhelm'', schon vor dem Krieg entwickelt und, aus 3 Teilen bestehend, von einer ganzen Reihe von metallverarbeitenden Firmen hergestellt. Er wurde zu günstigen Preisen auch an die Zivilbevölkerung verkauft. Ein Nachfolgemodell, produktionstechnisch vereinfacht aus einem Stück hergestellt, wurde gegen Kriegsende verfügbar. Im Jahresdurchschnitt wurde von 1939 bis 1943 jährlich etwa eine halbe Million dieser Luftschutzhelme gefertigt. Aussehen und Schutzfunktion waren umstritten, aber sie vermittelten ihren Benutzern ein Gefühl von Sicherheit und Autorität.

Reichswehr-Modell ca. 1931 mit RLB-Stern, (Reichsluftschutz-Bund) 1935

Feuerwehrhelmtyp 1934, für den Werkschutz, 1939

Stahlhelm, Modell TN (Technische Nothilfe) mit Luftschutz-Emblem, ca. 1942

Tschechischer Militär-Stahlhelm M34 als Beutehelm, neu lackiert im Luftschutz, 1943

Wehrmacht-Stahlhelm, Modell 1935, Version für die Luftschutzpolizei, 1944

Polnischer Militär-Stahlhelm M36 Beutehelm im Originalzustand, Luftschutz, 1943

Luftschutzhelme des Auslandes

Stahlhelm M35 der Motorisierten Verbände mit dem Emblem „DP" (Défense Passive) Zivilschutz, Frankreich, 1939

Armee-Stahlhelm, Typ Adrian M26 Défense Passive, Frankreich, 1939

Armee-Stahlhelm MKII, 1936, Emblem „AFS London" (Auxiliary Fire Service) Feuerwehr-Hilfsdienst/Zivilschutz, England, 1945

Armee-Stahlhelm M28, Emblem „NS" mit Nackenleder, Niederlande, ca. 1945

Zivilschutz-Stahlhelm, Emblem „CD" (Civil Defense), USA, ca. 1945

Der perfekte Feuerwehrhelm — gibt es ihn schon?

Polycarbonat-Helm, Hersteller: MSA, USA
Santiago, Chile, 1980

Hier und auf den nächsten Seiten werden „moderne'' Feuerwehrhelme gezeigt, d.h. Modelle, die nach 1945 getragen bzw. neu entwickelt wurden. Wie zu erwarten, werden mittlerweile fortschrittliche Fertigungstechniken und modernste Materialien auf breiter Basis eingesetzt. Überraschend ist aber, daß nach wie vor das traditionelle Design überwiegt, bzw., daß in vielen Fällen moderne Helmkörper durch allerlei „Zierat'' auf Tradition getrimmt werden, durch Hinzufügung von Metallkämmen, dekorativen Messingemblemen usw. Daß dieses Traditionsbewußtsein zu Lasten der Sicherheit geht, ist bedauerlich: ein Kunststoffhelm z.B. verliert durch zahlreiche Bohrungen und Verschraubungen für Metallkamm und Helmstern einen seiner wichtigsten Vorteile, die elektrische Isolierfähigkeit, außerdem wird er schwerer und teurer. Ist das nicht ein zu hoher Preis für falschverstandene Tradition?

Kunststoffhelm der Compania Primera
Santiago, Chile, 1980

Kunststoffhelm Paradeausführung
Chile, 1980

Vulkanfiberhelm Offiziershelm
Brasilien, 1980

Blechhelm, österr. Modell
Tschechoslowakei, 1945

Nachdem auch in Osteuropa in den ersten Nachkriegsjahren vorhandene Helme der Vorkriegszeit aufgetragen wurden, oder Militärstahlhelme der Kriegszeit mit geringfügigen Veränderungen an die Feuerwehren abgegeben wurden, treffen wir in den sechziger Jahren überall auf Versuche, den Feuerwehrmann optimal zu schützen. Das Design der Helme orientiert sich an traditionellen Formen, die aber bewußt modernisiert werden, und durch verbesserte Innenausstattungen und der Wahl von Kunststoff für die Helmschale einen erheblichen Fortschritt bedeuten. Bemerkenswert auch, daß das Klappvisier zum Bestandteil des Helmes wird. Jeder Feuerwehrmann, der einen Helm mit Visier getragen hat, weiß diesen Vorteil zu schätzen.

Militärstahlhelm, Typ Deutsche Wehrmacht 1935
Ungarische Produktion. Nach 1945 mit Spinne für den Feuerwehreinsatz, Ungarn, freiw. Feuerwehr

Kunststoffhelm, Phenol-Harz
Ungarn, 1970

Aluminium-Helm österr. Modell
Jugoslawien, 1970

Daß vertraute Militärhelmformen auf diese Weise leicht „up-to-date" zu bringen sind, beweisen die russischen und polnischen Feuerwehrhelme, letztere sogar mit integriertem Feuerwehr-typischen Kamm. Andere Länder, die sich keine eigene Neuentwicklung leisten wollten, haben sich an vorhandenen ausländischen Helmen orientiert, so z.B. Rumänien an französischen, und Jugoslawien an österreichischen Helmen, der in Originalform übernommen wurde.

Stahlhelm der Roten Armee für den Feuerwehreinsatz, UdSSR, 1960

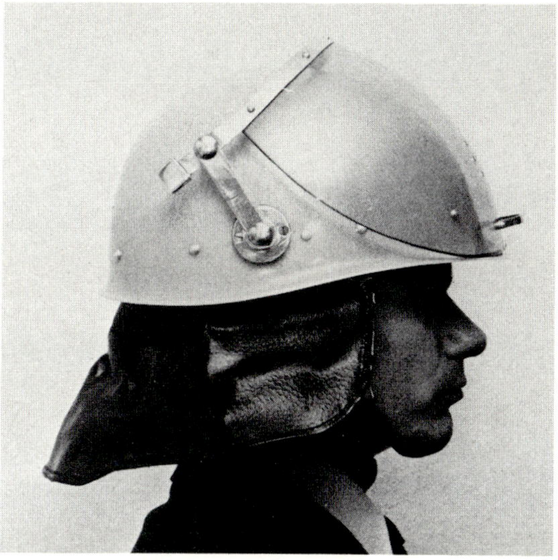

Kunststoffhelm mit Visier und Nackenschutz UdSSR, 1970

Kunststoffhelm, Phenol-Harz Polen, 1960

Metallhelm mit Kamm UdSSR, 1970

Militär-Stahlhelm, Typ Adrian
Franz. Modell M15, Offizier
Belgien, 1950

In Belgien wurden bis 1918 allgemein hohe napoleonische Messinghelme getragen, vereinzelt auch deutsche Modelle (z.B. in Flandern), danach wurden ausgemusterte belgisch/französische Militärhelme des Modells „Adrian" 1915 an die Feuerwehren ausgegeben; die Umrüstung war 1940 abgeschlossen. Sehr positiv aufgenommen wurde der Korkhelm aus belgischer Produktion, dessen Design von den englischen „Cromwell-Helmen" beeinflußt war; in etwas modernisierter Form ist er auch heute noch in Gebrauch. Um die Traditionalisten glücklich zu machen, wurde auch der alte Armeehelm vorübergehend in Kunststoff (Polyester) angeboten, erfreute sich aber verständlicher Weise keiner Beliebtheit, da er für den Feuerwehrdienst nicht optimal geeignet war. Seit etwa 1975 ist auch in Belgien der Kunststoffhelm nach amerikanischer Bauart in Verwendung, teilweise aus deutscher Produktion, mit Innenfutter nach DIN 4840.

Militär-Stahlhelm, Typ Adrian
Franz. Modell M33, Belgische Produktion
Belgien, 1950

Korkhelm
Belgische Produktion 1960

Kunststoffhelm
Belgien, 1980

Lederhelm, deutsche Produktion
Den Haag, Niederlande, 1950

Holländische Feuerwehren haben schon immer den größten Teil ihrer Helme aus Deutschland bezogen, allerdings in typisch niederländischen Ausführungen. Besonders große Berufsfeuerwehren (z.B. Amsterdam, Den Haag) haben sehr lange an Lederhelmen festgehalten, auch noch nach 1945 Lederhelme in Deutschland herstellen lassen. Als die Lederhelmproduktion hier eingestellt wurde, mußte man sich ,,notgedrungen'' um moderneres Material kümmern. Zuerst bot sich der bundesdeutsche Zivilschutzhelm an, der wohlfeil in großen Stückzahlen produziert wurde, und für Holland mit dem Alu-Kamm versehen wurde. Später kam als Alternativmodell noch ein Polycarbonat-Helm aus deutscher Produktion hinzu, der sich in der Form an die alten Lederhelme anlehnte.

Stahlblechhelm,
ehem. deutscher Luftschutzhelm
Niederlande, 1945

Kunststoffhelm, Polycarbonat
deutsche Produktion
Niederlande, 1975

Glasfaserhelm, Deutsches Zivilschutzmodell mit Kamm und neuer Innenausstattung, Niederlande, 1970

Kunststoffhelm, Polycarbonat, deutsche Produktion, mit Standard-Alu-Kamm, Niederlande, 1975

Aluminium-Helm in Anticorrodal
Schweizer Armeehelm-Form
Feuerwehr Basel, 1946

Kunststoffhelm für Feuerwehr-Elektriker
deutsche Produktion
Schweiz, 1970

Kunststoffhelm, Glasfaser, amerikanisches Modell aus deutscher Herstellung mit deutscher Innenausstattung, Schweiz, 1972

In der Schweiz wurden nach 1945 die alten Armeestahlhelme (Modell 1918) vorläufig weitergetragen. Da hohes Gewicht und mangelnder Tragekomfort immer wieder zu Klagen führten, wurde der Helm in der Nachkriegszeit auch aus Aluminium gefertigt, und in polierter Form bei einer Reihe von Feuerwehren eingeführt. Die Luftschutztruppen erhielten ihn in blau-schwarzer Lackierung. Da die Problematik eines Ganzmetallhelmes im Feuerwehreinsatz nicht wegdiskutiert werden konnte, wurde die Idee des ,,Feuerwehr-Elektriker-Helmes'', aus der Lederhelmzeit gut bekannt, neu aktiviert: ein deutscher Hersteller stellte den schweizer Militär- und Feuerwehrhelm aus Kunststoff her, allseits isoliert und ohne Metallteile. Langfristig konnte diese Lösung allerdings nicht befriedigen, und seit Beginn der siebziger Jahre werden die Feuerwehren mit amerikanischen Kunststoffhelmen (Polycarbonat oder Glasfaser) beliefert, die örtlich mit Stadt- oder Kantonwappen versehen werden. Obwohl diese Helme in Deutschland als ,,Cowboy-Hüte'' oft belächelt werden, finden sie offensichtlich auch in Europa immer mehr Freunde; in Südamerika, Australien und Asien z.B. sind sie schon seit vielen Jahren weitverbreitet.

Schweizer Armeestahlhelm
mit gegossenem Alu-Kamm
Lausanne, 1950

Aluminium-Helm, Wiener Form, innen beflockt, Mannschaftshelm für Oberösterreich, 1970

Alu-Helm
Tirol, 1970

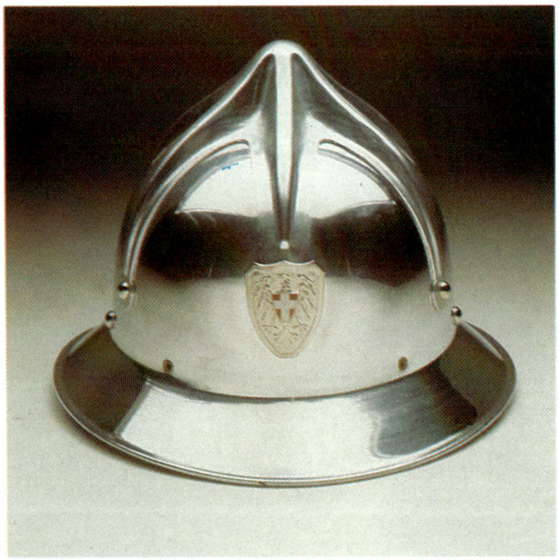

Alu-Helm, poliert
Berufsfeuerwehr Wien, 1980

Auch in Österreich: Der Weg vom Stahlhelm zum Kunststoffhelm

Österreichische Feuerwehren wurden nach dem „Anschluß'' mit großen Stückzahlen des Ganzmetallhelmes der deutschen „Feuerschutzpolizei'' beliefert, und mit oder ohne den traditionellen „blanken Kamm'' waren diese Helme bis in die sechziger Jahre im Einsatz. Noch heute wird in der Steiermark als Standardhelm ein im Schirm geringfügig geänderter Aluminiumhelm getragen, schwarz lackiert, mit innen aufgespritzter isolierender Beflockungsschicht, und noch immer mit Alu-Kamm. Auch auf das Landeswappen in Form eines emaillierten Metallschildes wird nicht verzichtet.

Parallel zum „deutschen'' Stahlhelm, der übrigens auch in Österreich seit dem 1. Weltkrieg aus eigenständiger Produktion zur Ausrüstung der Armee gehörte, entwickelten die österreichischen Feuerwehren auch nach 1945 wieder einen speziellen Einsatzhelm, der an eine Helmform anknüpfte, die teilweise schon vor 1938 verwendet wurde. Das Ergebnis war der Aluminiumhelm mit ziemlich breitem Schirm, und statt des Kammes mit einer „fünfbeinigen Spinne'' als spezieller Kopfschutz. Der Helm, in einer niedrigen und einer etwas höheren Form hergestellt, wird von Weinberger in Wien nach ÖNORM F 4045 produziert. Je nach Bundesland gibt es auch unterschiedliches Finish: Alu poliert, oder matt, oder schwarz lackiert; die Führungskräfte der BF Wien tragen ihn gelb-nachleuchtend gespritzt. Außerdem ist für jedes Bundesland das entsprechende Wappenschild vorgesehen, mit Lorbeerkranz-Umrandung für Offiziere.

Kunststoffhelm, Harz mit Fasern verstärkt, 1. Ausführung, Österreich, 1970

Kunststoffhelm, Polycarbonat, mit PVC-Emblem, Mannschaftshelm, Oberösterreich, 1975

Aluminium-Helm, innen beflockt, mit Alu-Kamm, Österreichische Produktion, Steiermark, 1975

Ende der sechziger Jahre wurden die ersten Versuche mit Kunststoffausführungen gemacht, z.B. mit faserverstärktem Kunstharz, unter Verwendung der Form des Metallhelmes, mit integrierter Kunststoffspinne. Der hier eingeschlagene Weg wurde offensichtlich als richtig erkannt, denn schon Mitte der siebziger Jahre wurde der Kunststoffhelm, in leicht geänderter Form und unter Verwendung besseren Plastikmaterials, auf breiter Basis eingeführt. Um einen der wichtigsten Vorteile des Kunststoffhelmes — die elektrische Isolierfähigkeit — auch konsequent zu nutzen, ging man vom Metallwappenschild ab und führte das selbstklebende PVC-Schild für das Helmwappen ein. Zur weiteren Kennzeichnung der Helme sind 3M Leuchtstreifen erhältlich.

Für Spezialeinsätze wird der schwedische „Trelleborg-Helm" mit Klappvisier benutzt.

Auch Frankreich hat den Weg vom Stahl- zum Kunststoffhelm gefunden. Nachdem bei kleineren Feuerwehren bis in die fünfziger Jahre noch Messinghelme (im Elsaß sogar deutsche!) oder schwarz gespritzte Poilu-Helme verwendet wurden, ging man vorerst auf breiter Linie zu dem verchromten Helm aus Nickel-Chromstahl über, mit aufgesetztem Kamm und Messingschild (bei Offizieren verchromtes Schild). Die gefährliche elektrische Leitfähigkeit dieses Helmes, die zu einer Reihe von Unglücksfällen führte, führte schon in den sechziger Jahren zu Versuchsmodellen aus glasfaserverstärktem Kunststoff. Das vorläufige Endergebnis ist ein preisgünstiger Kunststoffhelm von Petitcol-

Stahlhelm, verchromt, Fliegerhorst-Feuerwehr der Armeé de l'Air, Frankreich, 1960

Stahlhelm, verchromt, Offizier der Marins Pompiers, Marine Feuerwehr für die Stadt Marseille, Frankreich, 1970

Kunststoffhelm für Waldbrandfeuerwehren auf Korsika, Ajaccio, 1960

Kunststoffhelm, Glasfaser, Versuchsmodell, Chartres, Frankreich, 1960

Kunststoffhelm, Glasfaser
Modell Strassburg
Frankreich, 1970

lin, mit breiter Krempe, angelehnt an die traditionelle Form, der z.B. bei der BF Straßburg seit 1968 und beim Zivilschutz eingeführt ist.

Einsatzbesprechung bei der Brigade de Sapeurs-Pompiers de Paris, 1975

In Südeuropa — Tradition und Fortschritt

Die konservative politische Einstellung Spaniens nach dem Bürgerkrieg spiegelte sich auch in der Ausrüstung der Feuerwehren: noch heute tragen sie den Ganzmetallfeuerwehrhelm nach deutschem Muster. Das „Lokalkolorit" wird durch einen Messingkamm und ein großes dekoratives Helmschild, ebenfalls aus poliertem Messing, erreicht. Auch die freiwilligen Feuerwehren in der italienischen Provinz Südtirol, auf Eigenständigkeit und Abgrenzung zu Rom bedacht, halten am deutschen Stahlhelmtyp (nach DIN 14940) fest, garnieren ihn aber mit Kamm und großem Helmschild. Bis vor

Leichtmetallhelm
Deutsche Produktion, Messingkamm
Spanien, 1980

Leichtmetallhelm, Deutsche Produktion, Alukamm, Rabland, Süd-Tirol, Italien, 1972

Kunststoffhelm, Phenol
Hersteller: MISPA, Turin
Italien, 1972

Blechhelm, Messingkamm
Türkei, 1935

wenigen Jahren gehörte sogar ein Sturmband zu Schmuckzwecken dazu, mit hübschen Messing-Löwenköpfen außen am Helm befestigt, nach der Mode von 1870.

Die eigentliche Bebänderung entsprach jedoch der neuesten DIN, mit Fallschirmjägerkinnriemen. Die Kommunalen Feuerwehren Italiens benutzen einen sehr elegant wirkenden Kunststoffhelm, von MISPA in Turin aus Phenolharzgewebe in sehr guter Qualität hergestellt. Dazu wird ein Helmabzeichen aus Metall oder Weichplastik getragen.

Die Feuerwehren in der Türkei, die auf keine nationale Helmindustrie zurückgreifen konnten, benutzten bis vor wenigen Jahren noch Leder- oder Blechhelme österreichischer oder schweizer Bauart, rot lackiert mit Messingkamm und Kunststoffnackenleder. Mittlerweile sind italienische MISPA-Helme eingeführt worden, und parallel dazu findet man auch amerikanische Helme aus rotem Polycarbonat, von MSA.

Ebenfalls anzutreffen sind Kunststoffhelme im ,,military look'' nach Muster des Plastik-Innenhelmes für den amerikanischen Armeestahlhelm (NATO-Typ).

Kunststoffhelm
Modell MISPA, Italien
Istanbul, 1980

Kunststoffhelm, Polycarbonat
Modell MSA, USA
Istanbul, 1980

Kunststoffhelm, Phenol-Harz,
ähnlich NATO-Helm,
Offizier, Istanbul, 1980

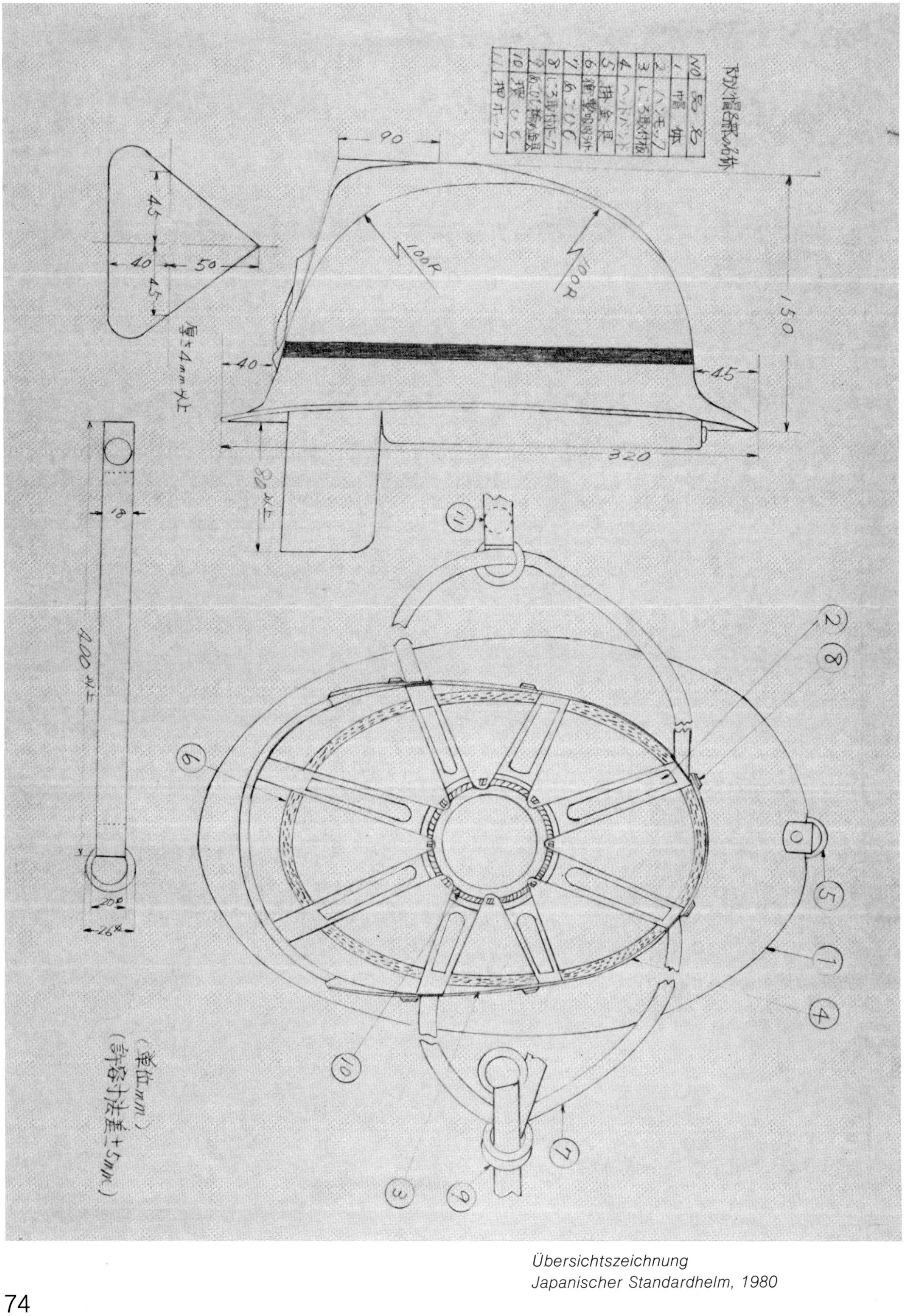

Übersichtszeichnung
Japanischer Standardhelm, 1980

Japan hat nach Kriegsende eine sehr modern konzipierte Helmgeneration entwickelt, in vollständiger Abkehr vom Ganzmetallhelm. Während es anfangs sehr leicht gebaute Helme waren, allerdings wieder mit einem traditonell bewährten, weit heruntergeführten Schulterschutz, der vorne zusammenknöpfbar war und nur einen Augenschlitz freiließ, ging man schon um 1955 auf Fiberglaskunststoffhelme über, in der Form eine Mischung von französischen und amerikanischen Designelementen. Auch hier wieder der angehängte Schulterschutz aus alubeschichtetem Gewebe (innen Wollstoff), und als großer Fortschritt, mit einem in den Helm einschiebbaren Visier (Acetatschild) als Augenschutz. Diese gelungene Lösung ist frappierend einfach, und es überrascht nicht, daß sie jetzt auch bei Helm-Neuentwicklungen, z.B. in Deutschland, beachtet wird.

Helmkörper aus Korbgeflecht
mit gepolstertem Baumwollüberzug
Feuerwehrmann, 1949—1954, Tokio, Japan

Glasfaser-Helm, Battalion Chief, 1955—1962
Tokio, Japan

Glasfaser-Helm, Wachvorsteher, 1970
Tokio, Japan

Kunststoffhelm, Polycarbonat
Tokio, Japan, 1980

Korkhelm, Modell Cromwell
Englische Produktion, Hongkong, 1980

Kunststoffhelm, Phenol-Harz, Englische Produktion, rechte Seite: Offiziersrangabzeichen, Hongkong 1960

Da die meisten Länder Asiens, bis auf Japan und die Volksrepublik China, keine eigene Helmproduktion haben (das Nachfragevolumen der wenigen organisierten Feuerwehren reicht hierzu nicht aus), lassen sich an den Feuerwehrhelmen frühere koloniale Traditionen oder Einflußsphären der Großmächte ablesen: der Hongkong Fire Service z.B. trägt seit seiner Gründung Helme aus England, mittlerweile das neueste Cromwell-Modell, während in Thailand der Einfluß der amerikanischen Berater unverkennbar ist: Kunststoffhelme vom Modell MSA sind bei den wenigen Berufsfeuerwehren anzutreffen.

Kunststoffhelm, Glasfaser, Modell MSA, USA
Bangkok, Thailand, 1975

Kunststoffhelm, Werkfeuerwehr
Volksrepublik China, 1978

Kommunale und Werkfeuerwehren in der Volksrepublik China verwenden z.Zt. einen sehr traditionell anmutenden Helm, der stark an die Topfhelme der japanischen Armee von 1932 erinnert. Aus Phenolharz mit Gewebeeinlagen hergestellt, werden diese Helme als ziemlich schwer empfunden; ein Nackenleder ist nicht vorgesehen. Trotzdem sich Fahrzeuge und Ausrüstung stark an russischen und deutschen Vorbildern orientieren, hat man sich beim Feuerwehrhelm offensichtlich von westlichen Vorbildern gelöst, obwohl der deutsche Stahlhelm eine gute Chance gehabt hätte, bei der Feuerwehr eingeführt zu werden: bis 1946 trug ihn die national-chinesische Armee als Standardhelm!

Wie schon in Asien, kann man auch an den Feuerwehrhelmen in anderen Ländern der Dritten Welt unschwer ablesen, welche westliche Industrienation beim Aufbau der Feuerwehr geholfen hat. Bedingt durch kleine Budgets und warmes Klima kommen oft die leichteren Modelle aus westlicher Produktion zum Einsatz, so z.B. der neue französische „Petitcollin" Kunststoffhelm in den ehemaligen französischen Kolonien, oder das flache englische Cromwell-Modell 3871 in Nigeria. Südafrika verwendet immer noch Lederhelme aus schottischer Produktion, während Indien, in Anlehnung an englische Vorbilder, eigene sehr stabile Lederhelme fertigt. Die wohlbetuchten Länder am Persischen Golf haben ihre Dollars teilweise in amerikanischen Feuerwehrhelmen angelegt, und in ehemals italienischen Einflußgebieten findet man Helme nach römischem Muster.

Lederhelm, Typ Bombay
Indien, 1970

Kunststoffhelm, Polycarbonat
Hersteller: Cairns, USA, Fliegerhorstfeuerwehr
Saudi Arabien, 1980

Lederhelm, englische Produktion, Offizier
Johannesburg, Süd-Afrika, 1960

Feuerwehrmänner mit Lederhelmen
Modell 1939—1962
Johannesburg, Süd-Afrika

Feuerwehrmänner mit schwerem Atemschutz,
Korkhelme von Cromwell
Oxford, England, 1965

Korkhelm, Cromwell Modell ,,LFB''
Station Officer, London Fire Brigade, 1965

Korkhelm, Cromwell Modell ,,Middlesex'',
Subofficer, Reading + Berkshire Fire Brigade 1965

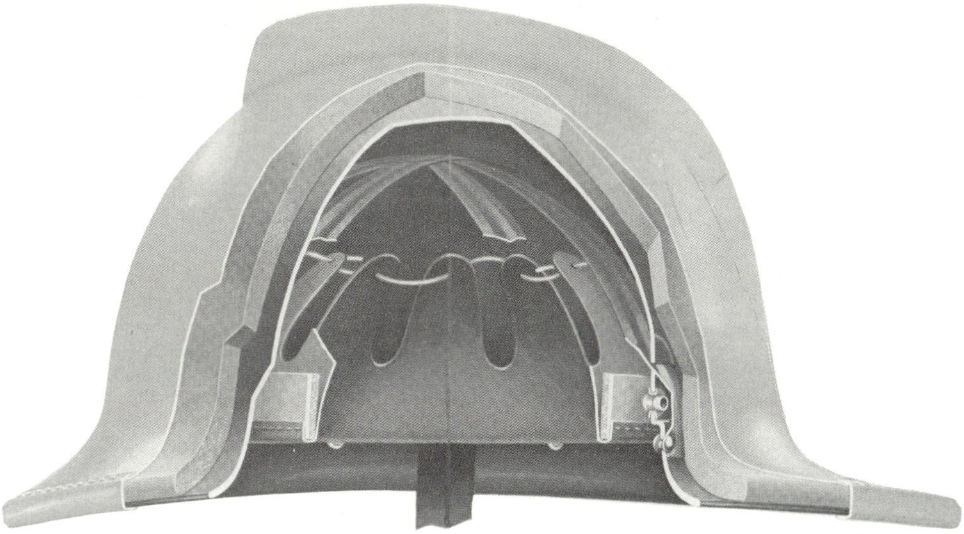

Korkhelm, Cromwell Modell ,,County'', England, 1980

Rangabzeichen London 1980

weißer Helm, schwarzes Band
Chief Officer

weißer Helm, schwarze Bänder
Deputy Chief Officer & Assistant Chief Officer

weißer Helm, schwarze Bänder
Deputy Assistant Chief Officer & Senior Divisional Officer & Divisional Officer

weißer Helm, schwarzes Band
Assistant Divisional Officer

weißer Helm, schwarzes Band
Station Officer

gelber Helm, schwarze Bänder
Sub Officer

gelber Helm, schwarzes Band
Leading Fireman

gelber Helm
Fireman

England — Ein gelungener Kompromiß von Alt und Neu

Um den mitleidig lächelnden Experten auf dem Kontinent den ,,Wind aus den Segeln'' zu nehmen, sei es gleich vorab gesagt: der z.Zt. in England getragene Korkhelm ist, vom Aspekt der Sicherheit gesehen, einer der besten auf der ganzen Welt! Über Design und Ästhetik kann man streiten, aber wenn es um Gewicht, Dämpfungseigenschaften und Durchdringungstests geht, schneidet der englische Korkhelm erheblich besser ab, als viele seiner ausländischen Konkurrenten.

Sie haben richtig gelesen: Kork wird mit Harz unter hohem Druck gepreßt und geformt, und diese ,,rohe'' Helmform wird mit einer Schicht aus Plastikmaterial (früher war es imprägniertes Leinen) ,,bezogen'' und verschweißt. Als Neuheit wurde auf der Interschutz 80 auch ein typisch englischer Helm aus Kevlar gezeigt, einem sehr hochwertigen Plastikmaterial der jüngsten Entwicklung.

Die relativ hohe Form ist Teil des Sicherheitskonzepts, wobei der ausgeprägte traditionelle Kamm in einer Anzahl von Varianten gefertigt wird, sodaß auch regionale Präferenzen berücksichtigt werden können. Als in den sechziger Jahren ein flaches Kork-Modell ohne Kamm vorgestellt wurde, ähnlich dem englischen Militärstahlhelm MK III von 1943/44, lehnten die englischen Feuerwehren diesen ,,teutonischen Panzerhelm'' als nicht feuerwehrgerecht ab. Feuerwehren wie Liverpool oder Luton, die ihn versuchsweise eingeführt hatten, musterten ihn bald wieder aus. Die Restbestände gingen nach Afrika, wo das Thema ,,Feuerwehrhelm'' nicht ganz so emotionsbeladen behandelt wird wie in Europa.

Cromwell Typ 3871
Luton, England, 1965

Cromwell Typ 3871, Offiziershelm
Liverpool, England, 1965

Armee-Stahlhelm MK3, Modell 1943/44
Feuerwehr-Hilfsdienst (AFS), 1965

Korkhelm, Cromwell Modell „NFBA"
Militärfeuerwehren, England, 1960

Kunststoffhelm, Phenol-Harz
Englische Produktion, 1960

Korkhelm, Cromwell Typ Middlesex
England, 1960

Kunststoffhelm, Glasfaser
für Fliegerhorst-Feuerwehren
England, 1960

Korkhelm, Cromwell Typ Arden
England, 1960

Kunststoffhelm, Glasfaser-Polyester, mit Alu-pigmentierter Oberfläche, für Flugplatz-Feuerwehren, Hersteller: Bristol Uniforms, England, 1980

Aluminium-Helm, Cairns Modell „Senator"
New Orleans Fire Dept. USA, 1960

Kunststoffhelm (Flint-Flex)
Cairns Modell „Continental"
Dreifacher Kamm (Regierungsmodell) USA, 1965

Kunststoffhelm (Flint-Flex)
Cairns Modell „Commando" USA, 1960

Kunststoffhelm (Flint-Flex)
Cairns Modell „Cavalier" USA, 1960

USA — unbegrenzte Möglichkeiten

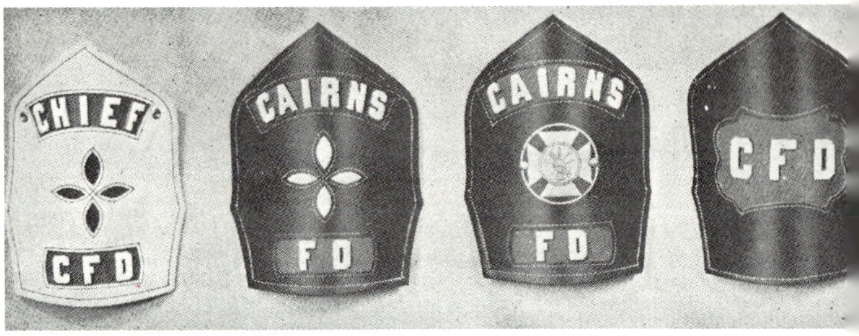

Diverse Schildausführungen in Leder, für US Helmtypen (Cairns)

Lederhelm, Cairns Modell „Volunteer"
Montgomery, Alabama, USA,

Lederhelm, Cairns Modell „New Yorker"
3M-Reflex-Markierungen, Odenten (Volunteer Fire Dept. Freiw. Feuerwehr) USA, 1970

Lederhelm, Cairns Modell „New Yorker"
Offizier, USA, 1970

Kunststoffhelm, Polycarbonat, Cairns Modell „Philadelphia", Drehleiter Nr. 8, Philadelphia, Pa. 1970

Für einen Designer von Feuerwehrhelmen sind die USA das ideale Arbeitsgebiet. Der Grund: die ziemlich eng gefaßten Normvorschriften der staatlichen und halbstaatlichen Fachausschüsse befaßten sich nur mit dem Leistungsstandard der Helme, nicht aber mit der Form oder dem Material. Das bedeutet z.B., daß der Helm elektrisch isolierend wirken muß, daß er bestimmte Minimalwerte der Schlagdämpfung erfüllen muß, daß er der Durchdringung spitzer Gegenstände an bestimmten Aufschlagpunkten widerstehen

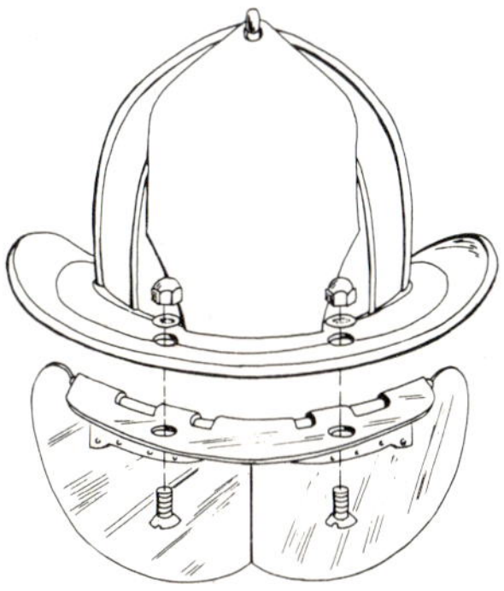

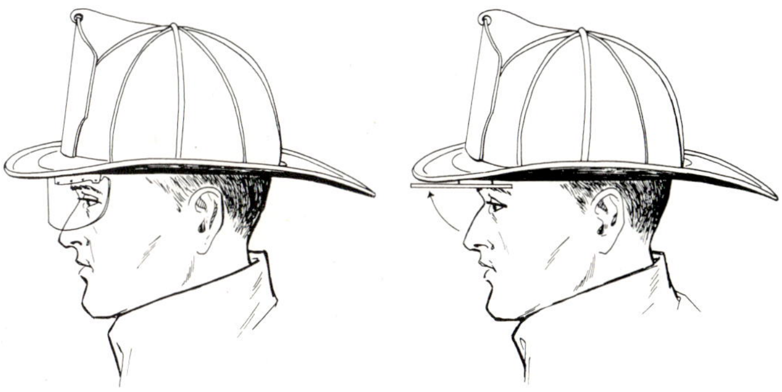

Augenschutz-Klappschild, System Bourke
Hersteller: Cairns, USA

Lederhelm, ähnlich Typ MSA
Deutsche Produktion
für U.S. Army Fire Department, 1946

muß. Solange diese Forderungen erfüllt sind, entspricht er den Arbeitsschutzbedingungen, und darf die Zulassung für sich beanspruchen. Diese sachliche und nüchterne Vorgehensweise hat dazu geführt, daß eine Vielzahl von Helmtypen angeboten wird, in allen gängigen Materialien, welche die Norm erfüllen. Da die Konkurrenz der Helmhersteller groß ist, werden neue Ideen schnell aufgegriffen und verwirklicht, nicht nur als Prototypen, sondern gleich in Großserie. Der Käufer, der örtliche Fire Chief, orientiert sich dann nur noch am Preis und der Form: wenn er traditionell gesinnt ist, wählt er einen Lederhelm von Cairns, oder den Kunststoffhelm von MSA; etwas modernere Feuerweh-

Phenol-Harz-Helm, Hersteller: MSA
Miami Beach Fire Dept. Florida, 1950

Kunststoffhelm mit Klappvisier, Hersteller: MSA
Taft Fire Dept., Florida, USA, 1960

Kunststoffhelm, Hersteller MSA
Helmschild als Abziehbild, USA, 1970

Glasfaserhelm, Kanadische Produktion für
Fliegerhorst-Feuerwehren der Royal Canadian Air
Force (RCAF), Kanada, 1960

Kunststoffhelm (PVC), Hersteller: Gentex Offiziersmodell, USA, 1970

Kunststoffhelm (Glasfaser) Hersteller: Gentex, USA, 1960

ren, die auch mehr technische Hilfsleistungen zu bringen haben, suchen sich einen Helm von Akron, Premier oder American Sports aus, mit kürzerem Nackenschutz und Klappvisier, und traditionsungebundene Feuerwehren nehmen gleich das non-plus-ultra, den „Astronauten-Helm'' von Federal, der sich laut Angaben Dietzenbacher Feuerwehrmänner, die 20 dieser Helme in einem fünfjährigen Trageversuch getestet haben, trotz seiner „massiven Form'' hervorragend bewährt hat.

Immer wieder faszinierend sind die farbigen ledernen Helmschilde an den traditionellen amerikanischen Helmen; bei neueren Modellen, deren Helmschale aus elektrischen Isolationsgründen nicht mehr durchbohrt werden darf, werden sie durch selbstklebende PVC-Aufkleber oder Abziehbilder ersetzt. Durch das „front-piece'' läßt sich der Träger sofort identifizieren, denn es zeigt in der Regel nicht nur den Namen der Feuerwehr (und evtl. die Nummer der Wache oder die Personalnummer des Feuerwehrmannes), sondern läßt auch die Funktion erkennen: Offiziere tragen weiße Schilde, Feuerwehrmänner eines Löschfahrzeuges schwarze, einer Drehleiter rote. Angehörige einer Rescue Company (Kombination von Rettungswagen / RTW und Rüstwagen) haben blaue Schilde, und Besatzungen von Flugfeldlöschfahrzeugen grüne. Mannschaften auf Sonderlöschfahrzeugen, großen Wasserwerfern usw. führen orangefarbene Schilde.

Kunststoffhelm (Glasfaser) Hersteller: Bullard für US-Militärfeuerwehren (FD = Fire Department) 1970

Kunststoffhelm, Hersteller: Bullard für Waldbrandeinsätze, USA, 1960

Die in Deutschland übliche nachleuchtende Heimlackierung ist in USA unbekannt, aber reflektierende Streifen von 3M, oder an die Helmkonturen angepaßte Dreiecke, sind mittlerweile akzeptiert. Interessant ist, daß viele Feuerwehren glauben, auf den Kinnriemen verzichten zu können, wenn sie jeden Helm einzeln verpassen lassen, und es ist in der Tat frappierend festzustellen, wie gut individuell angemessene Helme z.B. mit der traditionellen Cairns-Innenausstattung, sitzen können. Die neuesten Modelle jedoch tragen jetzt Kinnriemen, einige sogar mit einer Kinnkappe, die jedoch für den Feuerwehreinsatz generell ungeeignet ist.

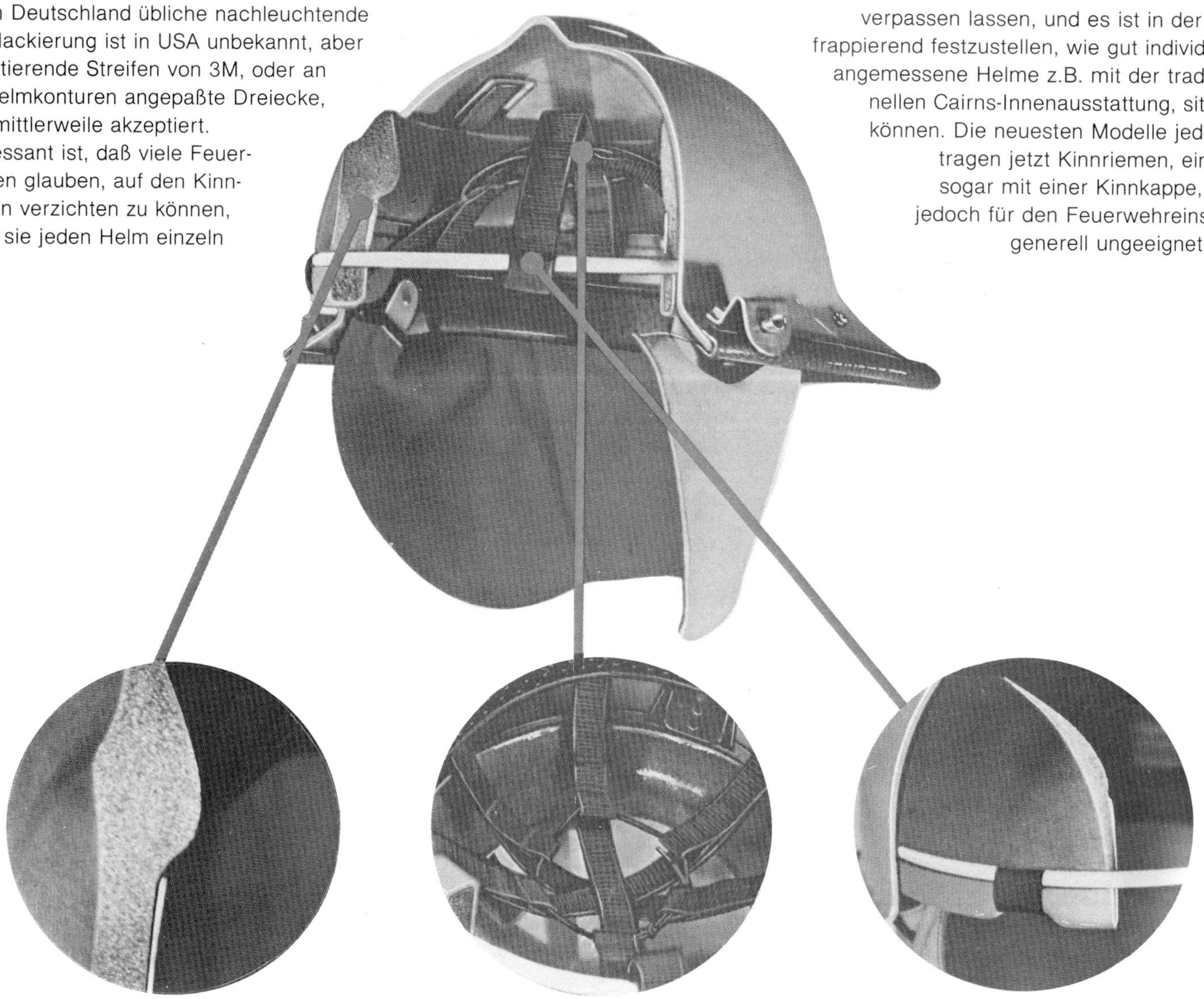

Schnittmodell Cairns ,,Phoenix'' USA, 1980

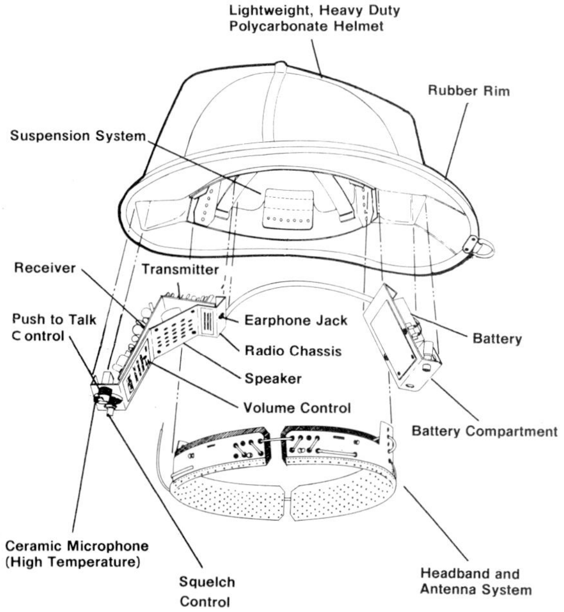

Comrad-Helm mit eingebautem Funkgerät (Helmschale von Cairns)

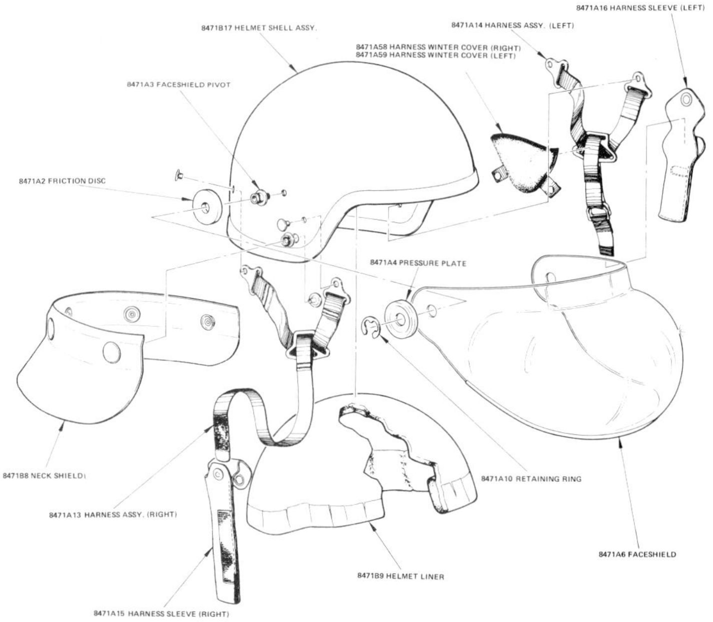

Federal ,,FH-1'', USA, 1965

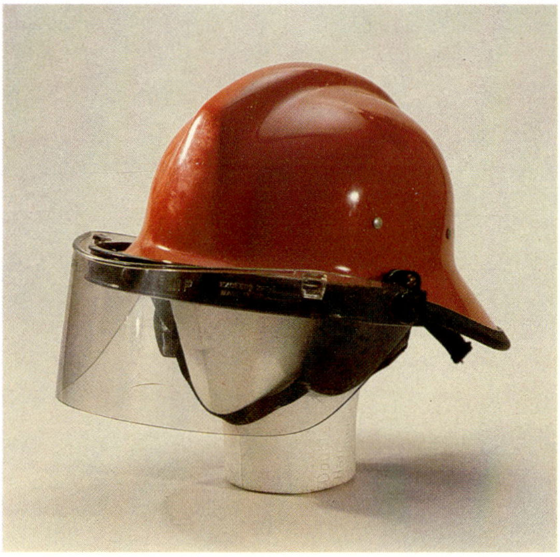

Kunststoffhelm (Glasfaser)
Modell „The Fireman''
Hersteller: American Sports, USA, 1970

Kunststoffhelm (Polycarbonat)
Hersteller: Akron, USA, 1974

Durchsichtige Klappvisiere finden wir an amerikanischen Helmen in begrenztem Umfang schon in den sechziger Jahren, riesige Ungetüme auf Acetat-Basis, die an die breiten Krempen aufgeschraubt werden konnten. Als Alternative entwickelte der New Yorker Feuerwehrmann Bourke einen kleinen klappbaren Schutz für die Augen, der auch nachträglich unterhalb der Krempe, besonders bei Lederhelmen der BF New York, montiert wurde. Hergestellt von Cairns in New Jersey, hat er in den übrigen Teilen der USA wenig Zustimmung gefunden. Seit einigen Jahren ist der Einsatz von Aluminium-Helmen in USA verboten, aufgrund ihrer elektrisch leitenden Eigenschaften; man findet daher heute nur noch Lederhelme (ziemlich weit verbreitet!) und Kunststoffhelme, deren Materialbasis mittlerweile, chronologisch gesehen, nach Phenolharz, Glasfaser und Polyvinylchlorid-PVC, bei Polycarbonat angelangt ist. Die Vor- und Nachteile der einzelnen Substanzen sind in der Fachwelt heiß diskutiert worden, und Insider behaupten, daß man vielleicht in absehbarer Zeit zu Phenolharz-Gewebe-Konstruktionen oder Glasfaser zurückkehren wird.
Helme mit 3 parallelverlaufenden Kämmen wurden von Cairns, Bullard und Gentex gebaut, übrigens ursprünglich speziell für die US-Regierung, die aus Sicherheitsgründen diese Verstärkung der Helmkuppe forderte für Feuerwehrmänner im Dienste der Regierung, d.h. bei Militärfeuerwehren, größeren Flughäfen, Rüstungsbetrieben usw. Helme der neueren Generation verzichten allerdings auf dieses Merkmal.

Kunststoffhelm, Federal Modell FH-1
Visier aus getöntem Butyl-Harz, USA, 1968

Kunststoffhelm (Polycarbonat) Federal Modell FH-2, Visier aus Polycarbonat, Versuchslackierung nach DIN 14940, USA, 1972

Aluminiumhelm ähnl. DIN 14940
mit Winterinnenausstattung
Deutsche Produktion für Finnland, 1970

Aluminiumhelm mit verchromtem Kamm
Oslo, Norwegen, 1960

Deutsche oder österreichische Feuerwehrhelme haben in Skandinavien schon im vorigen Jahrhundert einen guten Namen gehabt, der sich bis heute gehalten hat.

Finnland z.B. trägt heute Aluminiumhelme nach DIN 14940, aus deutscher Produktion. Früher waren es preußische Lederhelme, und, vorübergehend, französische Poilu-Helme des 1. Weltkrieges, die in den zwanziger Jahren kostengünstig verfügbar waren.

Die norwegischen Feuerwehren haben mit dem ,,deutschen Stahlhelm'' nur kurz geflirtet: ca. 1936 wurde eine Anzahl der Leder- oder Vulkanfiberhelme in Stahlhelmform nach Norwegen importiert und z.B. in Halden, bis 1945 getragen. Ansonsten orientierte man sich an früheren Lederhelmen des Typs Wien, und preßte sie zuerst in Stahl, mit deutschem Alu-kamm (ca. 1940), dann ab 1944 in Aluminium. Das Modell, in der Grundform dem jetzigen österreichischen Typ immer noch verwandt, scheint sich gut bewährt zu haben, denn es wird nun auch von TICON aus Kunststoff hergestellt, mit integriertem Plastikkamm, und dem traditionellen Metallhelmstern.

Dänemark bezieht nach wie vor Ganzmetallhelme aus deutscher Produktion; durch Messingkämme, verschiedenartige große Helmschilde und weitere Garnierungen sehen sie aber wesentlich dekorativer aus als das in Deutschland getragene ,,Urmodell.''

Kunststoffhelm
Norwegen, 1980

Stahlhelm ähnl. DIN 14940 mit Messingkamm und Messingschild und dänischer Kokarde (links)
Arhus, Dänemark, 1970

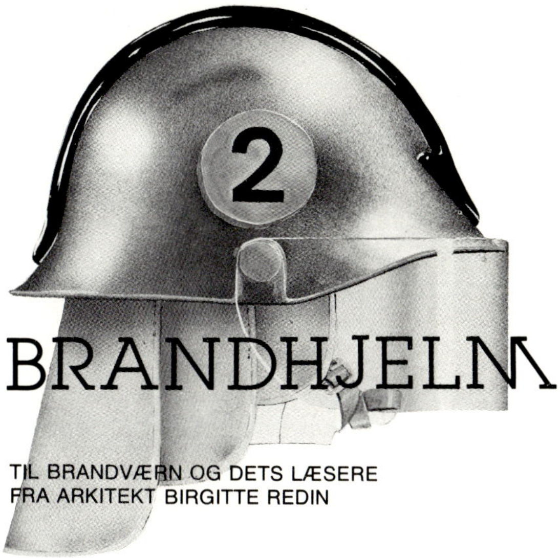

Dänischer Helmprototyp „Redin", 1979

Aluminiumhelm nach DIN 14940, mit Alukamm, deutsche Produktion, mit poliertem Helmkörper für Stockholm, Schweden, 1970

Glasfaserhelm, Hersteller: Brissman
Emblem 3 M Reflexfolie, Schweden, 1975

Da Schweden mehrere eigene Helmhersteller hat, gibt es dort eine größere Typenvielfalt bei Feuerwehren. Das Grundmodell ist auch dort der deutsche Stahlhelm, der seit den dreißiger Jahren in Ganzmetallausführung, aber auch als Vulkanfiberhelm getragen wurde. Der sehr aktive Helmfabrikant Brissman hat ihn schon 1936 als Versuchsmodell in Kunststoff kopiert, und 1938 tauchten die ersten Brissman-Helme in Aluminiumguß auf, in Stahlhelmform mit Kamm, aus einem Stück. Diese Helme fanden weite Verbreitung in Skandinavien. Überlegungen zum Thema Herstellungskosten und Sicherheit für den Feuerwehrmann führten dann 1958 zu Versuchsmustern aus Glasfasermaterial, mit relativ hohem integriertem Kamm, und seit 1960 gibt es den Brissman-Glasfaserhelm, mit etwas flacherem Kamm, in Großserie. Parallel dazu hat die Gummifabrik Trelleborg einen Glasfaserhelm mit Visier hergestellt, dessen Gesamtkonzeption sehr fortschrittlich ist; er gehört zu den modernsten Helmen Europas, und ist auch außerhalb Skandinaviens im Einsatz.

Glasfaserhelm (Polyester) mit Klappvisier (Polycarbonat) Nackenschutz aus Rayongewebe, alubeschichtet, Hersteller: Trelleborg, Schweden 1975

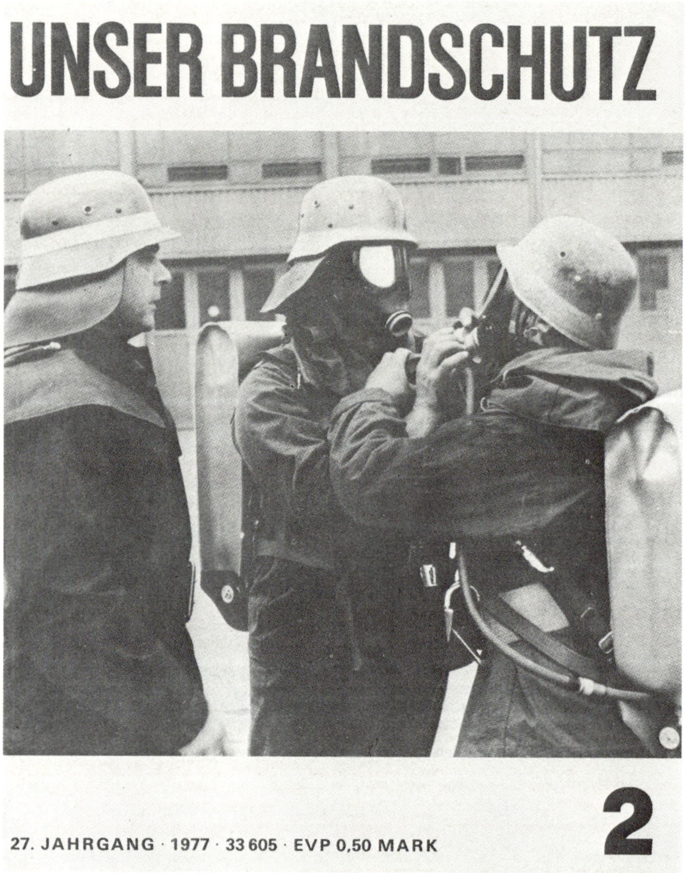

Kunststoffhelm (Glasfaser Polyester)
DDR, 1970

Kunststoffhelm (Glasfaser Polyester)
Arbeitsschutz mit Nackenschutz, DDR, 1977

Feuerwehrhelme in der DDR — eine bewältigte Vergangenheit

In der Nachkriegszeit wurden auch ,,drüben'' Feuerwehrhelme aufgebraucht, die das ,,Tausendjährige Reich'' übriggelassen hatte: Ganzmetallhelme der Feuerschutzpolizei, oder gar schwarz lackierte Helme aus Luftschutzbeständen. Nach der DDR-Norm TGL 121-940.01 (etwa vergleichbar mit der DIN 14-940) werden die Helmkörper, in Stahlhelmform, nach 1961 aus glasfaserverstärktem Polyester hergestellt, wesentlich leichter als Metallhelme, mit geringerer Wärmeaufnahme, und mit dem Bonus der besseren Isolierfähigkeit. Auch die Bebänderung wurde gegenüber den früheren Helmen in eine Vierpunkt-Aufhängung geändert.

Seit Ende der siebziger Jahre sieht man bei freiwilligen Feuerwehren häufig Arbeitsschutzhelme (,,Bauhelme'') aus Glasfaserkunststoff, mit Feuerwehr-Nackenleder. Es ist nicht bekannt, ob es sich hier um eine Übergangslösung handelt, bis ein Feuerwehrhelm nach Muster des NVA-Helmes vorliegt. Die Tradition wäre gewahrt, denn der jetzige Stahlhelm der Nationalen Volksarmee entstand schon 1944 in Thale, und wäre 1946 an die ,,Großdeutsche Wehrmacht'' geliefert worden.

Daß die Suche nach dem idealen Helm noch nicht abgeschlossen ist, zeigen einige Versuchsmodelle, seit 1980 im Tragetest.

Helm nach TGL 121-940.01

Stahlhelm der FSchPol, ca. 1940, mit aufgeschraubtem Schild für die Feuerwehren in Bayern, Mannschaftshelm, Schwandorf, Oberpfalz, 1955

*Militärstahlhelm Modell 1916
Bayern, 1950*

*Stahlhelm
mit Stadtwappen Büdingen/Hessen ca. 1950*

*Wehrmachtsstahlhelm Modell 1942, LS-Version
Bayern, 1960*

*Luftschutzhelm Modell 1944
mit bayr. Feuerwehrschild als Abziehbild
Unterfranken, 1960*

*Ehem. Luftschutzhelme im Einsatz
bei Feuerwehren in Hessen, 1970*

Seite 4 Entwurf DIN 14 940

6.2 Helmschale mit Anstrich und Reflexstreifen

6.2.1 Die mit * gekennzeichneten Maße und Abweichungen der Helmschale müssen Tabelle 2 entsprechen; Prüfung nach Abschnitt 8.3.6.

Nicht angegebene Einzelheiten sind zweckentsprechend zu wählen.

Allgemeintoleranzen DIN 7168 – 9

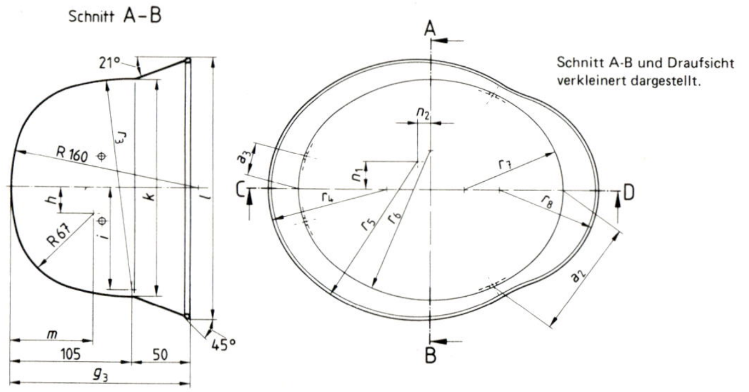

Feuerschutzhelme in der Bundesrepublik — „...man nehme, was man hat..."

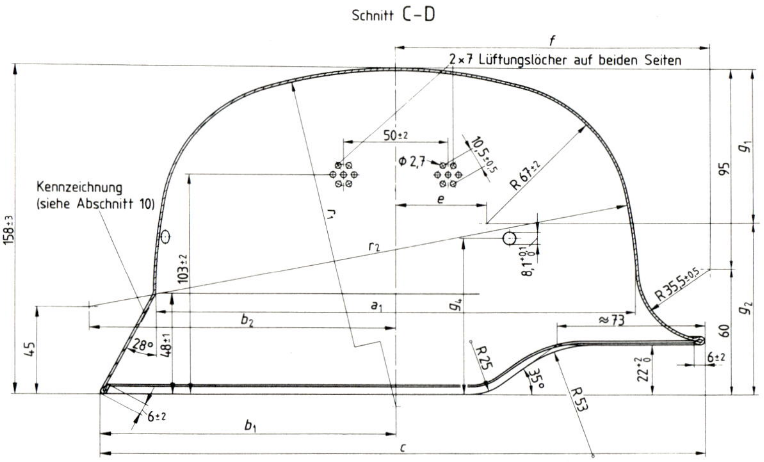

Bild 2. Helmschale

Auszug aus dem DIN-Blatt 14940 neueste Ausgabe 1980

Vielerorts mußten freiwillige Feuerwehren, auf Geheiß der örtlichen alliierten Kampfkommandantur, 1945 ihre Brandschutzhelme durch Bearbeitung mit der Spitzhacke unbrauchbar machen, wahrscheinlich, um Werwolfgelüste im Keime zu ersticken. Großzügigere Sieger bestanden lediglich darauf, „FIRE SERVICE" in großen weißen Lettern auf die Helme aufmalen zu lassen, damit alliierte Patrouillen nicht beunruhigt wurden. Eine Helmknappheit gab es nicht, denn Luftschutz-und Wehrmachtshelme lagen massenweise herum, und konnten durch einfaches Übermalen in den Feuerwehrdienst übernommen werden. Sie haben sich dort lange gehalten, wie ein Blick in die Gerätehäuser mancher Feuerwehren noch heute beweist.

Helm nach DIN 14940 der Berufsfeuerwehr Wiesbaden mit Rangabzeichen 1972

...und bei einer freiwilligen Feuerwehr in Baden, 1979

*Aluminiumhelm, Hochglanz-poliert
für zeremonielle Anläße, Hessen, 1975*

*Ehemaliger Luftschutzhelm im Einsatz
bei bayrischen Feuerwehren, 1968*

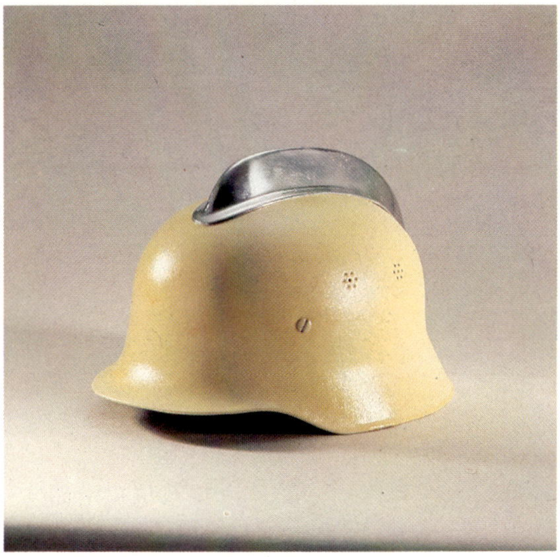

*Aluminiumhelm mit poliertem Alu-Kamm
Berufsfeuerwehr Darmstadt, 1980*

*Aluminiumhelm mit Kamm
Mannschaftshelm mit Schild
der bayr. Feuerwehren, Bayern, 1968*

Durch die Normungsarbeiten (DIN 14940) der sechziger und siebziger Jahre ist, zumindest auf dem Papier, Ordnung in die Typenvielfalt der Nachkriegszeit gekommen. Verbindliche Materialkriterien wurden festgelegt, die Innenausstattung wurde wesentlich verbessert, und Testanforderungen an die Sicherheit des Kopfschutzes wurden verabschiedet. Das war in der Tat ein großer Fortschritt, und der neuartige nachleuchtende Anstrich war nur sein äußeres Zeichen. Die Problematik jeder Normung liegt allerdings darin, daß man die Weiterentwicklung „festschreibt." Die Größe dieser Gefahr hängt davon ab, wieviele Einzelheiten in der Norm festgehalten sind, und die DIN 14940 ist in der Tat ein fast perfektes Meisterwerk, das dem Helmhersteller sehr wenig, dem Designer nicht den geringsten Spielraum läßt. Dies hat dazu geführt, daß das Thema „neuer Feuerwehrhelm" im Nachkriegsdeutschland nur sehr zaghaft angegangen wurde. Liegt es vielleicht daran, daß der Ersatz eines Helmmodelles, das sich in seiner Grundform schon in den Schützengräben Flanderns 1916 bewährt hat, gar nicht nötig ist? Oder ist die Industrie nicht interessiert, obwohl sie für ausländische Rechnung jeden beliebigen Helm produzieren kann? Oder ist es ein unbeabsichtigter Nebeneffekt der wohlgemeinten Normung, die gutdurchdachten Denkanstöße der vergangenen Jahre in Frage zu stellen?

Ein hoffnungsvoller Ausblick ergibt sich aus der Tatsache, daß 1981 bei einer Süddeutschen Berufsfeuerwehr ein groß angelegter Trageversuch mit Kunststoffhelmen nach DIN 14940 durchgeführt wurde (Stahlhelmform). Angeblich wurde der Test sehr erfolgreich abgeschlossen; Sieger war die Phenol-Harz Ausführung, die der Glasfaser-Version leicht überlegen war. Auf der Interschutz 80 als Prototyp gezeigt, ist die Kunststoffvariante jetzt produktionsreif.

In den vergangenen Jahren hat die DIN 14940 den Feuerwehren wenig Spielraum für individuelle Aspekte gegeben. Bei einigen Feuerwehren wird noch der Kamm getragen. Z.B. in Berlin und Darmstadt, wenige Wehren tragen ein Stadtwappen oder, bei Werksfeuerwehren, das Firmenzeichen. Die neuen Normvorschriften haben auch den rotlackierten bayrischen Helm verschwinden lassen. Lediglich bei Feuerwehraufmärschen sieht man hier und dort Unkonventionelles: verchromte oder aluminiumpolierte Helme (aus der Produktion für Schweden) verzieren die Köpfe der Fahnenträger oder der Ehrenwache.

Helm einer Texaco-Werksfeuerwehr, mit Firmenemblem, Hessen, 1974

Ersttagsumschlag, zur Erinnerung an die Herausgabe einer 30-Pfennig-Briefmarke, die den Bauarbeiterhelm nach DIN 4840 zeigt, ähnlich dem Jugendfeuerwehrhelm.

Arbeitsschutzhelme für die Jugendfeuerwehr nach DIN 4840

Die in der Bundesrepublik getragenen Jugendfeuerwehrhelme entsprachen durchweg den Sicherheitsanforderungen der jeweils gültigen DIN 4840, für Arbeitsschutzhelme. Da der offizielle, vom Deutschen Feuerwehrverband eingeführte Jugendfeuerwehrhelm, aus Polycarbonat, relativ teuer ist, wurden, mit regionalen Schwerpunkten, auch preisgünstigere Helme ähnlichen Designs nach DIN 4840 beschafft und mit Erfolg verwendet.

„offizielles" Modell für das Bundesgebiet...

Regional beschaffte Ausführungen für Hessen, 1972

...und für Bayern, 1970

Der neue Helm aus Bonn

Das Einsatzspektrum von Feuerwehr- und Zivilschutzhelmen weist schon von jeher einige Parallelen auf. Es ist deshalb interessant, auf Helmentwicklungen aufmerksam zu machen, die dem Bundesamt für Zivilschutz entstammen. Pate für das erste Nachkriegsmodell stand der Stahlblechluftschutzhelm des 2. Weltkrieges, der bis 1966, von zwei deutschen Herstellern, aus glasfaserverstärktem Polyesterharz hergestellt wurde. Auch die Innenausstattung wurde modernisiert: wie beim NATO-Stahlhelm der Bundeswehr wurde sie zentral an einer Schraube oben im Helmkopf befestigt. 1968 durchgeführte Versuche, die Helmschale aus Polycarbonat herzustellen, mißlangen; der Helm federte zu stark durch und gefährdete dadurch den Träger.

Schon 1971 wurde ein radikal geändertes Design vorgelegt, wieder in Polycarbonat, das jetzt aber die Forderung der DIN 4840 (Arbeitsschutzhelm) voll erfüllte. Die Helmform basierte auf den Modellstudien, die 1944 in Thale durchgeführt wurden, um die Helmproduktion rationeller zu gestalten, und um die Schutzfunktion der Helme zu verbessern. Ein großer Vorteil des Zivilschutzmodelles von 1971 war das einschiebbare Drahtvisier, eine Idee, die uns auch von einigen anderen Feuerwehrhelmmodellen vertraut ist. Produziert in kleiner Serie, wurden diese Helme erfolgreichen Trageversuchen unterworfen. Seine Einführung auf breiter Ebene scheiterte aber an dem Widerstand maßgeblicher Kräfte innerhalb der Feuerwehren, die diesen Helm, nicht zuletzt aus ästhetischen Gründen, ablehnten. Eine 1980 vorgestellte, stilistisch überarbeitete Form läßt eine Anlehnung an das traditionelle Stahlhelmmodell erkennen. Die Beschaffung dieses neuen Helmes aus Polycarbonat wurde vom Bundesamt für Zivilschutz inzwischen eingeleitet.

Helm aus glasfaserverstärktem Polyesterharz,
bis 1966 beschafft

Versuchshelm aus Polycarbonat,
mit geänderter Befestigung der Innenausstattung,
1968

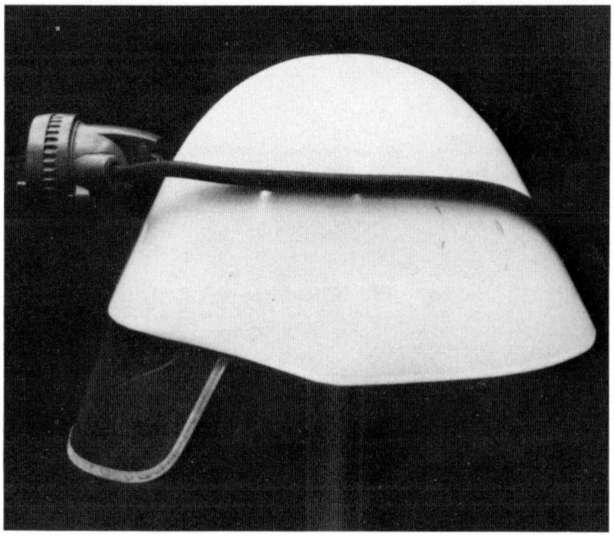

Helm in Nullserie aus Polycarbonat,
mit einschiebbarem Gesichtsschutzgitter, 1971

Neues Modell in Polycarbonat, 1980

Der neue Helm aus Berlin

Schon Mitte der sechziger Jahre begann bei der Berliner Auer Gesellschaft die Entwicklung eines neuen Konzeptes für einen modernen Brandschutzhelm. Unter Mitwirkung der Frankfurter Berufsfeuerwehr, die die Prototypen, auch einsatzmäßig, testete, entstand eine Vorserie von ca. 20 Helmen, in geringfügig unterschiedlichen Ausführungen. Gemeinsam war allen Versionen die zwei-Schalen-Konstruktion, d.h. der Innenhelm auf der Basis des Arbeitsschutzhelmes nach DIN 4840, und der Außenhelm, als abgewandelte Form des traditionellen „Stahlhelmes." Bei der Vorserie bestand dieser Außenhelm aus Polycarbonat; für die Serienproduktion wurde Aluminium-Druckguß erwogen. Eine Besonderheit des Frankfurter Auerhelmes war das einschiebbare Drahtgittervisier, das zwischen Außen-und Innenhelm montiert war und nach dem Prinzip eines Davy-Siebes das Durchschlagen von Flammen verhindern sollte. Die Befestigung des Innenhelmes am Außenhelm erfolgte bei den ersten Prototypen durch Klettenbänder, bei der Endversion durch einen Drahtklemmring. Die Grundüberlegung für dieses Zwei-Schalen-Konzept wurde aus der Gefahr entwickelt, die z.B. bei Explosions-Unglücken aufgetreten war: der normale Helm wurde vom Kopf der Träger abgerissen, die somit schutzlos wurden. Beim Zwei-Schalen-Helm kann sich der Außenhelm gefahrlos vom Innenhelm lösen, der dem Träger weiterhin Schutz bietet. Außerdem kann der Innenhelm als Schutzhelm bei Arbeitseinsätzen getragen werden, bei denen der große Außenhelm nicht nötig ist, oder störend wirkt.

Die Vorstellung des Frankfurter Auer-Helmes Anfang der siebziger Jahre löste eine ziemlich kontroverse Diskussion zum Thema ,,neuer Feuerwehrhelm'' aus, und Außenstehende konnten sich des Eindruckes nicht erwehren, daß Sachargumente sehr emotional behandelt wurden. Da sich bei den deutschen Feuerwehren keine gemeinsam getragene Entscheidung erreichen ließ, wurde 1972 die Einführung dieses oder eines ähnlichen neuen Helmes vorläufig ,,auf Eis gelegt.''

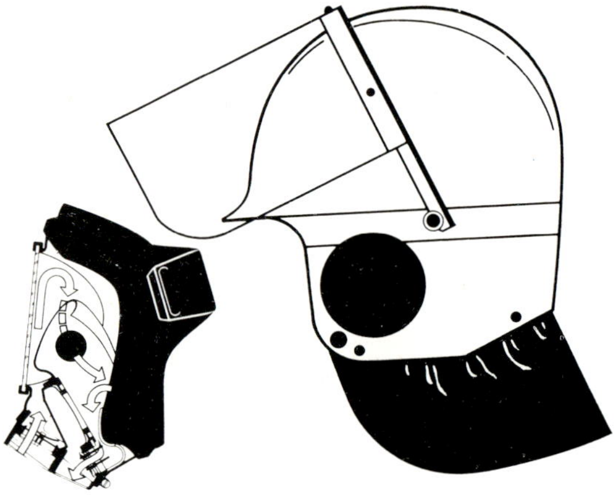

Der neue Helm aus Lübeck

Als das DRÄGERWERK in Lübeck, Mitte der siebziger Jahre, die Entwicklung eines neuen Feuerwehrhelmes bis zum Prototyp-Stadium durchgeführt hatte, staunte die Fachwelt. Der neue Helm aus Lexan (Polycarbonat) hatte eine bisher bei Feuerwehren nicht verwendete Form, die sich von der herkömmlichen Stahlhelmform weitgehend freigemacht hatte; der in die Helmschale integrierte flache Kamm deutete Feuerwehrtradition nur noch an.

Das Besondere an diesem neuen Helm war jedoch die mitgelieferte DRÄGER-Atemschutzmaske, die ohne störende Bebänderung in den Helm eingehängt werden konnte, und es somit dem Atemschutzträger ermöglichte, sich ohne zeitraubendes Helmabnehmen und Anpassen der Maske einsatzbereit zu machen. Das Geheimnis dieses Systems lag in zwei, seitlich am Helm eingebauten, Excenter-Befestigungen, die durch einfaches Drehen die Helmmaske fest an das Gesicht des Trägers preßten; Helm und Maske waren dadurch fest miteinander verbunden.

Daß dieses Helmkonzept keine theoretische Übung blieb, wurde durch die aktive Mitarbeit der Berufsfeuerwehr Den Haag gewährleistet, die den Helm, auch unter Einsatzbedingungen testete. Innenausstattung, Nackenschutz usw. wurden so gestaltet, daß die deutschen Helmnormen DIN 4840 und DIN 14940 nach Möglichkeit erfüllt wurden.

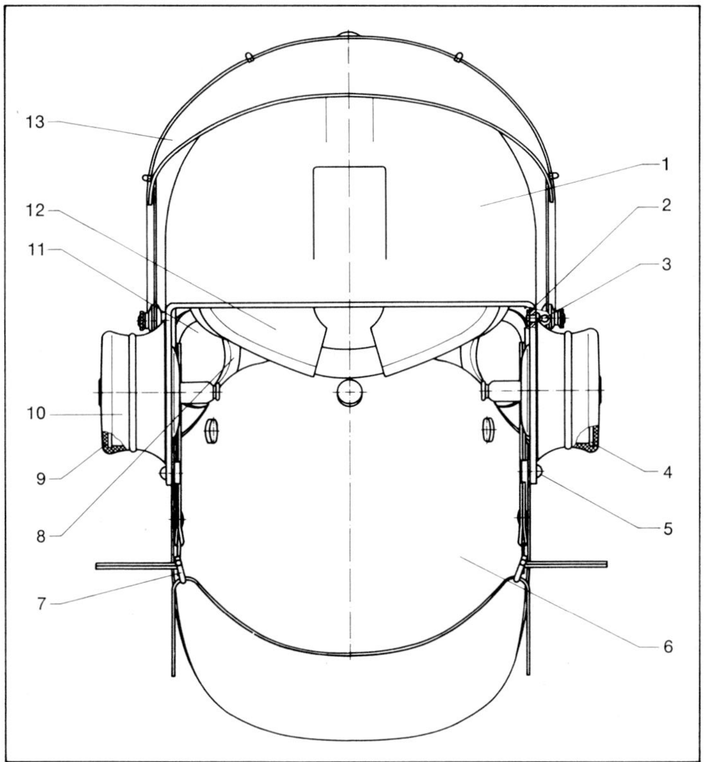

Darstellung des Atemschutzhelms

Was ist was?

1 Helm
2 Starlock-Schnellbefestiger
3 Tenaxknopf-Unterteil
4 Drehknopf, links
5 Gummiknopf
6 Nackenleder (Zubehör)
7 Kinnriemen, vollständig
8 Ring
9 Drehknopf, rechts
10 Maskenspannvorrichtung
11 Lochplatte
12 Innenausstattung
13 Visier (Zubehör)

Schnittdarstellung der Helmmaske RA

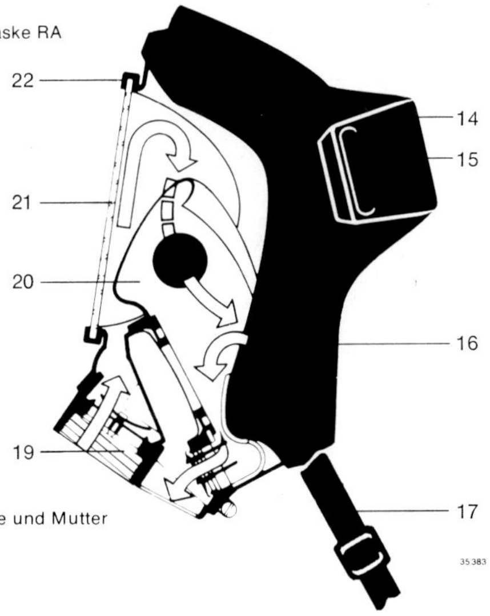

14 Zubehör (2 x Keil)
15 Maskenbeschlag
16 Maskenkörper
17 Nackentrageband
18 Spannband mit Schraube
19 Anschlußstück vollständig
20 Innenmaske vollständig
21 Scheibe (Plexiglas)
Scheibe (Triplex)
22 Spannrahmen mit Schraube und Mutter

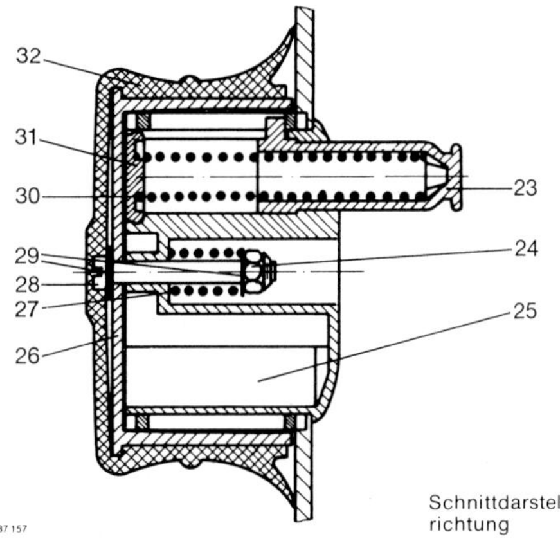

23 Bolzen
24 Mutter
25 Drehknopf, innen
26 Drehknopf links oder rechts
27 Schließfeder
28 Schraube
29 Scheibe
30 Feder
31 Deckel
32 Schutzkappe

Schnittdarstellung der Maskenspannvorrichtung

Auch die deutschen Berufsfeuerwehren hatten Gelegenheit, die ersten Prototypen, die zusätzlich mit Klappvisier ausgerüstet werden konnten, zu begutachten. Daß dem Helm in Deutschland bisher kein durchschlagender Erfolg beschieden war, liegt teilweise an der eng-definierten Feuerwehrhelmnorm DIN 14940, die keine Abweichung von der Stahlhelmform zuläßt. Darüber hinaus sind sich Fachkreise noch nicht einig, ob eine Zweipunktaufhängung der Maske, gegenüber der üblichen Fünfpunktaufhängung, für den Atemschutzträger einen ausreichend dichten Sitz der Maske gewährleistet. Anders als bei Pilotenhelmen, bei denen sich die am Helm befestigte Maske bewährt hat, da der Pilot in seinem Sitz angeschnallt ist und sich fast nicht bewegt, muß der Atemschutzträger bei der Feuerwehr harte körperliche Arbeit leisten, in fast allen Körperstellungen. Darüberhinaus muß im Feuerwehreinsatz damit gerechnet werden, daß der Helm von verschiedenen Richtungen Stöße erhalten kann, die seinen Sitz am Kopf verändern. Damit könnte sich automatisch auch eine am Helm befestigte Maske aus ihrer optimalen Lage am Kopf verschieben und dadurch undicht werden. Das Grundkonzept des DRÄGER-„Atemschutzhelmes" ist jedoch so interessant, daß es unbedingt weiter verfolgt werden sollte, denn das einsatzmäßige Anlegen der traditionellen, bebänderten Maske und des Feuerwehrhelmes nach DIN 14940, sowie die Trageeigenschaften dieses Helmes mit angelegter Maske beinhalten eine Reihe ungelöster Probleme.

KLAPPBARES VISIER.
FEURWEHRHELM.
18.5.78.

Der Helm, der den wichtigen Schutz des Kopfes übernimmt, wurde völlig neu konzipiert. Er besteht aus einer Polycarbonat-Helmschale mit Innenpolsterung. Ein vorklappbares Visier schützt die Augen vor Flammen, Verpuffungen und Säurespritzern. Im Helm integriert sind Wangenmikrofon und Ohrhörer, die bei Bedarf mit dem Funkgerät durch einen Stecker verbunden werden. Eine Stirnlampe läßt sich anbringen. Die Atemschutzmaske wird am Helm befestigt.

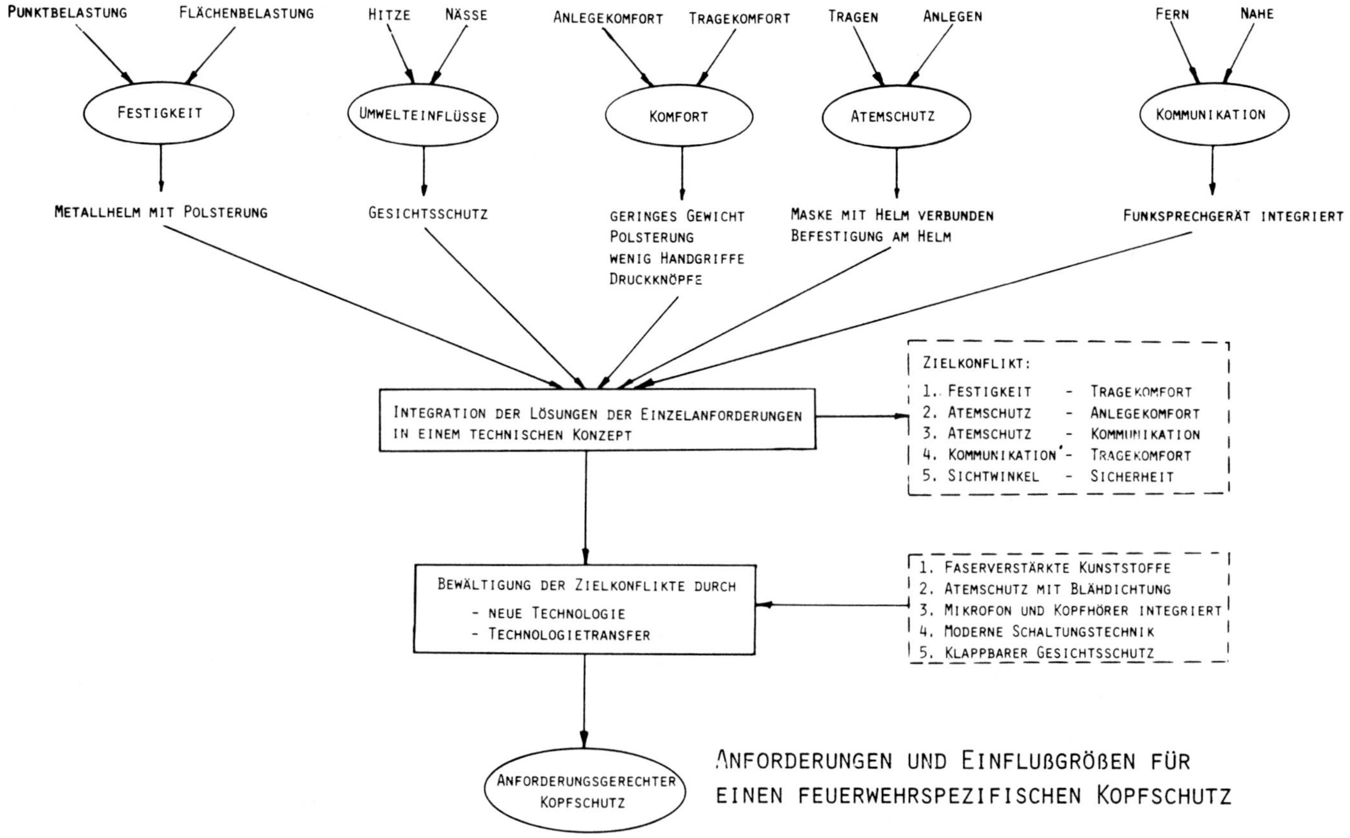

ANFORDERUNGEN UND EINFLUßGRÖßEN FÜR EINEN FEUERWEHRSPEZIFISCHEN KOPFSCHUTZ

Der neue Helm aus Stuttgart

Die Firma Porsche stellte 1978, in einer Designskizze, einen Integral-Helm für den Feuerwehreinsatz vor, der im Rahmen eines Forschungsauftrages der Bundesregierung als Ergebnis einer Perspekivstudie theoretisch entwickelt worden war, als Ergänzung zur neuen ORBIT-Fahrzeuggeneration, die in Feuerwehrkreisen mittlerweile zu heißen Diskussionen geführt hat.

Nachdem Anforderungen und Einflußgrößen für einen feuerwehrspezifischen Einsatz definiert waren, ergab sich als Resultat ein Polycarbonathelm im „Astronautenlook", mit vorklappbarem Visier und im Helm befestigter Atemschutzmaske. Kommunikationsprobleme bei diesem allseitig geschlossenen Helm sollten durch eingebautes Wangenmikrophon und Kopfhörer gelöst werden, die mit einem Funkgerät durch Stecker verbunden sind. Auch die Stirnlampe, schon immer auf der Wunschliste der Feuerwehren, aber bisher nie in wirklich brauchbarer Form realisiert, wurde nicht vergessen.

Nach anfänglicher Faszination über diesen „Helm des Jahres 2000", meldeten sich bald die ersten Kritiker und stellten die Frage, wie dieser Helm in der Praxis technisch zu verwirklichen sei, abgesehen von der bevorstehenden Kostenexplosion. Die Notwendigkeit, jedem Feuerwehrmann ein Funkgerät mitgeben zu müssen, wurde ebenso problematisch beurteilt, wie die Gefahr der möglicherweise undichten Atemschutzmaske, die, weil fest mit dem Helm verbunden, bei stoßartigen Einwirkungen auf den Helm ihren Sitz verändern wird. Es wäre trotzdem durchaus denkbar, daß der PORSCHE-Helm für spezielle Sondereinsätze ein ideales Ausrüstungsstück darstellen wird. Ob er in Produktion gehen kann, ist z.Zt. nicht bekannt.

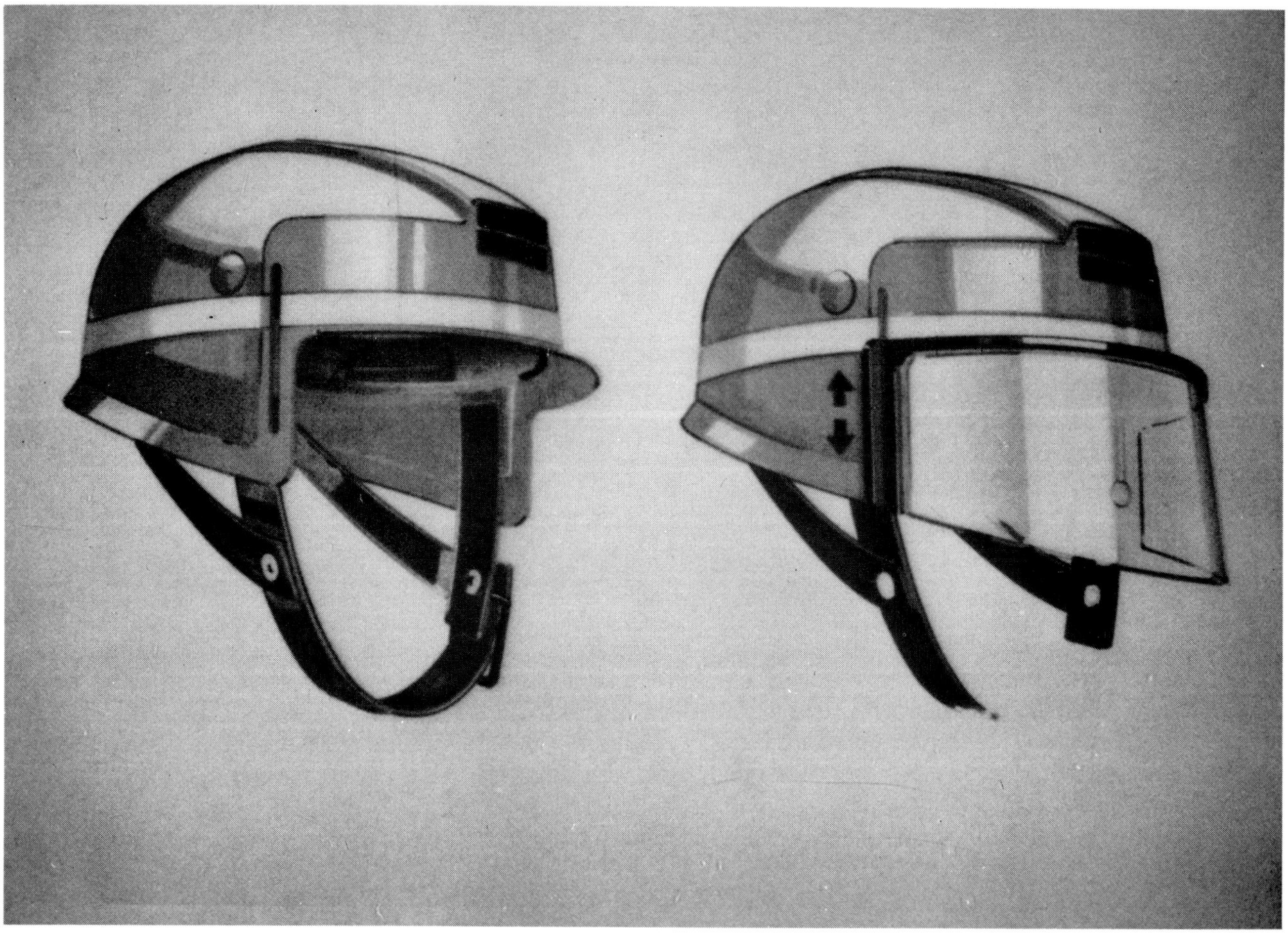

Der neue Helm
im TOTAL-Konzept

Als vorläufig neuester Beitrag zur Diskussion um einen Brandschutzhelm ist die Designstudie von TOTAL zu verstehen. Losgelöst von übernommenen Vorstellungen wurde hier versucht, vom Grundgedanken der Funktionalität geleitet, die Themen „Kopfschutz und Atemschutz" in einem Modular-System zu integrieren.

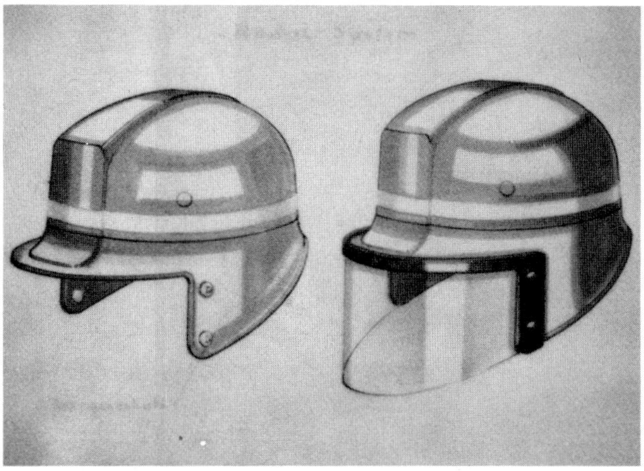

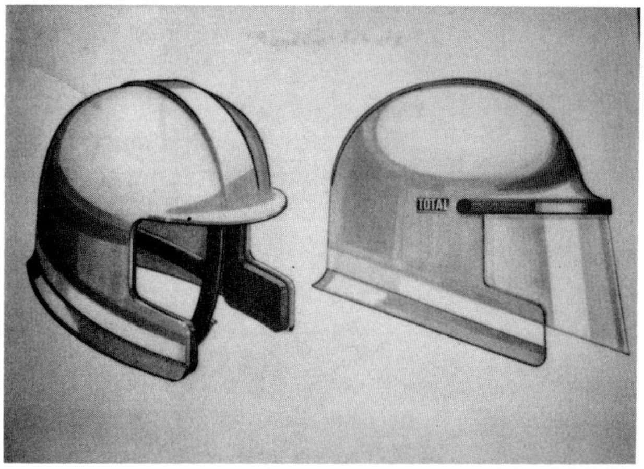

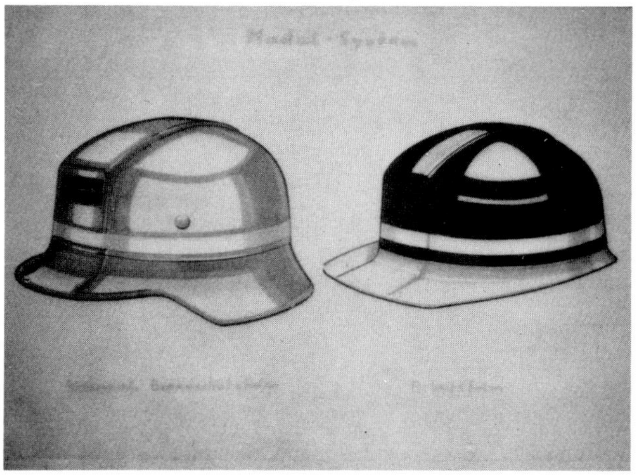

Die Helmschale, mit weit vorgezogenem Wangenschutz, ist in ihren Sichtfeld-Konturen so ausgelegt, daß sie sich an das Sichtfenster der Atemschutzmaske genau anpaßt. Um die bekannte Problematik der Bebänderung der Maske (schwieriges Anlegen bzw. Anpassen; Druckstellen am Kopf durch Aufsetzen des Helmes) zu lösen, wird eine Maske ohne Bebänderung vorgeschlagen. Durch eine flexible Kopfhaube aus porösem oder durchbrochenem, atmungsaktiven Material wird die Maske fest am Kopf gehalten, unabhängig von der Position des Helmes. Durch Aussparungen an den Ohrpartien der Maske wird die Kommunikation erleichtert. Falls die Atemschutzmaske nicht getragen wird, kann ein Visier am Helm angebracht werden.

Nachwort

Der Gedanke zu dieser kleinen Abhandlung ergab sich vor Jahren, als im Gespräch während einer langen nächtlichen Brandwache einer meiner Kameraden auf seinen Helm von 1940 zeigte und erklärte, daß es nur ,,einen deutschen Feuerwehrhelm'' gäbe; der Hinweis auf unsere anderen, neueren Helme überzeugte ihn nicht, und es stellte sich heraus, daß wir alle sehr wenig über unseren Helm wußten, trotzdem er das wertvollste persönliche Ausrüstungsstück jedes Feuerwehrmannes ist: es schützt seinen wichtigsten Körperteil, den Kopf, vor den Gefahren des Einsatzes.

Am Helm läßt sich ein Stück Kulturgeschichte ablesen. Technikgeschichte, wenn man die Herstellung und die verwendeten Materialien betrachtet. Sozialgeschichte, wenn man die Einflüße von Organisationsformen, Staat und politischen Denkweisen erkennt. Modische Trends sind festzustellen, mehr oder weniger ausgeprägte Geschmacksrichtungen zu definieren, aber auch schwer zu erklärende emotionale Einstellungen: kein anderes persönliches Ausrüstungsstück der Feuerwehren wird mit soviel Engagement, soviel Lob und Ablehnung behandelt wie der Helm. Es wäre wünschenswert, wenn die hier gebrachten Bilder und Erklärungen dazu beitragen, Zusammenhänge aufzuzeichnen, und hier und da Ansatzpunkte zu weiterführenden Diskussionen zu bieten.

Besonderer Dank gilt meinen Gesprächspartnern bei deutschen und ausländischen Helmherstellern, die meine vielen Fragen geduldig beantworteten, soweit es die Firmenpolitik zuließ; für die Interpretation der so gewonnenen Informationen bin ich allein verantwortlich. Dank auch meinen Kameraden bei freiwilligen und Berufsfeuerwehren in aller Welt, voran der Dietzenbacher Wehr, die mein Interesse an Feuerwehrhelmen voll Nachsicht und Verständnis duldeten. Besonders geholfen haben Allen Batchelor in London, John Christensen in Kopenhagen, Jean-Gabriel Jeudy in Paris, Marcel Nuijttens in Flandern, Paul Pribble in Kalifornien, Ken Smith in Croydon, Ken Snyder in Pennsylvanien und Wilhelm Wahrenberg in Stockholm.

Die meisten Farbbilder stammen aus der Kamera der Fotografenmeisterin Brigitte Harms. Helmut Raab von der Dietzenbacher Redaktion des ,,112 Feuerwehrmagazins'' verdanke ich eine ganze Reihe von Anregungen und konstruktiver Kritik. Der größte Teil der gezeigten Helme ist im Feuerwehrmuseum Salem zu besichtigen.

Thomas W. Herminghaus
Dietzenbach, im Mai 1982

Quellenverzeichnis

Soweit nicht besonders angegeben, stammen die Abbildungen aus dem Archiv des Verfassers.

Bücher:

L. Baer: Die Geschichte des Deutschen Stahlhelms, Eschborn, 1977
P. Connolly: Die Römische Armee, Hamburg, 1976
—: Courage et Dévouement, Paris, 1970
Currier & Ives: The American Fireman, Farbdruck, New York, 1858
K.H. Dunshee: Enjine! — Enjine!, New York, 1939
B. Fitzsimons (Ed.): Heraldry of War, London, 1973
J.v.d. Heiden: Slang-Brand-Spuiten, 2. Auflage, Amsterdam, 1735
J. Holaubek: Die Österreichische Feuerwehr, Wien, 1979
S. Holloway: London's Noble Fire Brigades 1833—1904, London, 1973
W. Hornung: Feuerwehrgeschichte, Stuttgart, 1981
Dr. Klietmann: Feuerschutzhelme der Feuerwehr bzw. Feuerschutzpolizei, Berlin, 1964
C.D. Magirus: Das Feuerlöschwesen in allen seinen Theilen, Ulm, 1877
Normenausschuß Feuerwehrwesen: Feuerwehrhelm, Entwurf DIN 14940, Berlin, 1979
Porsche: O.R.B.I.T. Studie, Stuttgart, 1979
C. Preyssl: The Firefighters of Vienna, Newbury, 1979
E. Radecke: Geschichte des Polizei-Tschakos, Hilden, 1981
G.M. Schulz: Helme und Mützen der Armee 1871—1945, München, 1978
F.J. Stephens and G. Maddocks: The Imperial German Army 1900—1918, New Malden, 1975
F. Thaler: Stahlhelme, München, 1978
Transfeldt- v. Brand-Quenstedt: Wort und Brauch im deutschen Heer, Hamburg, 1967
Wagner & Gelbert: Die Uniformen der Berufsfeuerwehr und der freiwilligen Feuerwehrmänner, Berlin, 1934

Zeitschriften:

E. Achilles: Frankfurts neuer Helm, in Fire International 26, 1972
P. Pribble: Trends in Helmet Design, in NFPA FIREMEN Magazine 2/1962
P. Pribble: Head Protection: a Top Priority, in FIRE COMMAND! 4/1976
P. Pribble: A Time, a Reason, a Need, Manuscript, 1978
B. Redin: Brandhjelm, in Brandvaern 5/80 (Abb.)
H. Schläfer: Der Feuerwehrhelm: Seine Geschichte, derzeitige Form und Weiterentwicklung, in Brandschutz 5/1971
G. Snively and C.O. Chichester: Evaluation of Firemen's Helmets, in NFPA FIREMEN Magazine 11/1968
—: Unser Brandschutz 2/1977; 4/1977 (Abb.)
E. Warnecke: Atemschutzmasken und Kopfschutzhelm, in Drägerheft 307, 1-4/1977

Quellenverzeichnis

Firmenkataloge:

Barnes, Boston, 1895
Beuttenmüller, Bretten, 1898, 1907, 1911, 1926, 1930, 1934, 1935, 1937
Cairns, Clifton, NJ, USA, ca. 1900, 1968, 1970, 1980
Ewald, Küstrin, 1924
Hayward, New York, 1902
Lieb, Biberach, 1886, 1887
Magirus, Ulm, 1874, 1884, 1885, 1906, 1907, 1913, 1914
Maury, Offenbach, 1911
Merryweather, London, ca. 1880
Meyer, Hagen, ca. 1925
Miller, New York, 1884
Mispa, Turin, ca. 1972
Müller, Offenbach, 1907
Olsen, New York, ca. 1890
Rannenberg, Hannover, ca. 1885
Rosenbauer, Linz, 1926, 1972
Waterous, Saint Paul, 1906
Woodhouse, New York, 1888-9

Informations- oder Bildmaterial haben zur Verfügung gestellt:

Die Berufsfeuerwehren Bonn, Frankfurt, Hongkong, Johannesburg, London, Tokio, Wiesbaden.

Das Bundesamt für Zivilschutz, Bonn.

Die Firmen Auer, Comrad, Cromwell, Dräger, Federal, Römer, Schuberth, Total, Vetter.

Die Herren W. Köcher, S. Ryan, H. Schröder, D. Seuffert

Fire Helmets

History and Development

Summary in English

EFB-Publishers
Breslauer Strasse 23
D-6450 Hanau/W. Germany
1982

Page 5

Historic reports about state or municipal institutions usually start with the 'Greeks and Romans.' The attempt fails however to trace the history of the fire helmet back to the days of antiquity, because organized and specialized firefighting forces were virtually unknown then. This does not mean that firefighting itself, even on a large scale, is an invention of modern times. An old work of art dating back to 850 B. C., an alabaster relief at a Ninive palace, depicts Assyrian soldiers fighting fires under wartime conditions, and a Chinese document of the year 564 B. C. gives orders to soldiers who are to take part in firefighting during conflagrations. Thus the oldest firefighters known to us, operating under unified command, were apparently military men, and firefighting was not the main part of their regular duty. We can also assume that they wore their regular military helmets when on firefighting duty. Surprisingly, the helmets worn by modern day firemen in many countries still look very much like military helmets.

P. Connolly in his book "The Roman Army" proves that already in ancient times much thought went into the development of helmets. Lacking a scientific approach, it took however several centuries to come up with real improvements, all over the Roman Empire.

Page 7

The Roman vigiles were the first real professional firefighters; unfortunately we do not know anything about their head gear. If they did wear a helmet, it was most likely of the military type.

The decline of the Roman Empire meant also, for some centuries, the end of an organized fire service. Only in the late middle ages did some of the emerging cities establish some sort of fire protection by hiring crews for the primitive manual fire pumps. These municipal crews wore some type of uniform garb. They were issued oil cloth caps or, towards the middle of the 18th century, stovepipe hats with the city's coat of arms (e.g. Vienna, Copenhagen) or with the crest of the insurance company (London) which organized the fire brigade. Hats of this type certainly did not offer much protection; in fact they were unsuitable for hard work, and were worn mostly for 'good looks', just like the colorful uniforms of the armies in those days. Firefighters at that period did not enter burning buildings, as a rule, because suitable hose material and breathing apparatus had not yet been invented, and although Jan van der Heyden, in his dramatic illustrations published 1735, shows firemen making interior attacks, they wear soft hats, according to then-current fashion.

Page 9

Head protection of firefighters attracted attention only later on, when firefighting became better organized. French Sapeurs-Pompiers in 1750 used a special cap of leather and linen with a big frontpiece of brass, showing the initial of the community, and after 1786 the Gardes-Pompes in Paris wear metal helmets with high combs. From now on fire brigades look to the military for helmets, and if we examine for example the German Army helmets of the past 150 years, we recognize that all models exist also in fire brigade versions.

There were good reasons for this development. The well organized military was admired by fire brigades which, being newly formed bodies, took some time to find their own style of drill and discipline. And military helmets, especially surplus stocks, were easy to find and inexpensive; this was important for the fledgling fire service, fighting for funds. Only in recent years have industry and government pressed on with scientific research for firefighter's head protection.

Apparently the task is quite complicated, as proven by the fact that the ideal fire helmet has not yet been invented. If and when it comes one day, its introduction into service should not be prevented by the 'high cost' argument. The helmet is the most important protective gear for each fireman, its quality can well decide over life and death in certain situations.

Page 11

Controversy in 1850: Brass Helmet versus Leather Helmet

In the 1850's, when many volunteer fire companies were founded, there was not much choice concerning helmet materials. Helmets made of felt were offered by the manufacturers, but they were not strong enough for everyday use. So, there was really only a decision to be made between leather helmets, as used by the military, or brass helmets, similar to the type used by the famous Paris fire brigade. Especially volunteer fire brigades in Southern Germany, partly due to their 'revolutionary' liberal political attitudes, looked to Paris for guidance, and consequently adopted brass helmets. C. D. Magirus, in his 1877 book, described the advantages of brass helmets:'' Although brass has to be treated (polished) more frequently than leather or felt helmets, it has the advantage of looking like new after cleaning; the lacquered leather- or felt helmets loose their lustre after a while. Looking at the brass helmet as a piece of uniform, it has the advantage of identifying a fireman at first sight, and even at night. In addition, a fire brigade, as a uniform body, looks very smart in brass helmets, this advantage should not be ignored.''

After all, the Germans were good patriots, and thus they opted for a shape different from the Paris helmets. In fact, local or regional pride resulted in a number of different helmet models, one for each province. However, in the founding days of the fire brigades, there were no binding uniform rules, and each fire chief could decide for himself which helmets to buy. Thus, we can find Bavarian helmet models in Northern Germany, and vice versa. Even within a fire fire brigade we can meet different helmet models, one for each company.

Page 14

A truly controversial item was the 'plume' to be attached to the comb, or to the side of the helmet. As a non-binding rule, chiefs wore a white plume, adjutants a white and red plume, after 1870. The advantage was good visibility of commanding officers, but the plume in many cases looked a bit ridiculous. Later on, when epaulettes were introduced as rank insignia, the helmet plumes were abolished.

Page 18

By 1850, when fire brigades in Germany were slowly being established, France had developed a highly organized system of fire departments. The Paris fire brigade had, at that time, introduced already their third helmet generation. French fire brigades started, around 1780, with padded cloth caps of very sturdy construction, featuring a long piece of cloth for the protection of the neck; the caps carried a brass front piece of generous dimensions with the number of the company. This type of cap was used by many French fire brigades; due to their solid design, quite a number has survived the past 200 years. At the beginning of the 19th century, a 'real' helmet made of brass and tin appeared, with a high comb, and a front emblem showing the imperial eagle, or a municipal coat of arms. This helmet was heavy and unattractive, and in 1812, a new model brass helmet was proposed, similar to the army fashion. It was much more expensive than the old model, but was soon adopted. It was the first fire helmet equipped with some sort of visor. The new helmet was an instant success: it was copied to some degree in Switzerland and in Germany, and later on served as a pattern even for the helmets of the London Fire Brigade. A somewhat lower model of this brass helmet type was worn in France until after World War II, mainly by volunteer fire companies. And when the trench warfare in 1915 required quick deliveries of a suitable steel helmet for the French army, the new military helmet was greatly inspired by the French fire helmet: it carried even the comb!
(See pages 60, and 70/71)

Page 21

In England, we find organized fire brigades after 1700. They were established by various fire insurance companies, and wore colorful uniforms and high hats, according to the fashion of those days. After 1833, these insurance brigades were combined into one fire fighting force, and a new brass helmet, similar to the French model, was introduced after 1866. At first, it carried the letters "MFB" (Metropolitan Fire Brigade), lateron a shield with LCC/LFB (London City Council/London Fire Brigade was adapted. Leather helmets were rejected by fire chief Shaw because the volunteers around London wore such helmets. A few years later, also the volunteer fire brigades had switched to the brass helmets...

This brass helmet model was used in London into the thirties, and when Australia in 1940, due to the war situation, could not receive brass helmet supplies from England, the Rider & Bell Co. of Sydney copied the design and produced such helmets for Australian fire brigades. They were retired only in 1964, after a "model life" of one hundred years!

Page 22

Brass did not only look good, but could be shaped easily, and this enabled many of the smaller European nations to specify their own brass helmet design. Firemen in Luxembourg, for example, for a while used a ball fitting on their helmets of South German manufacture. The Swiss accepted French helmets but replaced the high comb by a lower version.

Brass could be shaped easily, but was also susceptible to damage on the fire ground. Simple brass helmets would be made from one piece, but the more decorative helmets could be assembled only from many parts. These fittings had to be bolted, screwed or soldered onto the basic helmet shell, and this did not exactly add to the strengh of the helmet design. This problem, and electric conductivity as well as high production costs eventually led to the demise of the brass helmet: fire helmets were not bought any longer mainly for reasons of good look.

Page 25

Leather caps for head protection are known since the Middle Ages. On old illustrations, we see hunting caps made of leather, or simple head protection for miners. Since the middle of the nineteenth century, we find leather helmets with many of the newly established fire brigades, mainly in Northern Germany, but also in Bavaria. Often they were surplus military helmets which were readily available, or could be manufactured at low cost from existing tools. The spike or (artillery) ball was replaced by a brass or nickel comb, and the national eagle shield was substituted by a city front piece. Officers retained the metal chin strap, even though it had no function in fire service use. Surprisingly, the Prussians developed an alternative to the military style leather helmets. They created a relatively big "fire cap" of leather, with a built-in head cushion pad, and a long piece of leather sown into the helmet, for neck protection. This socalled "Berlin Cap" was adopted also by a number of other city fire brigades. The helmet shell did not carry the traditional comb, but a cross shaped protective fitting, to which a spike could be attached, for ceremonial purposes. Magirus in Germany and Shaw in England confirmed that a well built leather helmet compares well to the widely used brass helmets, at least by the primitive test standards of the day.

Page 27

Most fire helmets are equipped with some type of comb; this feature has become almost an international symbol of identification for firemen. For the helmets, this comb was not only a decoration, or replacement of military fittings, but was absolutely necessary to provide strengh to the helmet shell. Under actual fire ground use, it was found that heat tended to soften leather helmets, and humidity, e.g. from the nozzle spray, had a similar detrimental effect. Thus it was imperative to stengthen the crown of the helmets, and provide stability to the brim by attaching some sort of beading. To satisfy all regional and local tastes, combs in a variety of shapes were available; a well known helmet producer in Offenbach/Germany offered no less than 18 versions.

By the way, almost all leather helmets worn in Germany and neighbouring countries were manufactured in Offenbach near Frankfurt, still today the centre of German leather goods production.

Page 31

Leather helmets in Europe very often carried multi-coloured cockades, displaying flag colours of a province and/or nationality. These round metal discs, with a hole in the centre, were slipped under the chinstrip fasteners. Seen from the helmet wearer's postion, the national cockade (black, white, red in Germany) was to be worn on the right side of the helmet; the provincial cockade adorned the left side. In practical use, this rule was not always observed.

Page 33

Old fire brigade photographs often show the fire chief and other officers wearing the infantery spike or the artillery ball on top of their leather helmets. For a time, this corresponded to the Prussian fire uniform rules. However, this military ornament could also refer to the fire chief's former military branch in which he had done his national service. The eagle emblem of his military helmet was converted to a fire service emblem by placing a corresponding metal plaque over it (helmet with crossed axes).

Page 35

Leather helmets were available in a variety of regional models, differing in the shape of the helmet shell and for the type of comb. Nevertheless, many fire brigades ignored the model recommended for their region, and purchased whatever they liked. The typical Prussian military leather helmet was also worn in Bavaria, painted red and equipped with the full length Bavarian comb. On the other hand, traditional Bavarian leather helmets have been observed on the North German coast line. The easily recognizable Rhenish-Westphalian flat comb can also be found in parts of Hesse, and Saxonian leather combs were also employed elsewhere. Incidentally, all fire brigade electricians were issued helmets with leather combs; such helmets were not permitted to have any metal parts, for obvious reasons. Leather helmets were bought by German fire brigades up to the beginning of the thirties; many volunteer companies used them still during the second world war, especially for replacement personnel and junior firemen. In some cases, rural volunteer brigades had leather helmets as first line equipment until the sixties.

Page 39

European fire helmets of the 19th century followed military fashion rather closely. North America, however, preferred a different style. True, the first fire hats in America were similar to the stove pipe hats of the London insurance brigades, but after independance was declared in 1776, the development of the American fire helmet, as well, went independently from European influences. A leather helmet industry began to prosper mainly in the New York area and over the years arrived at a helmet design which basically is still in use today. Surprisingly, even leather helmets are still being produced, Cairns being the major manufacturer in this field. The typical front pieces and front piece holders exist in a great variety of styles.

Page 40

One cannot deny the Austrian influence when looking at fire helmets of the Balkan countries, in South East Europe. The countries belonging to the Austrian Empire before 1918 closely followed the Viennese pattern in regard to uniforms and helmets, even though local variations can be found. In a greater sense this rule applied also to Greece and Turkey; after all, Austrians modernized the fire service in these areas, and naturally introduced Austrian methods and equipments. The helmets shown here were used until recently.

Page 41

Leather helmets made in Offenbach were in use not only in Germany but also in Sweden, Denmark, Holland and Switzerland. Since the fittings could be easily changed, the Swedish helmet star with button insert could be attached as well as the typical Swiss chin strap holders, i.e a round disc with the Swiss cross instead of the German military lion-head holders.

Firemen liked their leather helmets, because constant wear, and softening by heat and moisture on the fireground made them comfortable like "an old shoe." By the way, most of the old helmets are too small for today's fireman heads. The reason: medical research has indicated that over the past 100 years, our bodies and heads have become bigger, at an average. It is open to question, though, whether we have become wiser...

Page 45

Progress into the 20th Century
— new material, new shapes —

Leather and brass helmets gave good service in the early days of the fire service, but problems arose which eventually lead to the search for new helmet materials. Production costs for those hand made helmets were high, and even rich fire brigades often could not afford to equip each fireman with his own helmet. Pumper crews, for example, often had to be content with simple wax cloth caps. As fire brigades grew, cheap surplus military helmets were often sold out, and consequently there were a surprising number of experiments with new helmet materials. As an alternative to the brass helmet, aluminum models were available; well liked for their lightness and shiny surface, they were even more expensive than brass helmets. Another new and inexpensive material of pressed fibres was praised in manufacturer's advertising brochures, but apparently fire brigades were not convinced so easily. Also helmets made of pressed felt were on the market, in a great many shapes corresponding to those of the traditional leather helmets, but they lacked strength; very few of these felt helmets have survived to our days.

Page 46

A real alternative to the hand made traditional helmets were, at that time, only those helmets produced of steel sheet in large quantity. These helmets were offered in all traditional shapes known from leather and brass helmets. to enable fire brigades to maintain their uniform appearance. The "new" helmets became accepted only after 1915, when even the Imperial German Army switched to felt or tin helmets, because leather was becoming non-available. Even the quality of the "new" tin helmets deteriorated as the war progressed; specimen made in the early twenties, during the economic crisis, were so thin and weak that they could hardly be considered protective headgear.

Around 1930 we notice a short comeback of the leather helmet, in excellent quality, and produced only for fire brigades. The Army had switched by then to the new model steel helmet (similar to the world war II model), and the police had adopted the Czako of the world war I machine gun detachments. After 1933, a new era began also for fire helmets.

Page 48

Also in Austria, the economic situation, the industrial progress and the severe leather shortage resulted in the manufacture of substitute leather helmets, lateron of sheet metal helmets. Production started with the old traditional helmet styles; also the old type helmet decorations were used. In order to the give helmet better side protection, a new spider-like helmet comb was developed. Later versions of this "helmet spider" are still worn on Austrian fire helmets today.

Page 49

Tin helmets were also widely used in Switzerland, especially by newly founded fire brigades, or as replacements for worn-out brass or leather helmets. When in the thirties air raid precaution became also one of the responsibilities of the fire brigades, the steel helmet of the Swiss Army was issued to the fire brigades. This heavy bullet-proof military steel helmet was developed in the US, for use in world war I trench warfare. As this new American helmet had a remote similarity to the German Army steel helmet of those days, it was quickly withdrawn from front line use. Swiss fire brigades retired this steel helmet only after 1970, when plastic fire helmets of US-pattern, similar to the MSA model, were introduced.

Page 51

In the 1920's and 1930's, also other foreign countries were searching for new fire helmet shapes and materials. We still find traditional shapes, but now manufactured from a simple piece of leather, or made of the brittle plastics of those days. We also notice light aluminum helmets (e.g. in USA), and this is also the period of introduction of the cork helmet in England, which replaced the old high brass helmet worn until the thirties. We can almost consider it a step backwards when in the thirties, the fire brigades in many European countries had to accept again versions of military steel helmets. Their material was certainly stronger than that of brass and leather helmets, but these military helmets were not built for fire department use. They were heavy and uncomfortable, and their scope of protection was not fireground oriented.

Page 52

After 1933 also the topic "fire helmet" became a political theme in Germany. Altho at first no uniform rules were laid down, active party members removed those emblems from their leather helmets which could have a remote connection with the church, or the free masons. And very keen members of the S.A. attached the party eagle emblem to their leather helmets, to tell the rest of the world of their great new ideas. We should not forget that the majority of the volunteer firemen throughout the war, guarded an attitude of political neutrality. This was in line with their humane mission, but did not make them very pupular with local party leaders. For reasons unknown today, even some of the old leather helmets received the decal shields reserved for the new steel fire helmet. This was strictly forbidden, but it shows the volunteers did not bother too much with these types of regulations.

Page 53

About 1930, the German Army conducted tests with the intention to replace the model 1916 steel helmet with a more modern helmet version. At the same time, the fire departments of Dortmund and Berlin experimented with fire helmets similar in shape to the German Army steel helmets; also, there were helmet prototypes made of aluminum, vulcan fibre, bakelite, and cloth-resin. The plastic helmets did not pass actual fireground tests, and the Prussian Ministry of the Interior, on February 6, 1934, issued the following instruction:

"Helmets should be made of leather or metal in the shape of the army steel helmet, with chin strap, buckle, comb, oak leaf decoration, detachable leather neck protection and helmet shield with the city coat of arms. Stocks of other head protective pieces can be used up."

The permission to use leather for the manufacture of the new fire helmet (army steel helmet shape) was a real surprise. Apparently, the government did not want to deprive the leather industry of a sizeable market. Indeed, some fire departments did order leather helmets of the new shape, but these helmets were difficult to produce, and were therefore expensive.

As expected, a few months later, in May 1934, leather helmets were deleted from the uniform code, and also the city shield with oak leaf decoration was withdrawn. The metal helmet was to be made only of steel or aluminum. The shape of the new fire helmet corresponded almost exactly to one of the experimental army helmets of 1933 (light version), which did not go into production for the army. In retrospect, it is useless to question this decision — the fire service had to do again with an army model helmet. After all, generations of firemen had to do with surplus Imperial Army leather helmets (spike removed), and it was common knowledge that those ex-army helmets had not provided adequate head protection to the firefighters. At least, the firemen in their new army type helmets provided a neat military spectacle, and this certainly was the declared aim of the new party leaders.

Hard to explain is the shifting of the helmet emblems which could be observed between 1933 and 1936.

While the new leather helmets (army shape) were equipped, after April 28, 1933, with a black/white/red shield on the right, and the black/white Prussian shield on the left (at first in straight upright position, lateron tilted), the new metal helmets were frequently redecorated:

—after April 26, 1934:
right: large white swastika
left: black/white/red shield (tilted)
—after July 10, 1934
right: black/white/red shield (tilted)
left: large white swastika
—after July 28, 1934:
right: red party shield with black swastika
left: police escutcheon shield, aluminum colour

These frequent changeovers annoyed the fire brigades. Still in the 1970's, some fire houses held stocks of helmets with the April 1934 emblems, the swastika crudely painted over.

The city shields, as front pieces, were deleted in 1934. It is not known if any such shields were ever fitted to the new steel fire helmets, although the first production batch had holes in the appropriate locations.

Another strange story evolved around the helmet comb for the new steel helmet "to be worn in the colour of the uniform buttons" (silver for firemen and local officers, and gold for high ranking fire chiefs, county officers etc). This regulation was issued on May 3, 1934, and was confirmed in summer of 1936 and in 1938; the rule was cancelled in November 1938 for full-time fire brigades, but was maintained for volunteers in January 1940. However, as the war progressed, only helmets without comb were delivered to the fire brigades, and all available combs had to be sent to the scrap collection points. The holes in the crown of the helmets had to be closed by splints especially made for this purpose.

Fire brigades and city administrations objected to the introduction of combs for steel helmets. Everybody agreed that such a comb was quite useful for the weak leather and brass helmets of former days, but it was obvious that the new steel helmet did not require such strengthening. On the contrary, the holes for the attachment of the comb tended to weaken the new steel helmet shell. Also, the procurement of sizeable quantities of the new aluminium comb was expensive, and also difficult, as this material was required for aircraft production, and was thus in short supply.

In 1935 and 1936, fire brigades and the Council of German Cities submitted several requests to abolish the combs, but to no avail, as it transpired that Hitler himself, in a decision of October 20, 1934, had ordered the new comb to be worn. Therefore, further discussion was useless, for the time being.

Page 57

Due to its stablity advantages in comparison to the old leather- and brass helmets, and because of its low price and easy availability, the new German steel helmet was used also by fire brigades in some neighbouring countries, in its original version or slightly altered. Even the comb on these helmets became internationally accepted, most likely as a fire brigade symbol. Denmark went even further by equipping the Falck fire service with steel helmets wearing the cross-shaped fixture which held the spike or artillery ball on the old Imperial Army leather helmets. This is to distinguish Falck firemen from their community-organized fellow firefighters, who, for good measure, still wear a brass comb on their steel fire helmets, of German manufacture.

Page 58

During the second world war, the many German organizations involved in firefighting (full-time fire departments, volunteer fire brigades, safety- and support-service S-H-D, airraid police, civil defence,

ARP, company brigades etc) used similar or identical equipment, in many cases. In particular towards the end of the war, new men called up as replacements were glad to get hold of any type of head protection, and fireground photographs show pumper and ladder crews with a mixed assortment of helmets, many of them captured from the enemy.

Most frequently found was the 'official air raid precaution helmet', joined together from three parts, and made by a number of companies since before the second world war. This helmet was sold also to civilians. An improved version, pressed from one piece of steel sheet only, became available towards the end of the war. Annual production of these helmets, between 1939 and 1943, amounted to about 500.000.

Looks of the ARP helmet and protective functions are open to discussion, but they provided their users with an air of authority, and some feeling of security, at least.

Page 61

The perfect fire helmet — does it exist?

Here and on the next pages we see 'modern' fire helmets, i.e. those worn or developed after 1945. As can be expected, modern production techniques and modern materials are being used, but surprisingly the design of fire helmets has remained mainly traditional. This also includes, in some instances, all kinds of unnecessary decorations, combs etc. This preference for traditional aspects can become a real problem, because holes and attachments will weaken even the most modern plastics helmet, and make it more expensive. This may be a high price to pay for preserving certain traditions.

Page 62/63

Eastern European countries, in the early post war years, used up existing stocks of prewar fire helmets, or employed World War II military helmets, with slight alterations. After 1960, however, we can notice attempts also in Eastern Europe to improve head protection of firemen. The design of those new fire helmets incorporates some traditional features, but the shapes of the helmet shells are modernized, the inner linings are improved, and plastics are selected for manufacture of the helmet shell. A visor becomes standard equipment for the fire helmet, i.e. a noteworthy improvement.

Russia and Poland, for example, modernized existing military helmets, the Poles even added the traditional comb. Other countries in Eastern Europe decided not to embark on a new development program, and instead adopted designs available in Western Europe; thus, the Rumanian fire helmet very much resembles the French model, and the Yugoslavs accepted the design of the current Austrian helmet. In each case, local manufacture results in some variations.

Page 64

Until 1918, firefighters in Belgium wore the napoleonic brass helmet with high comb; in some part of the country, for example in Flandres, also German fire helmets were used. After the first world war, surplus Belgian/French military helmets of the Adrian 1915 model were handed to the fire brigades; by 1940 all Belgian fire brigades were equipped with this helmet.

Since the 1950's, cork helmets of Belgian manufacture became available. Similar to the English Cromwell cork helmets, these new helmets were well received by the fire brigades; a modernized version is in production still today. To satisfy the more traditional minded fire brigades, the 1915 Adrian helmet was made for a while of polyester material, but did not gain wide acceptance, as the helmet was not well suited for the fire service.

After about 1975, many Belgian fire brigades have switched to the MSA-type American helmet design, made of plastic material. Some of these new helmets are made in Germany for the Swiss and Belgian markets, with inner linings according to DIN 4840.

Most fire brigades in the Netherlands have traditionally bought helmets of German manufacture, with typical Dutch trimmings, however. In particular, the big city full-time departments (e.g. Amsterdam, The Hague) have used leather helmets up to these days; even after the second world war did they order leather helmets from Germany. When leather helmet production ceased in Germany, new alternatives had to be found. At first, the German post-war civil defense helmet was introduced on a wide scale, and fitted with the German aluminum comb. Later on, a new polycarbonate helmet, made in Germany exclusively for Dutch brigades, was accepted; its design resembles that of the pre-war leather helmets.

Page 66

In Switzerland, the old army steel helmet (model 1918) continued to be used by the fire brigades after 1945. Its heavy weight and uncomfortable wear resulted in the production of some batches made of aluminum, in the post-war years. Air raid protection units received this helmet with dark blue paint. Steel helmets can create problems in some fireground situations, and therefore the idea of a special plastic helmet for fire brigade electricians was reactivated, well known from the days of the leather helmet. A German manufacturer undertook the production of this new helmet version, to the design of the Swiss army helmet.

However, in the long run this solution was not completely satisfactory, and since about 1970, the American MSA-type helmet has been declared standard helmet for Switzerland; each city or canton displays its own small metal shield on the front of the helmet, showing the coat of arms.

Page 68/69

Also in Austria: Changeover from steel helmet to plastic helmet.

When Austria was incorporated into the German Reich, before the outbreak of the second world war, Austrian fire brigades were soon equipped with the new German fire steel helmet. With or without comb, these helmets were used into the sixties. A slightly altered model is still today standard issue for the Steiermark province. After all, the typical German Army steel helmet was also produced in Austria after 1915, and was standard equipment of the Austrian Army, until 1945.

In the thirties, however, a completely different helmet was developed for the Austrian fire brigades, and a modernized version of this design was resurrected after World War II. It is nowadays the standard helmet of Austrian fire brigades, made of aluminum, and carrying a five-legspider comb to protect the helmet crown. The new helmet exists in a somewhat peaked version, and in a flat version, manufacturer is the Weinberger Co. in Vienna (national standard ÖNORM F 4045). Each province carries its own helmet shield; officers have a laurel wreath around the shield.

Since the late sixties, the new helmet is also made of plastic material. The metal front shield has been replaced by a small PVC-sticker. For better visibility, the helmets have 3M reflective bands.

For special jobs, the Swedish "Trelleborg" helmet with visor is used.

Page 70

France, too, has progressed from steel helmet to plastic helmet. Small fire brigades used brass helmets until the post war years (fire brigades in the Alsace region even used old German helmets!), or black-painted French military Adrian helmets. Therafter, the chrome helmet of the Paris fire brigade was adopted as standard. Firemen wear a brass front shield, officers have a chrome shield.

High cost of this helmet and its electric conductivity lead to search for improvement, and since the sixties, experimental models of glass fibre construction were tested. In 1968, the Strassburg fire department as well as the French Civil Defense introduced a new glassfibre helmet showing certain traditional design elements. Also the Paris fire brigade has toyed with the idea to introduce a plastic helmet.

Page 71

Most Asian countries, with the exception of Japan and the Chinese People's Republic, do not have helmet manufacturers within their borders. Thus, they import helmets from countries having influenced the development of their fire service; former colonial traditions or the influence of foreign advisers are clearly visible. The Hongkong Fire Service, for example, wears the newest English Cromwell fire helmet, whereas Thailand employs US-made helmets of the MSA-pattern.

Page 72

Southern Europe — helmets new and old

Spain's conservative attitude after the civil war is reflected also by their fire helmet; the fire brigades still wear the German steel helmet, i.e. the army model in fire service version. The local touch is added by a brass comb, and a fairly large decorative brass shield. Also the German speaking volunteer fire brigades of Southern Tyrol wear the German helmet, with comb and brass emblem; although situated in Italy, they are intent on maintaining their own traditions. The Italian Vigili del Fuoco employ an elegant black helmet of phenolic material, with a metal or soft plastic emblem. This helmet is made by MISPA of Torino, in excellent quality.

Page 73

Fire brigades in Turkey cannot fall back on a national helmet industry; until recent years, they used leather or tin helmets of Austrian and Swiss design, painted red, and sporting a typically Turkish brass comb. In the meantime, Italian MISPA helmets have been imported, as well as American MSA helmets, of red polycarbonate.

Also frequently encountered are white plastic inner liner helmets of the US-Army steel helmet.

Page 75

After the end of World War II, Japan began development of a modern fire helmet generation, completely doing away with metal helmets. The first versions of this new helmet family were of very light construction, but already in 1955 a switch was made to fibreglass helmets with the traditional neck and shoulder protection attached to the helmet. When closed, only a narrow slit for the eyes is left uncovered. The shape of the new helmet was influenced by French and American design elements, and as a special feature, an acetate transparent visor was incorporated into the helmet, fitting snug under the helmet shell when not in use. This simple solution has been followed also for recent five helmet prototypes of German development.

Page 77

City and works fire brigades in the Chinese People's Republic currently wear a rather traditional looking helmet which reminds of the 1932 pot helmets of the Japanese Army. Of phenolic construction, this Chinese fire helmet is considered to be rather heavy.

Fire vehicles and equipment show marks of Russian and German influence, but the fire helmet seems to be a purely national product. The German steel helmet had a good chance to be accepted by the Chinese fire service, because, widely unknown, it was part of Chinese Army equipment until 1946.

Page 78

When examining also the fire helmets of other new emerging nations, one can easily detect the influence of their European of American sponsors. Due to climatic conditions and small budgets, we encounter mostly the light models of western production. The new French plastic helmet from Petit-Collin is found in former French colonies, and the light Cromwell Model 3871, which was not adopted in England, is in use in Nigeria. South Africa still uses leather helmets of Scottish manufacture, while India builds sturdy leather helmets, similar to English models. The rich states on the Persian Gulf have invested their Dollars in good quality US-made helmets, and the former Italian colonies still buy their fire helmets in Italy.

Page 81, 82, 83

England — Successful compromise of tradition and progress

The cork helmet currently worn in England is, from the point of view of head protection, one of the best in the world! Design and external appearance of this helmet can be topics for discussion, but for criteria like weight, shock absorption, and resistance to penetration, this helmet surpasses most of its foreign 'competitors.' Its relatively high shape is part of the safety concept; the comb is offered in various styles, to satisfy regional requirements.

In the sixties, a model of low silhouette without comb was produced, somewhat similar to the British Army steel-helmet MK III of 1943/44. This helmet was rejected by fire brigades in England as not suitable for fire service; brigades like Liverpool and Luton which tested the new helmets withdrew them after a while. Remaining stocks were sold in Africa, where traditions were not so important. At the 1980 Interschutz Fire Exhibition in Hannover, a new version of the traditional high comb cork helmet was displayed, employing Kevlar, a new and very resistant synthetic material.

Page 84

USA — unlimited opportunities

For a designer of fire helmets, the United States are the ideal country, because helmet standards, which are fairly strict, are based on performance criteria only. Design of the helmet and material are left to the choice of the manufacturers, as long as they meet the prescribed performance. These standards prescribe limits for electrical isolation, shock absorption and penetration of the helmet shell; if a helmet meets these requirements, it is suitable for fireground use. This liberal approach has resulted in the production of a great variety of helmet models, of various materials, mostly synthetics. As competition is strong among helmet manufacturers, new ideas are quickly put into production, and progress of development is much quicker than in Europe.

The buyer, most likely the local fire chief, can select from a number of styles and price ranges. If he is traditional minded, he may select a leather helmet made by Cairns, or the standard model made by MSA. If he wants a progressive image, he may buy a helmet from Akron, Premier, or American Sports. Fire departments stressing a modern approach have in the past voted for Federal Helmets.

Fascinating, from the Eurpean point of view, are the leather front pieces on American fire helmets. As the shells of new plastic helmets, for reasons of electric conductivity, may no longer be pierced, self-adhesive stickers or decals/transfers are now used by many fire departments to replace the traditional front pieces. White shields identify an officer, pumper crews have a black front piece, ladder crews red, rescue companies blue, and airfield crash crews hav green front pieces. Firemen on special units sometimes have orange shields.

Helmets with three parallel combs were built by Cains, Bullard and Gentex for firemen employed by Federal Authorities; the additional combs were to provide greater strength to the helmet crown.

Page 90

Scandinavia

Already in the 19th century, German and Austrian fire helmets were often found in Scandinavian countries, as well.

Finland today imports German aluminum helmets built according to DIN 14940; before that, German leather helmets and French 'Adrian' steel helmets were used, from military surplus stocks liquidated in the twenties.

Norwegian fire brigades have used German steel helmets only for a short time: around 1936 a quantity of leather or fibre helmets of steel helmet shape were imported, and used for example by the Halden brigade until 1945. The standard helmet in Norway resembles the Austrian Vienna model, which was first pressed of steel (1940), later of aluminum (1944) and adorned with the German aluminum comb.

The basic model seems to have been successful, as it is produced nowadays, by TICON, of plastic material, with integrated comb, and decorated with the traditional helmet star emblem.

Denmark still buys metal fire helmets of standard German production, but the comb and other fixtures give these helmets a typically Danish appearance.

Page 91

Sweden has several helmet manufacturers, and thus a number of styles are available. The basic model is the German steel helmet, worn in metal and fibre versions. Brissman, a fire helmet manufacturer, produced a plastic prototype of the German style helmet already in 1936, and made a cast aluminum version in 1938; the comb was cast into the helmet crown. These helmets found wide acceptance in Scandinavia. For reasons of electric safety, a new plastic version of glass fibre was tested in 1958, with a relatively high comb; and a low silhouette version went into production in 1960, again with integrated comb. In parallel, the Trelleborg Rubber Co. made a very progressive fire helmet with full curved visor; this helmet is in use also in other European countries.

Page 92

Fire helmets in Eastern Germany

In the early post-war period, existing stocks of wartime fire and civil defense helmets were utilized. After 1961, according to the GDR-standard TGL 121 — 940.01 (somewhat comparable to the West-German DIN 14940), the standard fire helmet was made of glass fibre polyester construction. They were lighter than the metal versions, absorbed less heat, and gave better electrical insulation. The chin strap was attached to the helmet in four positions, giving better wearing comfort. In the past four years, standard workers helmets of glass fibre material have been issued to a number of volunteer fire brigades; a piece of soft material for the protection of the neck is attached to these helmets. Quite recently, a number of new helmet prototypes have started a prolonged test programme. Results are not yet known.

Page 93

West German fire helmets

In many German towns in 1945, the local military commanders of the allied troops ordered the German fire chiefs to destroy all fire helmets; apparently the fire helmets resembled the German military helmets to such a degree that misunderstandings and confusion could result, with detrimental effects for the German firefighters. When the allied invasion troops had established themselves, they required the white letters "FIRE" or "FIRE SERVICE" to be painted on the front of the fire helmets. Helmets were in generous supply, as civil defense helmets and army steel helmets littered the countryside. A simple coat of paint turned them into fire helmets, and they had a long service life: still today these heavy 'bullet proof' steel helmets can be found in German firehouses. Their outward appearance is very close to the current German fire helmet.

Page 94

In the sixties and seventies, the DIN 14940 standard was written, which at least on paper provided some rule of order. The materials to be employed for helmet manufacture were laid down, the inner linings of the helmet were improved, and test criteria for the level of protection to be afforded by the helmet, were developed. This was progress indeed, and the new night-glow paint of the helmet was the visible demonstration of an enlightened approach. Excessive standardization, however, can also be a block to further development, and as the DIN standard 14940 ist truly a masterpiece in most respects, it leaves very little freedom of choice to the helmet manufacturer, and no freedom at all to the designer, as the outward appearance of the helmet is fixed down to the most minute detail. This situation did not exactly encourage the discussion of new fire helmets in Western Germany.

Page 95

The DIN 14940 standard has done away with most individual 'decorations' of the West German fire helmets. A few fire brigades still wear the traditional aluminum comb, e.g. Berlin and Darmstadt; some fire brigades still have their city coat of arms on their helmets, or industrial fire brigades have a sticker displaying the company trade mark. In recent years, more and more brigades attach adhesive markings to the helmet, or stickers denoting breathing apparatus wearers (a miniature breathing mask). Some big fire brigades display the number of the fire station on the front of the fire helmets, or sometimes even a letter combination.

The most recent DIN 14940 supplements have abolished the red paint scheme of fire helmets for the Bavarian fire brigades; they, too, now have to wear the night-glow version. Only during firefighters conventions can one notice unusual helmet versions: chromed or aluminum polished models of the current helmet model are worn for ceremonial purposes. For funeral duty, some brigades keep a small stock of the old wartime fire helmets, painted black.

In 1981, at last, a South German fire brigade conducted wearing tests with the standard German fire helmet made in two versions, one of polycarbonate, the other of glass fibre construction. The latter won out, and in the near future we can expect introduction of this new helmet version.

Page 96

Junior firemen in Germany wear red plastic helmets conforming to the standard for construction workers' helmets, DIN 4840. Since the official helmet, introduced by DFV, the German Firefighter's Association, is made of high quality polycarbonate, and is thus relatively expensive, many fire brigades have purchased less expensive versions of the workman's helmet which differ in some details from the official junior firemen's helmet. All these helmets fulfill DIN 4840, however.

Page 97

The new helmet from Bonn

Fire helmets and civil defense helmets are built to do similar jobs, and it is interesting, therefore, to describe helmet development initiated by the German Office for Civil Defense in Bonn. Until 1966, the CD-helmet of World War II was produced, by two manufacturers, of glass fibre material. Tests conducted in 1968 of a polycarbonate version failed: the helmet was too flexible.

In 1971, a new completely different design was submitted, which fulfilled the safety aspects of DIN 4840. Made of polycarbonate, the helmet shape based on test series conducted in 1944 for a new German Army helmet to be introduced in 1946 (This new army helmet, incidentally, later entered production for the East German People's Army). A noteworthy feature of the new CD-helmet model was a wire mesh visor which, when not in use, could be retracted under the helmet shell. This helmet model was produced in a preproduction series, but was rejected by the fire brigades, mainly for reasons of outward appearance. A reworked design, presented in 1980, is in production now, exclusively for CD-units.

Page 99

The new helmet from Berlin

The Berlin based Auer Co., in conjunction with the Frankfurt Fire Dept., decided to submit a new helmet design, which was ready for testing after 1970. Twenty prototypes with slight variations were built, consisting of a polycarbonate outer helmet shell, with integrated wire mesh visor, and a separate inner liner helmet fulfilling the worker's helmet standard DIN 4840. For the production version, the outer shell was to be produced of injection cast aluminum. The basic design idea was to provide head protection to the wearer even if the outer helmet shell would be lost, by a heavy blow, or the blast of an explosion. The inner liner helmet could serve as head protection for those assignments where work in close quarters was necessary, traffic accidents, rescue, etc.

When the Frankfurt Auer helmet was publicly presented, a big discussion, often on an emotional basis, resulted. As the German fire brigades could not agree on requirements for a new helmet, production of this new helmet was shelved in 1972, for the time being.

Page 100

The new helmet from Lübeck

Around 1975, the Draeger Co. of Lübeck presented a new fire helmet made of polycarbonate; the Fire Dept. of the Hague in the Netherlands had cooperated in the development of this new helmet, which incorporated a breathing mask into the helmet design. Doing away with the traditional rubber straps which hold the breathing mask on the head of its wearer, the new Draeger mask was simply fitted to the front of the helmet, and attached to the helmet by two tightening fixtures, on the sides of the helmet. This indeed was a new and easy solution.

Critics pointed out, however, that if the helmet received a heavy blow, it would shift on the head of the fireman, and consequently the mask, firmly attached to the helmet, might shift also, resulting in possible leakage problems. The Draeger Co. tried as much as possible to fulfill the requirements of DIN 14940 and DIN 4840. It has to be seen whether the idea of an integrated mask can be realized on a safe basis.

Page 103

The new helmet from Stuttgart

In 1978, the Porsche design office of Stuttgard published the results of a helmet study based on theoretical assumptions. The report had been commissioned by the German Ministry of Research and Development, within the framework of a basic study to modernize the equipment of German fire brigades. Another result of this research was the proposed ORBIT vehicle series, again a purely theoretical concept at this time.

The new PORSCHE helmet was to be made of polycarbonate, with integrated visor, breathing apparatus mask, lighting fixture, and radio communications set built into the helmet. Aside from high cost of production also other problems of this theoretical concept have not yet been solved, including the question of leakage for the breathing mask, the energy source for lighting and radio, etc. A workable prototype has not yet been presented.

Page 105

A new helmet concept by TOTAL

The TOTAL Co., one of the foremost suppliers of fire brigade equipment in Germany, has undertaken a design study to integrate 'head protection' and 'breathing protection' into a modular fire helmet concept.

The helmet shell, with full side protection, is shaped in such a way to fit snugly around the window frame of the conventional breathing mask. The breathing mask itself does away with conventional rubber straps; a porous cover pulls it tightly over the head of the wearer, eliminating the fumbling with the straps, and the pressure points of the strap buckles on the head of the wearer.

If a breathing mask is not required, a transparent or wire mesh visor can be easily inserted into the helmet. The helmet shell, to be made of polycarbonate, or other suitable synthetic material, shows a comb like stiffening of the crown part. A space for an emblem, a sticker displaying the fire company number or similar, is provided on the forepart of the helmet. The chin strap is attached to the shell in four positions, and a piece of leather for neck protection is provided. This TOTAL helmet concept is an interesting contribution to the fire helmet discussions of the eighties.

Page 106

Epilogue

The idea for this book was born one night during a long fire watch, when discussion turned to fire helmets, and we all realized that most firemen knew relatively little about their fire helmets. Considering the fact that the helmet is probably the most important protective item a fireman owns, we felt that there was need for some more information about this piece of equipment.

The history of the fire helmet reflects also a section of cultural development. Looking at the technical aspects of this history, one comes upon the problems of new production methods and selection of new materials. Also discernible are steps in social history, affected by types of fire service organization, government philosophy, or prevalent political attitudes at one time or another. Fashion trends in fire helmet design become visible, and even the personal influences of great fire service leaders are reflected in some helmet designs. No other item for personal protection is viewed with so much enthusiasm or rejection than the fire helmet. This book attempts to explain the development of fire helmets to the present day, and to point to alternatives, as a basis for discussion.

In particular, I have to thank the manufacturers of fire helmets worldwide for helping me with a considerable amount of information, and for answering my many questions, as far as company policy permitted; for the interpretation of this information, only the author is responsible.

I have to thank also my fellow firefighters in all parts of the world, especially in the Dietzenbach volunteer fire department, who tolerated my special interest in fire helmets with understanding and forgiveness. Special assistance was provided by Allen Batchelor in London, John Christensen in Copenhagen, Jean-Gabriel Jeudy in Paris, Marcel Nuijttens in Flandres, Paul Pribble in California, Ken Smith in Croydon, Ken Snyder in Pennsylvania, and Wilhelm Wahrenberg in Stockholm.

The majority of the colour slides were taken by professional photographer Brigitte Harms; Helmut Raab of the Dietzenbach editorial office of 112 Fire Magazine provided valuable ideas and constructive criticism. Most of the helmets depicted in this book can be viewed in the Salem Fire Museum, in Southern Germany.

T. W. Herminghaus
Dietzenbach, May 1982